Lower Palaeozoic Rocks of the World

Volume 3 · Lower Palaeozoic of the Middle East, Eastern and
Southern Africa, and Antarctica
(with essays on Lower Palaeozoic trace fossils of Africa and
Lower Palaeozoic palaeoclimatology)

LOWER PALAEOZOIC ROCKS OF THE WORLD

VOLUME 1 · CAMBRIAN OF THE NEW WORLD

VOLUME 2 · CAMBRIAN OF THE BRITISH ISLES, NORDEN,
AND SPITSBERGEN

VOLUME 3 · LOWER PALAEOZOIC OF THE MIDDLE EAST, EASTERN
AND SOUTHERN AFRICA, AND ANTARCTICA

Frontispiece: Glacial striation from the upper Ordovician of southern Algeria. The striation is part of a very large surface, with striae from south-east to north-west (from bottom to top in picture). The locality is in the western part of Tassili de Tafassasset, on the east-north-eastern side of the Hoggar Massif, and has been described by IAP (1970, p. 94) as no. J 3.

Lower Palaeozoic
of the
Middle East,
Eastern and Southern Africa,
and Antarctica

with essays on Lower Palaeozoic
trace fossils of Africa
and Lower Palaeozoic palaeoclimatology

Edited by
C. H. HOLLAND

Department of Geology, Trinity College, Dublin

A Wiley–Interscience Publication

JOHN WILEY & SONS
Chichester · New York · Brisbane · Toronto

British Library Cataloguing in Publication Data:

Lower Palaeozoic of the Middle East, Eastern and
    Southern Africa, and Antarctica.—(Lower
    Palaeozoic rocks of the world; vol. 3).
    1. Geology, Stratigraphic—Palaeozoic
    I. Holland, Charles Hepworth
    II. Series
    551.7′Z   QE654   80-41688

    ISBN 0 471 27945 5

Printed in Great Britain by
J. W. Arrowsmith Ltd., Bristol BS3 2NT

PREFACE

The present volume of the series 'Lower Palaeozoic Rocks of the World' covers a substantial segment of the earth from the Middle East, through north-eastern and eastern Africa, to southern Africa, and then Antarctica. The Lower Palaeozoic rocks of north-western and west-central Africa are considered in Volume 4 of the series. In all these areas inaccessibility, a relatively late history of geological exploration, and the very nature of the rock successions themselves have variously combined to make a precise chronostratigraphy difficult to achieve. Accordingly, the various parts of these two volumes are each concerned with the Lower Palaeozoic as a whole rather than with a single Lower Palaeozoic system.

The Editor could write a short book on the history of preparation of Volumes 3 and 4: incidents, delays and difficulties have been many and the patience of some authors has been severely tried. This applies particularly to those who reasonably met an original deadline and those who entered the project at a later stage in order to save an impossible situation, and yet were themselves held up by other difficulties not of their own making. This is not, of course, to say that the delays have not been caused by unforeseen and unavoidable external factors. The result, however, is that some contributions were originally prepared in the early 1970s and, although there has been some opportunity for revision, addition, and subtraction, authors would clearly have preferred to write freshly and immediately. Nonetheless, the Lower Palaeozoic rocks of these vast territories are not well known to those not directly concerned with particular countries and the volumes now provide comprehensive new data and authoritative syntheses, inspiring to read. The Editor remains exceedingly grateful to all the authors for their kindly response to periods of harassment and periods of attempted consolation.

Charles Hepworth Holland

Trinity College,
Dublin.

CONTRIBUTING AUTHORS

T. P. Crimes Department of Geology, University of Liverpool, England

Eberhard Klitzsch Institut für Geologie und Paläontologie, Technische Universität, Berlin, Germany

M. G. Laird New Zealand Geological Survey, Department of Scientific and Industrial Research, Christchurch, New Zealand

R. M. Shackleton Department of Earth Sciences, The Open University, Milton Keynes, England

Nils Spjeldnaes Palaeoecology Department, Aarhus University, Denmark

I. C. Rust Department of Geology, University of Port Elizabeth, South Africa

Reinhard Wolfart Geological Survey of the Federal Republic of Germany, Hannover, Germany

CONTENTS

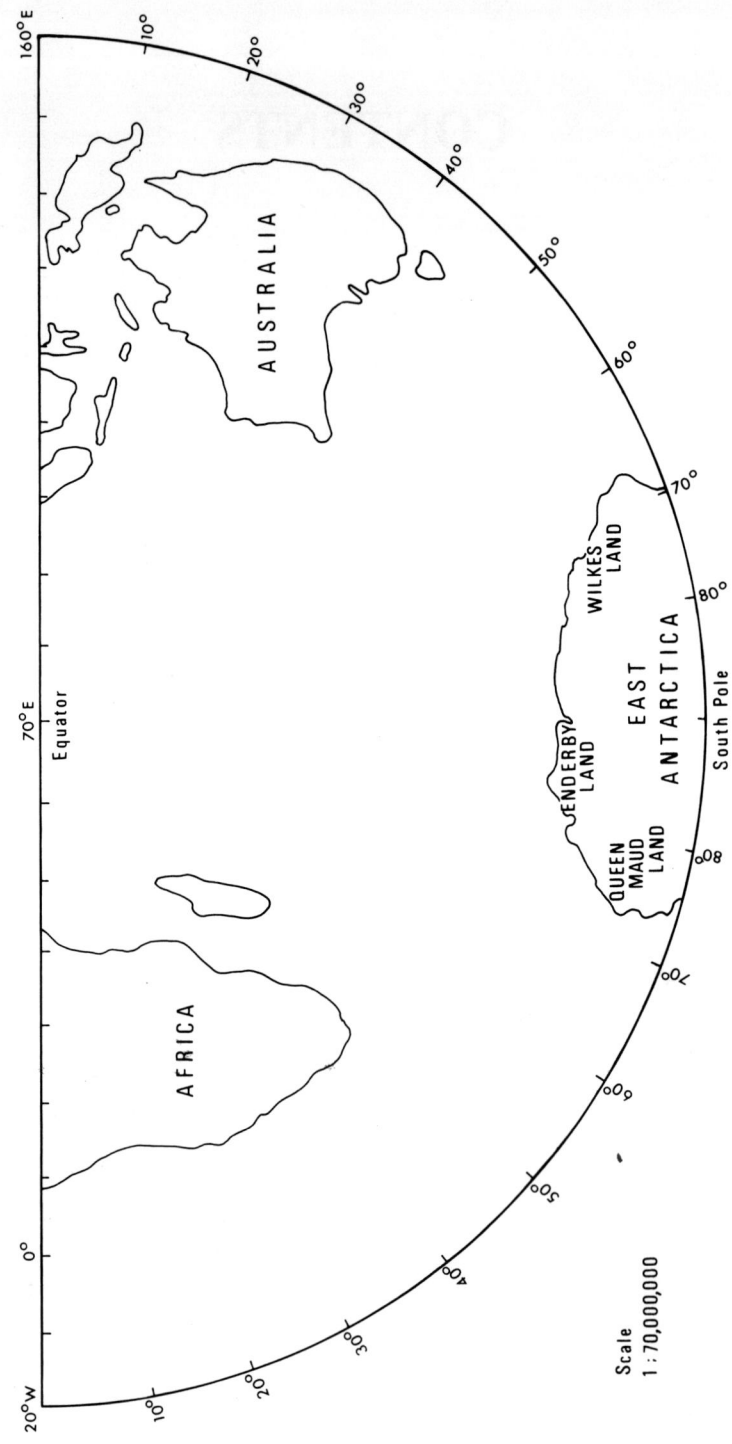

Scale
1 : 70,000,000

Above: Map of the Middle East and Africa showing political divisions. The area of Africa left unshaded is approximately that covered in Volume 4 of the series. Some countries are numbered as follows: 1, Lebanon; 2, Israel; 3, Jordan; 4, Kuwait; 5, Qatar; 6, United Arab Emirates; 7, Afars and Isaacs; 8, Rwanda; 9, Burundi; 10, Malawi; 11, Swaziland; 12, Lesotho; 13, Equatorial Guinea; 14, Dahomey; 15, Togo; 16, Liberia; 17, Sierra Leone; 18, Port Guinea; 19, Gambia; 20, Senegal; 21, Tunisia. Namibia is referred to as South West Africa in the fourth part of this book.

Left: Map to show east Antarctica in relation to southern Africa and Australia.

(*Maps drawn by Elizabeth Williams, Trinity College, Dublin.*)

Lower Palaeozoic of the Middle East, Eastern and Southern Africa, and Antarctica
Edited by C. H. Holland
© 1981 John Wiley & Sons Ltd.

INTRODUCTION

R. M. Shackleton

Department of Earth Sciences, The Open University, Milton Keynes, England

Throughout the Lower Palaeozoic periods, the present African continent was part of a much larger continental domain which was probably not very different from the Gondwanaland of the early Mesozoic. Arabia, much if not all of Iran, India, most of Australia, Antarctica, and South America formed part of this enormous continental area, within which Africa occupied an interior position. Whether the Iberian Peninsula, Brittany and southern Europe were part of the same continental mass (SMITH and others, 1973) or were separated from it by a Proto-Tethys (WHITTINGTON, 1972) is uncertain. This continental mass was itself presumably part of a still larger plate which included a rim of oceanic crust that was subsequently subducted. The early Palaeozoic Proto-Atlantic ocean on the site of the Caledonides in the North Atlantic region may have formed part of the oceanic rim.

The continental domain of which Africa formed part evolved as a result of episodes of intense deformation, metamorphism, and magmatism which occurred at intervals through the Pre-Cambrian. These processes are regarded as orogenic and are assumed to have given rise to mountain chains, although the depth of erosion is such that in many regions no post-orogenic molasse remains to demonstrate the former presence of high mountains. The successive orogenic belts form complex networks rather than linear belts and although in some places successive deformations follow the same course, more often pre-existing structures are crossed indiscriminately by later ones. Nearly every part of the continent has been involved in such deformations several times.

In considering the influence of the Pre-Cambrian basement on the subsequent evolution of the area, it is on the oldest and youngest of the Pre-Cambrian domains that attention must be focused. The oldest domains are the Archaean nuclei or cratons (Fig. 1) which have survived, essentially intact, since Archaean times, often with thin veneers of orthoquartzite which reflect repeated transgressions on to these stable areas. The shallow-water sediments of the Pongola Series in Swaziland have remained almost undisturbed, resting on the early Archaean Swaziland gneisses, for more than 3000 m.y.

The Archaean structures strike at the margins of the cratons so that it is clear that these cratonic nuclei are only small remnants of much larger structural domains and the Archaean basement is now being traced into the heart of the younger orogenic belts (HEPWORTH, 1972). Much if not all of the present African continental crust was already continental, in thickness and composition, in Archaean times.

The Archaean cratons do not stand at any constant or regular elevation. The present surfaces of the Rhodesian and Tanzanian cratons are about 1000 metres above sea level whereas in parts of the Kalahari, Congo, and West African cratons the basement surface is depressed below sea level. Thus not all the Archaean cratons have acted as positive areas during the Phanerozoic. The features which distinguish them are their stability or low tectonic activity and low heat flow, which is reflected in less magmatic activity than elsewhere.

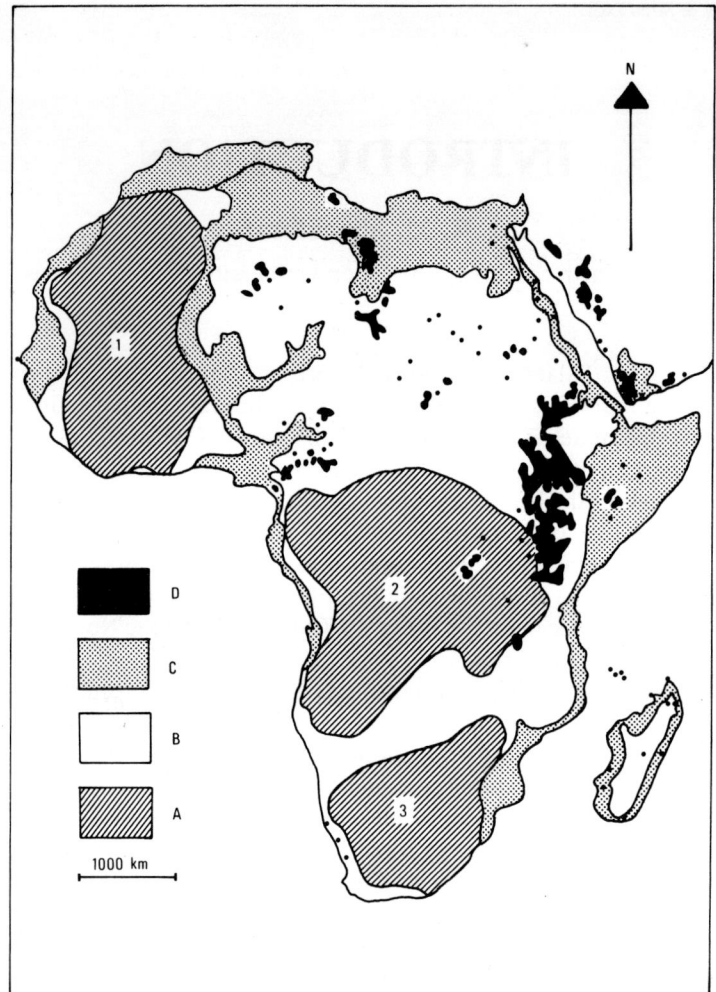

FIG. 1. Distribution of Cenozoic volcanism and Mesozoic and
Cenozoic marine areas in relation to fundamental basement
structure of Africa (after THORPE and SMITH, 1974, and
KENNEDY, 1965): (D) solid black, Cenozoic volcanic rocks; (C)
stippled, Mesozoic and Cenozoic marine areas; (B) blank and
stippled, domains yielding Pan-African ages; (A) oblique lines,
cratons yielding ages of more than 1000 m.y.—1, West African
craton; 2, Congo craton; 3, Kalahari craton.

   The most recent orogenic episode to affect the African basement was originally described as
the Pan-African thermo-tectonic episode (KENNEDY, 1964). Its effects were astonishingly
widespread. More than half of the African crust and equally extensive areas in the surrounding
continental area were involved. The Pan-African episode was initially recognised radiometrically, by widespread K/Ar dates of *ca.* 500 m.y. It has subsequently been established that there
are well defined peaks between about 650 and 450 m.y. with a mode of about 510 m.y., on
histograms of age determinations, whether these are based on K/Ar, Rb/Sr, or U–Th–Pb
(CLIFFORD, 1967). It has been argued that an earlier, Katangan episode about 620 m.y. ago can
be distinguished from a more widespread Damaran episode slightly more than 500 m.y. ago.

However, many of the younger dates represent post-tectonic magmatic rocks or biotite ages which reflect uplift and cooling; dates of syntectonic magmatic rocks tend to fall around 600 m.y. (CAHEN and SNELLING, 1966). Structural and metamorphic limits of the areas affected by the Pan-African episode have also been mapped. They show a similar pattern to that derived from the radiometric data, although micas in particular tend to show the effects of the episode outside the limits established from structure or metamorphism. The stratigraphical evidence shows that at least in north-west Africa (Morocco), north-east Africa (Egypt), and the Near East (Jordan, Saudi Arabia, south-west Iran), the Pan-African deformation, metamorphism, post-tectonic magmatism and also uplift and erosion of the deformed rocks, occurred before the Lower Cambrian, since in these regions the basement is covered unconformably by marine Lower Cambrian deposits. Continental equivalents of the marine Cambrian can be recognized within the African interior (CHOUBERT and others, 1968). It is possible that parts of the Pan-African domains were still being eroded while areas in the north of Africa were submerged under the Cambrian seas. Southwards from Egypt and Sudan, through Ethiopia into Kenya and Tanzania, there is a gradual change from high-level rocks with low-grade metamorphism, many high-level granite plutons, and typical molasse deposits to far more deeply eroded regions with widespread migmatites, high-grade metamorphism, and few discrete granite plutons. This deep erosion of the Pan-African domain in East Africa may have continued into the Lower Palaeozoic.

The effects of the Pan-African episode and the differences between the Pan-African domains and the major cratons (Kalahari, Congo, West Africa) were profound and long-lasting. Post-Viséan, pre-Jurassic dolerites are widespread in the West African craton but absent from the surrounding regions (BLACK and GIROD, 1970); the late Palaeozoic to Tertiary 'younger granites' of north-west Africa occur within the Pan-African domain but not in the cratons (KENNEDY, 1964; BLACK and GIROD, 1970); Mesozoic downwarping and marine sedimentation is confined to the Pan-African domains (KENNEDY, 1964); the Rift System developed mainly in circumcratonic areas; and diamondiferous kimberlites are widespread in the circumcratonic domains but very rare in the cratons (THORPE and SMITH, 1974). Cenozoic volcanism is mostly confined to the cratons (KENNEDY, 1964); the differences in behaviour between the cratonic and extra-cratonic areas must reflect long-lived differences not only in the crust but also in the attached upper mantle, since the Tertiary and Recent basalts, for example, originate from the upper mantle. It is not so clear how these differences were expressed during Lower Palaeozoic times but there can be no doubt that the same differences existed then. Because of their persistently greater mobility, the Pan-African domains must have been the sites of both erosion and deposition while the cratons maintained a lower relief. The main sources of sediment during the Phanerozoic thus lay within the Pan-African domains.

The motions of the plate of which Africa formed a part are known from palaeomagnetic data. The plate was moving—apparently as a single unit—continuously, extensively, and in irregular directions, throughout Pre-Cambrian and Phanerozoic times (MCELHINNY and others, 1968, 1974; PIPER and others, 1973) although the rate of movement varied.

The facies of the sedimentary cover is partly dependent upon climate (see SPJELDNAES, this volume) and thus on palaeolatitude. During late Pre-Cambrian and early Palaeozoic times the drift of the plate carried various parts of Africa across the pole. Different interpretations of the palaeomagnetic data have been proposed, ambiguity arising owing to the uncertainty in distinguishing between north and south poles. According to an interpretation which seems to be consistent with the geological as well as the palaeomagnetic evidence (MCELHINNY and others, 1974) the motions (expressed as movements of the pole relative to the continent) caused the pole to migrate northwards from South America (then joined to Africa) during the late Pre-Cambrian. If this is so, the widespread late Pre-Cambrian glacial deposits of western Africa should be diachronous, becoming progressively younger northwards. In Cambrian times when the motion was still rapid, the pole moved far away to the north of Africa and this is reflected in the warm-water Lower Cambrian deposits, including archaeocyathid reefs and red sediments. In the Ordovician the pole migrated southwards again across north Africa, resulting in the great late Ordovician Saharan glaciation (BEUF and others, 1971). By the end of the Lower Palaeozoic the pole had moved southwards as far as southern Africa.

Thus the Lower Palaeozoic deposits in Africa may be interpreted partly in terms of the long history of Pre-Cambrian orogenic episodes, partly in terms of the motion of the large plate of which Africa formed a part, and partly, naturally, in terms of the fluctuations in sea level which were themselves probably mainly caused by variation in the rate of formation of mid-ocean ridges.

References

BEUF, S., BIJI-DUVAL, B., STEVAUX, J., and KULBICKI, G. (1971). Les Grès du Paléozoique Inferieur au Sahara. *Publ. Inst. Fr. Pétrole*, No. **18**, xv + 464 pp., Technip, Paris.

BLACK, R. and GIROD, M. (1970). Late Palaeozoic to Recent igneous activity in West Africa and its relationship to basement structure, *in* CLIFFORD, T. N., and GASS, I. G. (Eds.), *African magmatism and tectonics*. Oliver & Boyd, Edinburgh.

CAHEN, L. and SNELLING, N. J. (1966). *The geochronology of Equatorial Africa*. North-Holland Publ. Co., Amsterdam.

CHOUBERT, G., FAURE-MURET, A., and CHARLOT, R. (1968). Le problème du Cambrien en Afrique nord-occidentale. *Rév. Géog. Phys. Géol. Dynam.* **2**, 289-310.

CLIFFORD, T. N. (1967). The Damaran Episode in the Upper Proterozoic–Lower Palaeozoic structural history of Southern Africa. *Spec. Pap. Geol. Soc. Am.*, **92**, 1-100.

HEPWORTH, J. V. (1972). The Mozambique orogenic belt and its foreland in north-east Tanzania: a photogeologically-based study. *J. Geol. Soc. Lond.*, **128**, 461-500.

KENNEDY, W. Q. (1964). The structural differentiation of Africa in the Pan-African (= 500 m.y.) tectonic episode. *8th Ann. Rep. Res. Inst. Afr. Geol.*, *Univ. Leeds*, 48.

KENNEDY, W. Q. (1965). The influence of basement structure on the evolution of the coastal (Mesozoic and Tertiary) basins in Africa, *in Salt basins around Africa*. Institute of Petroleum, London, 7-16.

MCELHINNY, M. W., BRIDEN, J. C., JONES, D. L., and BROCK, A. (1968). Geological and geophysical implications of palaeomagnetic results from Africa. *Rev. Geophys.*, **6**, 201-238.

MCELHINNY, M. W., GIDDINGS, J. W., and EMBLETON, B. J. J. (1974). Palaeomagnetic results and late Precambrian glaciations. *Nature, Lond.*, **248**, 557-561.

PIPER, J. D. A., BRIDEN, J. C., and LOMAX, K. (1973). Precambrian Africa and South America as a single Continent. *Nature, Lond.*, **245**, 224-228.

SMITH, A. G., BRIDEN, J. C., and DREWRY, G. E. (1973). Phanerozoic World Maps, *in* HUGHES, N. F. (Ed.), *Organisms and continents through time. Spec. Pap. Palaeont.*, **12**, 1-42. ·

THORPE, R. S. and SMITH, K. (1974). Distribution of Cenozoic Volcanism in Africa. *Earth Planet. Sci. Lett.*, **22**, 91-95.

WHITTINGTON, H. B. (1972). Ordovician geography and faunal provinces deduced from trilobite distribution. *Phil. Trans. R. Soc.*, B, **263**, 235-278.

Lower Palaeozoic of the Middle East, Eastern and Southern Africa, and Antarctica
Edited by C. H. Holland

LOWER PALAEOZOIC ROCKS OF THE MIDDLE EAST

Reinhard Wolfart

Federal Institute for Geosciences and Natural Resources, Hannover, Germany

Contents

1. Introduction

The Middle East is here taken to include Turkey, Syria, Jordan, Israel, the Peninsula of Sinai, Iraq, the Arabian Peninsula, Iran, and Afghanistan. It is part of the semi-arid belt of the northern hemisphere, exceptions being the mountainous regions in Turkey, Syria, Iran, Afghanistan, and Yemen, with rather high precipitation. Topographically, the Middle East is characterized by the west–east belt of Tethyan mountain chains in the northern part, which contrasts with the largely peneplained Arabian Peninsula in the south. The highest peaks of the Tethyan mountains rise about 4000 to 5000 metres above sea-level. Near its north-western boundary the peneplained Arabian platform is dissected by the Jordan Rift Valley with the Dead Sea surface 392 metres below sea-level.

The comparatively few complete Lower Palaeozoic sections exposed in the whole region provide information about sedimentation and faunal distribution over an area of 5 to 6 million square kilometres. In the Middle East environmental conditions were subjected to many changes during early Palaeozoic times. In general, they ranged from terrestrial through littoral to bathyal.

The Cambrian, Ordovician, and Silurian Systems are here described separately. The whole region is subdivided stratigraphically and environmentally, or simply politically (Fig. 1).

2. Cambrian

Cambrian rocks are broadly exposed along the eastern margin of the crystalline core of the Nubian–Arabian Shield in Saudi Arabia and southern Jordan. In the Tethyan mountain chains outcrops of Cambrian rocks are comparatively small and widely scattered. They mark uplifts of the Pre-Cambrian basement covered by thick post-Cambrian sediments in south-western Iran and the Persian Gulf region. In the latter, knowledge of Cambrian deposits is restricted to salt

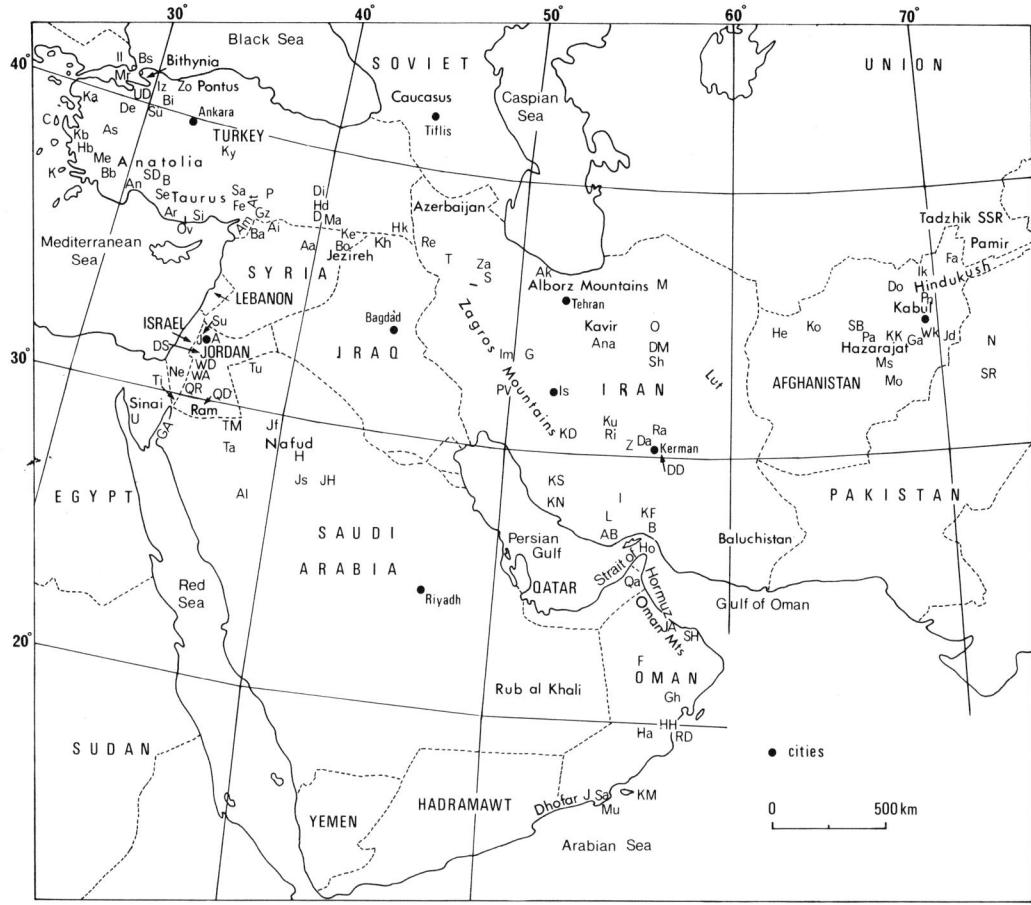

FIG. 1. Index map of the Middle East showing principal geographical features and names of outcrop areas: Aa = Abba; Ak = Alam Kuh; As = Alaşehir; AB = Al Buzah, Ras Bustaneh; Al = Al'Ula; Am = Amanos Mountains; A = Amman; Ar = Anamur; Ana = Anarak; An = Antalya; At = Antitaurus Mountains; Arghandab river basin (west of Moqur); Ai = Aril; Bb = Babadağ; Ba = Baflioun; B = Bandar Abbas; Bi = Bilecik; Bs = Bosphorus; Bo = Bouab; C = Chios; Chunchata (between Wardak and Sadmarda); Da = Dahu; Dašte Nawar (between Gadagak and Malestan); DS = Dead Sea; De = Demirkapi; DM = Derenjal Mountains; D = Derik; Di = Diyarbakir; Do = Doab; DD = Dorah Shah Dad; E = Eregli; F = Fahud; Fa = Faydzabad; Fe = Feke; Ga = Gadagak; Gz = Gaziantep; Gh = Ghaba; G = Golpaygan; GA = Gulf of Aqaba; Hb = Habibler; H = Ha'il; Ha = Haima; Hk = Hakkari; Hd = Handof; HH = Haushi-Huqf; He = Herat; Ho = Hormuz Island; Im = Imam Sayad Hassan; I = Irij; Is = Isfahan; Ik = Iškamyš; Il = Istanbul; Iz = Izmit; JA = Jabal Akhdar; Jabal-e-Kalchi (between Gadagak and Malestan); JH = Jabal Hanadir; JSa = Jabal Samhan; JS = Jabal Saq; Jd = Jalalabad; Jf = Jauf; J = Jordan Valley; Ke = Kamichlie; Kb = Karaburun; Ky = Kayseri; Ka = Kaz Dağ; Kk = Kerskhan; Kh = Khabour; Ko = Koh-i-Davindar; K = Kos; Ku = Kuhbanan; KD = Kuh-e-Dina; KF = Kuh-e-Faraghan; KN = Kuh-e-Namak; KS = Kuh-e-Surmeh; KM = Kuria Muria Islands; L = Lar, Aliabad, Kuh-e-Kurdeh; Ms = Malestan; Ma = Mardin; Mr = Marmora Sea; Me = Menderes Massif; M = Mila Kuh; Mo = Moqur; Mu = Murbat; Ne = Negev; N = Nowshera; Ov = Ovacik; O = Ozbakh Kuh; Pa = Panjaw; Pn = Panjir; PV = Pa Vashtah; P = Penbeğli-Tut; QD = Qa'a Disa; Qa = Qamar; Q = Wadi Quseib; Ram = Wadi Ram; RD = Ras Duqm; R = Ras en Nabq; Ra = Ravar; Re = Lake Rezayeh; Ri = Rizu; Sadmarda (north-east of Kerskhan); Sa = Saimbeyli; SR = Salt Range; Sare Pori (near Kerskhan); SH = Sayh Hatat; Se = Seydişehir; Sh = Shirgesht; Si = Silifke; S = Soltanieh Mountains; Spina Kada (near Kerskhan); Sü = Sülüklü-Eymir; SD = Sultan Dağ; SB = Surkh Bum; Su = Suweilieh; Ta = Tabuk; T = Takab; Ti = Timna; TM = Tubeiq Mountains; Tu = Turaif; UD = Ulu Dağ; U = Umm Bogma (= Abu Durba); WA = Wadi Araba; WD = Wadi Dana; Wk = Wardak; Za = Zanjan; Z = Zarand, Dahu, Desu; Zo = Zonguldak.

domes containing various rocks of the Lower Cambrian/?Pre-Cambrian ('Infracambrian')*
Hormuz Formation and even fossiliferous limestones of Cambrian age. From Iraq, Syria,
Lebanon, and the south-western part of the Arabian Peninsula Cambrian deposits are
unknown.

The thickness of Cambrian sediments which thus cover large parts of the Middle East vary
from a few hundred metres in the Arabian shelf region to more than 3000 metres in eastern and
northern Iran. After geosynclinal sedimentation in parts of the Tethyan belt during the
Mesozoic and early Tertiary, the Cambrian sediments of that region were strongly folded during
several Alpine phases. The Hormuz salt plugs of southern Iran pierced at random through the
Tertiary folds of the Zagros Ranges. They are found in every imaginable geometrical position,
the semi-arid climate revealing the complicated salt tectonics. In the Persian Gulf the salt domes
form small islands. On the Arabian shelf the exposed Cambrian deposits are in most cases
tectonically undisturbed and only slightly tilted eastwards.

2.1. History of Cambrian studies

BLANCKENHORN (1912) is the author of the earliest report on Cambrian fossils from the
Middle East. From the south-east corner of the Dead Sea, Jordan, he mentioned a small
collection of brachiopods and trilobites which he found in a limestone formation already
described by HULL (1886). Further descriptions of Cambrian faunas from the Dead Sea area
were published by DIENEMANN (1915), KING (1923), and PICARD (1942), whereas RICHTER
(1941) revised the collection of BLANCKENHORN. RICHTER recognized the West-Pacific
character of the Jordan fauna and postulated a Middle Eastern Tethys to be existent as early as
in the Cambrian. The first subdivision of the Cambrian deposits of southern Jordan was
proposed by QUENNELL (1951). Further contributions to the regional aspects of the stratigraphy
of the Cambrian System in southern Jordan were made by a group of German geologists
(BENDER, 1964, 1968).

BLANCKENHORN (1914) again was the first to mention the presence of Cambrian deposits in
the north-western part of the Arabian Peninsula. Extensive researches on the Cambrian rocks of
north-western Saudi Arabia were made much later by geologists of the Arabian–American Oil
Company and the United States Geological Survey who embarked upon the geological mapping
of the Arabian Peninsula (THRALLS and HASSON, 1956; STEINEKE and others, 1958;
BRAMKAMP and RAMIREZ, 1958; BRAMKAMP and others, 1963a–c; BROWN and others,
1963a,b; U.S. GEOLOGICAL SURVEY and ARABIAN–AMERICAN OIL COMPANY, 1963). This
enormous project was completed with publication of *The Geology of the Arabian Peninsula*
(POWERS and others, 1966), a source of much valuable information. In the Oman region, LEES
(1928) had already observed pre-Permian formations which were more precisely described as
probable Cambrian rocks by MORTON (1959) and especially by TSCHOPP (1967).

Another locality with Cambrian fossils was reported by BLAKE (1935) in the Timna area,
southern Negev, Israel, where stratigraphical knowledge of the Cambrian has been summarized
by BENTOR (1960). The latest discovery of fossiliferous Cambrian limestones within the

---

* In the Middle East, a barren group of cherty dolomites and red or variegated micaceous shales and sandstones up to
2 or 3 kilometres in thickness, having a wide distribution, is designated 'Infracambrian'. Upwards, these rocks pass
gradually into the biostratigraphically dated Lower Cambrian rocks. From the Pre-Cambrian basement complex, the
Infracambrian is separated either by a pronounced angular unconformity or by a less distinct (non-angular) disconformity. The Infracambrian is thus more closely linked with the Cambrian System than with the Pre-Cambrian basement
complex. At least partly, it may be still Pre-Cambrian in age (slightly modified definition of 'Infracambrian' of
STÖCKLIN, 1972).

Arabian platform was made by OMARA (1972), who found small archaeocyathids of Early Cambrian age in south-western Sinai.

In the Tethyan mountain chains, the northern part of the Middle East, LEES was the first to find Cambrian fossils in the salt plug Al Buza, Persian Gulf, Iran (see GREGORY, 1929). KING (1930, 1937) completed studies on the Cambrian fossils of Iran. Twenty-five years later came research on Cambrian rocks and faunas of Iran by HUCKRIEDE and others (1962) in the south-east Iranian Kerman area, and by the Iranian Geological Survey in the north Iranian Alborz Mountains (RUTTNER and others, 1968). The Cambrian deposits of the Alborz Mountains were further described by GANSSER and HUBER (1962) and ASSERETO (1963), who introduced part of a stratigraphical terminology of Cambrian rocks which is still valid. The second period of exploration of Cambrian deposits in Iran started by STÖCKLIN (1961) and HUCKRIEDE and others (1962) produced most important results concerning the age of the Hormuz Formation and the composition and stratigraphical correlation of Iranian Infra-cambrian deposits.

We owe the first references to Cambrian rocks of Turkey to TROMP (1941) and TASMAN (1949). They attributed sediments of more than 2000 metres thickness in the Mardin-Derik area, south-eastern Turkey, to the Cambrian or Cambro-Ordovician without being able to confirm that age by means of exactly determined fossils. It was STUBBLEFIELD (in TOLUN and TERNEK, 1952) who produced palaeontological evidence of Middle Cambrian age of part of these Lower Palaeozoic rocks. Geologists of the Shell Petroleum Company of Turkey discovered trilobites of Middle Cambrian age within an area of the Amanos Mountains where earlier geological maps show mainly undifferentiated Palaeozoic rocks (DEAN and KRUMMENACHER, 1961). KETIN (1966), HAUDE (1969), and FLÜGEL (1971) summarized existing knowledge of the Cambrian rocks of Turkey. Additionally, KETIN studied Cambrian outcrops in the Penbegli-Tut region and found there rocks with the Middle Cambrian genus *Paradoxides*. The latest contribution to knowledge of Cambrian rocks in south-western Anatolia, western Turkey, is by HAUDE (1972).

The first fossiliferous Cambrian deposits of Afghanistan were discovered in the central part of the country by WOLFART (1969) during a mapping campaign (1959–68) undertaken by the Afghan Geological Survey and a group of German geologists in central and southern Afghanistan. A preliminary synthesis of existing knowledge of the Cambrian System in central southern Asia (Iran to northern India), including the palaeontological description of south-east Iranian and Afghan Cambrian faunas, was completed by the author just before preparing this contribution (WOLFART and KÜRSTEN, 1974).

2.2. The Cambrian rocks of the Arabian Peninsula

Two areas of Cambrian sequences are known from the Arabian Peninsula, one along the eastern margin of the crystalline core of the Arabian Shield in north-western Saudi Arabia, southern Jordan, Israel, and south-western Sinai, and the other in the Oman region (Fig. 2). The first area is mainly characterized by relatively thin Cambrian deposits resting unconformably on Pre-Cambrian metamorphic and igneous rocks and conformably overlain by Ordovician rocks. In the second area, the Cambrian sequences are rather thick and unfossiliferous. They cover the Pre-Cambrian basement complex, the peculiarities of which are not known.

2.2.1. The north-western Saudi Arabian–Jordan region

In the north-western Saudi Arabian–Jordan region, autochthonous sequences crop out along the eastern margin of the Dead Sea rift valley extending southwards from approximately 31°30'

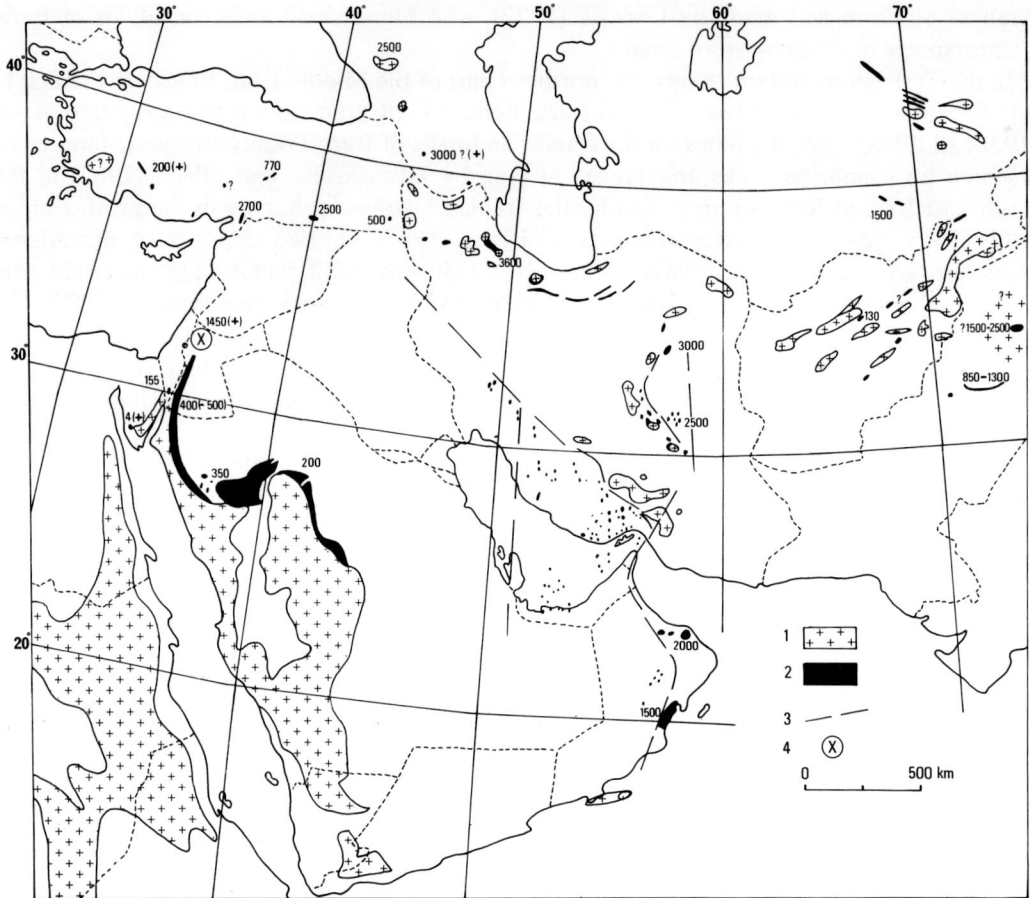

FIG. 2. Known areas of outcrops of Pre-Cambrian and Cambrian rocks. (1) Pre-Cambrian at surface;
(2) Infracambrian and Cambrian at surface; (3) trends governing Hormuz salt distribution; (4)
borehole. Total thicknesses of Infracambrian and Cambrian rocks in metres.

northern latitude into Saudi Arabia (QUENNEL, 1951; WETZEL and MORTON, 1959; BURDON,
1959; BENDER, 1964, 1965, 1968; HELAL, 1964, 1965). About 50 kilometres north of the Gulf
of Aqaba, the Cambrian outcrops turn south-south-east. Further Cambrian outcrops occur
along the southern margin of the Great Nefud and in a narrow band from Ha'il south-eastwards
until they disappear in the latitude of Riyadh under a Permian limestone formation. The Saudi
Arabian outcrop belt spans a distance of nearly 1200 kilometres (POWERS and others, 1966). In
the Middle East, the westernmost Cambrian outcrops of the Arabian Shield are small; they are
situated near Timna, southern Negev (BLAKE, 1935; BENTOR, 1960), and in south-western
Sinai (OMARA, 1972).

All of the sequences consist of a basal conglomerate or sandstone (Fig. 3) resting unconformably on the Pre-Cambrian basement complex of igneous rocks or metamorphics. This formation
is overlain by sandstones with shale or siltstone interbeds that are, in turn, followed by a
comparatively thin sequence of limestones and shales in part of the Jordan–Sinai region. The
limestone beds are overlain again by sandstones with shale interbeds. In north-western Saudi

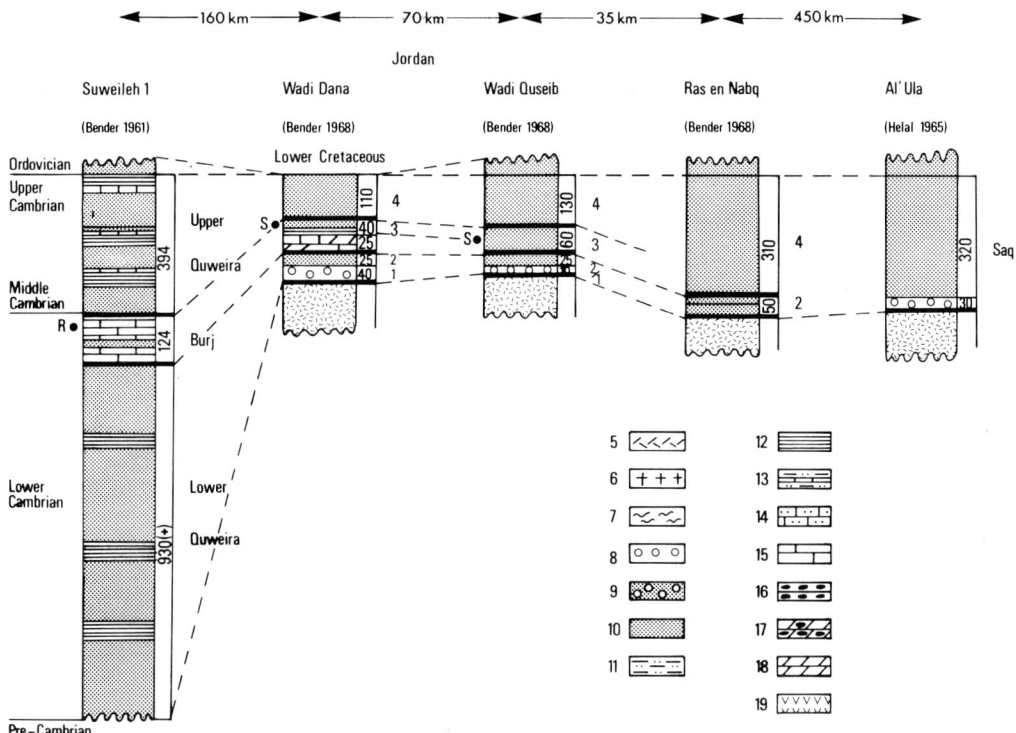

FIG. 3. Representative columnar sections and stratigraphical terminology for the Cambrian rocks of
Jordan and north-western Saudi Arabia. (1) Basal conglomerate; (2) Bedded Arkose Sandstone;
(3) White Fine-grained Sandstone; (4) Massive Brownish Weathered Sandstone; (R) *Redlichops*,
Palaeolenus, Obolus, Trematobolus, Orthotheca; (S) *Skolithos*. The faunal assemblage with *Redlichops* was found in the Burj Formation at the south-eastern end of the Dead Sea. Thicknesses in
metres. (5) Basement complex; (6) volcanics (lavas and tuffs); (7) phyllites and quartzites; (8)
conglomerate; (9) conglomeratic sandstone; (10) sandstone and quartzite; (11) siltstone; (12) shale
or marl; (13) interbedded siltstone; shale, and limestone; (14) sandy limestone; (15) limestone;
(16) chert; (17) cherty dolomite or limestone; (18) dolomite; (19) salt and other evaporites.

Arabia the entire Cambrian sequence is composed of clastic material. In all of the areas, the
Cambrian rocks are conformably overlain by sandstones of Ordovician age.

In Egypt and southern Jordan, the clastic sequences overlying the basement complex were
formerly called 'Nubian Sandstones' since RUSSEGGER (1847). The Nubian Sandstone,
however, consisted of deposits of Cambrian to Cretaceous ages, as could be shown by the
discovery of fossils. Consequently, the Nubian Sandstone which extends far into Saudi Arabia
was subdivided into several formations. The term 'Nubian Sandstone' has now been abandoned.

The basal conglomerate contains boulders, cobbles, and pebbles of white quartz and other
hard components derived from underlying basement rocks. BENDER (1968) observed in the
lowermost conglomeratic beds of southern Jordan boulders with a diameter of 1 metre.
Generally, however, the boulders are well rounded in south-west Jordan at a diameter of 20
centimetres. The matrix consists of arkosic sand. In south-western Jordan, the basal
conglomerates obviously filled basins and channels or valleys cut into the peneplained surface of
the Pre-Cambrian basement complex. Their thicknesses range from 0 to 50 metres on the
eastern bank of the Wadi Araba, partly reflecting original topographical irregularities on the

surface of the Pre-Cambrian terrain. This indicates tectonic activities in the Wadi Araba–Jordan Valley geosuture as early as late Pre-Cambrian times. In the eastern part of southern Jordan the surface of the Pre-Cambrian terrain is a peneplain.

In north-western Saudi Arabia, the basal conglomerates which are considered as a part of the 'Saq Sandstone' transgressed with a marked unconformity over a mature surface of Pre-Cambrian crystalline rocks. POWERS and others (1966) described the basement surface as a remarkably flat peneplain, the erosional relief of which seldom exceeds 10 metres, east and south-east of Ha'il. In that area the surface dips uniformly north-east under the sedimentary blanket with a dip of approximately 1.5°. In the range of the peneplain the basal conglomerates are composed of angular, more rarely subrounded, rock fragments varying much in size. The conglomerates are only locally well cemented. Sometimes they show steep dip, probably owing to initial depositional processes. According to BENDER (1968), the characteristics of the basal conglomerates are indicative of weathering under arid conditions. The morphometrical analysis of the clasts points to a fluviatile environment (LILLICH, 1963, *in* BENDER, 1968).

In north-western Saudi Arabia, the Saq Formation originally included the basal conglomerates as well as the overlying sandstone sequences of Cambrian and Ordovician ages (POWERS and others, 1966; THRALLS and HASSON, 1956; STEINEKE and others, 1958). HELAL (1964, 1965) proposed to confine the term 'Saq Formation' to Cambrian beds only, underlying the early Ordovician *Cruziana* Series. The lower part of the Saq Sandstone comprising sandstones of presumed Early Cambrian age only (POWERS, 1968) was named Siq Sandstone by BRAMKAMP and others (1963).

Saq lithology is strikingly uniform both vertically and laterally. According to POWERS and others (1966), the dominant rock type is buff to grey and white, commonly cross-bedded, poorly to well sorted quartz sandstone that is often friable and commonly hardened to a black, ferruginous quartzite surface. In places, some interbeds of hard, silty rocks occur as well as pebbly bands. The Saq Formation in the sense of HELAL (1964) amounts to a thickness of about 400 metres in the Hisma and Tabuk area. At the type section near 26°27′ northern latitude, the Saq thickness is more than 600 metres (Saq = Cambrian and Ordovician) and a few metres only near northern latitude 24°22′ owing to pre-Permian truncation which progressively cuts deeper into the Saq towards the south. No diagnostic fossils were collected from the Saq Formation. It can, however, be assigned with confidence to the Cambrian because it rests unconformably on the Pre-Cambrian basement and is followed conformably by fossiliferous Lower Ordovician beds. The outcrop surface of Saq rocks is a rough, hummocky plain with isolated wind-sculptured hills rising abruptly from the plain (POWERS and others, 1966).

Northwards, the uniform rocks of the Saq Formation of north-western Saudi Arabia can be traced laterally into various Cambrian deposits of southern Jordan and Israel, which are partly fossiliferous. The Saq Formation is an equivalent of QUENNELL's (1951) Quweira Sandstone in Jordan and of the Georgian rocks in Israel (BENTOR, 1960). BENDER (1964, 1968) published a detailed description of the Quweira unit which is lithologically more variously developed in the range of the Wadi Araba–Jordan Valley geosuture than in south-eastern Jordan. This fact is reflected in different subdivision of Cambrian rocks in different parts of southern Jordan. The next formation following above the basal conglomerate is the Bedded Arkose Sandstone. In places, this formation transgresses immediately over the Pre-Cambrian basement. Its thicknesses range between 20 and 50 metres in the southern part of southern Jordan and increase up to 200 metres towards the Dead Sea. At the base, the Bedded Arkose Sandstone consists of bedded, coarse, brown arkosic sandstones with scattered white quartz pebbles and well rounded granite gravel up to egg-size. The overlying medium to coarse-grained sandstones are rather well sorted, thickly banked, and of light brown to light violet colours. In the upper portion, strongly

FIG. 4. Stratigraphical correlation of the Cambrian rocks of north-western Saudi Arabia, southern Jordan, and southern Israel.

North-western Saudi Arabia

Powers 1966, 1968 / Helal 1964, 1965

- Tabuk — *Didymograptus* Shaly Member
- Ram and Umm Sahm — *Cruziana* Series
- Quweira Sandstone — Saq Formation
- Siq Sandstone
- Pre-Cambrian basement complex

Age column: Ordovician — Middle and upper / Lower (Saq Sandstone); Cambrian — Lower; Pre-Cambrian

Southern Jordan

Quennell 1951, Burdon 1959 / Bender 1964, 1968 / Lloyd in Selley 1972

- Ram and Um Sahm Sandstone — Graptolite Sandstone — Khreim Group
- Upper Quweira — Bedded Brownish Weathered Sandstone — Um Sahm
- Massive Whitish Weathered Sandstone — Disi
- Burj — Massive Brownish Weathered Sandstone — Ishrin
- White Finegrained Sandstone
- Dolomite Limestone Series
- Lower Quweira — Bedded Arkose Sandstone — Saleb
- Basal conglomerates
- Quweira Series — Slate-Greywacke Series
- Saramuj Series — Saramuj Conglomerates — Pre-Cambrian igneous basement complex
- Aqaba Granite Complex — Igneous rocks and metamorphics

Southern Israel

Bentor 1956, 1960

- Lower Variegated
- Nubian Sandstone
- Upper Shales, Zebra Sandstone, Dolomite Series
- Lower Shales
- Palaeozoic Nubian Sandstone
- Eilat Conglomerate
- Eilat Schist

ferruginous and bleached, fine-grained sandstones are interbedded locally with clay galls. According to BENDER's conclusion, the Bedded Arkose Sandstone was deposited under terrestrial environmental conditions the same as those for the basal conglomerate.

The next rock unit conformably overlying the Bedded Arkose Sandstone is the Massive Brownish Weathered Sandstone, which is recognized as a correlative unit of the upper portions of QUENNELL's Quweira Sandstone in the southern part of southern Jordan. The thicknesses of the formation vary between 250 and 350 metres in the Southern Jordan Ram area. The formation consists of massive or thick-bedded, frequently cross-bedded sandstones which form a distinct escarpment. The whole sequence is of medium to coarse-grained sandstone or brown, red–brown, dark violet, and yellowish colours. Arkosic layers, scattered gravel, and quartz pebbles are frequent. In the middle and upper portions there are silty, micaceous shale interbeds. No fossils have been found in the Massive Brownish Weathered Sandstone, the lithological characteristics of which suggest a terrestrial type of sedimentation. The age of the formation is determined by the underlying basement rocks and the overlying fossiliferous rocks of Lower Ordovician age as well as by quartz porphyries intruded into the lower portions of the Cambrian sequence. The radiometric age of the porphyries is about 530 (±50) m.y. The upper contact is of transitional character with gradual increase of white sandstone interbeds.

Towards the north-west, the lower part of the Massive Brownish Weathered Sandstone can be traced laterally into the various marine sediments of the Wadi Araba–Dead Sea area; the upper part, however, overlaps the marine formations of that area. In the central Wadi Araba area, the Bedded Arkose Sandstone is immediately overlain by the White, Fine-grained Sandstones of marine origin. Northwards, the lower portion of this formation passes into marine dolomites and limestones. The upper portion extends above these carbonates into the Dead Sea area. The White Fine-grained Sandstones—sometimes they are slightly greenish to reddish—are well sorted and up to 110 metres thick. Some beds contain, as do the terrestrial Cambrian formations east of the Wadi Araba area, scattered angular fragments of quartz porphyries which were interpreted by BENDER (1968) as volcanic ejecta. Along the central and northern part of the eastern bank of the Wadi Araba, the White Fine-grained Sandstones are copper-mineralized over a distance of about 70 kilometres. Locally, low-grade disseminated copper ores have been found in Cambrian quartz porphyries. Manganese ore is the main constituent of the mineralization in places. BENDER (1965) came to the conclusion that the syngenetic mineralization of the sandstones was caused by a contemporaneous quartz porphyry volcanism (Section 2.8). Beds with (?)Scolithus tubes are distributed over the whole area of the White Fine-grained Sandstones. The upper contact is marked by local erosional unconformities.

In the northern part of the Wadi Araba area, the Bedded Arkose Sandstone is overlain by a sequence of dolomites, limestones, and shales. From the Dead Sea area, the Cambrian carbonates have been known as 'Wadi Nasb Limestone' (HULL, 1886). QUENNELL (1951) named them 'Burj Limestone Group'; WETZEL and MORTON (1959) described them as 'Formation calcaire et marno-greseuse de Burj', consisting of 20 metres of sandy shales in the lower portion and massive dolomites and limestones with chert, of 30 metres thickness, in the upper portion. The thickness of the Burj Limestone Group increases northwards up to more than 124 to 135 metres, as proved by the borings Suweileh, about 20 kilometres north-west of Amman and Safra, about 40 kilometres east-south-east of Amman. The upper parts of the dolomites, some shaly layers, and some limestone horizons have yielded the trilobites *Redlichops*, *Palaeolenus*, *Hesa*, and *Protolenus*, the inarticulate brachiopods *Obolus*, *Trematobolus*, and *Micromitra*, and the hyolithid *Orthotheca*, which were described by KING (1923), RICHTER (1941), and PICARD (1942). RICHTER and PICARD assigned the fauna to the Lower to Middle Cambrian boundary or rather to the upper Lower Cambrian.

The uppermost Cambrian unit in the range of the Dead Sea area and north of it was named Upper Quweira Sandstone by QUENNELL (1951). It is the correlative unit of BENDER's upper portion of the Massive Brownish Weathered Sandstone and characterizes the transition from marine to continental environments in the range of the Wadi Araba–Jordan Valley geosuture. Its thickness varies from 110 metres south of the Dead Sea to nearly 400 metres at Suweileh, north of the Dead Sea. Upper Quweira lithology is of a marked terrestrial type at Suweileh. The formation consists of sandstones with some interbeds of shales and limestones (BENDER, 1961).

In the Timna area, southern Negev, Israel, BLAKE (1935) was the first to attribute the lower part of the Nubian Sandstone to the Cambrian on the basis of his finds of *Obolus*. BENTOR (1956, 1960) published more detailed information about the Nubian Sandstone sections in the Timna area, unconformably overlying the basement complex which is dominantly composed of granite, syenite, and monzonite. BENTOR tentatively subdivided the Cambrian part of the Nubian Sandstone into six units.

The lowermost unit consists of pinkish, cross-bedded fluviatile sandstones varying from 0 to 80 metres in thickness. It rests on a very irregular erosional surface of the basement complex. The second unit, termed Lower Shales, is composed of a basal conglomerate, red sandstone, and shale. It has a constant thickness of 12 to 13 metres and was interpreted by BENTOR to be of marine origin and Late Georgian age. The Dolomite Series originally considered to be Ordovician in age (BENTOR, 1956) has a maximum thickness of 23 metres and consists of well bedded dolomite, limestone, and dolomitic sandstone. Corresponding to its striped appearance, which is caused by alternating black and white layers, the fourth unit was called Zebra Sandstone. Thicknesses vary up to 16 metres because the formation rests on the very irregular surface of the Dolomite Series (Nimra Formation). COOPER (1976) described some new articulate brachiopods (*Trematosia, Glyptoria, Psiloria, Israelaria, Leioria*) from the late Early Cambrian Nimra Formation in the southern Negev (Israel). The upper shales are of olive-green, black, and red colours and are only 3 metres thick. Plastic flow of these shales has resulted in highly contorted strata in some places. Besides concretionary manganese ore, they carry a trilobite fauna which is indicative of a very early Middle Cambrian age according to an unpublished report by PARNES (*in* BENTOR, 1960). Part of the overlying terrestrial Nubian Sandstone may, therefore, still belong to the Cambrian System. To the present author an Early Cambrian age of the marine Timna units seems to be more probable than Middle Cambrian (*see also* PARNES, 1971). On the basis of the occurrence of *Strenuella, Timnaella, Kingaspis*, and *Myopsolenus*, PARNES assigned the fauna a late Early Cambrian age.

All of the Cambrian sequence, up to the Upper Shales inclusively, has a total thickness of 155 metres in the Timna area. With diminishing thicknesses it extends southwards to the Sinai border and beyond.

The first surface exposure of fossiliferous Cambrian rocks from the Sinai Peninsula was reported by OMARA (1972). In the Abu Durba area, south-western Sinai, he found a limestone bed underlying the huge 'Cambro-Ordovician' sandstone series near the Abu Durba granite. The limestone crops out as a narrow, isolated cuesta in juxtaposition to the granitic mass of Abu Durba. Its thickness is not accurately measurable but it may be about 4 metres. It is a creamish, white, hard, compact limestone mainly composed of a hard, compact calcarenite with minute laminated structures which have been identified as algal stromatolites. They are similar to Lower Cambrian stromatolites described by KRYLOV (1967) from southern Kazakhstan. Besides stromatolites, the main organic remains belong to archaeocyathids of a very small and primitive type. Consequently, OMARA assigned the limestone near Abu Durba to the Lower Cambrian. It is capped by a very hard, compact quartzite which yielded an indeterminable oboloid brachiopod faunule. The overlying Nubian facies sandstones with many shaly intercalations are about

600 metres in thickness. They are, in turn, overlain by Carboniferous rocks. The lower members, which constitute one quarter of the Nubian Sandstones, are predominantly reddish or brownish in colour, forming a complex of red beds. The upper three quarters are mostly buff in colour. OMARA left no doubt that he considers the Nubian Sandstone complex essentially to be Lower Palaeozoic in age. At least the red bed complex presumably must be assigned to the Cambrian System.

2.2.2. The Oman region

Comparatively little detailed information about the Cambrian rocks of the Oman region is available owing to difficult access until the years 1948–52. The first one to presume the presence of Cambrian rocks in the region was LEES (1928), who considered it possible that an arm of the Cambrian sea extended into what is now the great desert of Rub al Khali. He assumed the Cambrian shorelines to be governed by the strike of the Pre-Cambrian rocks. According to field observations and exploration wells, Cambrian sequences in this region are less homogeneous than those of the north-western Saudi Arabian–Jordan region (MORTON, 1959; TSCHOPP, 1967).

In the Oman Mountains and in the Haushi-Huqf area, clastics and carbonates were deposited in two sedimentary cycles during the Cambrian. At approximately the same time a thick series of salt, other evaporites, and clastics was formed in the desert plains. Continental sedimentation occurred in the late Cambrian to early Ordovician near the Haushi-Huqf area, passing westwards and northwards into shallow marine deposition. The two sedimentary cycles underlie Permian formations both in the Oman Mountains and in the Haushi-Huqf area. The observed total thickness of both cycles is more than 1800 metres in the Oman Mountains and more than 1300 metres in the Haushi-Huqf area. The crystalline basement complex has not yet been found in Oman. The nearest outcrops of basement rocks are known from Murbat, Dhofar, and the Kuria Muria Islands near the Dhofar Coast.

In the Haushi-Huqf area, Abu Mahara sandstones and gypsiferous shales at the base of the first sedimentary cycle were formed under fluviatile conditions. According to HENSON and ELLIOTT (1958), they are more than 516 metres thick. Khufai cherty dolomites and limestones are 289 metres thick and indicate a marine transgression bringing in tidal-flat conditions. A terminal regression caused evaporitic sedimentation. The second sedimentary cycle begins with the Shuram Formation consisting mainly of clastics. Its base is marked by a very sudden incoming of clastic sediments which may indicate an episodical uplift of the source area. HENSON and ELLIOTT described the Shuram and red and green mottled, silty, micaceous shales grading into laminated, dolomitic mudstones (271 metres). The following Buah Formation is composed of dolomitized tidal-flat limestones 252 metres in thickness. From Mukhaibah, near Ras Duqm, HENSON and ELLIOTT obtained some *Collenia*-type stromatolites occurring in the middle part of the Buah Formation.

TSCHOPP (1967) described the Mistal Formation at the base of the first sedimentary cycle in the Jebel Akhdar part of the Oman Mountains as conglomeratic clastics containing basement rock types and resembling tillites. The Mistal Formation is assumed to be synchronous with the Abu Mahara Formation and was also formed in a terrestrial environment. TSCHOPP published no information about the thickness of Cambrian formations in the Oman region. The basal Mistal Formation is overlain by the Hajir Formation, which is composed of shallow marine sandstones and thin tidal-flat limestones and sandstones. The second sedimentary cycle comprises the Miaidin and Kharus Formations, which correspond with the Shuram and Buah Formations of the Haushi-Huqf area. The Miaidin Formation is represented by a monotonous succession of siltstones, sandstones, and subordinate limestone bands. The Kharus Formation

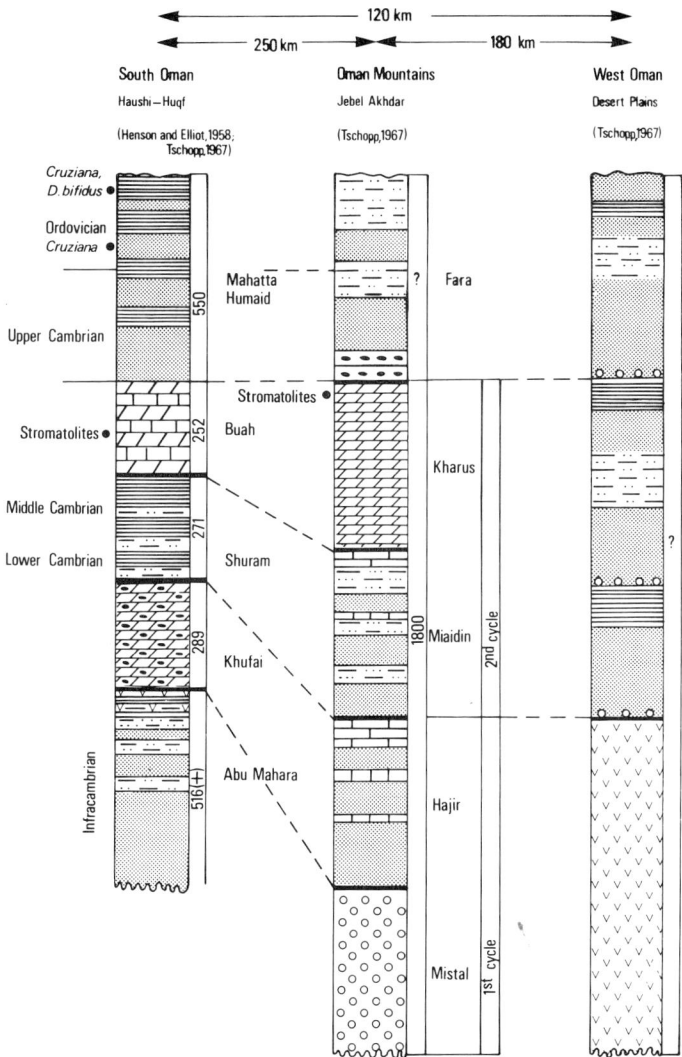

FIG. 5. Representative, partly schematic columnar sections and stratigraphical terminology for the Infracambrian and Cambrian rocks of the Oman region. Legend: see Fig. 3.

consists mainly of dolomites with silicified algal stromatolites in the uppermost part. According to TSCHOPP, the Hatat Metamorphics and Hiyam Dolomite are equivalents of (? parts) of the Huqf Series in the Sayh Hatat area, Oman Mountains. The metamorphics are a complex of green, highly sheared, epimetamorphosed phyllites and chloritic, sericitic schists, the base of which is not exposed. With a distinct lithological break, the Hatat Metamorphics are overlain by the light coloured Hiyam Dolomite, which is assumed to be synchronous with the Buah Dolomite.

In the Haushi-Huqf area as well as in the Oman Mountains, no diagnostic fossils are present in the rocks of the two cycles. Their age possibly ranging from Pre-Cambrian—'Infracambrian'—to Middle Cambrian can only be obtained by indirect arguments. TSCHOPP considered the absence of any shelly fauna in the sediments of the two cycles as an indication of Pre-Cambrian age. The overlying sequences, however, are graptolite shales of early Ordovician age encountered in the well Ghaba, north of Haushi-Huqf. These graptolite shales are, in turn, underlain by a monotonous clastic sequence with linguloid brachiopods tentatively dated as late Cambrian. In the Haushi-Huqf area the two sedimentary cycles are unconformably overlain by that Upper Cambrian–lower Ordovician sequence.

Large areas of Oman are occupied by evaporites which crop out in some piercement salt domes in southern Oman and have been encountered in wells drilled in western and southern Oman. As a result of geophysical research, they are also supposed to underlie several subsurface structures. TSCHOPP mentioned two explanations concerning the origin of the evaporites, the first of which seems to be the more probable. It considers the Oman evaporites as a correlative unit of the Iranian Hormuz Formation and interprets them as a thick sequence deposited in large basins which were bordered by highs. According to that interpretation the Huqf Series formed at approximately the same time on the positive elements bordering the basins. Based on the fact that wells in Dhofar encountered cherty dolomites and limestones beneath the salt horizon, the second explanation postulates, provided that the dolomites and limestones belong to an equivalent unit of the Buah Formation, the evaporites to be deposited during the course of the regression which commenced at the top of the secondary sedimentary cycle. The evaporites would thus be post-Buah in age.

		Desert Plains	Haushi-Huqf		Oman Mountains	
		Haima-Fahud			Sayh Hatat	Jebel Akhdar
Ordovician		clastics	Mahatta Humaid		Amdeh Quartzite	Fara
Cambrian	Upper					
	Middle	clastics	Huqf Series	Buah	Hiyam Dolomite	Kharus
	Lower			Shuram		Miaidin
Infracambrian		salt and other evaporites		Khufai	Hatat Metamorphics	Hajir
				Abu Mahara		Mistral

FIG. 6. Correlation of the Infracambrian and Cambrian formations of Oman (TSCHOPP, 1967).

In the Haushi-Huqf area and in the Oman Mountains uplift and erosion followed after the two sedimentary cycles. Subsequently in the Haushi-Huqf area the Mahatta Humaid Formation, about 550 metres in thickness, was deposited. It consists of sandstones with coloured chert grains and argillaceous siltstones. The lower portion of the Mahatta Humaid is characterized by poor sorting, coarse grain, cross-bedding, variegated colours, and absence of fossils. In the opinion of TSCHOPP it was probably formed in a continental environment, perhaps in sand dunes or large alluvial fans along the western edge of an uplifted Huqf massif. Westwards and northwards, the continental Mahatta Humaid sediments probably pass into sediments of shallow marine origin carrying the Upper Cambrian *Lingulella* cf. *nicholsoni*.

In the Jebel Akhdar part of the Oman Mountains, the Fara Formation may be the equivalent of the Mahatta Humaid. The Fara is composed of often laminated and brecciated, mainly black, chert in the lower part. The higher levels contain siltstone and quartz sandstone. In the Sayh Hatat area of the Oman Mountains, the Amdeh Quartzite is the correlative unit of the Mahatta Humaid. It is a thick, monotonous, uniform sequence of more or less silicified sandstone with thin epimetamorphosed shale. The degree of metamorphism is lower than that of the Hatat Metamorphics.

2.3. The Cambrian rocks of Turkey

The description of the Tethyan mountain region includes the Cambrian rocks in Turkey, Iran, and Afghanistan (Fig. 2). In the following, the southern provinces of the Soviet Union, the Cambrian outcrops of which are partly taken into consideration in Figs. 12–15 are not described in detail. Thus, the northern boundary of detailed description of Cambrian rocks generally coincides with the southern frontier of the Soviet Union. Turkey and Iran are mainly characterized by thick Cambrian deposits resting unconformably on the Pre-Cambrian basement complex, which is more exposed in Iran and Afghanistan than in Turkey. In Central Afghanistan, fossiliferous Cambrian sequences, in contrast, are very thin and up to the present precisely known from two regions only. Presumed Cambrian rocks of considerable thicknesses are known, however, from the Logar trough (Arghandab basin) and from the Badakhshan trough.

In Turkey, only a small number of limited outcrop areas from the Sultan Dağ Mountains, Amanos Mountains, Penbeğli-Tut, and Derik, west of Mardin, contain Cambrian sections with significant fossils which have already been described. The known Cambrian rocks of Turkey are confined to the south, i.e. to the two major tectonic units of the Taurids and the Border Folds. Especially in that part of the Taurus Mountains which borders the Mediterranean Sea, Lower and Middle Cambrian trilobites have been found in the Çal Tepe Formation near Beyşehir (DEAN, 1975). The trilobites are of western European and Mediterranean type. Middle Cambrian rocks in the Amanos Mountains and near Derik also include shales containing eastern North American faunal elements.

From the area south of Hakkari, in the south-easternmost part of Turkey, ALTINLI (1966) mentioned a sequence of thick limestones the Cambrian age of which has not yet been confirmed by fossils. DEAN (1975) classified the Hakkari succession as being of Arenig age in the upper part and late Pre-Cambrian ('Infracambrian') through Early Cambrian age in the lower part. Since TOLUN and TERNEK (1952) demonstrated for the first time the presence of fossiliferous Cambrian rocks in Turkey, only a few comprehensive studies on the rocks of the Cambrian System in Turkey have been made. The most detailed modern presentation of present knowledge about the Cambrian rocks in Turkey is given by DEAN (1975).

All of the sequences consist of an unfossiliferous lower part, of clastics attributed to the Infracambrian and Lower Cambrian which cover the Pre-Cambrian basement complex that is insufficiently known up to the present. The clastic formations are overlain by carbonates still partly assigned to the Lower Cambrian. In the Western Taurus Mountains, the upper portion of the carbonates yielded the oldest fossils from Turkey which are Early and Middle Cambrian in age (DEAN 1975). The carbonates are, in turn, followed by mainly clastic unfossiliferous formations again which are attributed to the Middle and Upper Cambrian. In some of the areas, the Cambrian rocks are conformably overlain by clastics of Ordovician age. Between Sultan Dağ in western Turkey and the area of Mardin in eastern Turkey, the correlative units of the sequences are amazingly uniformly developed over a distance of more than 800 kilometres. In spite of this fact, different local stratigraphical nomenclatures have been proposed for each outcrop area, to such an extent that Cambrian stratigraphy in Turkey seems to be rather complicated. It was HAUDE (1969) who recognized this first and tried to summarize present knowledge and to correlate the different local stratigraphies.

In the Sultan Dağ Mountains, the Cambrian section begins with the Ardiçli-Tepe Formation, which is about 100 metres in thickness (HAUDE, 1972). It consists of dark, pelitic shales grading into a greenish grey, fine-grained sandstone towards the upper portion. In the lower part, the

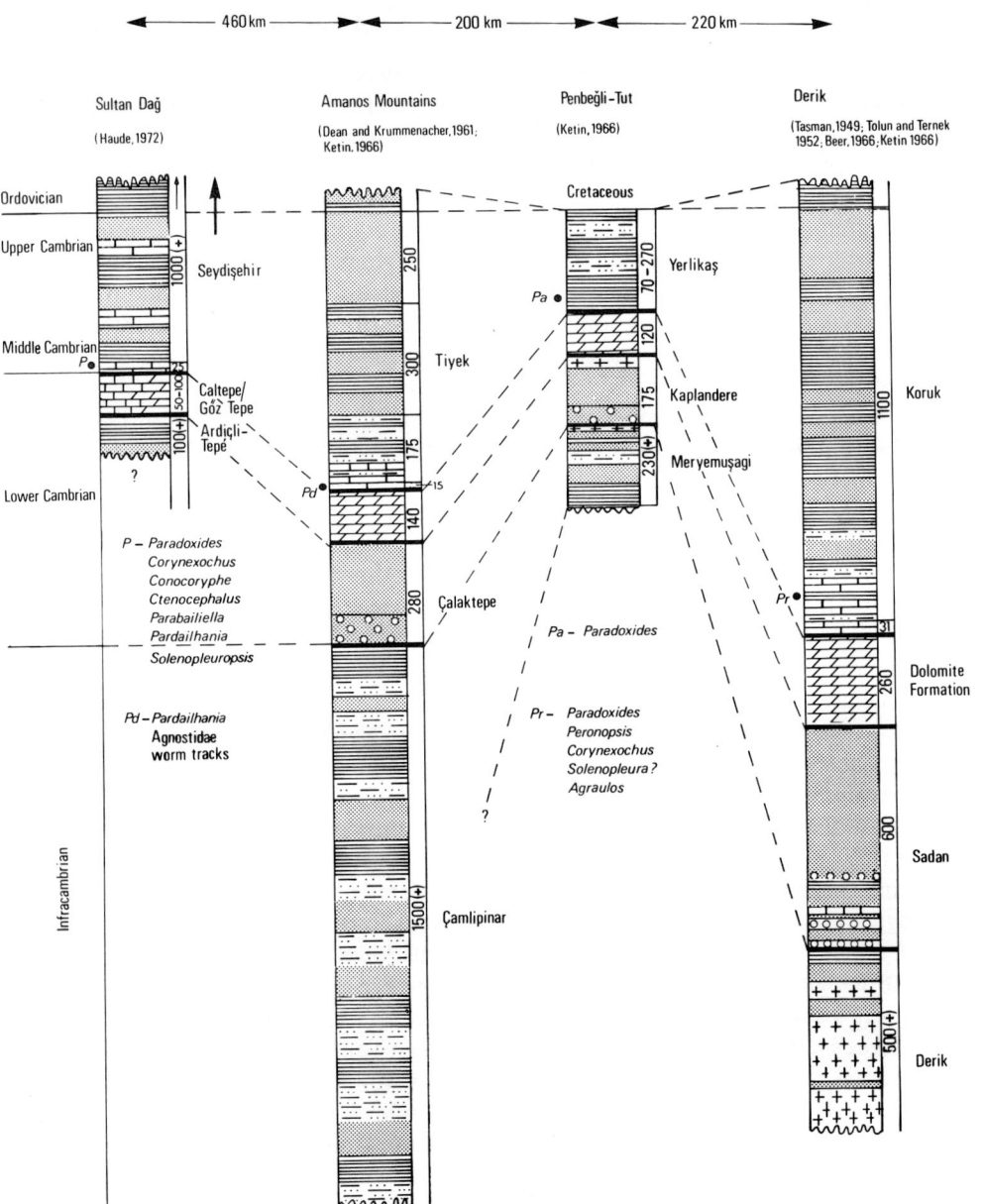

FIG. 7. Representative columnar sections and stratigraphical terminology for the Infracambrian and Cambrian rocks of Turkey. Legend: see Fig. 3.

shales are also slightly psammitic. The sandstone of the upper Ardiçli-Tepe is overlain by a limestone formation representing a key horizon of all of the Lower Palaeozoic sequences in southern and south-eastern Turkey. In the western Sultan Dağ area, the limestone is developed as a black, compact, crystalline rock that is only slightly dolomitized. Its thickness ranges between 50 and 100 metres. HAUDE (1972) named the western facies as the Çaltepe Formation corresponding to the section at the Çaltepe near Seydişehir, already described by BLUMENTHAL (1947) and DEAN and MONOD (1970). In the opinion of the latter authors, the Lower/Middle Cambrian boundary must be placed within the Çaltepe Formation. In the eastern part of the Sultan Dağ Mountains, especially north of Lake Beyşehir, the synchronous rocks of the Çaltepe Formation are in general totally dolomitized and ankeritized and dark brown in colour. They are named the Göz Tepe Formation, which is as thick as the Çaltepe. The rocks of the Ardiçli-Tepe exhibit a similar regional differentiation in facies, as do the overlying carbonates. In the western Sultan Dağ area the formation consists of non-metamorphosed shales and in the eastern part of phyllitic shales.

The following formation, the Phacoidal Limestone, is about 25 metres thick and also differentiated into western and eastern facies. In the western area it consists of red, nodular limestones with a red, shaly matrix. In the eastern area the limestones arc whitish to greenish; the shales are greenish. As a significant feature in the eastern Sultan Dağ, the Phacoidal Limestone is always overlain by a Middle Cambrian diabase which substitutes all of the limestone in places. In the upper part the Phacoidal Limestone grades into shales. In the transitional portion between limestone and shale HAUDE found the oldest fossil of the Sultan Dağ. After the preliminary identifications of SDZUY (in HAUDE, 1972), the fauna is Middle Cambrian in age and contains the following trilobites: *Paradoxides, Corynexochus, Conocoryphe, Ctenocephalus, Parabailiella, ?Jincella, Pardailhania,* and *Solenopleuropsis.*

Conformably above the Phacoidal Limestone follows the Seydişehir Shale (BLUMENTHAL, 1947; MONOD, 1967) in the western Sultan Dağ area and the Kuruderebasi Shale (HAUDE, 1972) in the eastern area. The two formations constitute a section of shales, fine-grained sandstones, and interbedded limestones which are more than 1000 metres in thickness. Both formations are differentiated only by the degree of metamorphism. Presumably they comprise rocks of Middle and Upper Cambrian, Ordovician, Silurian, and Devonian age. In the Seydişehir Shale HAUDE found a faunal association of Tremadoc age; MONOD (1967) and DEAN and MONOD (1970) found another association of Arenig age. The boundary between the Cambrian and the Ordovician could not yet be fixed.

The area south-east of Beyşehir, western Taurus Mountains, contains a succession of Cambrian and Ordovician rocks and faunas which is said to be one of the most completely documented in Turkey (DEAN, 1975). The complete lithostratigraphical succession is as follows:

Sobova Fm.	{ Grey Shale Member	*ca.* 20 m
	{ Limestone Member	0–10 m
Seydişehir Fm.	Shale, sandstone, limestone	>1000 m
	{ Red nodular limestone	40 m
	{ Light grey limestone	10 m
Çal Tepe Fm.	{ Black limestone	30 m
	{ Dolomite	>50 m
	{ (base of formation not seen)	

The lowermost part of the Çal Tepe Formation, the dolomite, is unfossiliferous, but the black limestone yielded the oldest fossils (Early Cambrian) known so far from Turkey. The lower part of the succeeding grey limestone contains Early Cambrian trilobites, too. The Lower/Middle Cambrian boundary is believed to lie within this light grey limestone. The red, nodular limestone marks a change in lithology. At two levels it contains corynexochids, paradoxidids, and solenopleurids, which are of Middle Cambrian age, as well as overlying shales of the Seydişehir Formation. The upper part of that formation yielded graptolites of the *Didymograptus extensus* Zone (lower zone of the Arenig Series). In the highest strata of the Seydişehir Formation are *Bactroceras*, *Proterovaginoceras*, *Protocycloceras*, and the trilobite *Symphysurus* suggesting a possible late Arenig age. In the succeeding Sobova Formation, which is definitely of late Arenig age, *Symphysurus* becomes more abundant. Since no evidence of Upper Cambrian or Tremadoc faunas has yet been found, the Cambro-Ordovician boundary cannot be drawn at the present.

ERENTÖZ (1966) reported a sequence of yellowish glistening shales with abundant trilobites, unconformably surmounting formations of probable Pre-Cambrian age in the mountains of Anamur between Antalya and Silifke. The shales are of Cambrian age according to ERENTÖZ, who did not provide detailed information about the fossils. They are, in turn, overlain by Ordovician rocks. This Lower Palaeozoic sequence presumably has already been referred to as Devonian by BLUMENTHAL (1947) from the southern Turkish coast between Silifke and Gilindir. According to DEAN (1975), the lowest strata seen in that region are silty, micaceous, brown–grey shales which yielded indeterminate asaphid trilobites and *Didymograptus* of extensiform type, suggesting an early Arenig age.

The next area of outcrop of Cambrian rocks to the east is situated in the Amanos Mountains where they are exposed in the central part near Hassa and Saylak. The Cambrian of the southern Amanos (or Giaour Dağ) was first studied by KRUMMENACHER (*in* DEAN and KRUMMENACHER, 1961; BAYKAL and ATAN, 1965). The Amanos succession is the thickest of the Infracambrian/Cambrian sequences in Turkey. It has been divided into four units which were called A to D in ascending order by KRUMMENACHER and named as Çamlipinar, Çalaktepe, Dolomite, and Tiyek Formations by KETIN (1966). Continuous successions can be followed only for short distances in the Amanos owing to the broken nature of its terrain. The best exposed section that has been examined lies west-south-west of Saylak.

The sedimentary sequence begins with the Çamlipinar Formation (=Formation A) which is more than 1500 metres in thickness and consists of alternating, well bedded, quartzitic sandstones, phyllitic shales, and greywackes, dark red to dark greenish brown in colour. The base of the Çamlipinar Formation is not exposed.

KRUMMENACHER reported that the beds of Formation B (=Çalaktepe Formation) are separated by a strong angular unconformity from the unfossiliferous rocks of the Infracambrian/Lower Cambrian Çamlipinar Formation. HAUDE (1969), however, suggested the unconformity to be simulated by a local fault, according to a cross section published by KETIN (1966). The Çalaktepe Formation is composed of massive, pinkish to red–violet quartzites, conglomeratic arkoses, and interbedded ferruginous, micaceous shales. Pebbles of pale quartzite, dark schist, and igneous material occur throughout, and cross-bedding is common. The rocks become more ferruginous and shaly towards the top. No fossils have been found but, because of its stratigraphical position and lithological characteristics, the formation can be recognized as a correlative unit of the Sadan and Kaplandere formations in the Mardin and Penbeğli-Tut regions. KRUMMENACHER reported a thickness of about 600 metres, but according to KETIN the average thickness amounts to only 280 metres.

There is a gradual passage to Formation C (=Dolomite Formation) the lower part of which consists of lenses of blue–grey limestone and coarsely crystalline, dark, dolomitic limestone in a ferruginous, shaly matrix. The upper part of the succession is formed by coarsely crystalline, dark blue to black, dolomitic limestones or dolomites with thin, shaly intercalations. According to KRUMMENACHER the thickness of the formation amounts to 185 metres whereas KETIN gives 100 to 150 metres. Being unfossiliferous, the Dolomite Formation has been assigned to the Lower Cambrian because it is conformably overlain by the Tiyek Formation, which has yielded Middle Cambrian fossils. For the same reason the underlying Çalaktepe Formation can confidently be attributed to the Lower Cambrian in the Amanos Mountains.

South of Tiyek, the Tiyek Formation (=Formation D) presumably comprises Middle and Upper Cambrian deposits and can be divided into three units of a total thickness of 750 metres, according to KETIN. KRUMMENACHER reported a much different total thickness of 250 metres from the area near Saylak. This corresponds to the observation that the rocks of Formation D are separated by a disconformity from the following Ordovician sequence. Near Saylak, Formation D consists of current-bedded, calcareous shales and shaly limestones in the basal part, whilst dark green, calcareous shales, both laminated and cross-bedded, prevail in the upper part of the succession. The basal beds yielded small shell fragments, worm tracks, and Middle Cambrian trilobites (*Pardailhania*, agnostidae) 5 metres above the base.

Around Tiyek, 2 kilometres west of Hassa, the lower part of the Tiyek Formation is 175 metres thick and consists of 10 to 15 metres of coloured, thin-bedded, nodular limestones and calcareous shales containing trilobite fragments and a monotonous series of shales. The middle part of the formation is represented by an alternation of well bedded quartzites and phyllitic shales 300 metres in thickness, whilst the upper part is mainly composed of coloured, thin-bedded quartzites 250 metres in thickness. In the Hassa region the Cambrian beds are capped by rocks of Mesozoic age.

The Lower Palaeozoic rocks in the middle part of the eastern Amanos Mountains were studied by LAHNER (1972), who described the Upper Cambrian Yuka Formation as the oldest rock unit exposed in that region. It is composed of fine-grained, light-coloured quartzites, greenish grey or violet sandstones, and siltstones. Occasionally cross-bedding can be noticed. Only *Lingula* and *Cruziana* were collected from the Yuka Formation, the thickness of which is about 250 to 300 metres. North of the region, BRYANT (*in* LAHNER, 1972) found the Middle Cambrian trilobite *Paradoxides* in rocks below the Yuka Formation which is, on the other hand, overlain by the Lower Ordovician Bahçe Formation. It was from this stratigraphical position that LAHNER concluded a Late Cambrian age for the Yuka Formation, which is an equivalent of the upper part of the Koruk Formation *sensu* HAUDE (1969).

In the Penbeğli-Tut region, the four lithological units of the Cambrian can easily be observed in the Kaplandere Valley, north of Penbeğli (KETIN, 1966). The base of the Infracambrian Meryemuşagi Formation is not exposed. The visible thickness of the formation is 230 metres. From bottom to top it consists of monotonous, dark shales; glauconite-bearing, green sandstones, greywackes, and coarse-grained clastics; dark green siltstones and shales; purple to wine-red sandstones, greywackes, coloured shales, and siltstones; and a diabase sill in ophitic texture at the top. The formation is unfossiliferous. Its age can be derived by comparison with the stratigraphical successions of the neighbouring areas.

The Kaplandere Formation, which is supposed to represent the oldest Cambrian rocks in the Penbeğli-Tut area, begins with thin-bedded quartzites and shales which are followed by reddish violet coloured, cross-bedded, sometimes conglomeratic sandstones and quartzites. The upper part of the formation is composed of fine-grained, thin-bedded, quartzitic sandstones with a diabase layer of 25 metres thickness at the top. The thickness of the unfossiliferous Kaplandere reaches 175 metres. The Dolomite Formation is about 120 metres thick and consists of unfossiliferous, thin- and thick-bedded dolomites and dolomitic limestones which are mostly crystalline and compact. The rocks of the Yerlikaş Formation are thought to be Middle and Late Cambrian in age. They are 70 metres thick in the western part of the Penbeğli-Tut area and 270

		Moses 1940 Tasman 1949	Kellogg and Kayar 1960	Schmidt 1965	Ketin 1966	Beer 1966	Haude 1969
Ordovician		?	Bedinan	Bedinan	Bedinan	Bedinan	Bedinan
Cambrian	Upper	(Telbesmi)	Sosink	Sosink	Sosink (Koruk)	Sosink	Koruk Fm.
	Middle						marl and nodular limestone
	Lower		Sadan	Koruk	Dolomite Fm.	Koruk	Dolomite Fm.
			Telbesmi	Zabuk	Sadan	Sadan	Sadan Fm.
				Sadan			
Infracambrian			Derik Volcanics	Derik	Telbismi	Derik	Derik Fm.

FIG. 8. Correlation of the Infracambrian and Cambrian formations of Mardin area, Turkey.

metres in its eastern continuation. The formation is composed of thin-bedded, grey siltstone and shale transgressed by Mardin Limestone of Late Jurassic to Early Cretaceous age. The limy shale beds yielded trilobites (*Paradoxides* s.l. and *Solenopleuropsis*).

Relatively large outcrops of Cambrian and Ordovician strata are found in the Saimbeyli region of the Taurus Mountains, between Pinarbaşi and Feke. Metamorphic rocks of the Emirgazi Formation (100 m) are assigned to the Pre-Cambrian and represent probably the oldest known rocks in the eastern Taurus. They are overlain by the Değirmentas Formation composed of 110 metres of grey and pink limestone which is lithologically similar to the Çal Tepe Formation and tentatively classified, therefore, as Lower to Middle Cambrian. The limestone is succeeded by the lower member of the Armutludere Formation (grey–green schist) which bears Tremadoc trilobites (*Macropyge taurina* Dean, eocrinoids).

In south-eastern Turkey [which corresponds to the major tectonic unit of the 'Border Folds' in the sense of KETIN (1966) and DEAN (1975)] at present three of the four different Infracambrian and Cambrian formations are named after the villages Derik, Sadan, and Koruk, situated within the outcrop areas, about 40 kilometres west of Mardin. The principal outcrops of Cambrian and Ordovician rocks form a small inlier bounded on the north by Cretaceous strata and on the west and south by Quaternary volcanics. Originally, TASMAN (1949) included the entire Cambrian sequence in the Telbesmi Formation, which he reported to be 2200 metres in thickness. But TASMAN's stratigraphical division (he had already found trilobite fragments) could not be maintained, and TOLUN and TERNEK (1952) did not use the term Telbesmi at all. They divided the Infracambrian/Cambrian sequence into éight unnamed formations of a total thickness of more than 1600 metres and believed the overlying rocks to be of Cretaceous age. Towards the top of their formation 8, TOLUN and TERNEK collected a rich Middle Cambrian fauna composed of trilobites, brachiopods, and eocrinoids which were determined by STUBBLEFIELD and later by DEAN (1975) as agnostids (*Peronopsis*), paradoxidids (*Eccaparadoxides*), dorypygids (*Dory-pyge*), and solenopleurids (*Solenopleuropsis*), Lingulidae, and *Echinocystites*. KELLOG and KAYAR (1960) were the first to divide the pre-Ordovician sedimentary sequence into four formations (in ascending order: Derik Volcanics, Telbesmi Formation, Sadan Dolomite, Sosink

Formation). According to DEAN (1975), the Middle Cambrian fossils have been recovered from the lowest portion of the Sosink Formation, and no fossils have been found in the underlying Dolomite Formation. SCHMIDT (1965) even introduced a division into five formations, whilst KETIN (1966) used the quadruple division again (Fig. 8). KETIN reported the thickness of the Infracambrian/Lower Cambrian Telbesmi Formation to be more than 2000 metres. Thus, according to KETIN, the entire pre-Ordovician sedimentary sequence would have a total thickness of more than 4000 metres. BEER (1966) and HAUDE (1969) assumed a total thickness of about 2500 metres, which seems to be more adequate than the figure given by KETIN. The meaning of the stratigraphical terms was always changing in the Derik area. Stratigraphical terminology, therefore, seems to be more complicated than it is in reality. In the following, the terminology used by HAUDE (1969) and later by DEAN (1975) is given preference.

The Derik Formation, which crops out around the villages Telbesmi and Kurtan near Derik, consists of about 500 metres of red fine- to coarse-grained andesites, porphyries, spilites, volcanic breccia, and tuffs, with interbedded reddish sandstone and shale. The thickness of the sedimentary layers forming eight different horizons varies from 2 to 30 metres. The Derik Formation is unfossiliferous. It is the lowest member of a conformable series with the fossiliferous Middle Cambrian Koruk Formation (=Sosink Fm.) at the higher level. From these facts, an Early Cambrian or Infracambrian age of the Derik Formation can be deduced.

With a slight disconformity, the Sadan Formation (=Telbesmi Fm. *sensu* DEAN, 1975) follows above the Derik Formation. It begins with an alternation of conglomerates containing volcanic pebbles and thin-bedded red sandstone, shale, and sandy limestone with red cherts. The Sadan Formation is unfossiliferous and has a total thickness of about 600 metres, half of which is made up of red coloured, cross-bedded, coarse-grained sandstone with high porosity. Thin-bedded red sandstone of the Sadan Formation gradually passes into dolomites and dolomitic limestones called the Dolomite Formation (or Sadan Dolomite). Its lower part consists of sandy dolomite with thin chert layers, the middle part of thick-bedded, compact dolomites, and the upper part of thin-bedded, dolomitic limestone. BEER (1966) observed undeterminable fossil fragments in the less dolomitized part of the sequence. The total sequence is about 260 metres.

The Koruk (Sosink) Formation is composed of Middle and (?) Upper Cambrian rocks cropping out near Koruk. The thin-bedded, dolomitic limestone of the Dolomite Formation passes into the coloured marl and nodular limestone of the transition beds which are 31 metres thick. Interbedded limestone, shale, and siltstone of more than 150 metres thickness yielded the oldest fossils of south-eastern Turkey, already mentioned by TOLUN and TERNEK (1952). According to KETIN the whole thickness of the Koruk Formation amounts to approximately 1100 metres. Ordovician black shale conformably overlies the Koruk Formation near Sosink.

DEAN'S (1975) interpretation of the correlation of the Cambro–Ordovician succession in the Derik region is rather distinctly different from that of earlier workers. According to DEAN, the Koruk (Sosink) Formation is disconformably followed by the Ordovician Bedinan Formation, consisting of shale and siltstone. The latter yielded an abundant trilobite fauna which indicates an approximately middle Caradoc age. There is neither faunal evidence for Ordovician strata older than middle Caradoc in the Bedinan Formation, nor is there any for Upper Cambrian strata in the Sosink Formation, the topmost 560 metres of which have yielded no fossils and comprise massive, cross-bedded sandstone of a shallow-water type.

According to DEAN'S (1975) interpretation (Fig. 9), the base of the Cambrian near Derik is conjectural and has not yet been defined faunally. Thus, it is not known whether the Derik Volcanics are of Pre-Cambrian or Early Cambrian age. Fossils show that the lower part of the Sosink Formation belongs to the Middle Cambrian and the unfossiliferous upper part may also

FIG. 9. Correlation of Cambrian and Ordovician rocks in Turkey (after DEAN, 1975, and others).

be of that age. Consequently, the disconformity between the Bedinan and Sosink Formations is a large-scale one ranging from Middle/Upper Cambrian to the early Caradoc. The top of the Bedinan Formation may involve another disconformity since there is no evidence of late Caradoc or Ashgill strata.

Information about Cambro–Silurian deposits in the south-easternmost part of Turkey, south of Hakkari, is still imperfect. ERENTÖZ (1966) mentioned dark grey quartzites below the Cambrian and limestone of probable Cambrian age above the quartzites. According to ALTINLI (1966), the Palaeozoic sequence begins with more than 500 metres of grey to black, unfossiliferous limestone overlain by grey or dark red quartzites with *Cruziana*. Calcareous shale, varicoloured marly shale, and tuff are intercalated in the limestone. The total thickness of this formation, known as the Giri Formation, is about 1000 metres. It may be of Late Cambrian and Ordovician age (FLÜGEL, 1971). DEAN (1975) considered the grey, unfossiliferous limestone to be of uncertain age (late Pre-Cambrian and/or Early Cambrian) and the overlying sandstone and quartzite with *Cruziana* to be of Arenig age (Fig. 9).

2.4. The Cambrian rocks of Iran

Two types of Cambrian outcrops occur in Iran. In southern Iran and the Persian Gulf area, knowledge of Cambrian deposits is restricted to salt domes containing various allochthonous extrusive rocks of Infracambrian and Cambrian age. This region is part of a broad transitional shelf zone between the Arabian Shield in the south-west and the Anatolian–Iranian range in the north-east. During late Proterozoic/early Cambrian time it was characterized by continental, lagoonal, and shallow marine environmental conditions with long-lasting sedimentation of evaporites. In the Alborz Mountains, eastern central Iran, and part of the Zagros Mountains, folded, sometimes overthrust Infracambrian and Cambrian rocks are exposed in more or less continuous sections. They consist mainly of clastics, carbonates, and some volcanics. Marine intercalations with significant fossils appear for the first time in the upper part of the Lower Cambrian sequence.

2.4.1. Persian Gulf area and southern Iran

Owing to the allochthonous nature of the Cambrian rocks in the salt domes piercing a thick sequence in the post-Palaeozoic rocks, the succession of Infracambrian and Cambrian deposits is only very incompletely known in southern Iran. The present state of knowledge of the salt domes in the Persian Gulf area and southern Iran is largely the result of work by STÖCKLIN (1961, 1968a, 1972) and KENT (1958, 1970). These investigations clarified and completed older work in this area by TAVERNIER (1642), BLANFORD (1872), PILGRIM (1908, 1922, 1924), RICHARDSON (1926, 1928), BÖCKH, LEES, and RICHARDSON *in* GREGORY (1929), HARRISON (1930, 1931), HIRSCHI (1944), SCHRÖDER (1946), O'BRIEN (1957), GANSSER (1960a), WALTHER (1968, 1972), and others.

BLANFORD (1872) introduced the name Hormuz Salt Formation for the entire complex of rock salt and associated sedimentary and igneous rocks building up Hormuz Island. PILGRIM (1908) applied the name Hormuz Series to the salt plug material of all the salt domes of southern Iran and the Persian Gulf islands. In general, the Hormuz Formation is composed of evaporites (rock salt and gypsum) and of blocks and contorted masses of black, laminated, fetid limestones, brown, cherty dolomites, red sandstones, and variegated shales, and of igneous material such as dolerite, basalt, quartz porphyry, keratophyre, and trachyte. Characteristic associated minerals are calcite, authigenic dolomite crystals, bipyramidal quartz, haematite, pyrite, ilmenite, epidote, apatite, sulphur, and others. The thickness of the formation is not exactly known in southern Iran and the Persian Gulf area. Estimates of the volume of the plugs indicating an average original salt thickness of the order of 300 metres were published by HARRISON (1931) and O'BRIEN. In the opinion of KENT (1970), however, the total original salt thickness could have been about 1000 metres.

For a long time, the original rock sequence of the Hormuz Formation has been a major objective of investigation (STÖCKLIN 1968a, 1972; KENT, 1970). According to KENT, the Hormuz Formation is made up of a cyclic series characterized by a repetition of salt, gypsum, coloured shales, and dark dolomite. From his observations at the salt plug of Chah Benu, east of Lar in Fars province, KENT drew the conclusion that thick salt sequences both underlie and overlie at least 1000 metres of non-evaporite beds. The further deduction is that there were at least two, and possibly several, major salt units interbedded with the clastic and dolomite rocks of the Infracambrian/Cambrian succession in southern Iran.

Very rare occurrences of basement rocks—metamorphic or plutonic boulders—are recorded from the Hormuz Formation. As an explanation, HARRISON (1930, 1931) suspected that the origin of the basement boulders might be connected with the linear distribution of the salt plugs independent of surface structures, which suggests control by basement faulting. Igneous rock debris often occurs in quantity in the Zagros and the Persian Gulf salt plugs. In the Zagros plugs, the dominant igneous rock is diabase of normal intrusive type. KENT regarded the intrusion as more or less contemporaneous with sedimentation which occurred at the beginning of the Cambrian. In contrast, the dominant igneous rocks of the Gulf islands are of extrusive type. Rhyolitic tuffs and tuffaceous sandstones are widespread. According to KENT, this extensive intrusive province marks at latest an Infracambrian/Cambrian episode.

A wide range of secondary minerals is another characteristic feature of the Hormuz Formation. Most abundant and ubiquitous is haematite, occurring in massive layers as well as a replacement in carbonates and igneous rocks. WALTHER (1968) proposed the following genetic interpretation. The iron was derived from Pre-Cambrian haematite–quartz–iron and was re-assorted at the time of salt deposition. During geological evolution, dissolution took place under hydrothermal conditions continuing up to the present time. After the salt plugs had appeared at the surface, formation of sedimentary ores happened repeatedly in the immediate neighbourhood of the salt plugs.

The age of the Hormuz Formation has long been discussed. The salt is now found in diapiric contact with Tertiary, Cretaceous, and, locally, Jurassic rocks and is, therefore, certainly older than Jurassic. Since the discovery, in 1925, of Late Cambrian trilobites in shales of the Hormuz Formation by LEES (BÖCKH, LEES, and RICHARDSON in GREGORY, 1929), a Cambrian age of the Hormuz salt has generally been accepted. Discussing the age of the Pakistanian Salt Range evaporites, which are considered to be synchronous with the Hormuz Formation (SCHRÖDER, 1946; SCHINDEWOLF and SEILACHER, 1955), SCHINDEWOLF ascribed an early Early Cambrian age to the Salt Range evaporites as well as to the Hormuz Formation. LOTZE (1957, 1969) came to the same conclusion on the basis of palaeogeographical considerations which show that evaporites of certain Pre-Cambrian age are not known up to the present. STÖCKLIN (1968a), on the other hand, postulated a greater age, arguing that because of the great thickness of the overlying (Lower Cambrian?) sandstones it would be doubtful whether the Hormuz should be included in the Lower Cambrian or should be assigned to the Proterozoic. KENT (1970) provided additional information by the discovery of rock-forming algae (stromatolites) at a number of salt plugs. *Conophyton* was obtained in quantity at Aliabad, *Cryptozoon* and possible *Solenopora* were found at Kuh-e-Kurdeh (identifications by Dr. W. J. Clarke of Sunbury). The mainland assemblage and that of the Gulf islands made up of salt domes could date from late Pre-Cambrian or Cambrian. Stromatolites have not been found associated with middle to late

Early or Late Cambrian fossils, nor are the latter so widespread. Thus, a pre-middle Early Cambrian age of the stromatolites is compatible with the palaeontological evidence for the Iranian and Persian Gulf area.

The assignment of the Hormuz Formation to the pre-middle Early Cambrian has been considered by KENT (1970) against the background of information from what is believed to be the equivalent of the Hormuz Formation in inland Iran, north-east of the Zagros thrust line (STÖCKLIN, 1961, 1968a,b; HUCKRIEDE and others, 1962). There, the Lower Cambrian Lalun and Zaigun Formations in the Derenjal area or Dahu and Kuhbanan Formations in the Kerman area, respectively, overlie dolomites and evaporites containing stromatolites with forms closely related to *Conophyton*. These beds are mainly conformable with the Cambrian but occur elsewhere as diapiric salt plugs. They are alternatively grouped by STÖCKLIN as doubtful Lower Cambrian or more probably Proterozoic. Such a date for the Hormuz Formation would be directly compatible with the position of the algal rocks in southern Iran in a sequence differing from that of the known Iranian Cambrian. Thus, it appears that the Hormuz Formation spans the boundary between Pre-Cambrian and Cambrian, and may extend into the early Lower Cambrian.

Hence, the concept emerging from the relative proportions of the salt plug rocks is that a series of about 3000 metres of salt interbedded with mixed normal sediments and igneous rocks, became mobilized, displaced the overburden, and gave rise to the great bulk of the Hormuz plugs. Consequently, most of the rocks formed a chaotic breccia by the process of translation from great depth.

The Hormuz Formation and equivalents in typical Hormuz development are known to extend from the northern Tabas area in eastern central Iran to southern Oman (STÖCKLIN, 1968a). Non-evaporitic or slightly evaporitic equivalents are believed to be present in the Infracambrian/Lower Cambrian formations of the Alborz Mountains and eastern central Iran, particularly in the Soltanieh Dolomite, the Barut Formation, the Rizu Formation, and the Ravar Formation. The Hormuz Salt Formation is further believed to correlate with the Punjab Saline Series of the Salt Range.

KING (1930, 1937) recorded Late Cambrian faunas from three Hormuz salt plugs of southern Iran (Kuh-e-Namak, Ras Bustaneh, and Irij), which have been re-interpreted by KOBAYASHI (1967). The richest fauna was collected at Kuh-e-Namak. It consists of the trilobites *Chuangina*, *Chuangia*, *Changshanocephalus*, *Maladioides*, and ?*Lioparia*, and the brachiopods *Westonia* and *Billingsella*, which are of early to middle Late Cambrian age. In some areas of southern Iran, grey limestones of 5 metres thickness have been observed cropping out between underlying greenish mottled sandstones 16 metres in thickness and overlying variegated sandstones 30 metres in thickness. The limestones yielded *Iranochuangia*, which indicates a middle Late Cambrian age. The sandstones are disconformably overlain by Carboniferous limestones. From these conditions terrestrial environments can be inferred for large parts of southern Iran during the Middle Cambrian and between late Cambrian and the Carboniferous Period.

2.4.2. Kerman area

Neighbouring the southern Iranian salt plug area north of the Zagros thrust line, numerous exposures of Infracambrian and Cambrian rocks are situated in the Kerman area (HUCKRIEDE, KÜRSTEN, and VENZLAFF, 1962) and, adjacent to the north, in the Derenjal Mountains (RUTTNER and others, 1968) and the Ozbakh Kuh area (FLÜGEL and RUTTNER, 1962). It is in this region that more or less continuous successions of Cambrian and Infracambrian rocks overlying the basement complex are exposed and have already been studied.

In the Kerman area, the Bafq Gneisses already known to STAHL (1911) were recognized by HUCKRIEDE and others (1962), BEHAIN (1970), and PILGER (1971) as highly metamorphosed, strongly folded rocks belonging to an early Pre-Cambrian orogenic phase, the Algomanian Orogeny. The Bafq Gneisses are overlain with a pronounced unconformity by basal conglomerates of late Algonkian age (PILGER, 1971).

Twenty kilometres west-north-west of Kerman, a monotonous sequence of sandstones, micaceous arkosic sandstones, quartzitic sandstones, and extremely fine-grained, sandy argillaceous shales of greenish grey, locally brownish red colour was named the Morad Series by HUCKRIEDE and others (1962). The rocks of the Morad Formation were first mentioned by GANSSER (1955). They contain intercalations of black, siliceous shales which are of marine origin, as indicated by the few organic remains of the Morad Formation including Radiolaria (Spumellaria), pellicles similar to *Laminarites*, elliptic membranes probably belonging to *Lophodiacrodium*, algae, and worm tracks of the *Sabellarifex* type. A Pre-Cambrian age is inferred from the stratigraphical relationships. In spite of the fact that the Morad Formation was subjected to the Assyntian Orogeny, it has since become classic evidence for non-metamorphic Pre-Cambrian sedimentary rocks in eastern Iran. At the type locality, the formation is more than 500 metres thick. Eighty kilometres north-west of the type locality, the Morad Formation shows the following sequence in ascending order: dark grey quartzite, red and green sandy shales, and slightly phyllitic green shales (thickness 400 metres); soft, yellow green, sandy shales with 1 or 2 metres of interbedded black siliceous shale containing radiolaria (thickness 300 metres). The second unit grades into 300 metres of white, hard quartzite. It is assumed that the Morad Formation is underlain by metamorphic complexes, particularly the Bafq Gneisses, though normal contacts have not been observed.

In the Kerman area, the Morad Formation is separated from the overlying Rizu Formation by a pronounced unconformity. The name was proposed by HUCKRIEDE and others (1962) for a unit of sandstones, dolomites, and volcanic rocks. Twenty kilometres west-north-west of Kerman, the Rizu Formation shows the following composition from bottom to top. The formation begins with 12 metres of breccia with up to head-sized components of siliceous shales and dolomites in a sandy dolomitic matrix. The next unit comprises 100 metres of alternating yellow weathered sandstones and dolomites, which are followed by about 500 metres of sedimentary–volcanic deposits of sandstones, sandy and ferruginous dolomites, quartzitic sandstones, conglomeratic layers, and red–brown and green tuffs. Pink and light-coloured cherts are characteristic of these beds. The Rizu Formation is capped by red–brown lavas and tuffs, in the upper part also light-coloured. The volcanics are of quartz porphyric to felsic-porphyric composition, but include also altered basic rocks. Organic remains comparable to *Spriggina*, *Dickinsonia*, and *Medusites* have been discovered about 90 kilometres north-west of Rizu (STÖCKLIN, 1968b). According to GLAESSNER (*in* STÖCKLIN, 1972) one specimen resembles the genera *Charnia* and *Rangea*. Radiometric dating of lead occurring in the fossiliferous beds and now generally considered to be a synsedimentary ore deposit yielded an age range of 595–760 ± 120 m.y. (HUCKRIEDE and others, 1962).

In the north-western Kerman area, east of Zarand, HUCKRIEDE and others named a saline sequence of complex lithology as the Desu Formation, which appears as chaotic, diapir-like structures. The formation consists of a mixture of sedimentary and volcanic rocks, the original stratigraphical relationships of which are only locally partly preserved. Near Dahu it is represented by an alternation of dolomites, rare black limestones, thick-bedded gypsum, and some red sandstones about 160 metres in thickness. According to HUCKRIEDE and others, the formation may reach a total thickness of 300 metres. HUCKRIEDE and others described the Desu rocks as appearing generally in the form of distorted complexes and disrupted blocks of all

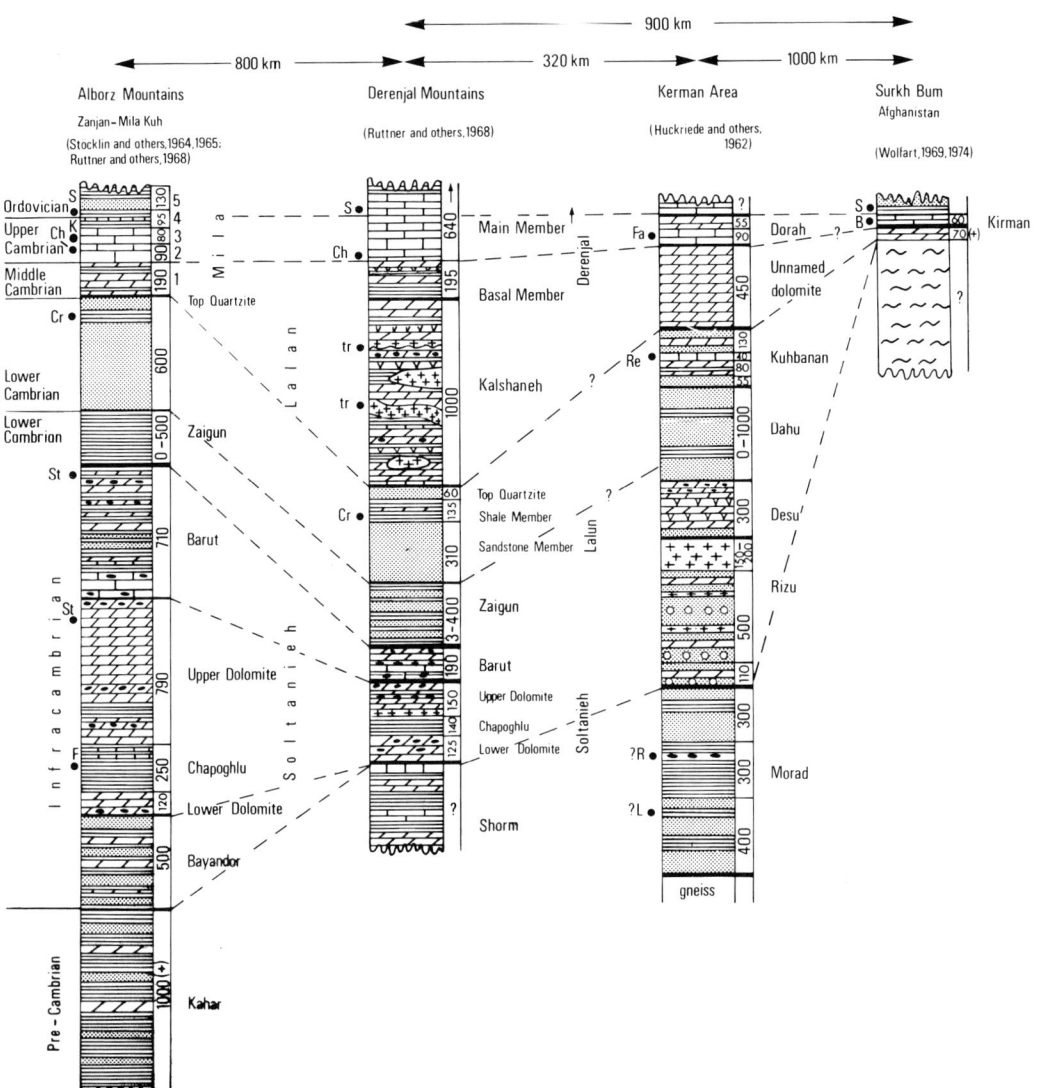

FIG. 10. Representative columnar sections and stratigraphical terminology for the Infracambrian and Cambrian rocks of Iran and Afghanistan. Alborz Mountains: (S) *Saukia* assemblage; (K) *Billingsella, Kaolishania, Anomocarella* s.l., cf. *Eochuangia,* Damesellidae, *Prochuangia*; (Ch) *Chuangia, Iranoleesia, Iranochuangia, Lioparella* ?, *Anomocarella* ?; (Cr) *Cruziana*; (St) *Stromatolites*; (F) *Fermoria*. Derenjal Mountains: (S) *Saukia* assemblage; (Ch) *Chuangia, Chelidonocephalus, Idahoia, Lioparella* ?; (tr) indistinct shells; (Cr) *Cruziana*. Kerman area: (Fa) *Farsia, Afghanocare, Pelagiella*; (Re) *Redlichia, Kermanella, Stoecklinia, Paragraulos* ?, *Latouchella* ?, *Orthotheca*; (?R) Radiolaria; (?L) Lebensspuren. Afghanistan: (S) Tremadocian *Saukia-Pilekia* assemblage; (B) *Blackwelderia, Pseudagnostus, Crepicephalus, Torifera, Afghanocare, Paracoosia, Farsia, Biaverta* ?, *Constrictella, Ariaspis,* Damesellinae. Lithological legend: see Fig. 3.

possible sizes floating rootlessly in a mylonite-like, gypsiferous dolomitic sandy mass. The sedimentary rocks include the following: gypsum mostly disintegrated to powder; whitish, grey and black, fetid, recrystallized limestones; violet–red, fine-grained, micaceous sandstones and sandy shales; and rare white quartzites. The volcanics are light-coloured to pink–grey quartz-porphyry, diorite or diorite–porphyry, quartz–diorite, green porphyry, ?gabbro–porphyry, augite–porphyry, tuffs, and tuffites. Basic rocks are most strongly chloritized and epidotized or altered to serpentine. Associated minerals are big haematite and apatite crystals. There are no fossils except questionable calcareous algae. From north of Dahu, a conformable contact between the Desu and the overlying Dahu Formation is known. Consequently, the Desu is definitely older than the Lower Cambrian Dahu Formation. HUCKRIEDE and others admitted the distinction between Rizu and Desu to be difficult. From more regional considerations, STÖCKLIN (1972) concluded that the Rizu and Desu formations now appear merely as different facies of about the same stratigraphical interval, with the Desu representing a transitional gypsum-rich facies between a salt facies (Ravar Formation) in the east, north of Kerman, and a marginal dolomite facies (Rizu) west of the Kuhbanan fault. Thus, the Desu Formation is believed to correspond broadly to the Infracambrian complex (Zaigun, Barut, Soltanieh, and Bayandor Formations) of northern Iran and the eastern part of central Iran as well as to the Hormuz Formation of southern Iran.

Still farther east, in the Ravar area, north of Kerman, the gypsum of the Desu Formation grades more and more into salt, and the rocks of the Ravar Formation (STÖCKLIN, 1961a) form a number of true diapirs and salt domes piercing through Palaeozoic, Mesozoic, and Tertiary formations. The main sedimentary rock types of the Ravar Formation are rock salt, gypsum, brownish recrystallized dolomites with abundant chert nodules, dark grey to black, fetid limestones, blue– to greenish grey limestones, wine-red, micaceous shales, marmorized limestones, and rare quartzites. Igneous rocks are mainly dark green 'diabases', partly altered to amphibolites and strongly epidotized, frequently coated with haematite. Crystals of bipyramidal quartz, dolomite, calcite, and haematite are present in large quantities. STÖCKLIN estimated the total thickness of the formation as several hundred metres.

Near the village of Dahu, 62 kilometres north-west of Kerman, the Dahu Formation (HUCKRIEDE and others, 1962) rests with a basal breccia conformably on the dolomites of the Desu Formation. In the Dahu area the formation consists of about 400 metres of barren, red to dark violet–brown sandstones with mudcrack fillings, mud rolls, and clayey intercalations. The greatest thickness of the formation amounts to 1000 metres. Near the classical Cambrian locality of Kuhbanan, the thickness of the Dahu Formation seems to be much less than elsewhere, because the white quartzite horizon usually overlying the Dahu Formation rests disconformably on a thick cherty dolomite of the Rizu Formation. STÖCKLIN, RUTTNER, and NABAVI (1964) and STÖCKLIN (1972) assumed the Dahu Formation to be a perfect equivalent of the Lalun Sandstone of northern Iran. They considered both of the formations to be capped by a synchronous Top Quartzite. An early Cambrian age was inferred for the Dahu Formation from the stratigraphical position. At Dahu, more than 100 metres above the Top Quartzite, a limestone horizon follows bearing trilobites of the *Redlichia* group which indicates middle (to late) Early Cambrian age.

WOLFART and KÜRSTEN (1974), however, came to the conclusion that the Dahu and Lalun Formations perhaps cannot be considered as perfect correlative units, because for bio-stratigraphical reasons, the *Redlichia*-bearing horizon of the Kuhbanan Formation might be synchronous with the *Cruziana*-bearing horizon of the Lalun Formation situated beneath the Top Quartzite. In RUTTNER and others (1968), SEILACHER compared the *Cruziana* from the Shale Member of the Lalun Formation in the Derenjal Mountains with identical ones from the Magnesian Sandstone of the Salt Range associated with *Redlichia noetlingi* (Redlich). The same species of *Redlichia* occurs in the Kerman area. It seems to be very improbable that in some regions of Iran marine horizons with *Redlichia* and *Cruziana*, respectively, which are synchronous fossils, are situated below and in another region above the Top Quartzite. An additional argument for the present author's view is that there are several quartzitic horizons in differing stratigraphical positions in the Kerman area. One quartzitic horizon, for example, follows above the limestones with *Redlichia* near Kuhbanan as well as in the areas between Rizu and Chabdjereh, about 120 kilometres north-west of Kerman.

The Lower Cambrian Kuhbanan Formation named 'Cambrian of Kuhbanan' by KING (1930) has been described by HUCKRIEDE and others (1962) from Dahu and Kuhbanan in its most complete succession. North of Dahu the formation is composed of about 80 metres of both thick- and thin-bedded, yellow, brown, and dark dolomites with interbeds of marl, dolomitic shale, and some chert in the lower portion. Towards Kuhbanan, some 80 kilometres to the north-west of Dahu, the dolomites of the lower Kuhbanan Formation gradually pass into brownish red, coarse- to fine-grained sandstones with some interbeds of brown to yellow dolomite. The thickness of the lower portion of the Kuhbanan Formation is about 80 metres near Kuhbanan village.

The middle part of the Kuhbanan Formation is formed by platy to thick-bedded, black to dark blue, partly dolomitic limestones with some intercalations of yellowish brown to reddish marl. The top of the limestone consists of thin-bedded, dolomitized beds with cherty layers. The limestones yielded the first significant Cambrian fossils discovered in east central Iran. The first description of the fauna was published by KING (1930) who identified *Redlichia* and *Hyolithes* and dated them as latest Early or Middle Cambrian. WOLFART and KÜRSTEN (1974) completed knowledge about that faunal assemblage and identified the brachiopods *Lingulella*, *Acrothele*, ?*Nisusia*, and *Diraphora*, the gastropod ?*Latouchella*, the calyptoptomatid *Orthotheca*, the trilobites *Redlichia*, *Kermanella*, *Stoecklinia*, and ?*Paragraulos*, and the eocrinoid ?*Gogia*. Corresponding to the results of KOBAYASHI (1967), LOTZE and SDZUY (1970), and SDZUY (1971) concerning the age of *Redlichia*, the fauna of the Kuhbanan limestones was considered to be middle (to late) Early Cambrian in age by WOLFART and KÜRSTEN.

The upper portion of the Kuhbanan Formation consists of about 130 metres of fine-grained, violet to wine-red, mainly massive sandstones with thick-bedded intercalations and lenses of dolomite north of Dahu. The upper boundary of the Kuhbanan Formation is sharply marked by the conformably overlying black to yellowish brown weathering, unfossiliferous dolomite which is 450 metres in thickness according to HUCKRIEDE and others. The unnamed dolomites were presumed to be of Middle Cambrian age by WOLFART and KÜRSTEN on the basis of their stratigraphical position and of palaeogeographical reflections. North of Dahu, the Middle Cambrian rocks are disconformably overlain by fossiliferous Lower Ordovician beds. Upper Cambrian beds presumably have been eroded in that region. Near Kuhbanan, the Kuhbanan Formation is disconformably overlain by Devonian rocks.

KÜRSTEN (*in* HUCKRIEDE and others, 1962) was the first to find a Late Cambrian faunal assemblage in a section near Dorah Shah Dad, 17 kilometres south-east of Kerman. The Dorah Formation, a correlative unit of part of the Mila Formation in northern Iran, consists of 90 metres of dark grey or blue to black, partly dolomitic, thin-bedded limestones the base of which is not exposed. The exact position of the fossiliferous horizon is not known. It has yielded some brachiopods (*Westonia*, Eoorthidae), the gastropod *Pelagiella*, and the trilobites *Afghanocare* and *Chelidonocephalus* ? (*Farsia*), which must be attributed to the *Crepicephalus* Zone (=*Glyptagnostus stolidotus* Zone) of late Kushanian age. There is no palaeontological evidence for the Late Cambrian age of the overlying dark, bedded dolomite which is 55 metres in thickness. The age of the following succession of white quartzites with interbedded dolomites, shales, and tuffites is not known. It is about 200 metres in thickness.

2.4.3. Zagros Mountains

From the Zagros Mountains, south-western Iran, comparatively little detailed information about Cambrian rocks has been published. KING (1930, 1937) provided some notes about Cambrian fossils and rocks from localities along the Zagros thrust line. However, it was SETUDEHNIA (1972) who gave some more detailed information about the Cambrian succession in south-western Iran obtained from the reconnaissance surveys which were conducted by J. V. HARRISON in the early and mid-1930s. In the Kuh-e-Dina area, about 530 kilometres west-north-west of Kerman, the most complete section of Cambrian rocks is exposed at Chal Parwari already mentioned by KING (1937). The total thickness of the outcropping Cambrian rocks is more than 1350 metres at Chap Parwari. The lower contact of the Cambrian rocks is not exposed, the upper contact with the ?Carboniferous sandstones is disconformable. HARRISON (*in* SETUDEHNIA, 1972) observed the following rocks in Kuh-e-Dina area which were tentatively correlated with the Cambrian sequence of the Alborz Mountains by SETUDEHNIA.

Lithological description	Thickness (metres)	Possible correlation with Alborz
Chocolate marls with trilobites	55	Mila
Mottled dark and yellow limestones interbedded with red shales	115	
White quartz sandstone, loosely cemented	6	Lalun
Deep red shales with a few white calcareous sandstones	55	
Red grit and conglomerate	30	
Variegated sandstones, mainly violet–red coloured with some beds of white sandstones and red shales	300	
Pink, laminated, medium- to coarse-grained quartz sandstone	200	Zaigun
Red shales with ribs of pink sandstone	85	
Red, micaceous shales	180	Barut
Red, micaceous shales with dark grey and yellow dolomites	25	
Red shales with intercalations of dark, coarsely crystalline dolomite and some bands of greenish grey shales	300	

According to SETUDEHNIA, the thickness of the Lower Cambrian sediments amounts to about 900 metres in the Zard Kuh area, whilst the thickness of Middle and Upper Cambrian rocks varies about 1200 metres. KING (1930, 1937) had already recorded the Early Cambrian *Redlichia* from Chal Parwara (= Parwari) in the Kuh-e-Dina area. The trilobites are preserved in hard, platy, finely crystalline limestones, the thickness of which is not known. The Upper Cambrian rocks were described by KING as thin-bedded, flaggy limestones, dolomitic limestones, and sandstones partly showing brecciation and mottling. KING estimated the thickness of

Upper Cambrian rocks to be up to about 700 metres. He assumed a gap between the *Redlichia*-bearing beds and the fossiliferous Upper Cambrian sequence during which no sedimentation took place. KING identified the following Late Cambrian fossils from the Zagros Mountains, revised by KOBAYASHI (1967). One horizon situated at least 700 metres below the Cambro–Ordovician boundary yielded four trilobites of Middle to early Late Cambrian age: *?Bonnia, Chelidonocephalus, ?Grandioculus,* and *Hundwarella,* which is not associated with the former three trilobites. The present author is not convinced of the generic identity of *Hundwarella,* which would be Middle Cambrian in age according to existing knowledge. Up to the present, there has been no palaeontological evidence for Middle Cambrian rocks in southern Iran. Possibly, the four trilobites are synchronous with the early Late Cambrian assemblage of Dorah Shah Dad. About 80 to 100 kilometres north-west of Kuh-e-Dina *Iranochuangia* has been found, which indicates an early to middle Late Cambrian age. The Upper Cambrian sequence is conformably overlain by limestones, shales, and siltstones which yielded *Saukia, Iranaspis, Briscoia,* and other trilobites attributed to the Daizanian or Fengshanian by KOBAYASHI (1967). In the opinion of WOLFART (1970, 1973), however, at least part of the genus *Saukia* must be assigned to the earliest Ordovician (= early Tremadocian).

2.4.4. Derenjal Mountains and Ozbakh Kuh Mountains

Owing to the work of RUTTNER and others (1968), the Cambrian sections of the Derenjal Mountains in the Shirgesht area along the eastern border of the Great Kavir, east central Iran, are fairly well known, although a detailed palaeontological–biostratigraphical analysis has not yet been performed. The Ozbakh Kuh Mountains, about 100 kilometres north of the Shirgesht area, represent another area with exposures of Cambrian rocks which are known only from the preliminary notes of FLÜGEL and RUTTNER (1962). More detailed information about the Ozbakh Kuh Mountains will be published by RUTTNER, NABAVI, and ALAVI in the near future.

The Derenjal Mountains, a solitary mountain group, proved to be an uplift of Infracambrian and Lower Palaeozoic rocks which are exposed in an almost complete succession. Two new Cambrian formations have been established in that region by RUTTNER and others, the Kalshaneh Formation of Middle Cambrian and the Derenjal Formation of Late Cambrian age. For the Lower Cambrian and the Infracambrian part of the rock sequence, the formational names used in the Alborz and Soltanieh Mountains, northern Iran, could be applied to the Derenjal area.

The oldest rocks of the Shirgesht area are the Pre-Cambrian Shorm Beds, consisting of thin-bedded, slaty limestones alternating in some places with greenish slates or yellowish, greenish, pinkish phyllitic shales, and a pink siliceous dolomite. The whole sequence has undergone slight metamorphism: the bedding planes of the slates and limestones are covered with a sericitic film, the limestone shows microfolding and B-lineation. The green slates intersected by calcite veins might be of tuffaceous origin. The thickness of the Shorm Beds is not known as their base is not exposed in the type area. Farther north, in the Ozbakh Kuh Mountains, the position of the Shorm Beds between highly crystalline rocks and the Soltanieh Dolomite is clearly visible. In spite of the lithological difference and the low-grade metamorphism, RUTTNER and others correlated the Shorm Beds with the Morad Series of the Kerman Area. STÖCKLIN (1972) considered the Pre-Cambrian low-grade metamorphics to merge vertically and laterally with the Pre-Cambrian non-metamorphics.

With a clear unconformity and a basal breccia, the Infracambrian Soltanieh Dolomite overlies the Shorm Beds in the Shirgesht area. The underlying Bayandor Formation, which has a thickness of several hundred metres, in northern Iran is obviously not present in the Derenjal Mountains. Concerning lithology and succession, the basal Infracambrian rocks of the Shirgesht area are an exact repetition of the Soltanieh Dolomite described by STÖCKLIN, RUTTNER, and NABAVI (1964) from the Soltanieh Mountains and from the central Alborz Mountains (ASSERETO, 1963). The Lower Dolomite Member is composed of black to grey, finely crystalline, bedded dolomite alternating with chert layers. The member is up to 125 metres thick. The Chapoghlu Member is 140 metres thick and consists of green, sandy to silty shales with intercalations of yellow and green sandstone, platy, cherty, black limestone and well bedded, cherty, black dolomite. Thin- and thick-bedded, cherty, light grey dolomites form the Upper Dolomite Member which is nearly 150 metres in thickness.

The similarity between the overlying rocks of about 190 metres thickness and the Barut Formation of northern Iran is, compared with the preceding formation, not quite so close. There are still striking similarities: black, fetid, thin-bedded limestone; thin-bedded, yellowish dolomites and limestones containing chert layers and ripple marks; and purple and green shales with dolomitic interbeds. On the other hand the intercalations of purple or green shales between the dolomite and limestone beds, otherwise so characteristic of the Barut Formation, are almost completely absent. No fossils have yet been found, either in the Soltanieh Dolomite or in the Barut Formation except for a few worm tubes and fucoid-like marks in the Chapoghlu Shales.

The discovery of typical Zaigun and Lalun Formations in the Shirgesht area, east of the Great Kavir, is evidence for their wide and uniform distribution. A wide, hilly and purple-coloured landscape is made up of those two formations. The Zaigun Formation consists of wine-red to pink shales at the base grading upwards into laminated, fine-grained, shaly sandstones, with intercalations of sandy or silty shales, mainly dark wine-red in colour, and containing plenty of mica on the bedding planes. At one place yellowish layers of conglomerate consisting of dolomite pebbles have been observed at the base of the Zaigun Formation. Whether the conglomerate is tectonic or sedimentary in origin cannot be decided. The total thickness of the Zaigun Formation may be estimated at about 300–400 metres. Compared with the Zaigun Formation in northern Iran, the correlative unit is more sandy in the Shirgesht–Ozbakh Kuh area and the boundary between the Zaigun and the overlying Lalun Formation is less distinct.

The Lalun Sandstone Formation can be divided into three units. The basal Sandstone Member is 310 metres thick and consists of purple, light red to pink, well bedded, partly quartzitic and cross-bedded rocks with ripple marks. Layers of coarse sandstone show graded bedding, beginning with a fine breccia with the fragments up to 1 centimetre in diameter, and gradually passing upwards into sandstone of normal Lalun type. The Shale Member is twice as thick (135 metres) as in northern Iran and consists of red shale with beds of red sandstone and arenaceous dolomite. Cross-bedding and ripple marks are common. At the basal contact, the bottom of some sandstone interbeds shows casts of footprints, digging tracks, and other traces caused by different animals. A. SEILACHER, Tübingen, identified *Cruziana*, which is identical with prints described from the Magnesian Sandstone of the Salt Range, Pakistan (SCHINDEWOLF and SEILACHER, 1955). There, the prints are associated with *Redlichia*. Thus, the Lalun Sandstone must be considered to be Lower Cambrian in age. The Lalun Formation is capped by the Top Quartzite which is 60 metres in thickness and well known from northern Iran.

In the Shirgesht area, the Lalun Sandstone Formation is overlain by a mixture of various sedimentary and volcanic rocks called the Kalshaneh Formation by RUTTNER and others (1968). The roughly estimated thickness of the sequence is up to 1000 metres. The formation is formed by a chaotic complex of different sedimentary and basic volcanic rocks, the original stratigraphical succession of which cannot be reconstructed. Nevertheless, the formation represents a distinct and persistent unit between the relatively undisturbed Lalun Sandstone and the overlying Derenjal Formation.

The sedimentary rocks are compact or finely crystalline, black and grey dolomites, locally with chert nodules and bands; black, fetid limestone and bluegrey, nodular limestone, rarely with chert; and dolomitic, partly sandy shales of mainly scarlet, subordinately yellow, green, and grey colours. Gypsum and possibly other evaporites are associated with the dolomitic shales. Crinoid-like structures and indeterminable traces of shells are the only organic remains found in the sedimentary components of the Kalshaneh Formation. A (?Middle) Cambrian age is inferred from its stratigraphical position between the Lalun Formation with *Cruziana*-bearing rocks of middle to ?late Early Cambrian age and the Derenjal Formation with fossils of Late Cambrian age. Volcanic rocks are green, diabase-like hornblende–augite–dolerite and black olivine–augite–dolerite. The volcanic components, however, cannot be regarded as part of the Kalshaneh Formation used in the sense of a stratigraphical unit because they are definitely younger than the sedimentary components, probably Silurian in age.

Typical Kalshaneh Formation is restricted to the type area in the Derenjal Mountains. The sedimentary portion of the formation seems to be perfectly correlative with Member 1 of the Mila Formation in northern Iran, consisting of dark dolomites and red and yellow dolomitic shales with salt pseudomorphs. It does not appear to be comparable to the Kuhbanan Formation of the Kerman area described by HUCKRIEDE and others (1962) and WOLFART and KÜRSTEN (1974), as suggested by STÖCKLIN (1972), for the Kuhbanan rocks contain fossils of definite Early Cambrian age which are probably synchronous with the *Cruziana* of the Lalun Formation.

In its lithological composition and disturbed appearance, the Kalshaneh Formation is similar to the diapiric Desu Formation (= Ravar Formation) of the Kerman area and to the Hormuz Salt Formation of southern Iran. RUTTNER and others (1968), however, came to the conclusion that the Kalshaneh Formation had been deposited after the Lalun Sandstone and that volcanic activity rather than salt tectonics caused the disturbed condition of the Kalshaneh. On the other hand, RUTTNER and others stated that an evaporitic Hormuz facies might be developed at two different stratigraphical horizons, one being late Pre-Cambrian to early Early Cambrian in age (= pre-Lalun rocks) and the other belonging to the Middle Cambrian Kalshaneh Formation (= post-Lalun rocks). Actually, the Hormuz Formation includes not only pre-Lalun but also post-Lalun Cambrian rocks.

In the Shirgesht area, the black dolomites of the Kalshaneh formation are overlain with sharp contact by the Derenjal Formation. This can be divided into a basal member, nearly 200 metres in thickness and a main member of more than 600 metres thickness. The basal member is composed of barren siltstones and marls with dolomitic intercalations, all of the rocks yellow to reddish grey in colour and displaying mud-cracks and salt-pseudomorphs. The main member is characterized by thin-bedded, flaggy limestones weathering yellowish to light brown, containing marly and silty intercalations and alternating with dark or black limestones and sandy lime-stones. There are several fossiliferous horizons, the lowest of which, at the base of the main member, yielded *Chuangia* and '*Lioparella*' s.l. Should *Lioparella* be identical with or closely related to *Farsia* from Dorah Shah Dad in the Kerman area, the basal portion of the main member must be attributed to an early stage of the Late Cambrian, possibly to the *Crepicephalus* Zone (= *Glyptagnostus stolidotus* Zone) which is of late Kushanian age (WOLFART and KÜRSTEN, 1974).

The next fossiliferous horizon found by RUTTNER and others is situated more than 230 metres above the basal horizon. It contains the trilobites *Iranaspis*, *Saukia*, *Idahoia*, and *Chelidonocephalus*, which were identified by WINSNES (*in* RUTTNER and others, 1968). The question is whether *Saukia* and its ally *Iranaspis* are really associated with *Chelidonocephalus*. The last is hitherto known only from Iran (KING, 1937) and Kazakhstan (KOBAYASHI, 1967), where it is of definite early Late Cambrian age. *Saukia*, on the other hand, has a wide distribution in the latest Late Cambrian/earliest Ordovician Trempealeauian and Fengshanian, respectively, of North America, south-eastern Asia, and Australia. Owing to the occurrence of *Saukia* 250 metres above the base of the main member of the Derenjal Formation, it is to be supposed that the Cambro–Ordovician boundary must be placed somewhere within the Derenjal Formation, perhaps some 300 to 400 metres below its top at the type section 3 kilometres south of Derenjal. Consequently, the uppermost fossiliferous beds of the Derenjal Formation with *Billingsella*, which was considered to belong to a late Middle to Late Cambrian assemblage by STÖCKLIN (1972), must be assigned to the Ordovician.

Information about succession, lithological composition, and faunas of the Cambrian rocks in the Ozbakh Kuh area is sparse and incomplete. FLÜGEL and RUTTNER (1962) reported the presence of calcareous shales with trilobites widely distributed east of Ozbakh Kuh. According to identifications by ERBEN (*in* FLÜGEL and RUTTNER), the trilobites indicate a Late Cambrian age. Besides the Upper Cambrian rocks, there are also dolomites and red sandstones (Lalun type) with *Cruziana*-like imprints comparable with similar tracks of the Lower Cambrian Purple Sandstone in the Salt Range area (SCHINDEWOLF and SEILACHER, 1955).

2.4.5. Northern Iran

Northern Iran, especially the Soltanieh Mountains and the central and eastern part of the Alborz Mountains, is the type area for a number of Infracambrian and Lower Cambrian formations similarly developed in the Shirgesht area of eastern central Iran and also in the Kerman area. The present state of knowledge of the Infracambrian and Cambrian rocks in northern Iran is the result of work by ASSERETO (1963), STÖCKLIN, RUTTNER, and NABAVI (1964), DEDUAL (1967), MEYER (1967), and ALAVI and AMIDI (1968). DIETRICH (1937) and GANSSER and HUBER (1962) should be mentioned as earlier workers.

The Kahar Formation is the oldest non-metamorphic formation in northern Iran. It is known from central and western Alborz, the Soltanieh Mountains, and southern Azerbaijan. DEDUAL (1967) gave the name Kahar to a very uniform sequence of argillaceous–sericitic to fine, sandy, micaceous, slaty shales of dull, green–grey colour, locally reddish grey, with subordinate intercalations of quartzitic sandstone, yellowbrown dolomite, dark, fetid, crystalline limestone, and some volcanics, which is 1600 metres in thickness in the central Alborz type area. The formation had already been recognized by STÖCKLIN, RUTTNER, and NABAVI (1964), GLAUS (1965), STEIGER (1966), and others. No fossils have been found in the formation, the base of which is not exposed. In the central Alborz type area, the Kahar Formation is overlain by the Soltanieh Dolomite; the contact is tectonically disturbed. A Pre-Cambrian age is thus inferred from the stratigraphical position.

Sharp limits between the Kahar Formation and metamorphics have nowhere been observed. Downward and lateral transitions from the Kahar slates into slightly metamorphic (phyllitic) schists are known from the Soltanieh Mountains. Nevertheless, STÖCKLIN (1972) considered the bulk of the Pre-Cambrian epi- and meso-zonal metamorphic rocks to be older than the Kahar. In his opinion, even the Shorm Beds of the Derenjal region are older than the Kahar. Everywhere the Kahar is overlain by Infracambrian formations: in the Soltanieh Mountains by the Bayandor Formation and in the Alborz by the Soltanieh Dolomite. Locally, a gentle unconformity has been suspected which is represented, perhaps, by the sharp lithological break between the Kahar and the overlying rocks. This suggests a period of non-sedimentation during which the Doran Granite intruded and partly metamorphosed the Kahar rocks. In southern Azerbaijan, ALAVI and AMIDI (1968) described a thick sequence of quartz porphyries and associated tuffs (Gharadash Formation) from between the Kahar and the Bayandor Formations. Lithologically the Kahar is comparable with the Morad Series in the Kerman area. The Morad Formation, however, is separated by a pronounced unformity, called 'epi-Assyntic' by STÖCKLIN (1972), from the overlying rocks.

In the Zanjan-Soltanieh area, a number of granite bodies was named Doran Granite by STÖCKLIN, RUTTNER, and NABAVI (1964). The granite is coarse-grained to slightly porphyritic, of whitish to pinkish colour, and almost completely lacking macroscopical dark minerals. The Doran Granite is truncated and transgressively overlain by the Bayandor Formation. It is thus definitely younger than the Kahar and older than the Bayandor, and was regarded as the youngest formation of the Pre-Cambrian basement complex. There are granites in other parts of Iran, e.g. in the Ozbakh Kuh Mountains, which are lithologically and stratigraphically comparable to the Doran Granite.

In the Takab area, southern Azerbaijan, a sequence of bedded quartz phorphyry and rhyolitic tuff alternating with green, black, and violet shale and subordinate micaceous shale was named the Gharadash Formation by ALAFI and AMIDI (1968). The formation is up to 1140 metres thick and includes a dolomite band in the middle part which is 130 metres in thickness. It conformably underlies the Bayandor Formation and is thus regarded by STÖCKLIN (1968b, 1972) as an equivalent of the Rizu Formation in the Kerman area and, further, to be the extrusive equivalent of the Doran-type granites intruded into the Kahar Formation.

STÖCKLIN, RUTTNER, and NABAVI (1964) described a series of purple sandstones and shales with thin dolomitic intercalations from south of Zanjan in the Soltanieh Mountains and called it the Bayandor Formation. It consists of dark purple sandstone and silty to fine sandy, strongly micaceous shales containing 15 intercalations of brown, recrystallized, partly ferruginous dolomite with numerous bands and nodules of chert. In the lower part of the section, greenish colours partly replace the purple ones. The basal bed consists of a 30 centimetres thick kaolinitic clay overlying an erosional surface of the deeply weathered Doran Granite. The first dolomite bed occurs only 2 metres above the base. STEIGER (1966) and ALLENBACH (1966) reported questionable stromatolites and archaeocyathid-like structures of probably inorganic nature. The Bayandor rocks reach a thickness of nearly 500 metres. STÖCKLIN (1972) inferred a late Pre-Cambrian (Infracambrian) age for the Bayandor Formation from its stratigraphical position below the Soltanieh Dolomite. The base of the Bayandor Formation has been selected as the boundary between the Infracambrian and the Pre-Cambrian basement complex. Equivalents of the type Bayandor are not present in the Alborz nor have they been recognized in Derenjal and in the Kerman area.

STÖCKLIN, RUTTNER, and NABAVI (1964) introduced the name Soltanieh Dolomite Formation for a thick dolomite sequence forming conspicuous cliffs in the mountains south of Soltanieh. Earlier, the formation was described by FURON (1941) as Silurian or Cambrian, by GANSSER and HUBER (1962) as Lower Hezarchal Formation, and by ASSERETO (1963) as Soltanieh Formation.

Three major divisions of the Soltanieh Formation can be distinguished: the Lower Dolomite Member, the Chapoghlu Member, and the Upper Dolomite Member. In the type section, the Lower Dolomite Member is 120 metres thick and consists of yellow, well bedded, recrystallized dolomite with many black and white chert bands up to 50 centimetres thick. A morphologically well expressed break in the uniform dolomite sequence is formed by the Chapoghlu Shale Member (247 metres) which consists of dark green, argillaceous, siliceous, and silty, micaceous, slaty shales; blue–black, thin-platy nodular, partly siliceous limestones and calcareous shales predominating in the uppermost part. The Chapoghlu shale contains a persistent horizon extending from southern Azerbaijan to the central Alborz with the oldest organic remains of northern Iran. These are 'fucoid' marks and abundant small, disc-like features variously compared to *Fermoria* (STÖCKLIN, RUTTNER, and NABAVI, 1964), *Chuaria* (EISENACK *in* ASSERETO, 1966), and *Beltanella* (STÖCKLIN, 1968). The Upper (Main) Dolomite Member is 790 metres thick and composed of massive, whitish or light grey, recrystallized dolomite. The uppermost part is more distinctly bedded and dark grey. Seams and nodules of black chert are common; two intercalations of green, micaceous slaty shale, 5 and 73 metres in thickness, are present in the type section. Stromatolites are fairly common in the Main Dolomite Member. According to MEYER (1967), they include *Hadrophycus* as well as *Collenia* in the central Alborz. ASSERETO (1966) referred to small, tube-like organisms of hyolithids and *Biconulites*. STÖCKLIN, RUTTNER, and NABAVI (1964) mentioned *Salterella*. Infracambrian (= latest Pre-Cambrian or earliest Cambrian) age of the Soltanieh Dolomite Formation was inferred from the fossils and from the stratigraphical position by STÖCKLIN (1972).

The Soltanieh Dolomite Formation is developed in a great part of northern and eastern central Iran, but it seems to be missing—like all Infracambrian and Lower Palaeozoic formations—on the Caspian side of the Alborz and in northern Azerbaijan. In the south-eastern Soltanieh Mountains and in the central Alborz, the Lower Dolomite Member is missing and the Chapoghlu Shale rests immediately on the Kahar. In the eastern central Alborz, the greater part of the Soltanieh Formation is changed to an alternation of variegated shales, sandstones, and dolomites; farther east, the typical cherty dolomite facies appears again. In the Kerman area, the dolomite contains gypsum and is in the lower part associated with sandstones, rhyolitic lavas, and tuffs. The Rizu, Desu, Ravar, and the Hormuz Salt Formations are thus broadly correlative with the Soltanieh Dolomite Formation as a non-evaporitic equivalent.

The name Zaigun Formation was established by ASSERETO (1963) to distinguish a more shaly lower part from the more sandy quartzitic upper part which is now called the Lalun Sandstone Formation, and formerly went under the term Old Red Sandstone. It consists of silty to fine sandy shales and fine-grained sandstones, cross-laminated, micaceous, of dark red colour with subordinate pink, greenish, and violet variations. It has a thickness of more than 450 metres at the type section. No fossils are known from the Zaigun Formation. An early Cambrian or latest Pre-Cambrian (= Infracambrian) age is indicated by the stratigraphical position. In the central Alborz type section, the contact between the Zaigun and the overlying Lalun Formation is gradational.

Except in the Kerman area, the Zaigun and Lalun Formations are nearly everywhere associated. In the Soltanieh Mountains the upper boundary is sharp, possibly marking a disconformity (STÖCKLIN, NABAVI, and SAMIMI, 1965). In the northern Golpaygan area, THIELE and others (1968) found the Zaigun facies to replace most of the Lalun Sandstone.

ASSERETO (1963) replaced the term Old Red Sandstone by the name Lalun Sandstone because the former became obsolete when the unit proved to be of Cambrian age. Prior to the discovery of Cambrian trilobites in rocks overlying the Lalun Formation, the name Old Red Sandstone, implying a Devonian age, was commonly used (e.g. STAHL, 1911; NATIONAL IRANIAN OIL CO., 1959; STÖCKLIN, 1960). FURON (1941) described the sequence as 'Dévonien gréseux' and GANSSER and HUBER (1962) as the Upper Hezarchal Formation.

In ascending order, the type section of the Lalun Formation in the central Alborz shows the following lithological composition: nearly 500 metres dark red to pink, medium-grained arkosic sandstones; 35 metres dark red and variegated shales with sandstone interbeds; 50 metres white subarkosic quartzite (= Top Quartzite). Cross-bedding and ripplemarks are frequent throughout the section. GANSSER and HUBER (1962) and ALLENBACH (1966) found *Cruziana*-like tracks in the Lalun of the central Alborz to be correlated with the trilobite footprints from the Pakistanian Salt Range which are middle (to late) Early Cambrian in age. The white Top

Quartzite of the Lalun is conformably, with sharp limit, overlain by Member 1 of the Mila Formation, which is of variable sandy or dolomitic composition.

The Lalun Sandstone is the most persistent Palaeozoic rock unit in Iran. Nearly everywhere it is crowned by the white Top Quartzite. In the Kerman area, however, the white quartzite overlying the Dahu Formation does not seem to be identical with the Top Quartzite of the Lalun (Section 2.4.2) as assumed by STÖCKLIN (1972). The Dahu and Lalun Formations are, therefore, not perfect equivalents. Northwards, the Lalun Sandstone pinches out against the Pre-Cambrian basement exposures along the northern foot of the Alborz and in the northernmost Azerbaijan. STÖCKLIN (1972) considered the Lalun Sandstone as the basal unit of the Cambrian System in Iran though the exact time–stratigraphical position of its lower limit is not known. The thick sequence of clastic and carbonatic rocks conformably underlying the Lalun is designated as Infracambrian. It is possible that this unit comprises Lower Cambrian as well as late Pre-Cambrian rocks.

STÖCKLIN, RUTTNER, and NABAVI (1964) described a sequence of dolomites, shales, and platy, nodular and sparry limestones with abundant trilobite fragments from the Mila Kuh in the eastern Alborz Mountains under the name Mila Formation. It was ASSERETO (1963), however, who used this term for the first time for part of the present Mila Formation. Earlier workers included the Mila Formation in younger Palaeozoic rocks. The Mila Formation can be divided into five members at the type locality which are described in ascending order.

Member 1 (189 metres) consists of barren dolomite with subordinate marl and shale. The basal bed is a thin, yellow marl which overlies the Lalun Top Quartzite with sharp but perfectly conformable contact. Locally, Member 1 contains salt pseudomorphs and can be considered as an equivalent of the partly evaporitic ?Middle Cambrian Kalshaneh Formation in the Shirgesht area. According to STÖCKLIN (1972), Member 1 is possibly also synchronous with at least that part of the Kuhbanan Formation which yielded the middle (to late) Early Cambrian faunas. The present author (*in* WOLFART and KÜRSTEN, 1974), however, suggested the *Redlichia*-bearing horizons of the Kuhbanan to be synchronous with the *Cruziana*-bearing horizons in the Lalun Formation.

Member 2 is 89 metres in thickness at the type section and composed of sparry and flaggy limestones, subordinate marls and siltstones containing the trilobites *Chuangia, Iranochuangia*, and cf. *Iranoleesia*, together with *Obolus* and *Hyolithes*; these according to ERBEN (*in* STÖCKLIN, RUTTNER, and NABAVI, 1964), who attributed the fauna to an early stage of the Late Cambrian. KOBAYASHI (*in* STÖCKLIN and others) identified trilobites of the *Lioparella* and *Anomocarella* group. He was inclined to assign some of Member 2 to the late Middle Cambrian whereas the present author prefers to consider all of the fauna to be of early Late Cambrian age. More recently, however, B. KUSCHAN (personal communication, 1972) has found *Dorypyge* in northern Iran which is significant as indicating a Middle Cambrian age.

Member 3, as the most conspicuous member of the entire formation, can be traced as a light band in the landscape throughout the Mila Kuh region. It is 82 metres in thickness and consists of a coarse, crystalline-sparry limestone, white–grey or white in colour with green or dark green spots and stripes due to a varying content of glauconite. The limestone yielded an abundance of trilobites and brachiopods identified by ERBEN (*in* STÖCKLIN and others, 1964) as *Prochuangia, Kaolishania*, cf. *Eochuangia*, and a new genus of the Damesellidae. Additionally KOBAYASHI and WINSNES (*in* STÖCKLIN and others) recognized trilobites of the *Anomocarella* group and the brachiopods *Billingsella* and ?*Apheoorthis*. The fauna indicates a middle stage of the Late Cambrian.

Member 4 is 96 metres in thickness and resembles Member 2 in containing numerous intercalations of sparry limestone which is in places entirely composed of trilobite fragments.

The principal rock type is a fine-grained, thin-bedded, calcareous sandstone or siltstone of greenish brown colour alternating with grey or greenish grey marls. Generally, the trilobite fragments are bigger in size than those of the preceding members. Some layers are full of *Hyolithes*. ERBEN (*in* STÖCKLIN and others, 1964) identified the following trilobites which appear to be a mixture of latest Late Cambrian and earliest Early Ordovician elements. The lower portion of Member 4 yielded *Quadraticephalus, Iranaspis, Sinosaukia,* ?*Kaolishania,* cf. *Eochuangia,* which are possibly of latest Late Cambrian age or partly already of early Ordovician age. The upper portion of Member 4 contains '*Briscoia*' (*in* KING 1937), *Saukia, Iranaspis,* ?*Changia,* and cf. *Eochuangia.* This assemblage seems to indicate an early Ordovician age. Consequently, Member 5 of the Mila Formation, which consists of barren shales and subordinate sandstones and limestones, belongs to the Ordovician.

The lithological succession, subdivision, and faunas of the Mila Formation are remarkably persistent in the entire northern region of Iran, being traceable from eastern Alborz to Azerbaijan. In the Shirgesht area, eastern Iran, the Upper Cambrian/lower Ordovician rocks are developed as perfect equivalents of the Mila Formation in northern Iran, but in eastern Iran they reach a thickness of about 3000 metres. This sequence has been subdivided into three formations, Kalshaneh, Derenjal, and Shirgesht, which form the Mila Group.

2.5. The Cambrian rocks of Afghanistan

The oldest known fossils of Afghanistan have been found near Surkh Bum, in the central part of the country, only within the past decade or so (WOLFART, 1969; WEIPPERT, WITTEKINDT, and WOLFART, 1970; WOLFART and KÜRSTEN, 1974). In Afghanistan, outcrops of fossiliferous Cambrian rocks are very limited according to present knowledge. The Cambrian sequence of Surkh Bum is lithologically divided into two parts, the barren lower dolomites (unnamed dolomites) and the fossiliferous upper limestones (Kirman Formation). The bedded, partly marly dolomites, the base of which is not exposed, are 70 metres thick. A ?late Middle Cambrian to early Late Cambrian age can be inferred from their stratigraphical position below the fossiliferous Upper Cambrian limestones. The latter are composed of 55–60 metres of dark grey–blue, crystalline and oolitic, well bedded limestones with some interbeds of yellow–grey, silty, dolomitic limestones. The number of the interbeds increases towards the top of the sequence. Barren, yellow dolomites of two metres thickness, with small pebbles of quartz in the upper part, form the top of the Upper Cambrian section which is disconformably overlain by lower Tremadoc limestones, marls, and slaty shales with abundant fragments of trilobites and some brachiopods. The Kirman limestones yielded the following faunas (WOLFART and KÜRSTEN, 1974). Brachiopods: *Westonia,* Eoorthidae; Monoplacophora: *Hypseloconus*; Trilobita: *Pseudagnostus, Crepicephalus, Torifera, Afghanocare, Paracoosia, Chelidonocephalus* ? (*Farsia*), ?*Biaverta, Constrictella, Ariaspis, Blackwelderia,* Damesellinae. Owing to the presence of *Pseudagnostus, Crepicephalus, Blackwelderia,* Aulacodigmatidae, *Paracoosia,* and ?*Biaverta,* the Upper Cambrian sequence of Surkh Bum must be considered to be correlative with the *Crepicephalus* Zone and the roughly equivalent *Glyptagnostus stolidotus* Zone, respectively, which corresponds to the late Kushanian of eastern Asia.

According to SLAVIN and MIRZAD (1969), up to 3500 metres of slightly metamorphosed rocks are interbedded between the crystalline basement complex and fossiliferous Devonian beds in the Badakhshan trough, northern Afghanistan (Section 3.6.3). The metamorphic sequence, called the Sijah-Darrah Formation by KOLČANOV and others (1971), possibly contains rocks of Cambrian age.

In the northern Logar trough between the upper Kabul valley and the upper Logar valley, DESPARMET and MONTENAT (1972) found the following sequence (from bottom to top): (1) 70 metres white and red limestones; (2) 70 metres red and green shales with sandy intercalations; (3) 170 metres greenish shales. This sequence is overlain by 370 metres of rusty brown shales and sandstones with *Afghanodesma*, assigned a Late Cambrian to Tremadoc age (Section 3.6.4). The formations 1–3 pinch out northwards.

Another occurrence of presumed Cambrian deposits in Afghanistan has been discovered by KARAPETOV and others (1971) in the south-western part of the Logar trough. In the Arghandab basin, west of Moqur and north-west of Qalat, the presumed Cambrian succession overlies with sharp angular unconformity the Pre-Cambrian basement complex consisting of slightly metamorphosed phyllitic shales, aleurolites, and sandstones of the Tschaman Formation showing grey–green and red colours. The upper portion of the Tschaman, which is 3500 to 4000 metres in thickness, contains tuffaceous materials.

KARAPETOV and others called the oldest non-metamorphic rocks of the Arghandab basin the Sargaran Formation. On the grounds of its stratigraphical position at the base of the Palaeozoic section and of algae of a Cambrian type, KARAPETOV and others (1971) came to the preliminary conclusion that the Sargaran must be, at least partly, Cambrian in age. The Sargaran is composed of the following members: (1) 25–70 metres of fine- to medium-grained, variegated conglomerates including fragments of the basement complex; (2) 50–60 metres of shales, aleurolites, and sandstones with an intercalation of dark, oolitic limestone; (3) 70 metres of bedded limestones with algae, at the surface brownish and cream coloured; (4) 50 metres of greenish grey and red aleurolites and sandstones; (5) 250 metres of grey, thick-bedded limestones mainly formed by algae; (6) 80 metres of yellowish, thin-platy sandstone. The seventh member of the Sargaran is composed of 120 metres of bedded, frequently dolomitic limestones with numerous small lenses and nodules of chert. KARAPETOV and others placed the Cambro–Ordovician boundary within this seventh member because 'the algae are positioned in the median members of the Sargaran' (Section 3.6.4).

2.6. Cambrian stratigraphy and correlation in the Middle East

In the Middle East region, the Cambrian sequences are mainly of barren deposits. Complete fossiliferous successions possibly may be present only in the Upper Cambrian of some Iranian regions. For this reason, it is very difficult or even impossible to subdivide the Cambrian rocks according to a classification of the Cambrian System based on faunal successions. Within the Middle East region, only a few attempts have been made to correlate Cambrian rock units of the various countries. Owing to the scarcity of fossils, those attempts were based more on lithological than on palaeontological characteristics. The faunal assemblages hitherto found in the various regions of the Middle East are of Middle Cambrian age in Turkey, of middle (to ?late) Early Cambrian age in Jordan, and of early Late Cambrian age in Afghanistan. Faunas of the middle (to ?late) Early Cambrian *and* of the Middle and Late Cambrian are present only in some regions of Iran.

In the past, additional difficulties in attributing the Cambrian rock units to one of the international stages arose from the various ages given to faunal assemblages with *Redlichia*. KING (1941), ÖPIK (1956, 1958), and others correlated the *Redlichia*-bearing horizons in the Salt Range and in northern Australia, respectively, with the early Middle Cambrian. Following KOBAYASHI (1967), it was SDZUY (*in* LOTZE and SDZUY, 1970) who, for the first time, proved the Lower Cambrian age for *Redlichia* in Spain. There, *Redlichia* is associated with the Lower Cambrian trilobites *Lusatiops*, *Triangulaspis*, *Andalusiana*, and *Strenuaeva*. In Spain, the *Redlichia* Beds are overlain by still further Lower Cambrian rocks with *Termierella* and *Hamatolenus*, and only after these genera, the Middle Cambrian *Paradoxides* appears. *Red-lichia* is widely distributed in eastern and southern Asia, Australia, and as far as Spain in the west (REPINA, 1969; ÖPIK, 1958; LOTZE and SDZUY, 1970). According to SDZUY, this fact indicates that morphologically so sharply defined a genus must be restricted to a narrow biostratigraphical range everywhere and, therefore, can be well applied as a guide fossil.

2.6.1. The lower boundary of the Cambrian

Owing to the lack of fossils, the problem of the Pre-Cambrian/Cambrian boundary cannot be discussed by the present author. Essential aspects of it have been treated by HUTCHINSON (1956), OKULITCH (1956), and ROZANOV (1967).

In the Middle East, the problem of the Pre-Cambrian/Cambrian boundary is an especially delicate one. Where fossiliferous Lower Cambrian rocks rest on the crystalline basement, there is of course no difficulty in placing the boundary, even though this boundary will not be synchronous in all places. However, where barren, non-metamorphic deposits of very great thickness overlie the crystalline basement and where the oldest significant fossils are middle to ?late Early Cambrian in age, as in the entire area of the Middle East, there is no definite solution as to where to draw the boundary.

In Iran, STÖCKLIN (1972) considered the Lalun Sandstone because of its stratigraphical position, great uniformity, and persistence, together with the sparse palaeontological evidence (Section 2.4), as the basal unit of the Cambrian System, although the precise time-stratigraphical position of its lower limit is not known. The thick sequence of non-metamorphic, carbonate, shaly, and sandy rocks (Zaigun, Barut, Soltanieh, and Bayandor Formations) conformably underlying the Lalun is grouped as Infracambrian. These rocks reach a thickness of up to 2800 metres in Iran and certainly comprise a Lower Cambrian as well as a late Pre-Cambrian portion.

The true Pre-Cambrian/Cambrian boundary is situated somewhere between the metamorphic basement complex and the oldest horizon with significant Cambrian fossils. The sedimentary sequence below this horizon is called Infracambrian and contains only few insignificant fossils as, for example, *Collenia* and *Fermoria*. The Infracambrian sequence includes, however, in some regions of the Middle East and adjacent areas (Oman, Persian Gulf area, southern Iran, eastern central Iran, and Pakistan Salt Ranges) the most conspicuous Hormuz Salt Formation or its equivalent Punjab Saline Series and Ravar Formation. Based on the very similar succession and nature of the deposits and on the comparable stratigraphical position of the *Redlichia*-bearing horizons, SCHROEDER (1946), SCHINDEWOLF and SEILACHER (1955), and ASRARULLAH (1961) postulated a late Pre-Cambrian to early Cambrian age for the Punjab Saline Series. SCHINDEWOLF supposed that the Purple Sandstone (= Dahu Formation) as well as the Punjab Saline Series could well have been formed during the pre-*Redlichia* portion of the Early Cambrian. LOTZE (1957, 1969) also preferred to assume an early Early Cambrian age for the Punjab Saline Series, bearing in mind the fact that evaporites of proved Pre-Cambrian age have not as yet been encountered. Consequently, it might be possible either to place the true Pre-Cambrian/Cambrian boundary at the base of the evaporitic formations, and their lateral equivalents the Barut Formation and part of the Soltanieh Formation in northern Iran, or to draw it at the base of the Lalun or somewhere between the two.

In the other regions of the Middle East, in the Arabian Peninsula, Turkey, and Afghanistan, the correlative units of the Iranian and Oman Infracambrian group are composed mainly of clastics and some volcanics, if they are developed at all. In the north-western part of the Arabian Peninsula and in Jordan, POWERS (1966, 1968) and BENDER (1964, 1968) placed the Pre-Cambrian/Cambrian boundary between the pre-Cambrian basement complex and the overlying clastic sequence, although the Cambrian age of the lower portion of this clastic sequence has never been proved. In that region as well as in Afghanistan, the stratigraphical gap between the Pre-Cambrian basement complex and the overlying sedimentary sequence might, however, be so comprehensive as to include the beginning of Cambrian times.

2.6.2. The upper boundary of the Cambrian

The upper boundary of the Cambrian System in the Middle East presents problems quite different from those of the lower boundary. NORTH (1971) emphasized that, if the Early Palaeozoic systems had been defined in eastern North America instead of in Europe, the early Ordovician of American usage (and not just the Tremadoc) would have been part of the system below it and not of the system above it. This partly applies also to the Middle East.

Especially in northern Iran, the carbonate facies continues uninterruptedly from the Upper Cambrian into the Tremadoc. As there is no lithological change, the Cambro–Ordovician boundary in that region is wholly palaeontological. In Afghanistan, the base of the Ordovician is lithologically more distinct than in Iran. The calcareous sequence of the lower Upper Cambrian is unconformably overlain by the calcareous–shaly Tremadoc Surkh Bum Formation which is lithologically obviously different from the Cambrian rocks. In the Arabian Peninsula and in Turkey, the Cambro–Ordovician boundary is mainly palaeontologically characterized. There, its biostratigraphical position has been clear since the first research upon it. In Northern and eastern central Iran, on contrast, it seems to be necessary to move the Cambro–Ordovician boundary from the top of Member 4 of the Mila Formation and from the lower portion of the Shirgesht Formation, respectively, into Member 4 itself and into the upper part of the Derenjal Formation, owing to the occurrence of saukiids. Possibly, the saukiids contain genera of late Late Cambrian and of Tremadoc age, respectively. In Iran, they are associated with some trilobite genera the relationships of which have not yet been sufficiently studied. Therefore, the problem of the Cambro–Ordovician boundary in Iran needs re-examination.

Before the work of ROBISON and PANTOJA-ALOR (1968) and WOLFART (1969, 1970a,b), all of the saukiids were considered to be of Trempealeauian (= late Late Cambrian) age. ROBISON and PANTOJA-ALOR were the first to find a mixed faunal assemblage in Mexico containing the Trempealeauian *Saukia* and Tremadoc genera such as *Parabolina*, *Asaphellus*, and others. Thus, they decided to correlate the *Saukia* Zone and the Early Tremadoc. In Afghanistan, *Saukia* is also associated with Tremadoc genera (WOLFART, 1969) and must, therefore, be assigned to the Ordovician System. Recently, JONES, SHERGOLD, and DRUCE (1971) described saukiids from western Queensland, Australia, and considered them to be of Late Cambrian age. They based their considerations, however, on conodont assemblage zones only, which are by no means comparable to the ranges of the conodont species in other regions, e.g. in Mexico and Afghanistan. The present author is not convinced that the saukiids were extinguished in Queensland at the end of the Cambrian, in northern America during the course of the Early Tremadoc, and in Afghanistan during the Late Tremadoc, as assumed by JONES, SHERGOLD, and DRUCE.

According to KUSHAN (1973), *Saukia* is associated with *Pagodia* in northern Iran also, and distinctly separated from older beds with the Upper Cambrian trilobites *Chuangia*, *Kaolishania*, etc. MILLER and ROBISON (1974) performed some further conodont researches on faunal assemblages from Oaxaca, Mexico. They stated that (1) at least part of the Lower Tremadoc rocks of Europe belong to the Lower Ordovician of North America; (2) the boundary between the lower and the upper Tremadoc is situated within the *Symphysurina* Zone; (3) the base of the Tremadoc is—concerning its age—nearer to the base of the North American Ordovician than has been assumed in the past; and (4) the *Saukia* of Oaxaca is probably of lower Ordovician age in the American sense.

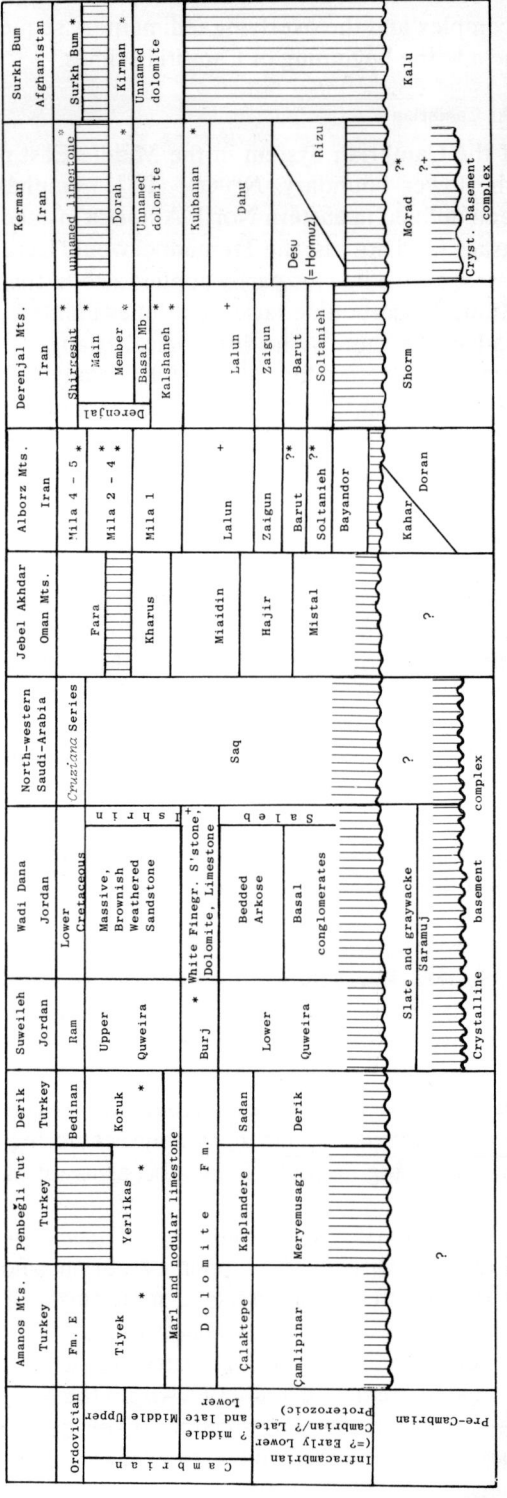

FIG. 11. Correlations of Cambrian rocks between selected areas in the Middle East. The wavy line indicates folding and metamorphism. * Fossils; + trace fossils.

2.6.3. Correlations within the Middle East

Correlations of Cambrian rocks between the various regions of the Middle East are summarized in Fig. 11. The Infracambrian rocks are correlated exclusively on grounds of lithological type and their stratigraphical position between the metamorphic basement complex and the formations bearing middle (to late) Early Cambrian and Middle Cambrian fossils, respectively. The basement complex was consolidated by orogenies of different ages. The last one, the Assyntic Orogeny, happened synchronously in the entire Middle East region (FLÜGEL, 1964; WOLFART, 1967b) and created a reliable time–stratigraphical feature. For lack of characteristic features, the Infracambrian rocks of Turkey and Iran cannot be correlated with synchronous units of southern Jordan and north-western Saudi Arabia. There are, perhaps, no equivalents at all. The successions of volcanics and red sandstones in the Kerman area (Rizu Formation) are very similar to those of the Derik area (=Derik Formation). Outside the Persian Gulf, southern Iran, and the Oman, evaporites of the Hormuz type have not been encountered in the Middle East. The Soltanieh Dolomite in northern and eastern central Iran and the red and green clastics of the Çamlipinar and Derik Formations in southern Turkey must be considered as lateral non-evaporitic equivalents of the Hormuz Salt Formation (KETIN, 1966; STÖCKLIN, 1972). On lithological grounds, the Turkish Çalaktepe, Kaplandere, and Sadan Formations and the Iranian Lalun and Dahu Formations seem to be correlative units. The carbonates of the Kuhbanan, Burj, and Dolomite Formations are widely distributed and yielded redlichiid faunas, the oldest significant fossils in the Middle East. They represent one of the most conspicuous guide horizons in large parts of the region. The problem also whether the Kuhbanan and the upper fossiliferous portion of the northern and eastern Iranian Lalun are contemporaneous is treated in Section 2.4.2.

In the eastern part of the Middle East (Oman, Iran, Afghanistan) the probable Middle Cambrian rocks are mostly of dolomitic character and must be regarded, on the grounds of their stratigraphical postion, as correlative with the clastic formations of the western part of the region. The latter are barren in north-western Saudi Arabia and southern Jordan; they yielded Middle Cambrian fossils in southern Turkey. Upper Cambrian rocks are composed of clastics in the western part and of carbonates in the eastern part of the Middle East. They are fossiliferous in Iran and Afghanistan only, and especially the early Late Cambrian faunas of Afghanistan and of the Kerman area show very close relationships.

2.7. Cambrian history of the Middle East

2.7.1. Tectonics and sedimentation

In the Middle East, the structure of the Pre-Cambrian basement was strongly affected by the various Pre-Cambrian orogenies, the last of which was the Assyntic orogeny. No folding happened during Cambrian times. Structural development and sedimentation during the Cambrian were largely governed by the Pre-Cambrian basement trends mainly showing three directions (N.W.–S.E., N.E.–S.W., and N.–S.). At least two significant episodes of uplift can be inferred from the gaps in the sedimentary record.

Both of the episodes comprised various lengths of time at different places. Uplift was invoked for all of the Middle East at the close of the late Pre-Cambrian Assyntic movements, followed by erosion or at least by a sedimentational gap. In Afghanistan, this gap comprises the time between the Cambrian Kalu Formation and Middle to Upper Cambrian dolomites. In Iran, Oman, and Turkey, sedimentation began as early as in Infracambrian times (= post-Assyntic late Pre-Cambrian). During the course of the Late Cambrian a further episode of uplift obviously affected Afghanistan, the Kerman region, part of Oman, and the northern part of southern Turkey only.

The sedimentary successions of the various regions of the Middle East are very complex and cannot be considered in detail. Therefore, only the major events are outlined here. In the Infracambrian and Cambrian, four major marine transgressions are indicated by marine faunas. The oldest is represented by the Infracambrian Hormuz Salt Formation and its carbonate equivalents in Oman, the Persian Gulf, and Iran. The presence of algal stromatolites (*Collenia*) and massive dolomites, relatively free of significant terrigenous detritus, strongly suggests a genesis in relatively shallow water. To the north, the salt basin of the Persian Gulf and southern Iran may well have been linked with a deeper marine trough through gaps in a system of swells (STÖCKLIN, 1968a). In the other areas of the Middle East, mainly sedimentary rocks of continental origin and volcanics were formed during this time.

The next marine transgression is of middle (to late) Early Cambrian age, as indicated by the faunal assemblage with redlichiid trilobites and *Cruziana* distributed all over Iran and as far to

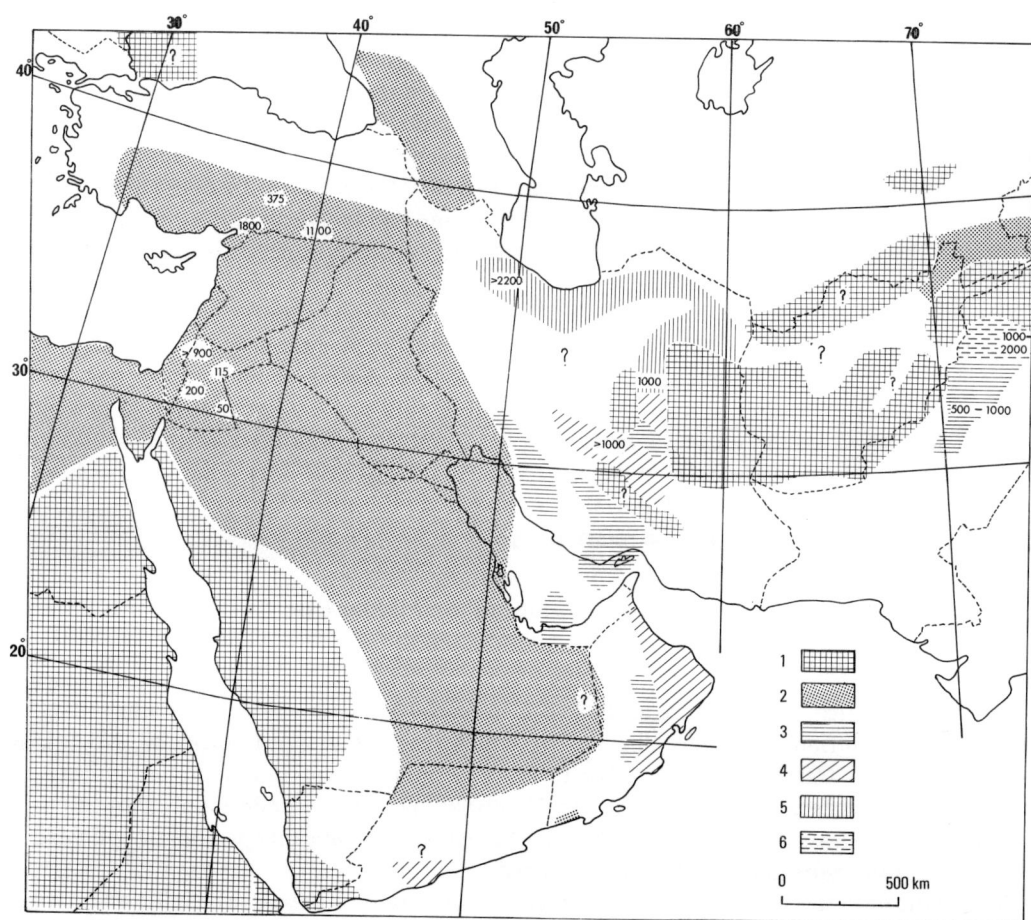

FIG. 12. *Palaeogeography*—late Pre-Cambrian to early Early Cambrian. 1, Potential source area; 2, continental to littoral environment (conglomerate, sandstone, shale, regionally volcanics); 3, lagoonal environment (shale, dolomite, gypsum, salt); 4, alternating continental and marine environment (conglomerate, sandstone, volcanics, limestone, dolomite, gypsum, salt ?); 5, mainly marine environment (dolomite, shale); 6, marine environment (shale). Thicknesses in metres.

the west as Jordan. Intercalations of oolites indicate shallow water conditions. Contemporaneously, in Arabia continental (to littoral) environments prevailed, and in Turkey non-fossiliferous carbonates and clastics were probably formed in marine environments.

The Middle Cambrian transgression mainly affected the southern part of Turkey and, perhaps temporarily, part of northern Iran. Middle Cambrian rocks have yielded the only marine fossils of the Turkish Cambrian. As can be inferred from dolomitic, gypsiferous rocks with salt pseudomorphs, in northern and eastern Iran and, perhaps, in Afghanistan, a temporarily lagoonal environment prevailed during the Middle Cambrian. Except for Oman, all over the Arabian Peninsula continental conditions continued.

Only in Iran and in Afghanistan, the Late Cambrian transgression left fossiliferous carbonates. Oolites, massive limestones, and shell layers indicate shallow water conditions. In Turkey unfossiliferous shales and sandstones were probably formed under marine conditions, while continental environments continued in the entire Arabian Peninsula.

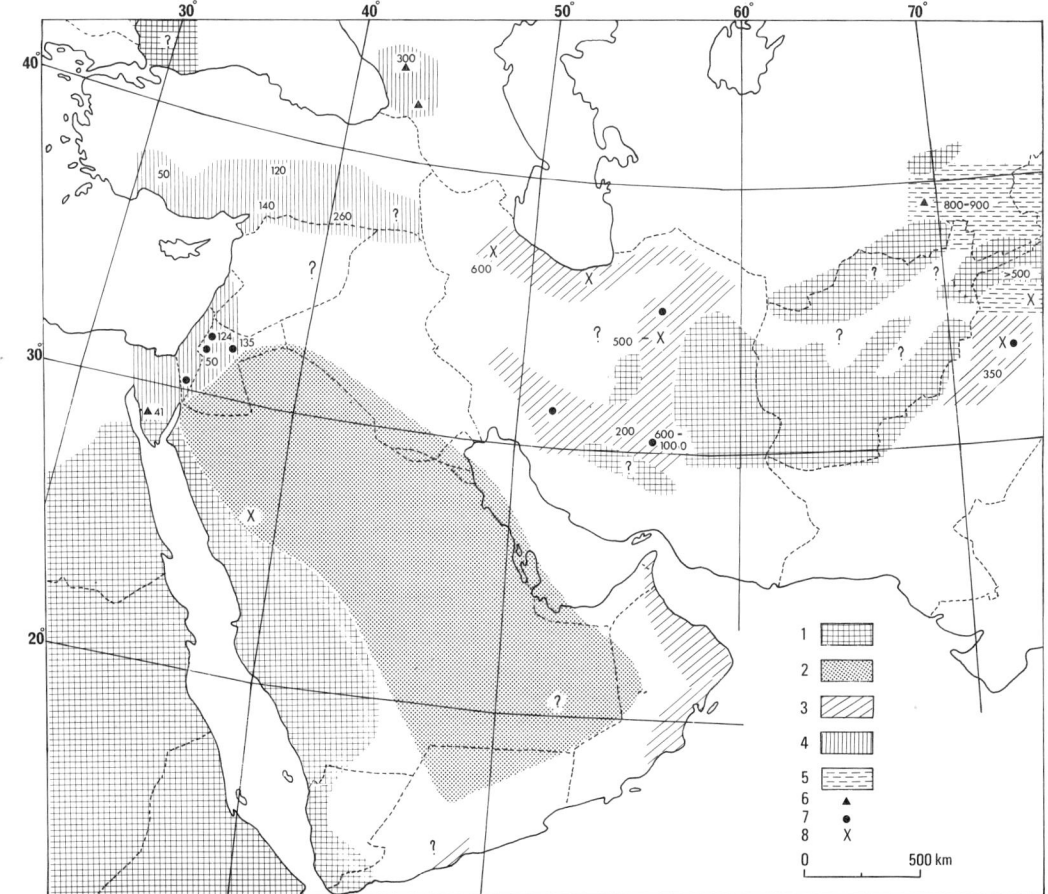

FIG. 13. *Palaeogeography*—middle to late Early Cambrian. 1, Potential source area; 2, continental to littoral environment (sandstone, shale); 3, marine following continental environment (sandstone, shale, dolomite, limestone); 4 and 5, marine environment; 4, dolomite, limestone, shale, sandstone; 5, shale, sandstone; 6, Archaeocyathidae (in Soviet Union associated with Siberian and south Asiatic faunas); 7, Redlichiidae, Ptychopariidae, *Orthotheca* (south Asiatic faunas); 8, *Cruziana, Rusophycus*, and other trace fossils. Thicknesses in metres.

In the Middle East, throughout the entire Infracambrian and Cambrian, the palaeogeographical configuration was governed by the huge Nubo-Arabian craton in the south-west and
by a large number of Pre-Cambrian archipelagous swells and highs in the middle and northern
part of the Middle East as demonstrated by Figs. 12–15. There is no clear evidence that the
stable block of the Arabian Shield and the Turkish–Afghan archipelago were separated by a
broad transitional zone striking north-west to south-east during the Cambrian as might be
suggested by present conditions. STÖCKLIN (1968a) assumed an additional high in that possible
transitional zone, the Qatar-Mund high, between the Arabian Shield and the Hormuz Salt basin.
During the Infracambrian and the Cambrian, a broad belt surrounding the Arabian Shield in the
east and in the north was covered with terrigenous sediments eroded in the cratonic regions. This
belt can be considered to be more or less identical with the stable shelf of the Arabian Shield.
Adjacent, to the north, the mobile shelf bordering the Arabian craton with the swells, serving as

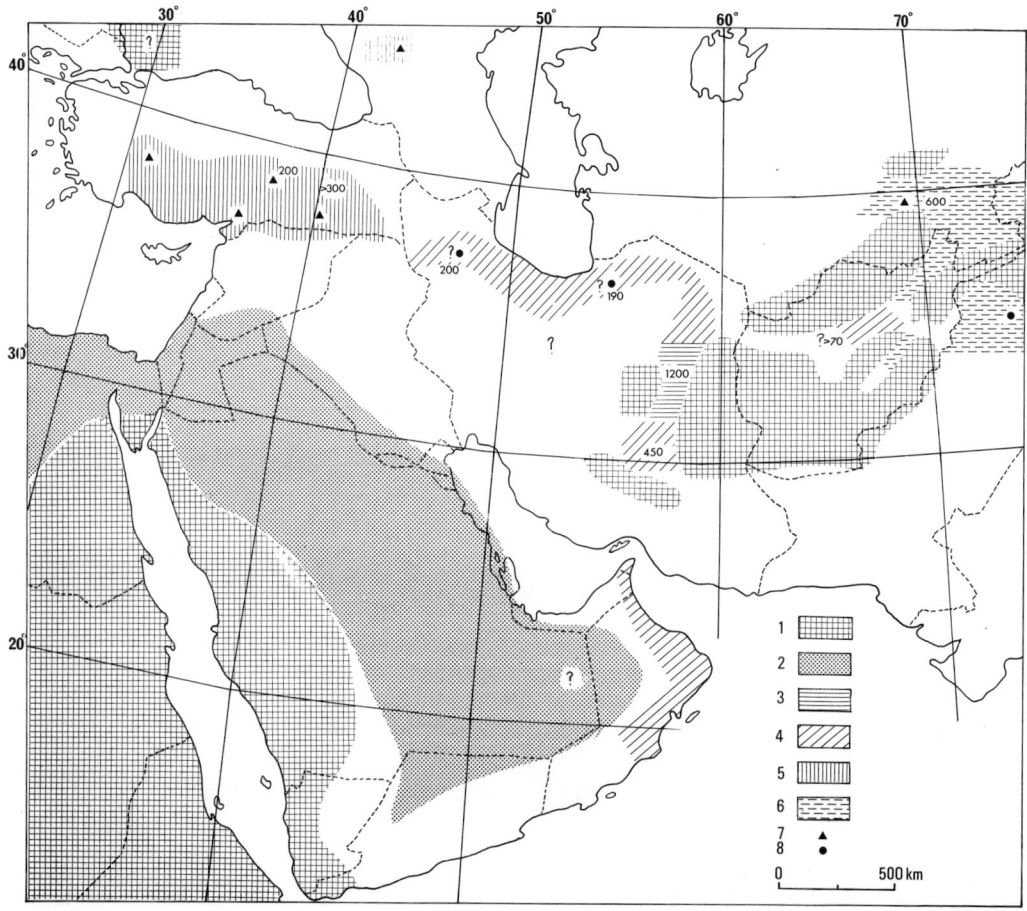

FIG. 14. *Palaeogeography*—Middle Cambrian. 1, Potential source area; 2, continental to littoral
environment (sandstone, shale); 3, lagoonal to marine environment (dolomite, dolomitic marl with
gypsum, limestone); 4–6, marine environment; 4, dolomite, dolomitic marl, limestone; 5, limestone, shale sandstone; 6, shale with intercalations of limestone; 7, faunas mainly composed of
Atlantic elements; 8, faunas composed of south Asiatic and Atlantic elements. Thicknesses in
metres.

source areas extended from Turkey to Afghanistan and Oman. It is in this region that sedimentational conditions frequently changed during the Early Palaeozoic and later.

In the entire mobile shelf region, the thickness of late Pre-Cambrian and early Cambrian deposits considerably exceeds that of the Middle and Upper Cambrian rocks. From this fact, and from the predominantly clastic nature of the Infracambrian and Lower Cambrian rocks, it can be inferred that during this time, immediately after the Assyntic orogenetical movements, a relatively strong relief still existed, on the one hand causing strong erosion and on the other accompanied by quick subsidence and sedimentation. In spite of the quick subsidence, the sea transgressed only once during the Early Cambrian episodically creating marine conditions (*Redlichia* transgression). During the earliest Cambrian (or latest Pre-Cambrian), however, short marine ingressions happened, coming from the north, west of the Afghan high as well as east of it, followed by lagoonal evaporitic sedimentation. During the remainder of the Cambrian (i.e. Middle and Late Cambrian), marine conditions increasingly spread into what are now Iran and Afghanistan causing carbonate sedimentation, while the deposition of clastics continued in

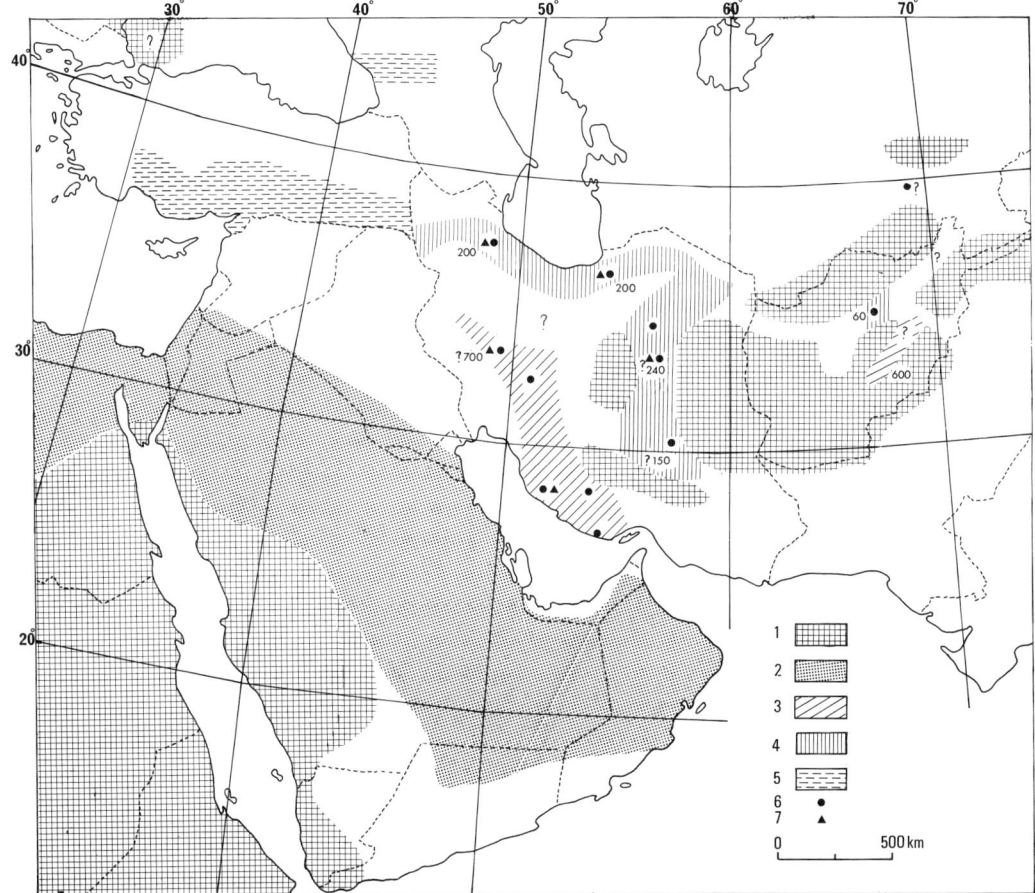

FIG. 15. *Palaeogeography*—Late Cambrian. 1, Potential source area; 2, continental to littoral environment (sandstone, shale); 3–5, marine environment, 3, sandstone, limestone, dolomite; 4, limestone, marl; 5, sandstone, shale; 6, early Late Cambrian faunas (Kushanian); 7, late Late Cambrian faunas (Changshanian to Daizanian). Thicknesses in metres.

the other regions of the Middle East. Everywhere, at the same time, however, subsidence was retarded. Consequently, the thickness of the Upper Cambrian deposits rarely exceeds a few hundred metres.

2.7.2. Zoogeographical features and climate

In the Turkish–Afghan west–east belt of mobile shelf conditions, the presence of Cambrian trilobites suggests the existence of a seaway which might be considered as an early Tethys. Since the work of NEUMAYR (1885) and SUESS (1893), a great equatorial sea of Jurassic age, assumed to extend from Mexico to the Himalayas, has been referred to as Tethys. SUESS recognized Tethys to separate two cratonic aggregations, Angaraland to the north and Gondwanaland to the south. According to RICHTER (1941), the name Tethys should be applied to Palaeozoic seas only if (1) they were situated in the range of a European–Asian Mediterranean Sea, (2) they communicated with the Pacific, and (3) they are defined by uniform zoogeographical features.

The middle (to late) Early Cambrian trilobite genus *Redlichia* is widely distributed in Australia, China, Korea, Siberia, India, Pakistan, Iran, and Spain. Consequently, SDZUY (1967) came to the conclusion that the Early Cambrian Tethys was a marine connection between Europe (Spain), south-eastern Asia, and Australia, serving for faunal migration between the Atlantic Province to the west and an oriental province to the east. COWIE (1971) distinguished two Early Cambrian trilobite world realms, the redlichiid realm and the olenellid realm, the latter of which was divided into Acado–Baltic and Pacific Provinces.

According to SDZUY, in the Middle and Late Cambrian, Tethys as an open seaway for faunal exchange was replaced by a path of migration through northern Siberia. Middle Cambrian faunas of the Mediterranean region, which was zoogeographically isolated from northern Europe in the Middle and Late Cambrian, can be traced into Turkey. Obviously, they did not penetrate any farther east, although the Turkish and northern Iranian marine areas could well have been linked. The distribution of Middle Cambrian faunas was probably controlled by unfavourable lagoonal facial conditions predominating in Iran. The Late Cambrian trilobite faunas of Iran and Afghanistan with *Crepicephalus*, *Paracoosia*, and *Blackwelderia* show affinities pointing strongly to faunas of eastern Asia, Australia, and even of North America (WOLFART and KÜRSTEN, 1974). There existed no connection between the Middle East and Europe during the Late Cambrian.

In the Middle East, the land was, during the Early Cambrian, probably a tropical or subtropical desert without vegetation. Consequently, the surface was unprotected from erosion. Early Cambrian (or late Pre-Cambrian?) saline and dolomitic sequences suggesting warm, arid evaporating conditions are present in large parts of southern Iran, the Persian Gulf, and Oman. Other indications of tropical or subtropical climate are plenty of red sediments which are widely distributed in the surrounding regions of Turkey and the Arabian Peninsula, and archaeocyathids which have been found in Tadzhikistan (MARKOVSKIJ, 1959), north of Afghanistan, and in Sinai (OMARA, 1972). By the occurrence of archaeocyathids, the ecological conditions of the Early Cambrian are characterized fairly precisely (HILL, 1965). Archaeocyathids, as benthonic organisms, preferred shallow water conditions of between 20 and 50 metres; this especially applies to the large specimens of Tadzhikistan. For archaeocyathid bioherms, as for coral reefs, optimal temperatures ranged between 20 and 25 °C. Lagoonal conditions excluded the occurrence of archaeocyathids. COWIE (1971) presumed that areas with evaporitic sedimentation were situated near the Early Cambrian palaeoequator derived from palaeomagnetic evidence. Trilobites, brachiopods, and other phyla were widely distributed in Early Cambrian times. From this, it might be inferred that the seas were universally habitable. Thus, the evidence might suggest that the quiet Early Cambrian times had a uniform climate which contrasted with the widespread glaciation of the late Pre-Cambrian. Therefore, the provincial distribution of the faunas could, perhaps, have been controlled by salinity or other factors rather than by temperature. PALMER (1972), however, came to the conclusion that the contrasting faunas with European, American, and Asiatic affinities around the margins of Gondwanaland seem to be controlled by Cambrian palaeolatitudes related to a pole near north-west Africa. The Cambrian faunal provinces and their origin were discussed by LOCHMAN–BALK and WILSON (1958) on the basis of environmental diversification of the faunas on a geotectonical background. PALMER (1968, 1969) re-evaluated this concept by additional studies.

As can be inferred from the predominantly chemical sedimentation during the Middle and Late Cambrian, the Middle East must have had a warm climate during that time as well.

2.7.3. Early Cambrian palaeogeography and continental drift

Owing to the concepts of sea floor spreading, tectonic plate movements, and continental drift, knowledge of the structural history of the earth has advanced rapidly. At present, it is not possible to explain the complicated relationships between the continents without Alfred Wegener's hypothesis of continental drift, which was modernized on the basis of numerous results of geological and geophysical research. In the opinion of modern authors (PICHON, 1968; DIETZ and HOLDEN, 1970; SMITH and HALLAM, 1970; COWIE, 1971; CRAWFORD, 1971; SMITH, 1971), Gondwanaland, comprising South America, Africa and the Arabian Peninsula, Antarctica, Australia, and India, existed as a more or less uniform southern continent—a large tectonic plate—as late as the Permian, through early to middle Mesozoic. By tectonic plate movements, parts of Gondwanaland drifted northwards during Cretaceous and Cainozoic times. According to this hypothesis, India, south Afghanistan, Iran, and the Arabian Peninsula were separated from the Asian continent by a funnel-shaped Tethys Sea of several thousand kilometres width, opening eastwards (WOLFART and WITTEKINDT, 1980). TAKIN (1972) stated that 'it is now generally agreed hat the Tethyan Ocean, a few thousand kilometres wide, existed in the present area of the Middle East in the Mesozoic and probably in the Palaeozoic or even earlier.'

COWIE (1971) considered it a basic assumption that the continents were positioned in the Early Cambrian, about 550 million years ago, at much the same position as can be deduced for Late Palaeozoic or later times, 300 to 150 million years ago. Arguments for the position of the continents, arranged as a 'Pangea', during the Permian and the pre-Permian were supplied by palaeomagnetic evidence, by the remains of the late Pre-Cambrian or Permo-Carboniferous glaciation, and by biogeographical evidence as, for example, the extension of the Malvinokaffric Province during the Early Devonian or the distribution of the Gondwana flora of the Permo-Carboniferous (MARTIN, 1969).

Opinions concerning the beginning of the split-up of Gondwanaland are divergent, partly on the basis of contradictory observations (MARTIN, 1969). According to DIETZ and HOLDEN (1971), the northward drift of India and Africa inclusive of the Arabian Peninsula, began during the Triassic or even later during early Cretaceous times (WOLFART and WITTEKINDT, 1980), and the Middle East Tethys was closed during the course of the Cainozoic. There are some observations (WOLFART, 1973; WOLFART and KÜRSTEN, 1974) about Cambrian conditions in the Middle East which confirm the assumptions of DIETZ and HOLDEN (1971), COWIE (1971), and others.

(1) Following PETRUSHEVSKY (1971), WOLFART (1973) stated the Cambrian deposits of middle south Asia to be platform sediments because of their facies, thicknesses, and faunas. From this fact and from the nature of the middle south Asian troughs recognized as near

fault troughs in Afghanistan (WEIPPERT, WITTEKINDT, and WOLFART, 1970) and in the Himalaya (PETRUSHEVSKY, 1971), the latter concluded that before the Neogene most of the Himalaya was presumably part of the Proterozoic Indian platform, at the northern edge of which the Himalayan trough of Tibet had already formed in the Cambrian. Concerning the north-eastern edge of the Arabian Shield, FALCON (1967) favoured similar conclusions. According to him, the north-eastern boundary of the Arabian Shield coincides with the Zagros thrust zone with its position apparently already defined in the Pre-Cambrian.

(2) The Arabian Peninsula and Iran were positioned in the range of lateral facies changes during the phase of saline sedimentation in the early Early Cambrian. A broad belt of continental clastics surrounded the Nubian–Arabian Shield and graded eastwards into the saline formations of Oman and the Persian Gulf/southern Iran, replaced by dolomites in eastern central and northern Iran. On the basis of those conditions, STÖCKLIN (1968) concluded that the northern sea (Tethys) must have been linked with the salt basin in the south (Section 2.7.1). Thus, evaporitic sedimentation in the south was apparently causally connected with marine invasions from the north. Platform sediments and closely connected lateral facies zones seem to indicate that, in the Early Cambrian, the Arabian Peninsula and Iran were part of the Gondwana continent which was separated from Angaraland by the Tethys Ocean several thousand kilometres in width.

In the Salt Range, east of the Afghan high, facies zones were not developed as in the Iranian–Arabian region. Nevertheless, there are indications that in the Salt Range area, the lowest Cambrian evaporitic sedimentation is causally connected with marine invasions from the north: (a) marine sedimentation in northern India and Kashmir during long periods of the Early and Middle Cambrian; short-time marine and continental sedimentation or non-sedimentation, respectively, in the Salt Range area during the same time; and (b) general decrease in thicknesses from north to south.

(3) Corresponding to the conditions described above—existence of Lower Cambrian lateral facies zones and marine invasions from the north—the regressions always began, remarkably early, in the south, i.e. in the areas of southern Iran and the Salt Range. In the latter area, Lower Cambrian deposits are relatively limited in thickness, and the Middle Cambrian is not present. In southern Iran, fossiliferous rocks of the Lower and Middle Cambrian have not been discovered up to the present, while marine limestones of the Upper Cambrian are locally only 5 metres in thickness, intercalated between continental clastics. All of the observations give rise to the presumption that a large continental hinterland extended south of the Cambrian sedimentational regions of Iran.

(4) The occurrence of the Lower Cambrian trilobite species *Redlichia noetlingi* (Redlich) in eastern central Iran, Spiti (northern India), and in the Salt Range indicates close geographical relationship of the three areas. Correspondingly, the very close relationship of the early Late Cambrian faunas of Surkh Bum, central Afghanistan, and Dorah Shah Dad, eastern central Iran, suggests the combined existence of old seaways in the Late Cambrian and even in early Ordovician times (WOLFART, 1970, 1973). Possibly, these seaways already served as migration paths in earlier Cambrian times.

(5) According to SAHASRABUDHE and MISHRA (1966), palaeomagnetic investigations of Cambrian or late Pre-Cambrian quartzites yielded a northern latitude of 14° referred to Nagpur, India. This result corresponds to the palaeomagnetic evidence gained from Pre-Cambrian igneous rocks indicating a very old position of India in northern latitudes (ATHAVALE and others, 1963). The same result, however, contradicts the palaeomagnetic characteristics of numerous younger rocks (ATHAVALE and others, 1970a). ATHAVALE and others (1970b) explained the contradictory results by assuming repeated north–south

Plate 1. Wadi Ram, southern Jordan. Cambrian sandstone overlying aplite granite. (*Photograph by Prof. Dr. F. Bender, Hannover*).

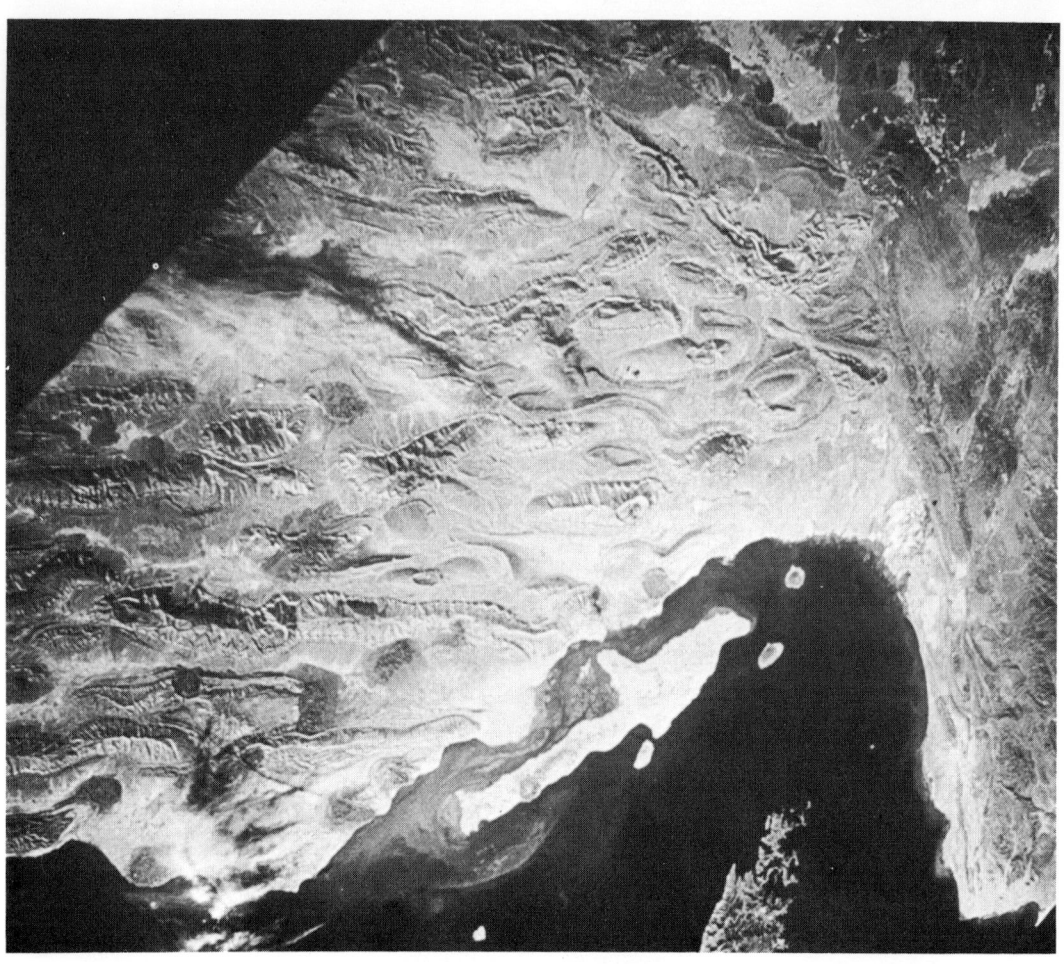

Plate 2. The Strait of Hormuz which connects the Persian Gulf with the Gulf of Oman and the hinterland of Bandar Abbas, southern Iran, showing anticlines and salt plugs formed by the Hormuz Formation. (*Photograph by NASA S-66-63082, Gemini XII, from about 250 kilometres vertical height*).

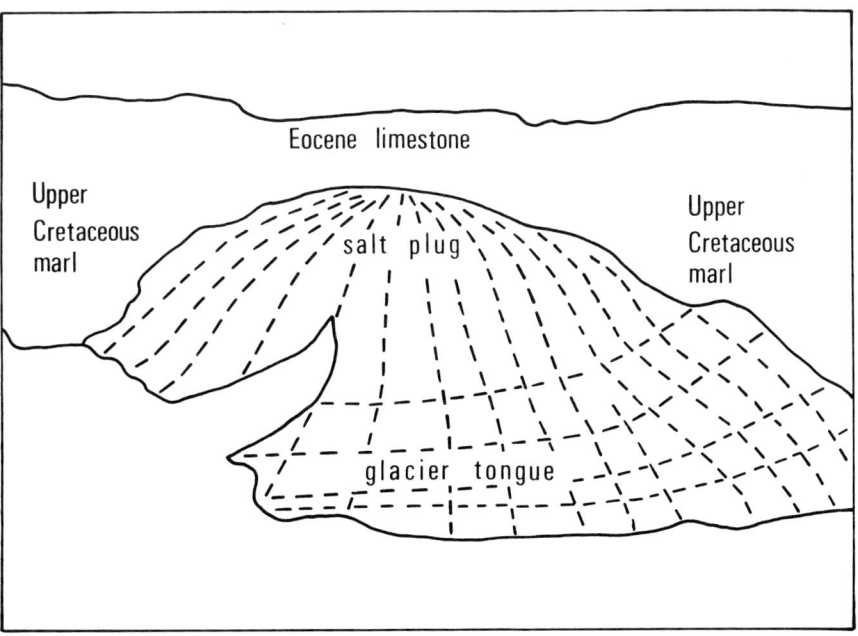

Eocene limestone

Upper
Cretaceous
marl

salt plug

Upper
Cretaceous
marl

glacier tongue

Plate 3. Salt glacier of Kuh-i-Anguru, south-eastern Iran. The true salt plug is
morphologically marked. Towards the left and in front of the salt plug, flowing salt
masses and the glacier tongue are distinctly recognizable. (*Aerial photograph by
Dr. H. W. Walther, Hannover, from about 1900 metres above sea-level*).

Plate 4. Qa'a Disa, southern Jordan. Lower Ordovician sandstone. (*Photograph by Prof. Dr. F. Bender, Hannover*).

Plate 5. Surkh Bum valley, north of Kotale Kirman, west of Panjaw, central Afghanistan. Foreground: Upper Cambrian limestones overlain by lower Ordovician (lower Tremadoc) limestones and shales. Background: Cretaceous limestones and conglomerates. (*Photograph by the author, 1968*).

Plate 6. Tangi Saidan in the upper Logar valley, eastern Afghanistan. Ordovician quartzites folded by orogenic movements during Oligocene. (*Photograph by Dr. H. Wittekindt, Hannover*).

movements of the Indian subcontinent. According to this concept, India would have drifted from a northern position in the Cambrian to about 50° southern latitude during the Palaeozoic. In the course of the Mesozoic and Cainozoic, India drifted northwards again.

2.8. Economic geology

In the Cambrian sequences of the Middle East, only small mineralized areas of little significance are known at present, such as south-western Jordan, southern Israel, and Umm Bogma in the Sinai. BENDER (1965, 1968) described copper and manganese ores from along the central and northern east bank of the Wadi Araba, south of the Dead Sea, whereas BENTOR (1956) reported occurrences of similar ores from Timna, 25 kilometres north of the Israel Red Sea port of Eilat.

At Timna, manganese and copper ores occur in various types of deposits, especially in the Cambrian Upper Shales which overlie the Zebra Sandstone (Section 2.2.1). Four types of manganese occurrences are distinguished: manganiferous Zebra Sandstone, disseminated manganese oxides, concretionary ore, and manganiferous veins. At Timna, the evidence points to a syngenetic sedimentary origin of the manganese and copper ores which have been formed by leaching from dilute sources in the Pre-Cambrian basement.

In south-western Jordan, the manganese and copper ores are interpreted by BENDER (1965) to be of syngenetic origin and to have been derived from a contemporaneous quartz porphyry volcanism. It seems to be probable, however, that the Jordan ores also are of sedimentary origin. Mining activity has been recorded from both sides of the Wadi Araba even as early as 2000 years BC. At present, there is only some exploitation of copper ores at Timna.

In the arid regions of the Near East, as has been reported, for example, from north-western Saudi Arabia (POWERS and others, 1966), the Cambrian clastics sometimes attain significance as an aquifer for domestic water supply.

3. Ordovician

Rather broad outcrops of Ordovician rocks extend for more than 1000 kilometres from southern Jordan south-eastwards into north-western Saudi Arabia. They mostly join the Cambrian outcrops to the east, normally showing a dip between 1° and 5° towards the east or north-east. Frequently the outcrop surface represents a mature stage of an arid inselberg landscape covered with the debris of arid weathering. Usually, the landscape is of low relief, although a few isolated, precipitous hills rise abruptly from the plain. In north-western Saudi Arabia, the more prominent features are between 150 and 300 metres high; in the mountainous region of Ram, southern Jordan, they tower up to 600 metres above the crystalline basement.

In the Tethyan mountain chains Ordovician outcrops are generally small and widely scattered, marking uplifts of the Pre-Cambrian basement. In Syria, Iraq, and the Persian Gulf region, the Ordovician rocks are overlain by a thick cover of later sediments and knowledge is restricted to deep boreholes. From Lebanon, Palestine, and Sinai, no Ordovician deposits are known hitherto, but it can be assumed that they are regionally present under a large cover of younger sediments.

The thickness of Ordovician deposits in the Middle East are mostly less than those of the Infracambrian and Cambrian rocks, with the exception of Afghanistan. It is only in the latter region that the Ordovician deposits are generally much more widely distributed and thicker than the Cambrian ones. Especially in Iran, the Ordovician is comparatively thin. Together with the Cambrian and younger rocks, the Ordovician rocks are strongly folded in the Tethyan belt

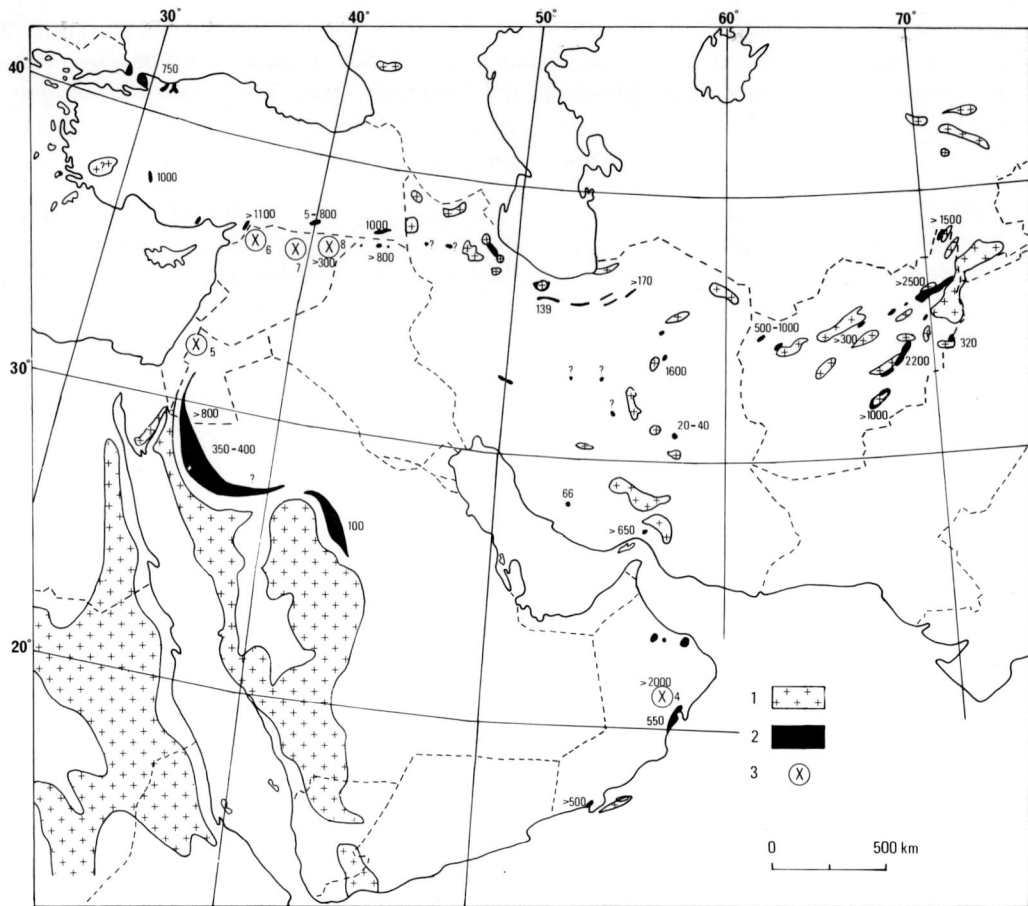

Fig. 16. Known areas of outcrops of Ordovician rocks. (1) Pre-Cambrian at surface; (2) Ordovician at
surface; (3) borehole; (4) Ghaba; (5) Suweilieh; (6) Baflioun; (7) Abba; (8) Kamichlie. Total
thicknesses of Ordovician rocks in metres.

during several Alpine phases and, in north-western Turkey, during the Variscan orogeny, also.
On the Arabian Shelf, the outcropping Ordovician rocks are in most cases tectonically undisturbed and only slightly tilted towards the east or north-east.

3.1. History of Ordovician studies

BENDER and HUCKRIEDE (1964) were the first to recognize the presence of fossiliferous
Ordovician rocks in Jordan. Previous authors, such as QUENNELL (1951), BURDON (1959), and
WETZEL and MORTON (1959), presumed the same formations, from which BENDER and
HUCKRIEDE identified Ordovician fossils to be entirely of continental origin and considered
them to be of Palaeozoic to even Triassic or Jurassic age. Originally, BENDER and HUCKRIEDE
(1964) attributed the *Sabellarifex* and *Conularia* Sandstones (Fig. 17) to the Silurian System;
WOLFART (1968) interpreted them as middle and late Ordovician age on the basis of new and
nicely preserved fossils.

BLANCKENHORN (1914) and DIENEMANN (1915) were the first to report from northern Saudi
Arabia sandstones with black, bituminous silicified shales bearing *Diplograptus*. Much later, the

Arabian–American Oil Company and the United States Geological Survey carried out their geological mapping project (see Section 2.1) on the Arabian Peninsula, providing much valuable information about the Ordovician deposits. Present knowledge was completed by the studies of HELAL (1964, 1965). In the Dhofar region, the first information on rocks which were later assigned to the Ordovician by BEYDOUN and GREENWOOD (1968) is due to CARTER (1852). Geological research in Oman began rather late. The only published information on fossiliferous Ordovician rocks was provided by MORTON (1959) and TSCHOPP (1967).

Exact knowledge of Ordovician deposits in Turkey is comparatively recent. PAMIR (1960) cited only two very uncertain occurrences of Ordovician rocks from the Taurus Mountains and the south-eastern part of Anatolia. Earlier, some scattered observations were made by FUCHS (1902) and by BROILI (1911), who described the Ordovician *Phycodes circinnatus* from the Antitaurus Mountains. FRECH (1917) found an Ordovocian trilobite of the genus *Dalmanitina*, according to R. and E. RICHTER (*in* DUBERTRET, 1936), originally assigned to the Silurian genus *Acaste* by FRECH. FLÜGEL (1964) reported this trilobite to belong to *Dalmanitina* (*Chattiaspis*) *kegeli* (R. and E. RICHTER). The first systematic work on the Ordovician rocks of north-western Turkey was by PAECKELMANN (1925, 1938) and, earlier, by PENCK (1919). During and after the Second World War, BLUMENTHAL (1941, 1947) discovered some Palaeozoic rocks in south-western Anatolia, at present partly attributed to the Ordovician, but it was during the 1960s that the modern period of Ordovician research work began all over Turkey. The most outstanding studies were done by HAAS (1968) in north-western Turkey, by DEAN and KRUMMENACHER (1961), MONOD (1967), DEAN and MONOD (1970), DEAN (1971), and HAUDE (1972) in south-western Anatolia, and by TOLUN (1960), KELLOG (1960), SCHMIDT (1966), KETIN (1966), ALTINLI (1966), and DEAN (1967) in south-eastern Anatolia.

In north-eastern Syria, the only knowledge of Ordovician rocks was provided by deep boreholes drilled during the 1950s. The results were published by SUDBURY (1957), DANIEL (*in* DUBERTRET and others, 1963), WEBER (1963), and FLÜGEL (1963). At nearly the same period, WETZEL (*in* BELLEN and others, 1959) and SEILACHER (1961, 1963) described for the first time Ordovician rocks from northern Iraq.

On a palaeogeographical map of Iran, KOBAYASHI (1934) characterized knowledge of the early Ordovician by a large question mark. The first Ordovician fossils from Iran, a small fauna with *Symphysurus palpebrosus* (Dalman), were found by BOBEK (1934) in the northern Iranian Alam Kuh 'area and described by DIETRICH (1937). Under the name Lashkerak Formation, GANSSER and HUBER (1962) included beds predominantly of Ordovician age between the Cambrian Lalun Sandstone and the younger Devonian–Carboniferous formations. Additional Ordovician fossils from the Alam Kuh area were collected by GLAUS (1965). Further studies on Ordovician rocks and faunas were made by HUCKRIEDE and his associates (1962) in the Kerman area. The faunas from these scattered occurrences proved to be partly of early, partly of late Ordovician age. Thirty years ago, Douglas (1950) published some remarks on a thick formation of clastic rocks in the Kuh-e-Faraghan area, southern Iran. The lower part of this formation must be of Ordovician age as it is overlain by a sequence bearing early Silurian graptolites. New data on the Lower Palaeozoic rocks in the Mila Kuh area, northern Iran, were published by STÖCKLIN and others (1964). Cambrian and Ordovician rocks were described under the name Mila Formation, the fifth and partly also the fourth member of which have been assigned to the Ordovician. ASSERETO (1963) used the name Mila Formation for the first time; he included, however, rocks belonging to the Mila in his descriptions of younger Palaeozoic rocks. RUTTNER and others (1968) introduced the name Mila Group for the thick Cambro–Ordovician rocks in eastern Iran. This was subdivided into three formations, the youngest of which—the Shirgesht Formation—includes rocks of Ordovician age only. An early Ordovician age must be ascribed to the uppermost part of RUTTNER's Derenjal Formation containing the trilobite *Saukia*, which is closely related to the Afghan *Saukia*. Earlier, KING (1937) mentioned *Saukia* from the south-western Iranian Zagros Mountains.

As late as 1964, FESEFELDT collected the first Ordovician fossils in the eastern Afghan Logar trough. These were identified by WOLFART. Important knowledge of the Ordovician rocks of northern Afghanistan is due to MIRZAD and others (1968) and KOLČANOV and others (1971).

Many valuable scattered observations on Ordovician deposits in eastern Afghanistan have been published by French geologists (LAPPARENT, BLAISE, DESPARMET, LAVIGNE and others, 1962–71). DÜRKOOP (1970) and KARAPETOV and others (1971) studied the Ordovician rocks in the south-western part of the Logar trough. Many studies on the Ordovician System in eastern, central, and western Afghanistan were made by German geologists (1959–68). Their results were compiled by WOLFART (1970), who elaborated the first detailed palaeogeographical map of the Ordovician System in Afghanistan.

3.2. The Ordovician rocks of the Arabian Peninsula

From the Arabian Peninsula, two areas of marine Ordovician rocks are known. One lies along the eastern margin of the crystalline core of the Arabian Shield in north-western Saudi Arabia and southern Jordan and the other in the Oman region (Fig. 15). In the first area, the Ordovician rocks consist mainly of sandstones grading downwards into Cambrian and upwards into Silurian sandstones. In the Oman region, the Ordovician rocks are sandstones and shales reaching a greater thickness than in the north-western Arabian region, in spite of the succession being incomplete. In southern Israel, west of the Wadi Araba–Dead Sea rift valley, no strata of proved Ordovician age are known. In the southern Negev, the fossiliferous late Lower Cambrian beds are overlain by continental sandstones, formerly named 'Nubian Sandstones', which comprise rocks of Cambrian to Cenomanian age.

3.2.1. The north-western Saudi-Arabian–Jordan region

In the north-western Saudi Arabian–Jordan region, autochthonous sequences of the Ordovician are exposed east of the Cambrian outcrops extending from about 31°30′ north latitude in Jordan, south of the Dead Sea, southwards into Saudi Arabia (QUENNELL, 1951; WETZEL and MORTON, 1959; BENDER, 1964, 1968; HELAL, 1964, 1965; POWERS and others, 1966; WOLFART and others, 1968). In Saudi Arabia, exposures of Ordovician rocks extend south-east from the Jordan border until they disappear partly overlain by Silurian rocks, under the blanket sands of the Great Nafud, west of Ha'il. South-eastwards, they are gradually overlapped by Permian beds and wedge out completely near latitude 24°20′ N (POWERS and others, 1966).

All of the Ordovician sequences consist of silicate clastic rocks following the Cambrian sandstones. In Saudi Arabia, the Ordovician clastics are up to 400 metres in thickness. They are composed of units of purple, olive-green, and grey graptolitic shales which alternate with massive beds of reddish brown to grey, partly cross-bedded, commonly micaceous sandstones. The Saudi Arabian sequence extends without significant lithological change into southern Jordan. In both of the areas the lithological nature of the Ordovician rocks certainly reflects extensive shallow-water conditions, graptolite bearing shales and fossiliferous sandstones record several marine transgressions. In Saudi Arabia, unfossiliferous and cross-bedded, massive sandstones sometimes containing abundant bivalve moulds indicate intermittent emergence and near-shore conditions.

In north-western Saudi Arabia, POWERS and others (1966) and POWERS (1968) considered the Saq Sandstone, previously published as Cambrian (?), to be Cambrian and Ordovician in

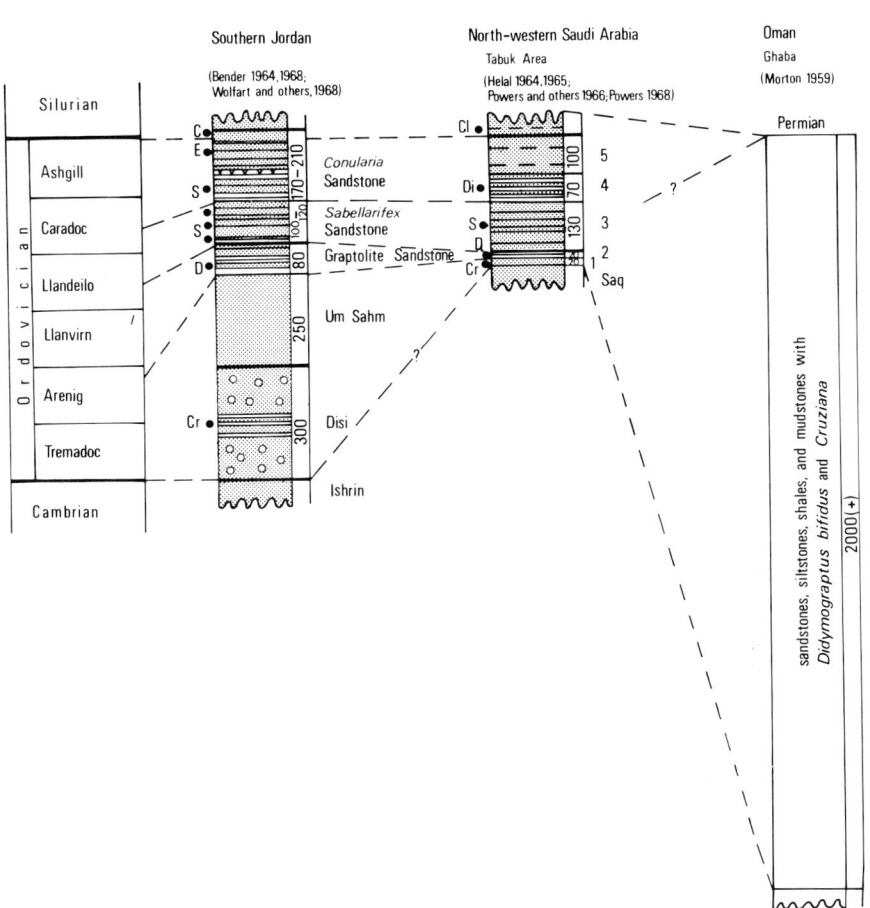

FIG. 17. Representative columnar sections and stratigraphical terminology for the Ordovician rocks of southern Jordan and the Arabian Peninsula. For legend of petrographical characteristics, see Fig. 3 (5–18). (1) *Cruziana* Series; (2) *Didymograptus* Shaly Member; (3) Lower Tabuk Sandy Member; (4) *Diplograptus* shaly Member; (5) Upper Tabuk Sandy Member. (C) *Climacograptus scalaris* cf. *scalaris* (Hisinger), *Distomodus kentuckyensis* Branson and Branson; (Cl) *Climacograptus*; (Cr) *Cruziana*; (D) *Didymograptus bifidus* (Hall); (Di) *Diplograptus*; (E) *Exoconularia, Trematis, Orbiculothyris, Brongniartella, Diplograptus*; (S) *Sabellarifex, Arthrophycus, Rusophycus*.

age. Saq lithology is strikingly uniform both vertically and laterally (Section 2.2.1). The dominant rock type is buff to grey and white, commonly cross-bedded, poorly to well sorted quartz sandstone that is often friable and commonly case hardened to a black, ferruginous quartzite surface. West of the Great Nafud, the lower Saq presumably is continental in origin, whereas thin lenses of purple shale with *Cruziana* indicate that the upper portion is at least partly marine, P. E. CLOUD, JR. (*in* POWERS and others, 1966) referred the trilobite tracks to *Cruziana* cf. *furcifera* d'Orbigny and *C. huberi* (Meunier) of probable early Ordovician age

		Graptolite zones of the British Isles Williams et al. 1972	Southern Jordan: Quennell 1951 Burdon 1959	Southern Jordan: Bender 1964, 1968 Wolfart et al. 1968	Southern Jordan: Lloyd in Selley 1972	North-western Jordan: Powers 1966, 1968	North-western Jordan: Powers 1966, 1968	Saudi Arabia Tabuk area: Helal 1964, 1965	Dhofar: Beydoun and Greenwood 1968	Oman (Haushi-Huqf, Ghaba): Morton 1959	Oman (Haushi-Huqf, Ghaba): Tschopp 1967
			↑		↑	Tabuk	Tawil Mb. / Member 6	Jauf	Cretaceous	Permian	Permian
Silurian				sandstones			Members 4 - 5 sandstone, shale	Tawil / Sharawra			
Ordovician	Ashgill	*D. anceps*	Ram and Um Sahm Sandstones	Nautiloidea Sandstone				Upper Tabuk Sandy Member (Tabuk)			
		D. complanatus									
	Caradoc	*P. linearis*		*Conularia* Sandstone			sandstone, shale with graptolites (3)	*Diplograptus* Shaly Member			
		D. clingani									
		C. wilsoni		*Sabellarifex* Sandstone			sandstone with *Scolithus* (2)	Lower Tabuk Sandy Member			
		C. peltifer									
	Llandeilo	*N. gracilis*								?	?
		G. teretiusculus		Graptolite Sandstone	Khreim Group		Hanadir Member (1)	*Didymograptus* Shaly Member	Murbat	Ghaba: sandstones, shales with *Didymograptus*	Haushi-Huqf
	Llanvirn	*D. murchisoni*									
		D. bifidus									
	Arenig	*D. hirundo*		bedded brownish weathered sandstone	Um Sahm	Umm Sahm Ram	Saq	*Cruziana* Series	Sandstone	Huqf: sandstone, shales with trilobites	Mahatta
		D. extensus									
	Tremadoc	*C. tenellus* / *D. flabelliforme*		massive whitish weathered sandstone	Disi				?	dolomite, shale, sandstone	Humaid
Cambrian			Quweira	massive brownish weathered sandstone	Ishrin	Quweira		Saq			

FIG. 18. Stratigraphical correlation of the Ordovician rocks of southern Jordan and the Arabian Peninsula.

(about Arenig). HELAL (1965) obviously separated the Ordovician part from the Saq Formation calling it the *Cruziana* Series and assigned it to Tremadoc? to Arenig.

The *Cruziana* Series is conformably overlain by the Tabuk Formation (STEINEKE and others, 1958; BRAMKAMP and others, 1963a,b) which, in its original definition, comprised rocks of Ordovician to Lower Devonian age. According to POWERS and others and HELAL (1965), the *Cruziana* Series can easily be distinguished from the overlying Tabuk Formation because the contact is very sharp between the graptolitic shale of the Tabuk and the coarse-grained, cross-bedded sandstones of the *Cruziana* Series. Within the Tabuk area, the *Cruziana* Series consists of sandstones with minor amounts of sandy shales about 20 metres in thickness.

HELAL (1965) confined the Tabuk Formation to rocks of Ordovician age and divided it into four members. Within the type region near Tabuk, the Tabuk Formation is about 320 to 340 metres thick. There it crops out in many areas, forming mesas, low shale benches, or slopes between the massive sandstones that overlie or underlie it. Most characteristic rock types are variegated sandstones and micaceous siltstones. The *Didymograptus* Shale Member (= Hanadir Member *in* POWERS, 1968) consists of about 20 metres of dark, faintly calcareous, silty shales and dark, tan-weathered siltstones. The shales are evenly bedded and usually poorly exposed. They yielded *Didymograptus protobifidus* Elles and *D. bifidus* (Hall) occurring abundantly in shales which rest directly on the *Cruziana* Series. The graptolites indicate a late Arenig to Llanvirn age (zone 6 of ELLES, 1925). The *Didmyograptus* Shaly Member covers large areas in north-western Saudi Arabia and extends over a distance of more than 700 kilometres to the south, as far as to the Jabal Hanadir, latitude 24° N.

Upwards the dark grey graptolitic shales grade into a sandstone sequence of dull, yellowish grey to brownish colours, 130 metres in thickness. Helal named the sequence the Lower Tabuk Sandy Member. It is composed of impure, arkosic, micaceous, partly glauconitic sandstones with interbeds of siltstones and grey–green shales. No diagnostic fossils have been collected from the Lower Tabuk Sandy Member; only *Sabellarifex* is widely distributed. These tubular moulds have been known within the Tabuk area since the work of AULER PASCHA (1908), BLANCKENHORN (1914), and KOBER (1919).

The *Diplograptus* Shaly Member of the Tabuk Formation includes about 70 metres of medium to dark grey, soft shales following directly over the Lower Sandy Member. The lower contact is fairly abrupt but apparently conformable. In the Tabuk area Helal collected from the richly fossiliferous shales *Diplograptus* sp., which dominates the other genera, and poorly preserved fragments identified by R. F. ROSS (*in* POWERS, 1968) as: *Climacograptus* sp., *Cl.* cf. *brevis*, *Orthograptus* aff. *O. calcaratus priscus*, *Glyptograptus*, and *Rectograptus*. From Saudi Arabia, diplograptids have been known since the work of BLANCKENHORN (1914).

The Tabuk Formation ends with the Upper Tabuk Sandy Member which includes 100 metres of marine sandy beds consisting mainly of greyish, moderately hard, thin- to thick-bedded, fine-grained, micaceous, cross-bedded sandstones with ripple-marks and interbeds of sandy siltstones and some shales. It grades into the underlying *Diplograptus* Shaly Member and is rather sharply separated from the overlying Silurian Sharawra Formation. HELAL assumed a hiatus at this boundary. No fossils other than *Sabellarifex* were found in the homogeneous sandstones.

In the Jabal Saq–Jabal Hanadir area, about 220 kilometres south-south-east of Ha'il, the Ordovician rocks include only the *Cruziana* Series, with *Cruziana furcifera*, *C. goldfussi*, and *C. huberi*, and the *Didmyograptus* Shaly Member, both of them (?) 100 metres in thickness (HELAL, 1965). No younger Ordovician rocks could be traced in that area.

In southern Jordan, the Ordovician rocks are in general of similar lithological nature as in north-western Saudi Arabia. Concerning lithological subdivision and fossil content there are differences between the regions, as has been shown by BENDER (1964) and HELAL (1964, 1965). It was BENDER and HUCKRIEDE (1964) who recognized for the first time in southern Jordan, east of the Dead Sea–Wadi Araba rift valley, that the 'Nubian Sandstones' (Section 2.2.1), formerly comprising the entire sequence of clastic sediments above the crystalline basement complex and below the Cenomanian limestones, could be divided both lithologically and by means of fossils. The lower Ordovician rocks are characterized by an alternation of continental, littoral, and neritic sandstones showing increasing marine influence northwards. Between Llanvirn and Llandovery times the fine clastic sedimentation continued to be marine only. There seems to be no indication of a hiatus between the Ordovician and Silurian Systems as suggested by HELAL in north-western Saudi Arabia with the exception that the Ordovician–Silurian boundary should coincide with the rock salt horizon in the middle portion of the *Conularia* Sandstone (Section 4.2). In southern Jordan, the entire Ordovician sequence amounts to more than 800 metres in thickness.

BENDER (1964, 1968) subdivided the Ordovician rocks into five units and used the lithological characteristics for formational names. Later, LLOYD (*in* SELLEY, 1972) renamed part of them (Fig. 18). The Massive Whitish Weathered Sandstone (Disi Formation of LLOYD) gradationally follows the Cambrian Ishrin Formation, which is composed of the Massive Brownish Weathered Sandstone. The Disi Formation (up to 300 metres; BENDER, 1968) represents the greater portion of the 'Ram Sandstone' (QUENNELL, 1951), originally considered to be unfossiliferous and of terrestrial origin. This rock unit is composed of massive, strikingly whitish weathering sandstones of white, beige, rosy, and pale violet colours. Predominantly coarse- and medium-grained sandstones contain much quartz gravel and scattered quartz pebbles up to 5 centimetres in diameter. There are two or three intercalations of grey–brown, silty, micaceous shales up to 170 centimetres in thickness. In the upper portion some thick-bedding can be observed. Cross-bedding appears frequently throughout the unit. Besides the colour, rounded, smooth weathering shapes of the sandstones are typical which can thus be clearly distinguished from the units beneath and above. In one of the shaly intercalations, the oldest identifiable fossils of the southern Jordan sandstones have been found: *Cruziana* of the group *furcifera* Orbigny, *C. goldfussi* (Rouault), *C. vilanovae* Saporta, and *Bilobites*. On the basis of the trace fossils, HUCKRIEDE (*in* BENDER, 1964) assigned an Arenig age to the Disi Formation, which thus seems to be an equivalent of the lower part of HELAL's *Cruziana* Series in north-western Saudi Arabia. Fluviatile transport is indicated by the nature of the sedimentary components of the Disi Formation the deposition of which, however, happened, at least partly, in a near-shore shallow marine sea.

HELAL (1965) considered the following unit, the Bedded Brownish Weathered Sandstone (BENDER, 1964), as a correlative of the upper portion of his *Cruziana* Series. BENDER himself and LLOYD (*in* SELLEY, 1972) correlated this formation with the Um Sahm Sandstone (QUENNELL, 1951). Its lower contact is marked by the beginning of parallel-bedded, pale red sandstone above the white, massive sandstone. In its lower portion, the unit consists of thick-bedded, pale red, beige, and brownish, dominantly coarse-grained sandstones with some scattered, fine gravel. Layers of medium- and fine-grained sandstones and subordinately, micaceous, argillaceous fine sandstones are intercalated. The upper portion of the Um Sahm is formed of well sorted, homogeneous, thick-bedded, locally cross-bedded fine sandstone of white and pale beige colours. The total thickness is about 250 metres. Surficially, the rocks of the unit are marked by the dusky, brown colour of a thin desert varnish thus differentiated from the units beneath and above. In the north-western area of outcrop, both the Disi and the Um Sahm Sandstones are transgressively capped by Mesozoic sandstones. No fossils have so far been found in the Um Sahm Sandstone, which is assumed to be of lower Ordovician age on the basis of its position between rocks of Arenig and Llanvirn age, respectively. As to the origin of the Um Sahm rocks, BENDER (1964) suggested the lower portion to have been deposited in a continental or marine littoral environment. From grain shapes, size, and distribution, BENDER concluded the upper portion to be of aeolian origin.

In north-western Saudi Arabia as well as in southern Jordan, the Graptolite Sandstone (= *Didymograptus* Shaly Member) is developed as a significant sequence of fossiliferous marine rocks, which was discovered by BENDER (1964). In Jordan, it is composed of 70 to 90 metres of vari-coloured, hard shales, flaggy and quartzitic fine sandstones, and grey–green and brown, micaceous, sandy shales and fine sandstones. At its lower contact, the beginning of the unit is marked by a distinct lithological break from bedded fine sandstones below to the vari-coloured shales above. Besides indeterminable traces of organisms, *Didymograptus* cf. *bifidus* (Hall) has been collected in fine sandstones as well as in shales. It is the key fossil of the early Llanvirn (= zone 6 of ELLES, 1925).

Adjoining to the east, the *Sabellarifex* Sandstone (BENDER, 1964) conformably overlies the Graptolite Sandstone, both of them unconformably overlain by Mesozoic sandstones in the area of the north-western Ordovician outcrops of Jordan. The *Sabellarifex* Sandstone is composed of shaly, micaceous, flaggy sandstone, bedded fine sandstones with innumerable, mainly vertical tubes of *Sabellarifex*, medium-grained sandstones with manganese spots, and sandstones with ripple-marks and tracks of organisms. The *Sabellarifex* bed itself is up to 20 metres in thickness and is intercalated between sandstones which are 25 metres in thickness below and 65 to 75 metres above. The total thicknesses of the unit range between 100 and 120 metres. Lithology and fossil content characterize the unit as deposited in a marine littoral environment. *Sabellarifex dufrenoyi* (Rouault) (= *Tigillites dufrenoyi R.*) has long been known from north-western Saudi Arabia, near Tabuk, by the reports of AULER PASCHA (1908), KOBER (1919), and BLANCKENHORN (1914), the sticks being called 'worms of Hiob' by the bedouins. Besides *Sabellarifex*, *Harlania alleghanensis* (Harlan), and, in the upper part of the unit, *Rusophycus* and poorly preserved moulds of modiolopsid shells have been found, besides *Cruziana* identified by A. SEILACHER as of early Caradoc age. According to the biostratigraphical position of the under- and overlying units—Graptolite Sandstone and *Conularia* Sandstone—a Llandeilo to early Caradoc age must be assigned to the *Sabellarifex* Sandstone (WOLFART and others, 1968).

In southern Jordan, BENDER (1964) established the *Conularia* Sandstone, the lower contact of which is marked by 12 to 15 metres of greenish brown shales overlying the *Sabellarifex* Sandstone. The following part of the unit is mainly composed of an alternating fossiliferous sequence of fine sandy shales, flaky fine sandstones, *Sabellarifex* sandstones which are about 10 centimetres in thickness, and sandstone flags with ripple-marks and traces of organisms. This sequence is repeated in small cycles of about 15 to 20 metres. It contains intercalations of hard layers re-occurring more or less regularly and forming the characteristic landscape of table mountains in this area. Approximately 70 metres above the base of the *Conularia* Sandstone, laminated, brown and green, fine sandy shales of about 4 metres thickness follow, containing up to 2 centimetre layers and lenses of rock salt. Above the salt-bearing beds again fossiliferous sandstones appear with the same lithological characteristics as the sequence below. Calcareous fine sandstone layers as well as millimetres-thick gypsum laminae and fibrous gypsum in joints are frequent in the upper portion of the unit. The fossil content and type of sediments indicate a fully marine, littoral environment. Total thicknesses of the *Conularia* Sandstone are estimated to range between 170 and 210 metres. In places, the unit is unconformably overlain by massive, white sandstones of the Mesozoic. The contact between the *Conularia* Sandstone and the overlying Nautiloidea Sandstone is masked by the Quaternary sediments of the plain of

Muddawwara. The boundary between the two units is tentatively placed about 80 metres above the salt-bearing horizon. Originally, HUCKRIEDE (*in* BENDER, 1964) assigned a middle and upper Silurian age to the *Conularia* Sandstone on the basis of *Rhipidomella*-like brachiopods which turned out to be the new inarticulate brachiopod *Orbiculothyris costellata*, which is associated with the Caradoc *Trematis tenuiornata* according to WOLFART and others (1968). *Exoconularia?*, *Lingulella?*, *Rostricellula*, *Modiolopsis*, *Brongniartella*, and Nautiloidea complete the fossil assemblage of the *Conularia* Sandstone. Most of the unit can, therefore, be attributed to the Caradoc and Ashgill. In the upper 30 metres of the unit, BENDER found a faunal assemblage which is clearly different from the usual assemblage of the *Conularia* Sandstone and must already be assigned to the late early to middle Llandovery (Section 4.2). Consequently, it has been proved that the upper part of the unit grades upwards into the Silurian.

3.2.2. The Oman region and south-western Arabian Peninsula

Little information is available on the Ordovician rocks of the Oman region. According to MORTON (1959) and TSCHOPP (1967), continental sedimentation occurred during the Cambro–Ordovician near Haushi-Huqf, passing westwards and northwards into shallow marine sedimentation. In the whole Haushi-Huqf area, uplift and erosion, with possibly some tectonic disturbances, followed the deposition of the Huqf Series which is Pre-Cambrian to Early and Middle Cambrian in age. Subsequently, the Mahatta Humaid Formation (TSCHOPP, 1967) was deposited, lying unconformably on Cambrian rocks. MORTON compared the Mahatta Humaid with the sandy–shaly Rann Beds of Qamar. The ill-exposed formation reaches a thickness of about 550 metres. It consists partly of fluviatile sedimentary rocks which probably pass westwards and northwards into shallow marine deposits. MORTON (1959) reported the red sandstones and mottled shales to contain trilobites and *Cruziana*, but in the Haushi-Huqf area, the Mahatta Humaid is described by TSCHOPP as an unfossiliferous sequence of sandstones with coloured chert grains and argillaceous siltstones in the lower part characterized by poor sorting, coarse grain, cross-bedding, and variegated colours, perhaps being formed in sand dunes and/or in large alluvial fans along the western edge of an uplifted Huqf massif.

To the north-west, at Ghaba, lower Ordovician sandstones, siltstones, shales, and mudstones, more than 2000 metres in thickness, have been drilled without reaching the base. Geophysical evidence suggested that salt-doming occurs at depth. The former rocks yielded *Didymograptus bifidus* (Hall) and *Cruziana* and must be assigned, therefore, to the early Llanvirn. Over the entire Oman region, erosion prevailed after the Ordovician until the Permian transgression initiated an almost continuous period of carbonate sedimentation which lasted until the end of the Cenomanian.

In the Sayh Hatat area, Oman Mountains, the Cambro–Ordovician Amdeh Quartzite has a sharp, conformable contact with the underlying Hiyam Dolomite. It is considered to be an equivalent of the Mahatta Humaid and consists of a thick, monotonous, uniform sequence of more or less silicified sandstones with thin interbeds of epimetamorphosed shales. The degree of metamorphism is lower than that of the Pre-Cambrian Hatat Metamorphics. In the Jebel Akhdar area, the Fara Formation is considered by TSCHOPP to be the correlative unit of the Mahatta Humaid. In its lower part, this formation comprises mainly black chert, often laminated and brecciated. The higher levels are composed of siltstones and quartz sandstones.

From behind Murbat, in Dhufar, CARTER (1852), LEES (1928), FOX (1947), and BEYDOUN (*in* BEYDOUN and GREENWOOD, 1968) described micaceous sandstones and argillaceous beds named the Murbat Sandstone by LEES. In the type section here, LEES reported the formation as reaching a total thickness of 540 metres, consisting, in the basal part, of coarse grits, arkose sandstoncs, sandy, micaceous shales, and some pebble beds composed entirely of detritus from the basement rocks. The series becomes finer upwards, and thinly bedded, brown, green, and purple shales and sandy shales appear. Locally, these become white, yellow, and red with gypsum veins and some sulphur or carbonaceous zones, but the predominant colour is red. The formation rests unconformably on Murbat crystalline rocks and is overlain by Cenomanian marls and limestones with slight angular discordance.

In the western part of the Jebel Samhan, BEYDOUN (1964, 1966, 1968) found the total thickness of the Murbat Sandstone to be 1116 metres. The succession is predominantly sandy in the lower and upper parts and shaly–silty in the middle part. Conglomerates with basement pebbles occur both in the basal part and in the upper 100 metres. As to the age, no diagnostic fauna has yet been found in the succession. The Murbat Sandstone is unaffected by folding or metamorphism and must be younger, therefore, than the folded, veined, and peneplained beds cropping out in the neighbouring region. It resembles, however, the non-folded lower Ordovician rocks of Oman, and BEYDOUN (1964) suggested its age to be Ordovician, although there is room for time gaps in the upper part, so that a younger age for that part is not improbable. For that reason, LEES (1928) considered the formation to be similar to the 'Nubian Sandstone'.

3.3. The Ordovician rocks of Turkey

In Turkey, Ordovician rocks are exposed in two regions, the northern Anatolian geosyncline and the mobile shelf of southern Turkey (WOLFART, 1967; FLÜGEL, 1971). Around Istanbul, Ordovician rocks are mostly of continental type; they are composed of basal conglomerates, arkosic sandstones, and quartzites which indicate the presence of a source area in the immediate neighbourhood (HAUDE, 1969). The Ordovician rocks of southern Turkey are rather different from those of northern Turkey and consist of marine sandstones, shales, and limestone interbeds.

In Anatolia, Ordovician rocks are more widely distributed than Cambrian rocks. BRINKMANN (1971) reported the westernmost outcrop of Ordovician rocks to be on the Greek island of Kos. There, the sequence consists throughout of marine siltstones and sandstones with a brachiopod fauna of a Bohemian type. DEAN (1975) tentatively considered the sequence to be part of the Seydişehir Formation of the Taurids. Most of the Taurus Mountains is composed of Mesozoic and Tertiary formations; but inliers of Palaeozoic rocks occur at intervals.

From the Taurus Mountains, near Beyşehir, MONOD (1967), DEAN and MONOD (1970), and DEAN (1967, 1970, 1971, 1973, 1975) described the Seydişehir Formation introduced by BLUMENTHAL (1947). HAUDE (1972) traced the formation farther to the north, in the southern Sultan Dăg area. DEAN and MONOD subdivided the sequence surmounting the fossiliferous Middle Cambrian rocks into the Seydişehir and Sobova Formations. The Seydişehir Formation is composed of the Seydişehir Shales and the Upper Greywackes. Lithologically the Seydişehir Shales consist essentially of a succession of psammitic quartzites alternating with silty micaceous shales occupying a total thickness of roughly 1000 metres. Most of the succession is barren; fossils occur sporadically throughout the upper half of the Seydişehir Shales and become more frequent towards the top. They were collected from brown, sandy limestone interbeds in the sandstone and comprise trilobites (*Geragnostus, Ptychopyge, Megistaspis, Paramegalaspis, Taihungshania, Symphysurus, Colpocoryphe, Neseuretus*), dalmanellid brachiopods, molluscs (*Redonia*), and graptolites (*Tetragraptus* cf. *reclinatus* Elles and Wood, *Didymograptus deflexus* Elles and Wood, and *D.* cf. *nitidus* Hall). Most of the fossils indicate an early Arenig age. The trilobites are mostly of Tethyan type, but some genera exhibit Balto–Scandinavian affinities (DEAN, 1971).

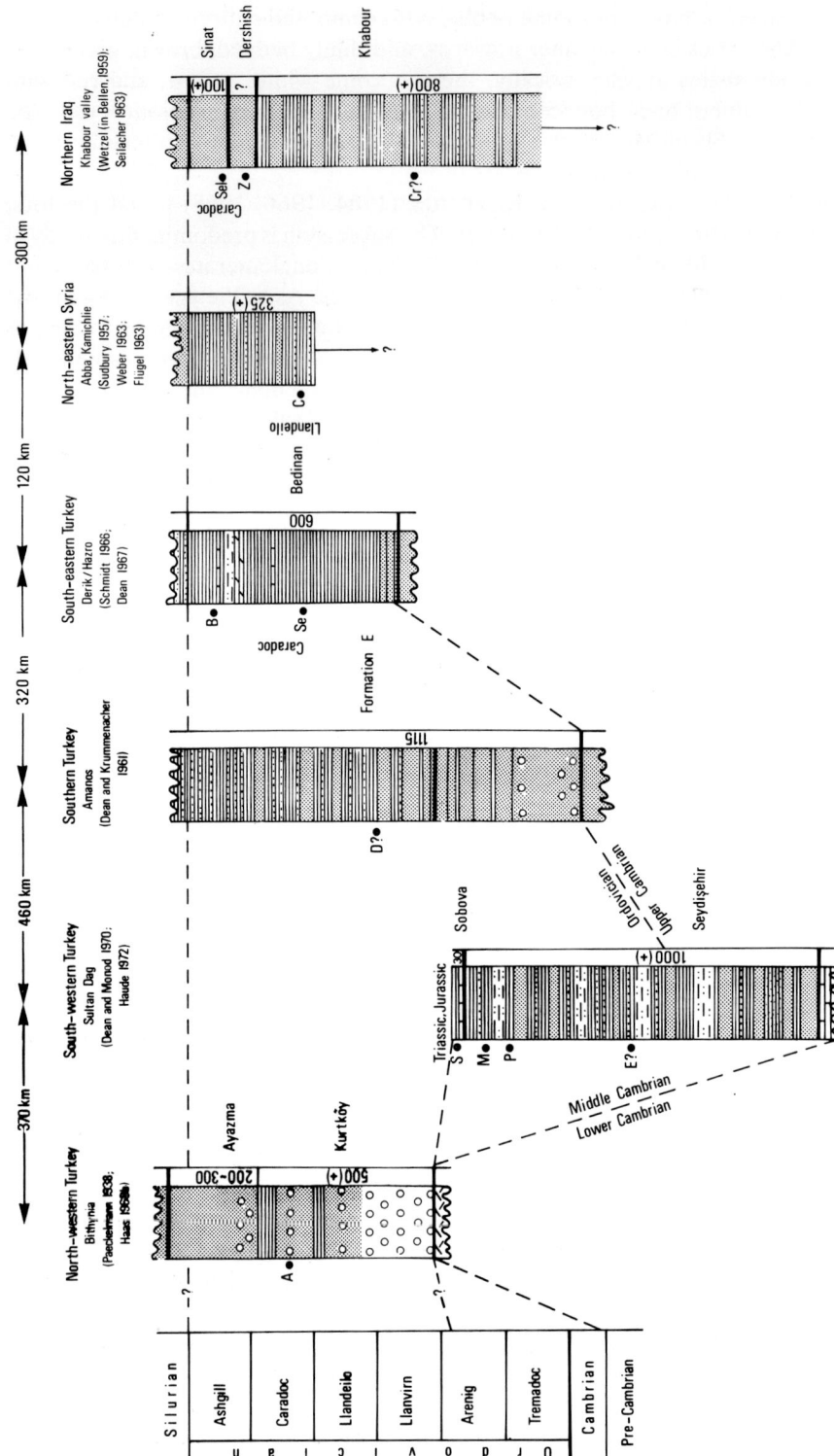

FIG. 19. Representative columnar sections and stratigraphical terminology for the Ordovician rocks of Turkey, Syria, and Iraq. For legend of petrographical characteristics, see Fig. 3 (5–18). (A) Asaphids, *Exoconularia*; (B) *Brongniartella, Kloucekia, Dalmanitina, Cryptolithus* ?, *Marrolithoides*; (C) *Colpocoryphe, Pseudobasilicus, Diplograptus spinolosus* Sudbury; (Cr) *Cruziana, Billingsella*; (D) *Dalmanitina*; (E) *Euloma, Pseudagnostus, Asaphellus, Symphysurina, Macropyge*; (M) *Megistaspis, Neseuretus*; (P) *Paramegalopsis*; (S) *Symphysurus, Illaenus, Niobe, Carolinites, Agerina*, and others; (Se) *Selenopeltis, Ampyx, Colpocoryphe, Dalmanitina, Kloucekia, Cryptolithus* ?, *Marrolithoides, Neseuretus*; (Sel) *Selenopeltis buchi* (Barrande); (Z) *Zoophycos*; (?) exact position in section uncertain.

In the southern Sultan Dağ area, HAUDE (1972) succeeded in finding the following trilobite assemblage in the Seydişehir Shales, indicating a Tremadoc age: *Geragnostus*, *Pseudagnostus*, *Euloma* (*Proteuloma*), *Asaphellus*, *Symphysurina*, *Macropyge*, and *Lichapyge*. HAUDE took into consideration that the lowermost part of the Seydişehir Shales could well be of Upper or even of Middle Cambrian age. Eastwards, the Seydişehir Shales pass into slightly metamorphosed shales called the Kuruderebaşi Shales by HAUDE. The Upper Greywackes of the Seydişehir Formation comprise approximately 20 metres of coarser beds.

The Seydişehir Formation is conformably overlain by the Sobova Formation, the lower portion of which consists of the Sobova Limestone which is 0 to 10 metres in thickness. The limestone is well bedded to massive, grey to pink, mainly unfossiliferous. Some beds contain crinoid or cystoid ossicles, small brachiopods (*Eodalmanella*), and trilobites representing the most prolific Ordovician trilobite fauna in the district (*Geragnostus*, *Euloma*, *Carolinites*, *Apatokephalus*, *Niobe*, *Symphysurus*, *Cyclopyge*, *Illaenus*, *Ampyx*, *Neseuretus*?, *Agerina*). The Sobova Formation is terminated by a group of grey shales passing laterally into red shales and sandstones of approximately 20 metres thickness. This yielded *Symphysurus* only. Thus, the entire formation is not significantly younger than the Seydişehir Shale. It is transgressively overlain by Triassic to Upper Jurassic limestones.

An isolated occurrence of Ordovician shelly fauna was discovered in black shales near Kemer, south-west of Antalya by M. MARCOUX (*in* DEAN, 1975). The trilobite fauna was only poorly preserved (*Placoparia* sp., cyclopygids, illaenids), and its age could not be exactly determined. The trilobites exhibit Bohemian affinities and are probably younger than those of the Seydişehir and Sobova Formations. The black shales of Kemer are quite distinct from any recorded beds from Beyşehir or in the Anamur/Silifke area. Since the black shales lie within a region of tectonic nappes, they may have been transported to their present position from a location farther south.

ERENTÖZ (1966) recorded the presence of Ordovician/Silurian sandstones and shales with abundant Silurian graptolites following the Cambrian rocks in the area of Silifke/Anamur, some 200 km south-east of Beyşehir. BLUMENTHAL (1947) has already described the sequence as Devonian on the southern Turkish coast, between Silifke and Gilindir. No fossils have yet been described or illustrated from that region, and the only published account deals with the coastal section near the Bay of Ovaçık, west of Silifke. YALÇINLAR (1964) described the succession for the pre-Devonian strata, but it turned out to be somewhat confused (DEAN, 1975), and the latter considered silty, micaceous, brown–grey shales with poorly preserved asaphids and *Didymograptus* of the extensiform type to be the lowest strata. DEAN suggested an early Arenig age for them and correlated them with part of the Seydişehir Formation. The shale is succeeded by massive, yellow-weathering sandstone with bands of conglomerate, which are considered to represent the local base of the Silurian resting with a marked unconformity upon Arenig strata.

Rather early Ordovician rocks were reported from the Antitaurus Mountains. FUCHS (1902) described *Phycodes circinatus* from red–brown, platy sandstones with fucoids in the area south of Saimbeyli. In the same region, KETIN (1963) found Cambro–Ordovician sandstones with conglomeratic interbeds below fossiliferous Silurian rocks. DEAN (1975) interpreted the rocks with *Phycodes circinatus* as a probable local equivalent of the upper Armutludere Formation or part of the Seydişehir Formation.

Large outcrops of Ordovician strata are present in the Saimbeyli region of the Taurus Mountains. In that region, Lower to Middle Cambrian limestones are succeeded by the lower member of the Armutludere Formation, consisting of grey–green schist which yielded the trilobite *Macropyge taurina* and eocrinoids (Tremadoc). *Symphysurus*, asaphids, *Didymograptus*, and *Tetragraptus* indicate an Arenig age for the upper member of the Armutludere

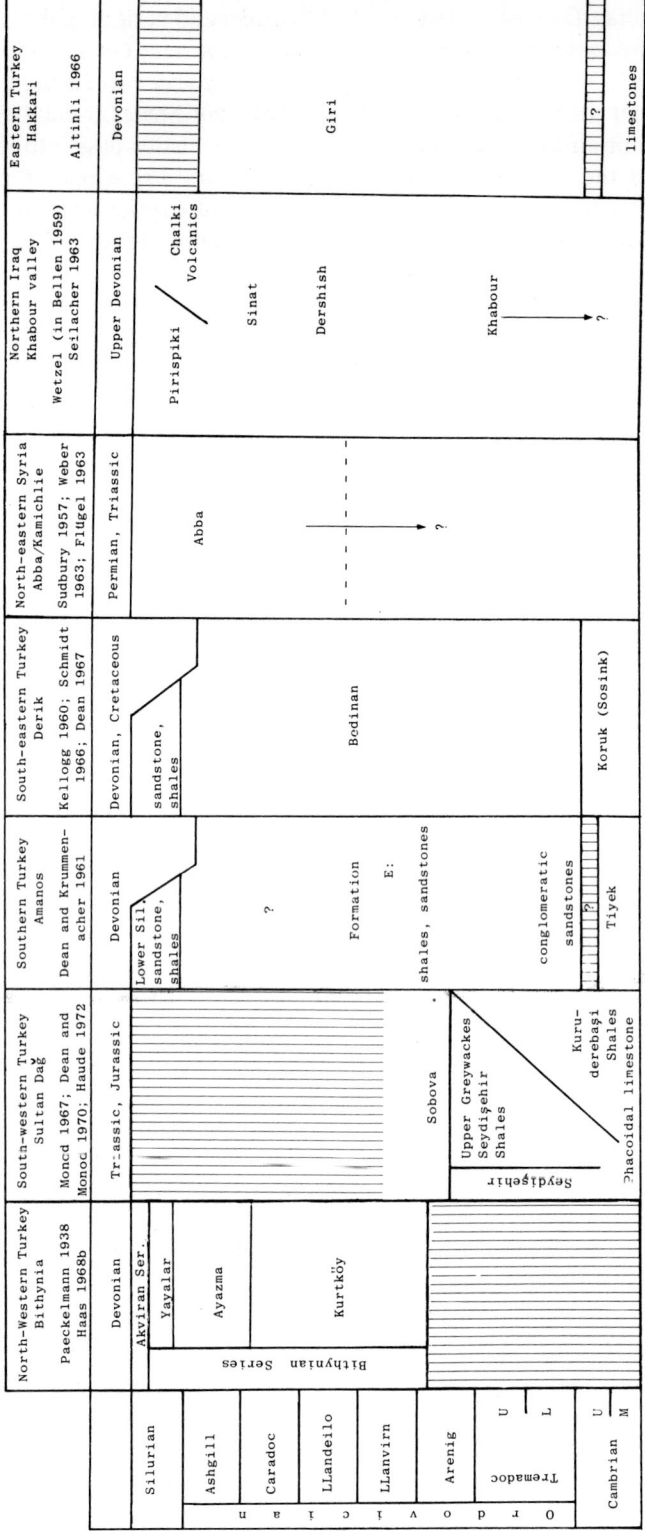

FIG. 20. Stratigraphical correlation of the Ordovician rocks of Turkey, Syria, and Iraq.

Formation conformably overlying the lower member and made up of hard, brown–grey shale. In the Taurus Mountains south of Saimbeyli, *Taihungshania* has been found associated with *Niobella* and '*Ampyx*'. The Cambro–Ordovician boundary cannot be drawn in the Saimbeyli region since neither Upper Cambrian nor Lower Tremadoc faunas have yet been found there. Possibly there is a hiatus between the Armutludere and Degirmentas Formations.

In the southern Amanos Mountains, Cambro–Ordovician rocks are exposed near Saylak, 12 kilometres north by east of Kirikhan. There, the fossiliferous Middle Cambrian rocks are separated by a disconformity from the overlying Ordovician Formation E which is up to 1115 metres thick (DEAN and KRUMMENACHER, 1961). Formation E is formed by massive, pale, coarse to conglomeratic, quartzitic sandstones in its lower part. Higher in the succession, green, sandy, micaceous shales alternate with nodular, quartzitic sandstones. The topmost strata comprise a few layers of soft, sandy shales of purple colour. Slump structures occur in the upper part of the section, where worm tracks and the trilobite track *Fraena* were encountered. The only stratigraphically useful fossil yet found in this formation is *Dalmanitina* (*Chattiaspis*) *kegeli* (R. and E. Richter) collected from a brownish shale by FRECH (1917). Formation E is transgressively overlain, though without angular unconformity, by carbonates held to be Devonian in age.

In the middle part of the eastern Amanos Mountains, LAHNER (1972) distinguished the lower Ordovician Bahçe Formation and the upper Ordovician Kizlaç Formation, which reach thicknesses of 500 and 300 metres, respectively. The Bahçe Formation is of restricted distribution. It is composed of quartzites and grey–green or brownish sandstones with shaly interbeds. On the bedding planes, *Cruziana furcifera* d'Orbigny is preserved which corresponds to *Fraena* (FRECH, 1917). Additionally, *Scolithus* and *Vexillum* are present in the rocks. According to SEILACHER (*in* LAHNER, 1972), the fossils indicate an Arenig age. Lahner noticed that there is a gradual transition between Cambrian and Ordovician rocks in the middle Amanos region, thus excluding tectonic movements during that time as assumed by DEAN and KRUMMENACHER in the southern Amanos.

The competent Bahçe Formation is overlain by the incompetent Kizlaç Formation consisting of sandstone, siltite, and shale of grey–brown to greenish brown colours. From the occurrence of *Onnia ornata* (Sternberg) and a certain *Cruziana*, LAHNER deduced a Caradoc age. The Kizlaç Formation is conformably overlain by the Dedeler Formation, the basal part of which contains lower Silurian brachiopods (Section 4.3.2). South-eastwards, the Kizlaç Formation passes into the upper Ordovician part of the Fevzipaşa Formation, consisting of a uniform grey to olive-coloured shale sequence with interbeds of sandstones and thin pebble layers. The Fevzipaşa Formation reaches a thickness of about 800 metres. Its upper part is assumed to be Silurian in age. It was in the Ordovician part of the Fevzipaşa that FRECH (1917) found *Dalmanitina*.

In the Border Folds region of south-eastern Turkey (Derik–Mardin area), the principal outcrops of Cambrian and Ordovician rocks are situated within an inlier of 22 kilometres length and up to 5 kilometres width. TOLUN and TERNEK (1952) were the first to give short account of the Cambrian outcrops near Derik, and the highest part of the Cambrian succession as shown by them proved to be Ordovician in age. Later, TOLUN (1960) noted the occurrence of Silurian rocks (*sensu lato* including Ordovician) in south-eastern Turkey and mentioned a succession of 900 metres of marly–sandy, fossiliferous beds underlying Cretaceous limestones at Bedinan. He stated also the similarity of the Bedinan Ordovician rocks to others found in boreholes in northern Syria. The most important work on the Lower Palaeozoic rocks in the Mardin area was done by KELLOGG (1960), who recognized already the middle Ordovician age of the Bedinan fauna. SCHMIDT (1966) gave a detailed description of the lithological composition of the

Bedinan Formation which comprised, according to this author, beds of Ordovician and Silurian age up to 600 metres in thickness.

DEAN (1967) reported the Bedinan Formation to rest unconformably on the Cambrian Sosink Formation in the type region between Derik and Mardin, south-eastern Turkey. BEER (1966), however, considered the Cambrian rocks to be conformably overlain by the Bedinan Formation, the total thickness of which in his opinion ranges between 500 and 800 metres. For the most part, the Ordovician rocks are composed of grey–green mudstones and shales with some intercalations of calcareous siltstone. Higher in the succession, the beds become more arenaceous and pass upwards conformably into a series of current-bedded sandstones. The shaly parts of the formation are sometimes highly fossiliferous containing particularly trilobites. All the other animal groups (graptolites, brachiopods, etc.) are in a minority in comparison with the trilobites. DEAN identified the following trilobites: *Ampyx, Marrolithoides, Cryptolithus?, Dionide, Dalmanitina, Kloucekia, Neseuretus, Brongniartella, Platycoryphe?, Colpocoryphe, Selenopeltis* and the brachiopods *Lingula, Schizocrania, Trematis, Svobodaina, Aegiromena.* The graptolites are represented by *Diplograptus, Climacograptus,* and *Lasiograptus.* This faunal assemblage indicates a late Caradoc age and exhibits marked Bohemian/Tethyan affinities. The higher, more massive, arenaceous strata of the Bedinan Formation yielded only occasional fragments of inarticulate brachiopods. In the lower sandstones, *Dalmanitina* and *Kloucekia* still occur rarely indicating a Caradoc age. There is no faunal evidence for Ordovician strata older than middle Caradoc in the Bedinan Formation, nor is there any for Upper Cambrian in the Sosink Formation. DEAN's (1975) conclusion was that there must be a large-scale disconformity between the Bedinan and Sosink Formations. The top of the Bedinan Formation may involve, according to DEAN's interpretation, yet another disconformity, since there is no evidence of late Caradoc or Ashgill strata. The Lower Palaeozoic rocks are unconformably overlain by Devonian, Cretaceous, and Lower Tertiary limestones.

From the Hakkari district, south-eastern Turkey, ALTINLI (1966) described the Giri Formation consisting of about 1000 metres of grey and purple, coarse-grained, evenly bedded quartzites with some lenses of variegated marl, shale, tuff, and limestone. The *Cruziana*-bearing quartzites are probably correlative with the Ordovician quartzites of northern Iraq. They overlie conformably or disconformably the presumed Cambrian grey to black limestones and are, in turn, disconformably followed by fossiliferous Devonian limestones and shales.

In north-western Anatolia, between Instanbul and the Zonguldak area, the Ordovician rocks comprise a red-bed sequence resting on older crystalline rocks. The Ordovician sequence had already been described by PENCK (1919) and PAECKELMANN (1925, 1938), but it was HAAS (1968b) who named the various formations. The Ordovician rocks of northern Turkey have also been referred to by KAMEN-KAYE (1971), BRINKMANN (1971), and FLÜGEL (1971). They are composed of variegated, conglomeratic greywackes and feldspathic sandstones. The Ordovician portion of the Lower Palaeozoic rocks of the Bithynian Peninsula, north-western Turkey, comprises the Kurtköy and parts of the Ayazma Formations according to HAAS (1968b).

The Kurtköy Formation, which is a correlative unit of the conglomeratic and arkosic beds of PAECKELMANN (1938), consists in the lower part of conglomerates 100 to 200 metres in thickness. OKAY (1947) observed quartz, chert, quartzite, arkosic sandstone, mica schist, granite, etc., to form the conglomerate. The base of the Kurtköy Formation is not exposed. The upper portion of the Kurtköy is composed of more than 300 metres of variegated, mainly violet coloured, arkosic sandstones and shales with intercalations of conglomeratic layers. In a chamosite–oolite bed in the uppermost part of the Kurtköy, SAYAR (1962, 1964) discovered numerous specimens of the Ordovician genus *Exoconularia* a.o. (*Archaeoconularia fecunda* (Barrande), *Exoconularia exquisita bohemica* (Barr.), *Metaeonularia consobrina* (Barr.)). The same locality yielded asaphids (YALÇINLAR, 1956) which indicate a middle Ordovician age.

The Ayazma Formation which corresponds to the 'Hauptquarzit' (PENCK, 1919; PAECKEL-MANN, 1925, 1938) is composed of 200 to 300 metres of red–brown to whitish quartzites, the lower portion of which is thick-bedded with interbeds of conglomerates; the upper portion is thin-bedded. Neither the arkosic sandstones of the Kurtköy nor the Silurian Yayalar Formation following above the Ayazma Formation can always sharply be separated from the quartzites, which locally appear to be only lenticular intercalations in the arkosic sandstones. The Ayazma quartzites are unfossiliferous. HAAS (1968b) assigned most of them to the upper Ordovician on the basis of their stratigraphical position between the middle Ordovician Kurtköy and the lower Silurian Yayalar Formations.

The bulk of the Ordovician rocks of north-western Anatolia are of continental origin and must have been eroded in a nearby source area of igneous and metamorphic rocks. BRINKMANN (1971) came to the conclusion that this source area most probably was located to the north, in the area of the present-day Black Sea. He postulated a 'Pontic Land' as the northern frame of the Anatolian geosyncline.

East of Istanbul, west of Izmit, YALÇINLAR (1959) found ?*Dictyonema flabelliforme sociale*, Dendrograptidae, and Conulariidae in light grey shales which are, thus, of early Ordovician age. Outcrops of red and grey quartzites, sandstones, and conglomerates both in Çamdağ and Bolu Dağ eastern part of the Bithynian Peninsula, were correlated with the Ordovician red clastics of the Kurtköy and the Ayazma Formations by BRINKMANN (1966). In the Bolu Dağ area, a volcanic series is intercalated between the red clastics and the Pre-Cambrian gneisses. BRINKMANN regarded the volcanics as of Cambro–Ordovician age. From the region south-west and south of the Bithynian Peninsula, from the Ula-Dağ and Kaz Dağ area, and from south-east of Bileçik, KAADEN (1959) and HÖLL (1966) considered green schists and marbles as equivalents of the Ordovician red clastics described by HAAS. The green schists rest unconformably on meso- to katazonal gneisses.

3.4. The Ordovician rocks of Syria and Iraq

In Syria, Ordovician rocks are not exposed at the surface. They have only been encountered by boreholes in the Jezireh, north-eastern Syria, near Abba, Kamichlie, and possibly also Baflioun. The Ordovician rocks are composed of greyish black, thinly bedded, micaceous shales with thin beds of fine-grained sandstone. The entire thickness of the Ordovician is not known. At Abba, they yielded a faunal assemblage of brachiopods, trilobites, and graptolites from a depth of 3099 to 3124 metres (DANIEL *in* DUBERTRET and others, 1963; WEBER, 1963). According to SUDBURY (1957), and STUBBLEFIELD (*in* SUDBURY), *Colpocoryphe arago* (Rouault), *Pseudobasilicus* cf. *nobilis* (Barrande), and *Diplograptus spinulosus* Sudbury indicate a Llandeilo age. *Plumulites*, *Lingula*, and *Sowerbyella*? complete the assemblage. The fossiliferous beds are overlain by more than 300 metres of rocks of similar lithological composition which are unfossiliferous. Their age is held to be Ordovician or ?Silurian. The base of the Ordovician rocks is not known at Abba.

From Kamichlie, WEBER (1963) reported the Ordovician rocks to be of the same lithology as at Abba and to be more than 318 metres in thickness. NICOLAUS (*in* WEBER, 1963) and FLÜGEL (1963) identified the following faunas: *Lingula*, *Sowerbyella*, *Onniella*; *Ctenodonta*, *Hyolithes*; *Onnia ornata* (Sternberg), *Dalmanitina socialis* Barrande; *Turrilepas*; *Tomaculum problematicum* Groom; *Climacograptus*, several species of *Diplograptus*, *Amplexograptus*, *Glyptograptus*, and *Rectograptus*. Especially on the basis of the presence of *Amplexograptus* cf.

perexcavatus (Lapworth), FLÜGEL considered the Ordovician rocks of Kamichlie to be Llandeilo in age. At Kamichlie, the overlying beds are probably Silurian.

According to WETZEL (*in* BELLEN and others, 1959), the lithology of the Ordovician rocks in the area of the Turkish–Iraquian boundary is rather different from that in Syria. In the latter area, the Ordovician rocks are of fully marine origin, while the former are shallow water or littoral deposits. There are only few marine horizons with *Orthoceras*, *Cruziana*, and linguloids, but without any graptolites. At the type section in the Khabour valley, northern Iraq, the Ordovician rocks are more than 800 metres in thickness, with the base not exposed. They were named the Khabour Quartzite-Shale Formation by WETZEL. The formation is composed of alternations of thin-bedded, fine-grained sandstones, quartzites, and silty, micaceous shales, olive-green to brown in colour. The quartzites are generally cross-bedded and white. They show fucoid markings, infilled trails, and burrows. Metamorphism is very slight in the thin-bedded shales and almost unnoticeable in the thicker shale beds. SEILACHER (1963) considered the Khabour Formation to be of early to middle Ordovician age, mainly on the basis of *Billingsella* and *Cruziana*. He designated the formation as a geosynclinal sequence owing to thickness and vertical succession of both biological and sedimentary facies. In his opinion, the Khabour Quartzite represents a typical *Cruziana* facies. In south-eastern Turkey, the Khabour Formation is exposed in much greater thickness than in Iraq. It is regarded as perhaps comparable with the ?Giri Quartzites (Tasman, 1949) of the Telbesmi area, near Derik, and with parts of the 'Hakari complex' consisting of black, micaceous shales (TÜRKÜNAL, 1953).

The Khabour Quartzite is overlain by the Pirispiki Red Beds Formation. The contact is apparently conformable and gradational (WETZEL *in* BELLEN and others, 1959). The entire Pirispiki Formation is unfossiliferous and presumed to be Silurian in age. It is overlain by the Upper Devonian Kaista Formation. Additionally, SEILACHER (1963) described the Dershish Sandstone and the Sinat Greywacke from between the Khabour and the Pirispiki Formation. The Dershish Formation contains *Zoophycos*; the Sinat Formation is more than 100 metres in thickness and yielded *Selenopeltis buchi* (Barrande) which is probably of Caradoc age.

The Chalki Volcanics, flows or sills of basalts and dolerites laterally wedging out into red beds of the Pirispiki, are of much smaller thickness in Iraq than in south-eastern Turkey where the Pirispiki may partly be represented by igneous rocks. According to WETZEL, the Chalki Volcanics with associated conglomerates provide the most definite evidence of a ?Caledonian break somewhere between the Upper Devonian Kaista Formation and the Ordovician Khabour Formation. This break might be placed below as well as above the Pirispiki Formation.

3.5. The Ordovician rocks of Iran

In general, Ordovician rocks are restricted to the same regions from which the Cambrian has been described. At several localities in northern, eastern central, and south-western Iran, the largely carbonate rocks of the Upper Cambrian pass upwards into the Ordovician without visible break. The boundary between the two systems can only be drawn by means of palaeontological evidence (Section 2.6.2). In the Kerman area, early Upper Cambrian rocks are unconformably overlain by very thin fossiliferous Ordovician limestones and shales. The thickness of the Ordovician sequence is generally of a few hundred metres only, except in the Shirgesht area, where it amounts to more than 1200 metres.

3.5.1. South-western Iran and the Kerman area

In southern Iran, at Kuh-e-Faraghan and Ku-e-Gahkum, north of Bandar Abbas, Ordovician rocks are likely to be present in more than 500 metres of barren shales, siltstones, and sandstones full of fucoids and worm-tubes. They underlie fossiliferous Silurian deposits according to

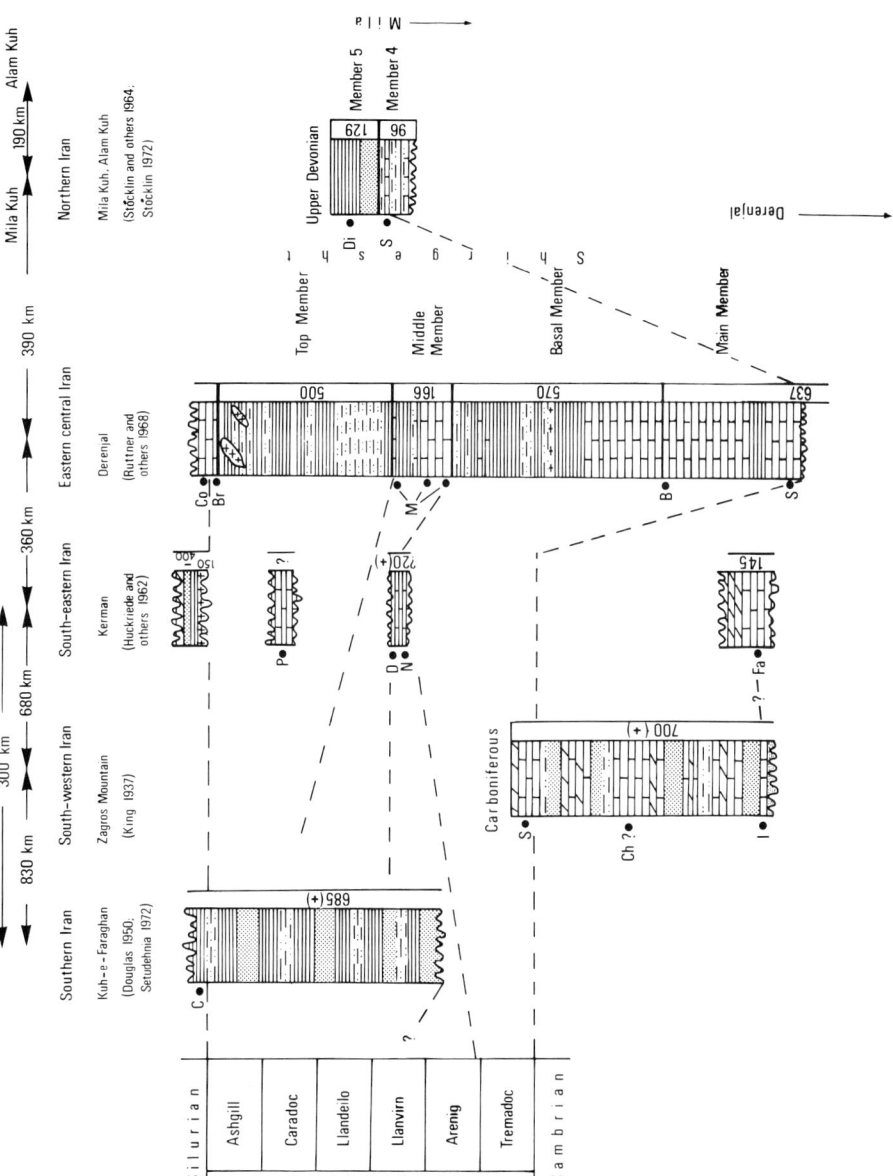

FIG. 21. Representative columnar sections and stratigraphical terminology for the Ordovician rocks of Iran. For legend of petrographical characteristics, see Fig. 3. (B) *Billingsella*; (Br) Brachiopods; (C) *Climacograptus* cf. *scalaris normalis* Lapworth, *Monograptus* aff. *incommodus* Tornquist; (Ch) *Chuangia*; (Co) corals, brachiopods; (D) *Didymograptus bifidus* (Hall); (Di) *Didymograptus, Illaenus*; (Fa) *Farsia, Afghanocare, Pelagiella*; (I) *Irania, Chelidonocephalus*; (M) *Megalaspides, Illaenus, Symphysurus, Hystircurus*, etc.; (N) *Nanorthis, Drepanodus arcuatus* Pander; (P) *Platystrophia*; (S) *Saukia, Iranaspis*.

HARRISON (1930) and DOUGLAS (1950). At a higher Ordovician/?Silurian horizon, abundant crinoid remains are reported. Near the top of the group, an early Silurian fauna with *Monograptus* aff. *incommodus* Tornquist, *Climacograptus* cf. *scalaris* Lapworth, etc. was obtained (DOUGLAS, 1950; SETUDEHNIA, 1972). Thus, the lower part of the group could be assigned to the Ordovician. SETUDEHNIA (1972) reported Ordovician rocks also to be exposed in Kuh-e-Surmeh, Fars Province, about 350 kilometres west-north-west of Bandar Abbas. There, 66 metres of olive-green to grey shales and interbedded thin micaceous sandstones, with 6 metres of finely crystalline, ferruginous dolomite at the top, yielded *Dalmanitina*, *Eohomalonotus*, and *Climacograptus*, which are of middle to late Ordovician age.

From two localities in the Zagros Mountains, about 150 to 200 kilometres west of Isfahan (Imam Sayad Hassan and Pa Vashtah), KING (1937) described an Upper Cambrian sequence of more than 700 metres thickness consisting of limestones, shales, and sandstones. The uppermost part of the sequence yielded *Saukia*, *Iranaspis*, and *Briscoia* attributed to the Daizanian or Fengshanian by KOBAYASHI (1967). In the opinion of WOLFART (1970a,b, 1973), however, at least part of the genus *Saukia* must be assigned to the early Tremadoc (Section 3.6.1). No additional information about the Ordovician part of these sections is available.

HUCKRIEDE, KÜRSTEN, and VENZLAFF (1962) identified thin, fossiliferous rocks in the north-western Kerman area as lower Ordovician. Near Dahu, they found grey limestones with abundant brachiopods (*Nanorthis*, *Syntrophina*) and conodonts (*Drepanodus arcuatus* Pander, *D.* cf. *subarcuatus* Furnish) indicating an Arenig rather than Tremadoc age. The limestones are overlain by 10 metres of yellowish brown shales with *Didymograptus bifidus* (Hall), which is early Llanvirn in age. The shales are, in turn, overlain by tuffaceous, red sandstones and shales considered to be partly Ordovician in age by HUCKRIEDE and others. At various localities in the Kerman area, limestones and dolomites with tabulate corals and brachiopods have been found which could be Ordovician as well as Silurian in age. Several hundred metres of variegated shales, sandstones, and dolomites above the fossiliferous limestones are barren. Their age ranges between late Ordovician and Carboniferous.

3.5.2. Derenjal Mountains and Ozbakh Kuh Mountains

In eastern central Iran, the fossiliferous limestones of the Derenjal Formation pass from the Upper Cambrian into the lower Ordovician as indicated by *Saukia*, which was obtained from the upper part of the Derenjal. The Derenjal Formation is conformably overlain by the Shirgesht Formation (RUTTNER and others, 1968), which is more than 1200 metres thick. The latter covers only a comparatively small area in the southern part of the Derenjal Mountains, forming hills with smooth slopes. From the extremely folded and faulted Ozbakh Kuh Mountains only a single sample of shale containing *Dictyonema* has become known, pointing to the presence of Ordovician rocks in the northern prolongation of the Derenjal Mountains (RUTTNER and others, in preparation).

RUTTNER and others (1968) subdivided the Shirgesht Formation into three members. The basal member is up to 570 metres thick and consists of nodular limestones at the base and marls at the top. Colours are grey–green and red. A few fossils were discovered in the very base of that member only. The middle member is 166 metres thick and characterized by fossiliferous, red, sandy limestones and greenish marls. The faunal assemblage is composed of bryozoans (*Stictoporella*, *Strombopora*, *Escharopora*, and others), brachiopods, gastropods, cephalopods, trilobites (*Megalaspides*, *Illaenus*, *Symphysurus*, *Hystricurus*, *Ningkianolithus*, *Marrolithus*?), and crinoids, indicating a middle Ordovician age, possibly including also early Ordovician. From the stratigraphical position, an early Ordovician age is inferred for the basal member of the formation. The top member is composed of 500 metres of barren, greenish grey marls, shales,

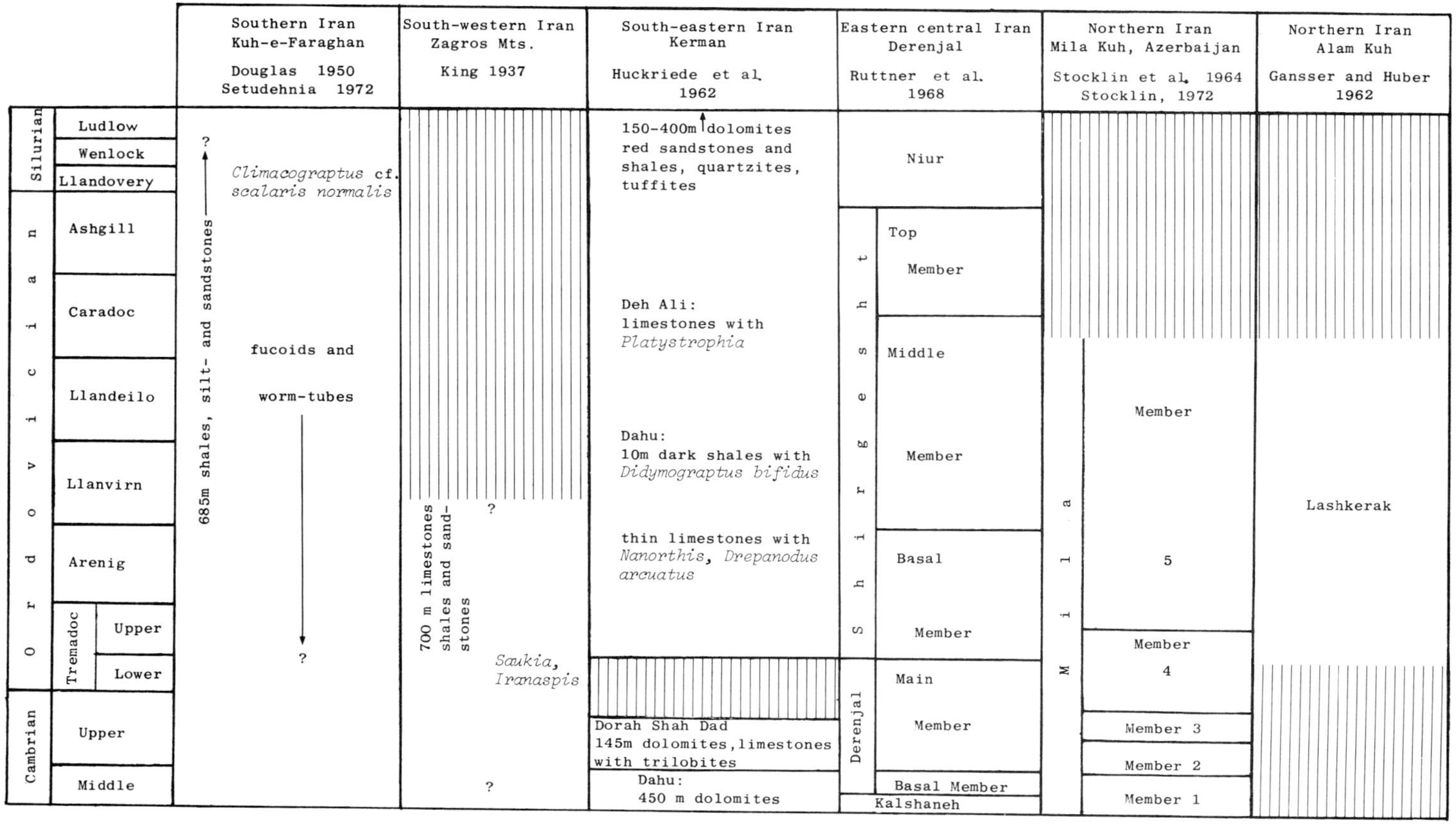

FIG. 22. Stratigraphical correlation of the Ordovician rocks of Iran.

and siltstones. The formation is conformably overlain by the Silurian Niur Formation. Altogether, the Ordovician rocks of the Derenjal area reach a thickness of approximately 1600 metres.

Lithologically and partly also in its faunal content, the Shirgesht Formation is comparable to the Lashkerak Formation of the western central Alborz, the thickness of which is, however, only one fifth that of the Shirgesht; and the trilobites seem to be slightly older, essentially early Ordovician in age. According to STÖCKLIN (1972), the Shirgesht Formation, perhaps exclusive of its calcareous lower part, is correlative with member 5 of the Mila Formation in northern Iran.

The presence of marine Ordovician rocks has also been reported from south-east of Anarak in central Iran by DAVOUDZADEH and others (1969). In order to allow comparisons, the Ordovician fauna of Anarak needs further studies.

3.5.3. Northern Iran

In the eastern Alborz Mountains, northern Iran, fossiliferous limestones pass from the Upper Cambrian into the lower Ordovician as it is known from the Derenjal Mountains, eastern central Iran. These limestones are part of member 4 of the Mila Formation named after Mila Kuh, a mountain in the eastern Alborz, by STÖCKLIN and others (1964). The Mila Formation is widely distributed in northern Iran and traceable from the Mila Kuh type area through the central Alborz, the Soltanieh Mountains, and the Takab area as far as the Lake Rezayeh area in Azerbaijan. Its lithology is remarkably persistent all over the northern region. The Mila Formation is divided into five members. Member 5 and part of member 4 are Ordovician in age and correlate with the Shirgesht Formation of eastern central Iran and with the Lashkerak Formation in the Alam Kuh area, western central Alborz. In the type area, member 4 is composed of siltstones, sandstones, sparry glauconitic limestones, and marls which yielded *Quadraticephalus*, *Iranaspis*, *Saukia*, *Sinosaukia*, '*Briscoia*', *Kaolishania*?, *Chuangia*?, cf. *Eochuangia*, and *Hyolithes*. Genera associated with *Saukia* must at least partly be considered to be early Tremadoc in age (Section 3.6.1). In the type area, member 5 consists of barren shales, subordinate thin sandstones and limestones, and a basal quartzite of 129 metres thickness altogether. Beds containing a Late Devonian fauna overlie the Mila Formation with a basal conglomerate and a sharp, non-angular disconformity. In the western central Alborz and in the southern Azerbaijan, shales, sandstones, and limestones corresponding to member 5 of the Mila yielded an early to ?middle Ordovician fauna including *Tetragraptus*, *Didymograptus*, *Illaenus*, *Paurorthis*, *Clitambonites*, etc. (STÖCKLIN, 1972).

From the Alam Kuh area, western central Alborz, GANSSER and HUBER (1962) described the Lashkerak Formation which is a correlative unit of part of member 4 and member 5 of the Mila Formation. The Lashkerak Formation follows with conformable contact upon the Top Quartzite of the Lower Cambrian Lalun Formation. It is separated from Carboniferous limestones by a sharp, non-angular disconfirmity. In the type area, the basal part of the Lashkerak comprises 59 metres of barren shales, sandstones, and thick, yellow to red, cherty dolomites. The barren basalt member is overlain by 25 metres of red, nodular limestones and marls with trilobites, cystoids, and brachiopods found by BOBEK (1934) and described by DIETRICH (1937). Additional observations were made by GLAUS (1965). The faunal assemblage contains the following trilobites: *Illaenus*, *Asaphus*, *Panderia*?, *Symphysurus*, *Presbynileus*, *Ptychopyge*, *Megalaspis*, and *Ogygites* which are early to ?middle Ordovician in age. The upper member consists of 55 metres of barren sandstones and shales with a few thin interbeds of limestones.

3.6. The Ordovician rocks of Afghanistan

The Ordovician sedimentary rocks of Afghanistan were deposited in three marine arms which were probably relatively sharply limited. The Transafghan trough predominantly running east–west connected the Iranian and the Pakistanian–Indian seas of the Ordovician. In eastern Afghanistan, the Badakhshan and Logar troughs branch off towards the north-east and south-west, respectively (Fig. 24). There are two different facies in the Ordovician sequences of Afghanistan. An alternation of limestones and shales with red and green colours and rich faunas is restricted to a small area in central Afghanistan called Surkh Bum, near Panjaw. In the opinion of the present author, these rocks are Early Tremadoc in age. Sandy–shaly rocks with interbeds of sandy and pure limestones with predominantly grey–blue colours contain comparatively poor faunas. They are distributed all over the Ordovician sedimentational areas of Afghanistan. Either these rocks were formed during the entire Ordovician Period or in post-early Tremadoc times only.

3.6.1. Surkh Bum area

WEIPPERT and WOLFART (1968) and WOLFART (1969, 1970a,b) described marine sedimentary rocks—the Surkh Bum Formation—with early Tremadoc faunas from Surkh Bum in the central part of the Transafghan trough. From the same area, the unique occurrence of richly fossiliferous Afghan Cambrian rocks is known. The lower Tremadoc rocks are separated from the early Upper Cambrian by a non-angular unconformity. They are 50 metres in thickness and consist of three lithological divisions. The lowest division is composed of nearly 7 metres of red limestones with interbeds of greenish and reddish shales or marls. It is overlain by about 11 metres of olive-green shales bearing large trilobites of more than 20 centimetres length. The upper division is represented by an alternation of about 32 metres of red and light grey, greenish and reddish, partly sandy limestones and greenish shales. Very rich faunas are concentrated in the pure limestones of red and greenish white colours. The shales are comparatively poor in fossils. The faunal association is predominantly composed of abundant trilobites, besides some articulate and inarticulate brachiopods, numerous conodonts, and primitive ?gastropods. WOLFART (1970a) identified the following: Brachiopoda: *Lingulella*, *Paracraniops?*, *Apheoorthis*; Trilobita: *Geragnostus*, *Trinodus*, *Afghancephalites*, *Onchonotellus*, *Pagodia* (*Wittekindtia*), *Saukia*, *Macropyge*, *Hazarania*, *Panderia*, *Harpides*, *Pilekia*. In the limestones of the Surkh Bum Formation, more than 30 conodont species have been identified by Stoppel (1970): *Acodus* cf. *campanula* Mound, *A*. cf. *housensis* Miller, *A*. cf. *sevierensis* Miller, *A*. cf. *tetrahedron* Lindström, *Acontiodus* cf. *curvatus* Mound, *A*. ? *falcatus* Ethington, *A. rectus* cf. *sulcatus* Lindström, *Cordylodus* aff. *angulatus* Pander, *C. prion* Lindström, *C. rotundatus* Pander, *Drepanodus arcuatus* Pander, *D*. cf. *concavus* (Branson and Mehl), *D. conulatus* Lindström, *D*. ? *gracilis* (Branson and Mehl), *D. homocurvatus* Lindström, *D. incurvus* Pander, *D. numarcuatus* Lindström, *D. proetus* Lindström, *D*. cf. *subarcuatus* Furnish, *D. suberectus* (Branson and Mehl), *O*. cf. *excelsus* Stauffer, *O*. aff. *forceps* Lindström, *O. inaequalis* Pander, *O. multicorrugatus* Harris, *O. parallelus* Pander, *O. pseudomulticorrugatus* Mound, *O. scalenocarinatus* Mound, *Panderodus* ? *lineatus* (Furnish), cf. *Ptiloncodus* sp., *Scandodus pipa* Lindström, *S*. cf. *rectus* Lindström, *Scolopodus cornutiformis* (Branson and Mehl), and *S*. cf. *insculptus* (Branson and Mehl).

The trilobite faunas represent a unique assemblage of Pacific and Atlantic genera which is considered to be of early Tremadoc age on the basis of the association of *Saukia*, *Pagodia* (*Wittekindtia*), *Harpides*, *Pilekia*, and *Panderia*. In south Asia, this is the first palaeontological evidence for a Tremadoc age of (?part of) *Saukia*. ROBISON and PANTOJA-ALOR (1968)

	Transafghan trough			Badakhshan trough
	Herat — Bluemel 1967	Surkh Bum — Weippert and Wolfart 1968	Jalalabad — Wittekindt 1967	Durumbak — Mirzad et al. 1968
Silurian	Devonian / ?	?	20m limestones and black shales	?
Ordovician — Ashgill / Caradoc	500–1000m shales, calcareous shales, and quartzites (middle Ordovician conodonts have been found in Devonian limestones)	>250m slightly phyllitic shales and sandy limestones with early to middle Ordovician conodonts	shales and sandstones	Sijah Darrah Fm. up to 2500m shales and phyllites with interbeds of marmorized limestones and *Basilicus?*
Ordovician — Llandeilo			200m quartzites	
Ordovician — Llanvirn / Arenig	blue-grey limestones with *Prooconodontus*		120m algal sandstones	
Ordovician — Tremadoc U / L		Surkh Bum Fm. 50m red and green limestones and shales with *Saukia, Pilekia, Harpides*		
Cambrian — U / M / L	?	60m limest. with trilob. 50–70m dolomites	?	
Pre-Cambrian			basement complex	

	Kherskhan Lapparent et al.1968	Gadagak, Sadmarda Blaise et al.1971 Fesefeldt in Wolfart 1970	Chunchata Desparmet et al. 1971	Jabal-e-Kalchi Duerkoop 1970	Malestan Wolfart 1970	Argandab river basin Karapetov et al.1971
			L o g a r t r o u g h			
Silurian	Middle Devonian	Devonian		Early Devonian	Devonian	Devonian
	20m black shales 25m limestones 20m black shales	20m black shales		up to 1500m limestones, shales, sandstones	20m dark limestones	Badokalaj Dewalak
Ashgill	250m sandy shales with Crypto-lithus, Diacalymene	Logar Fm. (partim) Cryptolithinae	?		200m quartzites, shales, and sandy limestones with *Pharostoma*	Nawdesh Fm. 800–1000m shales and sand-stones with interbeds of limestones
Caradoc	50–100m coarse sandstones with *Cruziana furcifera* d'Orb.			2200m greenish brown to black shales and greywackes with interbeds of algal limestones (*Dasy-porella norvegica*)		
Llandeilo		*Didymograptus murchisoni. Cruziana* e.g. *furcifera*	brown shales with graptolites			
Llanvirn		*Hesperonomiella*	200m quartzites			
Arenig		830m quartzites, sandy shales and sandy lime-stones with	370m redbrown shales and sandstones with *Afghanodesma, Cruziana*			Sargaran Fm. 600–670m dolomitic limestones with chert interbeds; basal conglomerates
Tremadoc U		320m quartzites	300m red, green shales and limestones	42m conglomerates		
Tremadoc L		240m sandy shales, quartzites, and sandy limestones				
Cambrian U/M/L						
		P r e – C a m b r i a n b a s e m e n t c o m p l e x				

Fig. 23. Stratigraphical correlation of the Ordovician rocks of Afghanistan.

recorded *Saukia* to be associated with the Atlantic genera *Parabolina*, *Pharostomina*, and others in southern Mexico. *Saukia* and *Pagodia* are known as typical elements of Pacific faunas. They are genera of the eastern Asian Fengshanian and of the North American Trempealeauian, respectively, hitherto attributed to the latest Late Cambrian. The findings at Surkh Bum yielded the evidence for a Tremadoc age of at least part of the Trempealeauian and Fengshanian stages. The opinion of JONES, SHERGOLD, and DRUCE (1971), who understood the Australian occurrences of *Saukia* to be of Late Cambrian age, is discussed in Section 2.6.2.

KOBAYASHI (1971) also considered the Afghan faunal assemblage with *Saukia* to be of early Tremadoc age. He suggested the Afghan species of *Saukia* to belong to some relic species which have been reported from the lowest Ordovician rocks in North and Central America and eastern Asia. According to KOBAYASHI, the early Tremadoc faunas of Afghanistan cannot be correlated with the Fengshanian and Trempealeauian faunas because the latter do not contain typical Tremadoc trilobites. This interpretation seems to be inconclusive because there is still the possibility that the real problem is rather of a zoogeographical nature (WOLFART, 1970).

3.6.2. Transafghan trough

The widely distributed main part of the Afghan Ordovician deposits is predominantly composed of grey–blue clastics. In general, the sequences consist of thick, sandy shales with interbeds of sandy limestones and thick-bedded, light grey quartzites up to several hundred metres thick. Exposed quartzites are usually coated with a dark brown film of iron and manganese oxides. Within the Transafghan trough, Ordovician rocks are known from numerous exposures. Only by means of rare occurrences of conodonts could the age of the otherwise unfossiliferous rocks be determined. The Transafghan trough can be subdivided into three parts showing differences in lithology and faunas.

From the western part of the trough, south of Herat, BLÜMEL (1967) described a slightly metamorphosed formation of shales with intercalations of pure and sandy limestones and quartzites. The formation is up to 1000 metres thick. Compared with the Ordovician rocks of the Badakhshan and Logar troughs and of the rest of the Transafghan trough, the rocks are more calcareous south of Herat. Limestones positioned about 40 kilometres south of Herat yielded *Proconodontus mülleri serratus* Miller and *P. notchpeakensis* Miller (STOPPEL, 1970). According to MILLER (1969), both of the species are associated with the Trempealeauian and early Canadian trilobite genera *Euptychaspis* and *Symphysurina*, respectively, in North America. Therefore, the Afghan finds of *Proconodontus* probably must be regarded to be of Tremadoc age.

About 60 kilometres east-south-east of Herat, middle Ordovician conodonts (*Acodus*, *Gothodus* or *Tetraprioniodus*, *Ligonodina*, *Scandodus*?, *Scolopodus*, *Stereoconus*) have been found in Devonian limestones with spiriferids. Ordovician shaly rock types as known from other Afghan regions are not preserved in that area. It seems to be probable, therefore, that the cited conodonts originally were part of an Ordovician shaly, calcareous sequence eroded in pre-Devonian or Devonian times.

From the Koh-i-Davindar, 60 to 80 kilometres east-north-east of Herat, MIRZAD and others (1968) reported an unfossiliferous alternation of sandstones, phyllites, and limestones with intercalations of acid effusive rocks more than 2500 metres thick. The entire sequence was tentatively considered to be of Cambrian to Devonian age.

In the central part of the Transafghan trough, 6 kilometres south-west of Surkh Bum, WEIPPERT (1967) observed slightly phyllitic shales and sandy limestones which are more than 200 metres thick. STOPPEL (1970) identified the following conodonts of early to middle

Ordovician age: *Acontiodus, Coelocerodontus* or *Gothodus, Cordylodus* or *Paracordylodus, Hibbardella prima* (Walliser), *Oistodus, Scandodus,* and *Tetraprioniodus.*

From the eastern part of the Transafghan trough, two regions are known with outcrops of Ordovician rocks. In the Panjir valley following the north-eastern continuation of the axial trend of the Logar trough, variously metamorphosed shales, quartzites, and sandy limestones several hundred metres thick are exposed. Up to the present, no fossils have been found in this formation, which is tentatively understood as Ordovician owing to its facies. East-south-east of Jalalabad, WITTEKINDT (1967) reported a sequence of several hundred metres of shales, sandstones, and quartzites. The formation is overlain by fossiliferous limestones and black shales of late Silurian age. The lowest portion is composed of about 100 metres of calcareous sandstones with abundant concentric algae of a diameter of 2 cm (*Solenopora, Malacostroma*). The middle part is formed by an alternation of 200 metres of cross-bedded, whitish grey quartzites and quartzitic shales. The upper part contains slightly metamorphosed, light grey to greenish brown shales and calcareous, grey–green sandstones. Owing to its facies and stratigraphical position, the entire formation has been considered to be of Ordovician age.

3.6.3. Badakhshan trough

During the Ordovician, the Badakhshan trough was probably the largest area of Afghanistan with marine sedimentation. The Ordovician sequence is up to several 1000 metres in thickness and consists mainly of clastic rocks which are very poor in fossils. The differentiation between rocks of Cambrian and of Ordovician age is, therefore, not yet possible.

In the south-western part of the Badakhshan trough, in the area of Doab-Išpušta, 150 kilometres north-west of Kabul, Soviet geologists (KOLČANOV and others, 1971) attributed the Sijah-Darrah Formation, a sequence of unfossiliferous, quartzitic sandstones, blastopsammitic schists, and phyllites rarely with interbeds of marmorized limestones, to the ?Upper Proterozoic to Lower Palaeozoic (Cambro–Ordovician). The formation is up to 2500 metres in thickness. It is conformably overlain by Silurian rocks and separated by a hiatus from the Pre-Cambrian basement complex. KOLČANOV and others compared the Sijah-Darrah Formation with the Wishar Formation of the northern Pamir, assumed to be Late Proterozoic to Ordovician in age on the evidence of acritarchs.

SEMENOW and others (*in* MIRZAD and others, 1968) described Lower Palaeozoic rocks from the central part of the Badakhshan trough comparable to similar rocks in the central Pamir. The lower portion consists of more than 700 metres of dolomites, limestones, sandstones, and shales and, though unfossiliferous, was attributed to the Cambrian. An Ordovician age for the lower portion cannot, therefore, be excluded. The upper part of the Lower Palaeozoic rocks is composed of 300 metres of sandstones containing ?*Basilicus nobilis* Barrande of Llandeilo age.

In the north-eastern part of the Badakhshan trough, north-eastern Hindukush, near Iškamyš, Soviet geologists observed unfossiliferous phyllitic shales with intercalations of basic effusive rocks more than 1500 metres in thickness. They designated those rocks which are ?unconformably overlain by fossiliferous Devonian rocks as Cambro–Ordovician. According to the geological map (1 : 500,000) of Afghanistan (BRATASH and others, 1964), Ordovician rocks are also present 125 kilometres north-north-east of Faydzabad. Their age can only be inferred by comparison with Ordovician rocks in the neighbouring northern Pamir described by BARKHATOV (1863). There, *Trinucleus* and other fossils indicate a middle to late Ordovician age.

3.6.4. Logar trough

In the centre of the Logar trough situated in the eastern Hazarajat, the Ordovician sequence seems to be fairly complete and, being comparatively rich in fossils, different from the widely distributed grey–blue, clastic Ordovician rocks which bear only conodonts. It was FESEFELDT (1964) who recognized the Ordovician age of part of the Logar Formation nearly 1400 metres in thickness on the basis of *Cruziana furcifera* d'Orbigny. The Logar Formation comprises all of the Lower Palaeozoic rocks as well as early Lower Devonian. It represents, at least partly, a synonym of the Kotandar Series (LAPPARENT, 1962) consisting of unfossiliferous, greenish black shales and sandstones which overlie unconformably the metamorphic basement complex in the region west of Kabul.

From the area of Kherskhan, in the northern part of the Logar trough, LAPPARENT and others (1968) and DESPARMET (1969) reported 50 to 100 metres of coarse-grained, well bedded sandstones overlying chloritic gneisses which are probably equivalents of the Pre-Cambrian Kalu Formation. The sandstones contain *Cruziana* and *Rouaultia* considered to be Arenig to Llandeilo in age. Near Kherskhan, lenses of brownish limestones are intercalated between the gneiss and the overlying sandstone. According to STÖCKLIN (*in* DESPARMET, 1969), these limestones are reminiscent of similar rocks in Iran. The basal sandstones are overlain by 250 metres of sandy, sericitic shales bearing trilobites in the upper part. *Cryptolithus* and *Diacaly-mene* indicate a middle to late Ordovician age. They mark a horizon which is presumably correlative with the Cryptolithinae-bearing horizon near Gadagak.

In the Sadmarda valley, north-east of Kherskhan, BLAISE and others (1971) discovered brownish black, sandy shales with graptolites attributed to *Didymograptus murchisoni* (Beck) which indicates a late Llanvirn age. Near Sadmarda, the trace fossils *Cruziana furcifera*, *Tigillites*, and *Vexillum* must be assigned to the Arenig or Llandeilo. From Chunchata between Sadmarda and Wardak, DESPARMET and others (1971) described a sequence of about 900 metres thickness. The lower portion is formed by unfossiliferous, red and green shales with red and white limestones in the basal part. It is more than 300 metres thick and must presumably be assigned to the latest Cambrian or early Ordovician. The overlying formation is composed of 370 metres of red–brown shales and sandstones containing three horizons with *Afghanodesma*, besides *Cruziana* in the lowermost horizon. Besides a Late Cambrian to Tremadoc age for *Afghanodesma* as assumed by DESPARMET and others, in the opinion of the present author an Arenig age must also be considered. The sequence with *Afghanodesma* is followed by 200 metres of unfossiliferous quartzites designated as Arenig by DESPARMET, and by brown shales with graptolites of late Llanvirn age.

The Gadagak section in the upper Logar valley exposes the Ordovician part of the Logar Series fairly completely. According to FESEFELDT (1964), it begins with 240 metres of unfossiliferous, sandy shales, sandy limestones, and quartzites which might be considered to be of late Cambrian as well as of early Ordovician age. The following unit consists of 320 metres of bedded quartzites, also unfossiliferous. For lithological reasons, the present author is inclined to ascribe an early Ordovician age to the entire lower portion of the Gadagak section rather than a late Cambrian age. The quartzite is overlain by an alternation of sandy shales, sandy limestones, and quartzites more than 800 metres thick. The alternation contains at least three fossiliferous horizons, the lowest of which is situated about 60 metres above the base. It yielded *Hesperonomiella*, *Goniophorina*?, and *Iocrinus*? and must be considered, therefore, to be Arenig to Llandeilo in age. Fossils of nearly the same age—*Cruziana furcifera* d'Orbigny and *Receptaculites*—have been found 300 metres above the base of the alternation. The next fossiliferous bed with Cryptolithinae of middle to late Ordovician age was discovered about 750 metres above the base. With a sharp boundary, the Ordovician sequence is disconformably overlain by lithologically different black shales of middle to late Silurian age.

The thickest Ordovician sequence (2200 metres) of the Logar trough was investigated by DÜRKOOP (1970) near Jabal-e-Kalchi between Gadagak and Malestan. There, 42 metres of conglomerates, mainly consisting of quartz pebbles, unconformably overlie the slightly metamorphosed Pre-Cambrian basement complex. Greenish brown to black shales and grey-wackes with ripple-marks, cross-bedding, Lebensspuren, and interbeds of algal limestones represent the main part of the entire sequence. The age of the Jabal-e-Kalchi Formation follows from its stratigraphical position and from the occurrence of *Dasyporella norvegica*, which is probably an Ordovician species.

From the south-western part of the Logar trough, near Malestan, WOLFART (1970a) described 200 metres of quartzites and shales with intercalations of sandy limestones containing middle Ordovician faunas (late Llanvirn to early Caradoc): *Moorephylloporina*, *Sowerbyella*, *Quadrotheca*, and *Pharostoma*.

In the southernmost part of the Logar trough, the Arghandab basin west of Moqur, the Ordovician deposits are still rather thick (800–1000 metres). According to KARAPETOV and others (1971), the Ordovician section comprises the uppermost portion of the Sargaran Formation and the Nawdech Formation. In the Arghandab basin, the lower boundary of the Ordovician System was tentatively placed within the seventh member of the Sargaran Formation, consisting of 120 metres of bedded, often dolomitized limestones with numerous small lenses and nodules of chert. KARAPETOV and others found ?Cambrian algae in the middle members of the Sargaran and attributed, therefore, a lower Ordovician age to the seventh member. The Nawdech Formation is typically developed in the valley of the Argasu River. It is composed of dark aleurolites, being greenish in the upper portion and containing interbeds of fine-grained sandstones. Two intercalations of limestones yielded brachiopods of the *Stropheodontacea* and *Orthida* and some undeterminable crinoids. Chiefly based on the conformably overlying lower Silurian sandstones, a middle to upper Ordovician age is assigned to the Nawdech Formation.

In the Logar trough, the Ordovician rocks are rather uniformly composed of shales and quartzites with intercalations of sandy limestones. The bedded, massive quartzites of the lower part of the Ordovician sequence generally pinch out south-westwards. From Jabal-e-Kalchi southwards, thicknesses of Ordovician rocks diminish rapidly.

3.7. Ordovician stratigraphy and correlation in the Middle East

3.7.1. Stratigraphy of the various regions and correlations

In the Middle East, the biostratigraphy of the Ordovician deposits is based on different groups of fossils. Particularly the non-trilobitic fauna was made more profuse by the development of graptolites, brachiopods, cephalopods, and conodonts, so that the monopoly in biostratigraphical classification and palaeozoogeography held by the trilobites during the Cambrian can no longer be maintained for the Ordovician System. Many of the Ordovician successions, however, are poor in fossils or include only few fossiliferous horizons; many of them have as yet been insufficiently studied, and no definitive synthesis, therefore, can be presented.

The Ordovician faunal assemblages are of Arenig to Ashgill age in north-western Saudi Arabia and southern Jordan and of early Llanvirn age in the Oman area. In addition to graptolites, the assemblage includes conulariids, brachiopods, nautiloids, trilobites, and associated trace fossils which can clearly be attributed to the Ordovician. Lithologically and palaeontologically, the different Ordovician formations of Jordan and Saudi Arabia can be correlated, though fossils generally diminish towards the latter country, and the lower portion of the Jordan Ordovician succession is scarcely developed in it. Most widely distributed are the Llanvirn beds with *Didymograptus* which can be traced from the Arabian Peninsula through Iran to eastern Afghanistan. In the uppermost (?Ashgill) part of the Ordovician succession of southern Jordan, the trilobite *Brongniartella* is associated with an assemblage of typical Ordovician fossils (Section 3.2.1). According to BENDER (1964) and WOLFART and others (1968), younger *Brongniartella*-bearing beds are underlain by a faunal assemblage of early Llandovery age in the same region. Consequently, *Brongniartella*, hitherto ascribed only to a late Ordovician age, must have ranged from the late Ordovician into the early Llandovery. In south-eastern Turkey DEAN (1967) assigned a late Caradoc age to a faunal assemblage with *Brongniartella*.

From north-western Turkey, Tremadoc graptolitic facies and a small middle Ordovician fauna have been described by YALÇINLAR (1956, 1959). No comparable occurrence of the Tremadoc graptolite facies has been encountered in the remaining part of the Middle East. In south-western Turkey, the Tremadoc is represented by a shaly facies with trilobites of Acado–Baltic type only. In Iran and central Afghanistan, the lower Tremadoc rocks consist of limestones containing the Pacific trilobites *Saukia* and *Pagodia*, as well as Acado–Baltic forms. Arenig to Llandeilo faunas are widely distributed in southern Turkey, north-eastern Syria, Iran, and Afghanistan (Fig. 24). In comparison with the trilobites, all the other animal groups such as brachiopods, graptolites, and conodonts are frequently in a minority, particularly in Turkey. Caradoc and Ashgill faunas have not yet been found in Iran; in south-eastern Turkey they are mainly represented by trilobites, some graptolites and brachiopods, and in the Logar trough, Afghanistan, by some trilobites.

3.7.2. The upper boundary of the Ordovician

The problem of the Ordovician/Silurian boundary is very different in the various regions of the Middle East. There are only a few sections of Syria, Turkey, and southern Jordan where the Ordovician rocks pass without visible break into the Silurian. In Afghanistan, the problem can be studied only in the eastern part of the country, especially in the Logar trough. On the basis of some observations of abrupt changes in facies between Ordovician and Silurian formations, gaps in the stratigraphical record, occurrences of conglomerates, etc., WOLFART (1970) postulated a sedimentational break between the two systems. DÜRKOOP (1970, Table 1) came to the same conclusion; KARAPETOV and others neglected the problem. In north-western Saudi Arabia, HELAL (1964, 1965) stated that the Ordovician rocks are separated from the overlying Silurian formations by a noticeable hiatus combined with a sharp lithological change. In Afghanistan and Saudi Arabia, the Ordovician/Silurian boundary coincides with the described sedimentational gaps because the rocks below contain Ordovician and the rocks above Silurian fossils.

In southern Jordan, Turkey, and eastern Iran, it is more difficult to fix the Ordovician/Silurian boundary than in the preceding regions. In southern Jordan, the problem is (WOLFART and others, 1968) that a faunal assemblage with graptolites, conodonts, and brachiopods of definite early Llandovery age, according to BENDER (1964), is followed by the trilobite *Brongniartella* which has hitherto always been assigned to the Ordovician. The same trilobite has been observed to occur below the early Llandovery assemblage. Provided that the observations concerning the succession of beds are correct, the Ordovician/Silurian boundary must be placed below the Silurian faunal assemblage in southern Jordan, and *Brongniartella* must be ascribed a late Ordovician to early Silurian age.

There are only two regions in Iran where Ordovician rocks are immediately followed by Silurian: in the eastern Iranian Derenjal Mountains and in the southern Iranian Kuh-e-Faraghan area. In the latter region, the Ordovician rocks are unfossiliferous; they are conformably overlain by lower Silurian beds with *Monograptus*. The Ordovician/Silurian boundary has tentatively been placed at the base of the graptolitic formation. In the Derenjal Mountains, the upper part of the Ordovician succession is also unfossiliferous. The top of the Ordovician System is obviously marked by a sudden change in facies; marly, shaly Ordovician rocks of several hundred metres thickness are overlain by richly fossiliferous Silurian limestones. In the remaining regions of Iran, upper Ordovician rocks have not yet been found.

In southern Turkey, a continuous section of fossiliferous upper Ordovician and lower Silurian rocks is not present. From north-eastern Syria, WEBER (1963) reported the Ordovician rocks to be conformably followed by Silurian beds which are nearly of the same facies as the former. As the upper portion of the Ordovician succession is unfossiliferous, and as the Silurian deposits yielded only chitinozoans and hystrichosphaerids, it is impossible to fix the Ordovician/Silurian boundary exactly in that region. In north-western Turkey, easy of Istanbul, the unfossiliferous upper Ordovician succession is followed by an early Llandovery faunal assemblage.

3.8. Ordovician history of the Middle East

3.8.1. Tectonics and sedimentation

The Ordovician Period was a quiet time in the Middle East, and no major orogenic events are recorded in the sedimentary sequences. Gradual subsidence in the regions of marine sedimentation is indicated by the stratigraphical record within the entire region, with a significant and widely distributed reversal in parts of the Middle East towards late Ordovician times. In detail, Ordovician palaeogeographical evolution was very complex, particularly in the basinal regions of the Turkish–Afghan west–east belt of mobile shelf conditions. The formation of local basins with a considerable thickness of sediments which have been deposited rather rapidly may be taken as evidence of tectonic activity. Another indication could be of acid and basic volcanics of possible Ordovician age which were reported from western and northern Afghanistan (BLÜMEL, 1967; KOLČANOV and others, 1971). From the other areas of the Middle East, no Ordovician volcanics have been described. Stable shelf conditions with more uniform geological development prevailed in the surroundings of the Arabian Shield. As in the Cambrian, structural development and sedimentation were largely governed by the Pre-Cambrian basement trends, showing three main directions (N.W.-S.E., N.E.-S.W., N.-S.). Consequently, the Ordovician regions of sedimentation were nearly identical with those of Cambrian time.

In some regions of the Middle East, e.g. Surkh Bum, central Afghanistan, Kerman in south-eastern Iran, and possibly also in the Amanos Mountains, southern Turkey, the transition from Cambrian to Ordovician time is marked by gaps in the sedimentary record of the late Upper Cambrian. In these regions, the beginning of the Ordovician more or less coincides with marine transgressions growing even more intensive during the further course of the period. In north-western Turkey, east of Istanbul, Ordovician sedimentation must have begun later, possibly in early middle or late early Ordovician time. In other regions, such as the stable shelf of the Arabian Peninsula, eastern and northern Iran, and probably some areas of southern Turkey, the Cambrian rocks pass without visible break into Ordovician deposits. Major regressive events at different places in the Middle East are indicated by gaps of various size in the sedimentary succession of the upper Ordovician and Silurian. Gaps comprising upper Ordovician and Silurian time exist in south-western Turkey, northern and south-western Iran, and Oman. The remaining regions of the mobile shelf, such as Afghanistan and Turkey, were probably also affected by an episodic, short-lived uplift at the end of the Ordovician and the beginning of the Silurian.

In the Middle East, the palaeogeographical pattern of the Ordovician land masses is very closely related to that of the Cambrian. In the south-west, it is dominated by the large Nubo–Arabian craton, in the northern and central part by a great number of Pre-Cambrian swells and highs. The lithofacies in the stable shelf region around the Arabian craton is

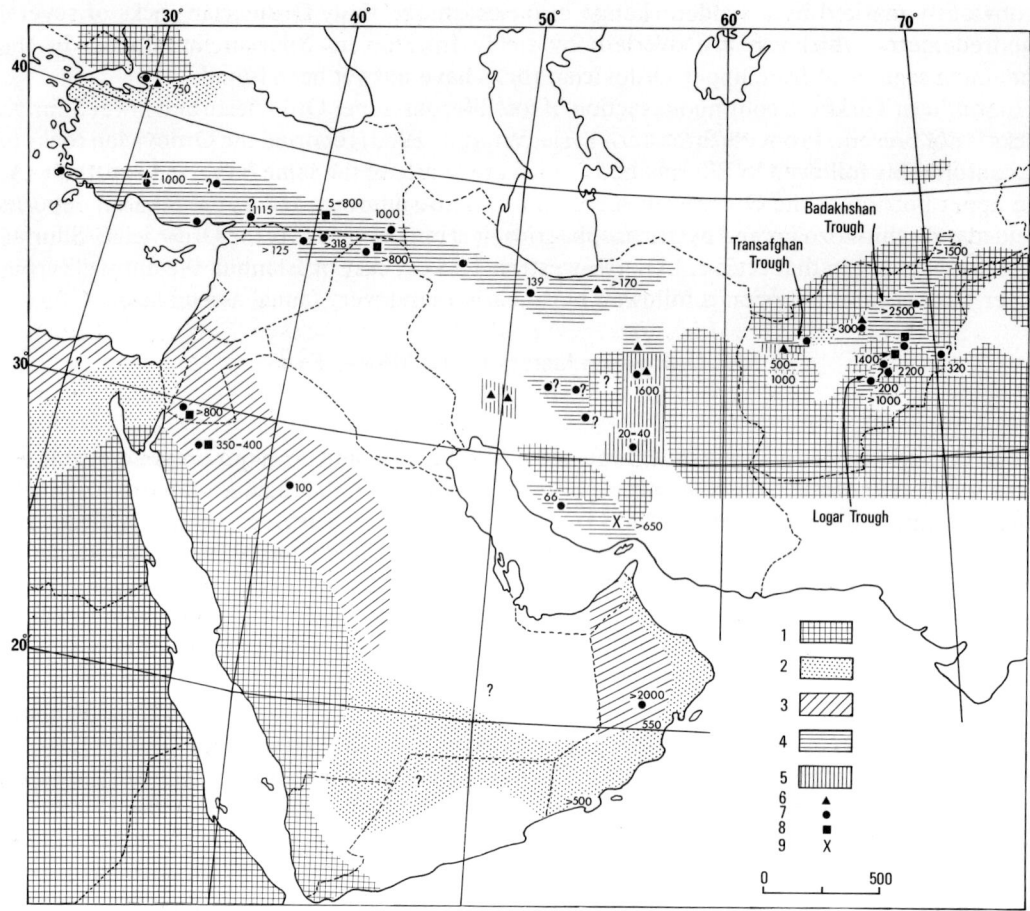

FIG. 24. Ordovician palaeogeography. (1) Potential source area; (2) continental to littoral environment (conglomerate, sandstone, shale); (3)–(5) marine environment; (3) sandstone, shale; (4) quartzite, sandstone, shale, sandy limestone; (5) predominantly marl and limestone; (6) Tremadoc faunas; (7) Arenig to Llandeilo faunas; (8) Caradoc to Ashgill faunas; (9) trace fossils. Total thicknesses in metres. In north-western Saudi Arabia, the boundary between Cambrian and Ordovician rocks is uncertain.

composed only of clastic, detrital material derived from the craton. In the mobile shelf region, particularly in Afghanistan, northern Iraq, and in Turkey, thick sequences of conglomerates and quartzites yield abundant evidence of numerous swells which served as sources for the detrital sediments. In most parts of the mobile shelf region comprising Afghanistan, southern and northern Iran, north-eastern Syria, and southern Turkey, the Ordovician rocks are predominantly composed of grey–green or blue–grey, partly graptolitic shales, (regionally calcareous) sandstones, and quartzites. Only in the Derenjal region, eastern Iran, limestones and marls make up the entire Ordovician sequence.

Lower Tremadoc rocks consist of partly red and green limestones and marls in central Afghanistan and all over Iran. As indicated by coquina-like limestones, conditions of accumulation in the region of Surkh Bum, central Afghanistan, are believed to be near-shore, at any rate very shallow water conditions during early Tremadoc time. This phase was terminated abruptly by a flood of terrigenous shaly-sandy sediments which continued nearly all over the Ordovician of Afghanistan.

3.8.2. Zoogeographical features and climate

As pointed out by DEAN (1967b) and WOLFART (1967), the Middle Eastern part of Tethys in the Ordovician played an important role as a seaway connecting the Far East and the Mediterranean area. This situation is comparable to the palaeogeographical conditions of the Lower Cambrian (Section 2.7.2). In the eastern part of the Middle East, in Iran and Afghanistan, the early Tremadoc trilobite fauna is represented by an assemblage of Pacific genera (*Saukia* and *Pagodia*) which are associated with genera of Acado–Baltic type (*Harpides, Macropyge, Pilekia, Panderia*) in Afghanistan. WHITTINGTON (1966) considered the Atlantic genera to belong to the *Ceratopyge* or *Pharostomina* fauna, respectively. From south-western Turkey, HAUDE (1972) reported a Tremadoc trilobite fauna of Acado–Baltic type only.

After early Tremadocian times the zoogeographical situation changed thoroughly in the Middle East. The post-Tremadoc deposits of the Ordovician bear trilobite faunas of a southern province comprising the Andes, England, central Europe, China, Tethys, and Australia (WHITTINGTON, 1966). Concerning provinciality of Ordovician trilobite faunas, WHITTINGTON and HUGHES (1972, 1973) outlined that in the lower Ordovician (Arenig–Llanvirn) four trilobite provinces existed. According to these authors, Turkey was part of one of them, of the European *Selenopeltis* province. The rest of the Middle East must be presumed to exhibit also rather close faunal relations to the *Selenopeltis* province. The post-Llanvirn history of distribution appears to be one of decreasing provinciality in trilobites (and in brachiopods, according to WILLIAMS, 1973). By Caradoc times only the *Selenopeltis* province is distinct from the rest of the world including England, the Baltic region, and Australia. In the late Ashgill, an essentially cosmopolitan fauna began to emerge.

Following WHITTINGTON and HUGHES (1973), provinciality of faunas originated from the physical characters of seas between the continental masses, such as width, depth, pattern of circulation and temperature difference. Progressive merging of faunal provinces during the course of the Ordovician is held to be due to relative movements of continental blocks by means of which barriers of migration between the single faunal provinces were removed. According to a suggestion of WHITTINGTON and HUGHES (1972, 1973), the *Selenopeltis* fauna inhabited cool waters on the margins of Gondwanaland. This is possibly supported by the assumption of SPJELDNAES (1961) that polar ice-caps existed during the Ordovician and that one of the poles was in or just west of Central Africa. The latter assumption is supported by the results of palaeomagnetic studies. There are more observations which are in favour of a cold Ordovician climate in the Middle East. LINDSTRÖM (1972) came to the conclusion that in the early Ordovician great parts of Europe were at times fairly cold, and ROGNON and others (1972) reported a late Ordovician glaciation in northern Africa.

Summarizing present knowledge, MCCLURE (1978) supposed that late Ordovician glaciation is indicated by the presence of glaciated boulders and tillites in the Arabian Peninsula. He concluded that the broad continental ice sheet which covered much of North Africa at that time (TUCKER and REID, 1973) extended eastwards across the Arabian portion of the Afro–Arabian Pre-Cambrian craton. Glacial activity in Arabia apparently took place within the uppermost part of the Caradoc or immediately following. A hiatus possibly associated with glacial tie-up of an enormous continental ice mass and consequent marine regression may account for the apparent absence of Ashgill sediments. The Ordovician and Silurian sediments above and below the glaciated horizon contain significant faunal elements.

4. Silurian

By far the largest Silurian outcrops of the Middle East extend for more than 900 kilometres from southernmost Jordan south-eastwards into north-western Saudi Arabia. The Silurian clastic deposits belong to the huge blanket of sedimentary rocks covering the north-eastern slope of the Arabian craton, the original inclination of which seems to be reflected by the dip of the bedded rocks. Usually, the landscape is of an arid inselberg type.

In the Tethyan mountain chains, the Silurian outcrops are very small, widely scattered, and even less numerous than those of the Ordovician. In north-eastern Syria and large parts of south-eastern Turkey, the Silurian rocks are covered by thick post-Silurian sediments and knowledge of them is restricted to deep boreholes. From western Afghanistan, western Iran, Iraq, Lebanon, Israel, Sinai, and the south-eastern part of the Arabian Peninsula, no Silurian deposits are known hitherto.

The thicknesses of the Silurian deposits are about the same as or somewhat less than those of the Ordovician. Variscan folding possibly only affected the Silurian deposits of Bithynia and some regions north of the central Anatolian crystalline massif, as well as northern Afghanistan

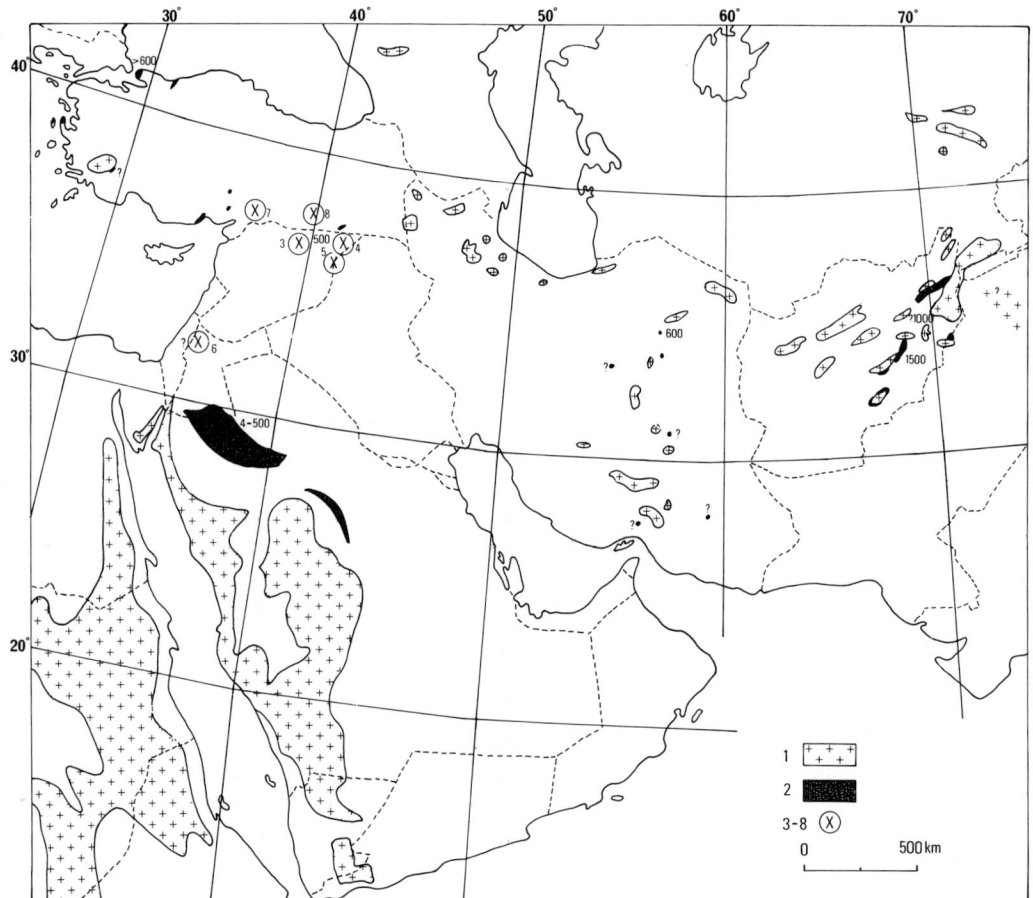

FIG. 25. Known areas of outcrops of Pre-Cambrian and Silurian rocks. (1) Pre-Cambrian at surface; (2) Silurian at surface; (3–8) boreholes; (3) Abba; (4) Kamichlie; (5) Bouab; (6) Suweilieh; (7) Aril-l; (8) Handof-l. Total thicknesses of Silurian rocks in metres.

and very small regions near the margins of tectonically particularly exposed blocks in southern Afghanistan. Within the Tethyan mountain chains, all of the Lower Palaeozoic rocks were strongly folded during the Alpine orogenesis.

4.1. History of Silurian studies

In the early times of geological exploration of the Arabian Peninsula, THRALLS and others (1956) and STEINEKE and others (1958) assigned some rocks of the former 'Tawil Sandstone' to a Silurian age. In the following period, the Silurian rocks were collectively treated with other Lower Palaeozoic deposits as the Tabuk Formation by BRAMKAMP and others (1963) and BROWN and others (1963), as well as by POWERS and others (1966) and POWERS (1968). HELAL (1964, 1965) restricted the term Tabuk Formation to the Ordovician rocks and divided the Silurian into Sharawra and Tawil Formations. BENDER (1964) was the first to discover fossiliferous Silurian rocks in southern Jordan as part of the former Nubian Sandstone. The fossils were described by BENDER and HUCKRIEDE (1964) and WOLFART, BENDER, and STEIN (1968).

Until about two decades ago, Silurian rocks were entirely unknown in southern Turkey and north-eastern Syria. Evidence for their presence in these regions was published by FLÜGEL (1955), YALÇINLAR (1955–73), IMANDT (1962), and WEBER (1963). In north-western Turkey, Silurian rocks were already described by PAECKELMANN (1925, 1938) and re-studied by HAAS (1968). KAUFFMANN (1965), HÖLL (1966), and BESENECKER and others (1968) were the first to recognize the Silurian rocks of Chios and Karaburun Peninsula, west of Izmir.

The first reports on lower Silurian rocks were published by LEES (1938) and DOUGLAS (1950). Later, FLÜGEL and RUTTNER (1962), and in particular RUTTNER, NABAVI, and HAJIAN (1968), completed knowledge of fossiliferous Silurian deposits in eastern central Iran, as did HUCKRIEDE, KÜRSTEN, and VENZLAFF (1962). We owe the first note on Silurian rocks of Afghanistan to FESEFELDT (1964). More important work, however, was carried out by DESPARMET (1969), DÜRKOOP (1970), PLODOWSKI (1970), and KARAPETOV and others (1971).

4.2. The Silurian rocks of the north-western Arabian Peninsula

The outcrops of Silurian rocks are restricted to the north-western part of the Arabian Peninsula, along the eastern margin of the crystalline core of the Arabian shield, east of the outcrops of Cambro–Ordovician rocks. They cover the eastern part of southernmost Jordan and the Tabuk area of north-western Saudi Arabia, west of the Great Nafud. The extensive Tabuk exposure of Silurian rocks forms a broad planar surface, in places broken by north-west–south-west structural ridges and drainage channels (POWERS and others, 1966). Differential relief averages about 50 metres, individual hills rising higher above the general ground level. Alternating hard and soft clastics of the Sharawra weather to a relatively smooth surface in comparison with the rough, hummocky surfaces formed by the strongly cross-bedded sandstones of the Cambrian Saq below and the late upper Silurian Tawil above. From the southern part of the Arabian Peninsula, especially from the Oman region as well as from the Negev in southern Israel and from the Peninsula of Sinai, no Silurian rocks are known up to the present. In the southern Negev and near Abu Durba, Peninsula of Sinai, fossiliferous late Lower Cambrian beds are overlain by the continental Nubian Sandstones, which might contain a portion of Silurian age.

Autochthonous sequences of the Silurian extend from about 30° northern latitude in Jordan southwards into Saudi Arabia until between about 25° and 26° northern latitude (Fig. 25), wedging out completely as well as the Ordovician rocks. All of the Silurian sequences consist of sandstones and shaly rocks overlying the Cambro–Ordovician sandstones.

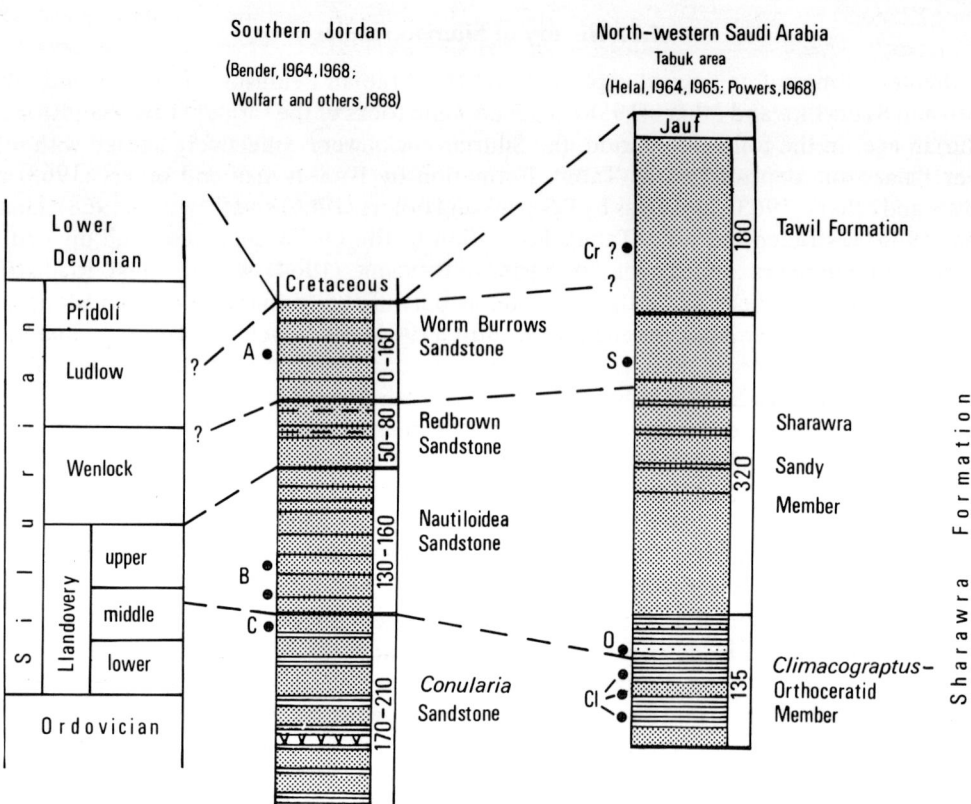

FIG.26. Representative columnar sections and stratigraphical terminology for the Silurian rocks
of Jordan and north-western Saudi Arabia. (A) *Arthrophycus, Rusophycus, Onchus*;
(B) *Brongniartella, Rostricellula*; (C) *Climacograptus scalaris* cf. *scalaris* (Hisinger), *Climacograptus innotatus jordaniensis* Stein, *Demirastrites*, e.g. *triangulatus* Harkness, *Distomodus
kentuckyensis* Branson & Branson; (Cl) *Climacograptus* sp., (Cr) *Cruziana*; (O) orthoceratids; (S) shelly faunas and 'worm burrows'. For lithological legend, see Fig. 3.

In Saudi Arabia, west of the Great Nafud, the Silurian rocks reach a thickness of more than
650 metres. They are composed of shales and sandstones divided into the Sharawra and the
Tawil Formations by HELAL (1964, 1965). According to the latter, the Tabuk Formation of
BRAMKAMP and others (1963), BROWN and others (1963), POWERS and others (1966), and
POWERS (1968), originally comprising Ordovician to Devonian sediments, must be restricted to
the Ordovician System, and the Silurian rocks cannot be collectively treated together with other
Lower Palaeozoic sediments as the Tabuk Formation (Fig. 27).

HELAL divided the Sharawra Formation into two members, the *Climacograptus*–Orthoceratid Member at the base and the Sharawra Sandy Member at the top. The lower unit consists
of coloured sands and siltstones with sandy intercalations and overlies the Ordovician Upper
Tabuk Sandy Member unconformably. The richly fossiliferous *Climacograptus*–Orthoceratid
Member is about 135 metres thick. According to THRALLS and others (1956) and STEINEKE and

others (1958), the *Climacograptus*-bearing shales are lowermost Silurian in age. On the grounds of *Monograptus* sp., and *Rastrites* sp. collected in the Qusayba escarpment (POWERS, 1966, 1978), BERRY and BOUCOT (1972) considered the fossiliferous beds to be approximately middle Llandovery in age. The *Rastrites* beds of Qusayba scarp may well be the *Climacograptus–*Orthoceratid Member of Tabuk area as defined by HELAL. From Turaif water well in northernmost Saudi Arabia, HEMER (*in* BERRY and BOUCOT, 1972) described identifiable chitinozoans and acritarchs found in grey to brown, fissile, pyritic, carbonaceous shale with common plant scraps. HEMER stated that the chitinozoans and acritarchs from Turaif are well comparable with similar fossils from the Wenlock Shales. If so, the fossiliferous formation of Turaif may, perhaps, be correlative with part of the Sharawra Sandy Member as defined by HELAL (1964, 1965).

With gradational contact, the Sharawra Sandy Member overlies the *Climacograptus* Member forming the main body of the Sharawra plateau. The member is mainly composed of pastel-shaded, pink, green, and grey, fine- to coarse-grained sandstone showing wool-sack structures in the upper part and intercalations of red and greyish green shales. The Sharawra Sandy Member is about 320 metres thick; its upper boundary is everywhere marked by iron concentrations at the sharp and unconformable contact with the overlying Tawil Formation. Marine fossils are found only in the upper portions of the member. Bivalves and gastropods represent the dominant part of the fauna which inhabited a shallow environment corresponding to a neritic–littoral lithotope. In addition to the molluscs, tubular moulds, worm trails, and arthropod tracks were also noticed within the Sharawra Sandy Member, the age of which was assumed to be Silurian by HELAL because of the occurrence of lower Silurian rocks below it.

The name of the following Tawil Formation was first introduced by THRALLS and others (1956), who applied the term to Silurian sandstone of the obsolete Upper Uyun Group. HELAL (1964) raised the Tawil to formational status. According to his definition, the Tawil Formation is chiefly distinguished by its purplish red to black colour and by haematite-coated quartz pebbles and grains occurring in some horizons. At the type locality, the Tawil Formation consists of 180 metres of brown sandstone and minor amounts of siltstone and shale. The sandstones are haematitic, fine- to coarse-grained, cross-bedded, and arkosic. Partly very micaceous, quartz pebble zones are common. Shaly interbeds are commonly red. Iron concentrations are a conspicuous feature at the base and the top of the Tawil Formation. HELAL considered them as an indication of the change from transgressive to regressive conditions. The unconformity at the base of the Tawil is marked by pastel, light-coloured, thin-bedded sandstone of the uppermost Sharawra and by massive-bedded sandstone with iron concentrations and massive quartz pebble beds of the basal Tawil. Within the Tabuk area, the Tawil Formation is unconformably overlain by marly limestones of Upper Cretaceous and/or Lower Eocene age. In the Jauf area, however, the Tawil is superposed by the Devonian Jauf Formation. The contact is non-gradational and shows a marked unconformity. No diagnostic fossils were noticed in the Tawil Formation besides worm trails and arthropod tracks (*Cruziana*). An upper Silurian age is suggested for the Tawil Formation as it is overlain by the Jauf Formation which is of definite Lower Devonian age.

In southern Jordan, lithology of the Silurian rocks is generally similar to that in north-western Saudi Arabia. In the basal part of the Silurian sequence of Jordan, in comparison with the Tabuk area, a minor amount of shaly beds was noticed. Subdivision and fossil content were slightly different in the two regions. It was BENDER and HUCKRIEDE (1964) who recognized for the first time rocks of Silurian age in southern Jordan, these being part of the obsolete Nubian Sandstones (Section 3.2.1). BENDER (1964) and BENDER and HUCKRIEDE (1964) divided the Silurian into several units originally including also the *Sabellarifex* Sandstone. In addition to the latter, BENDER and HUCKRIEDE (1964) and BENDER (1968) distinguished the *Conularia*

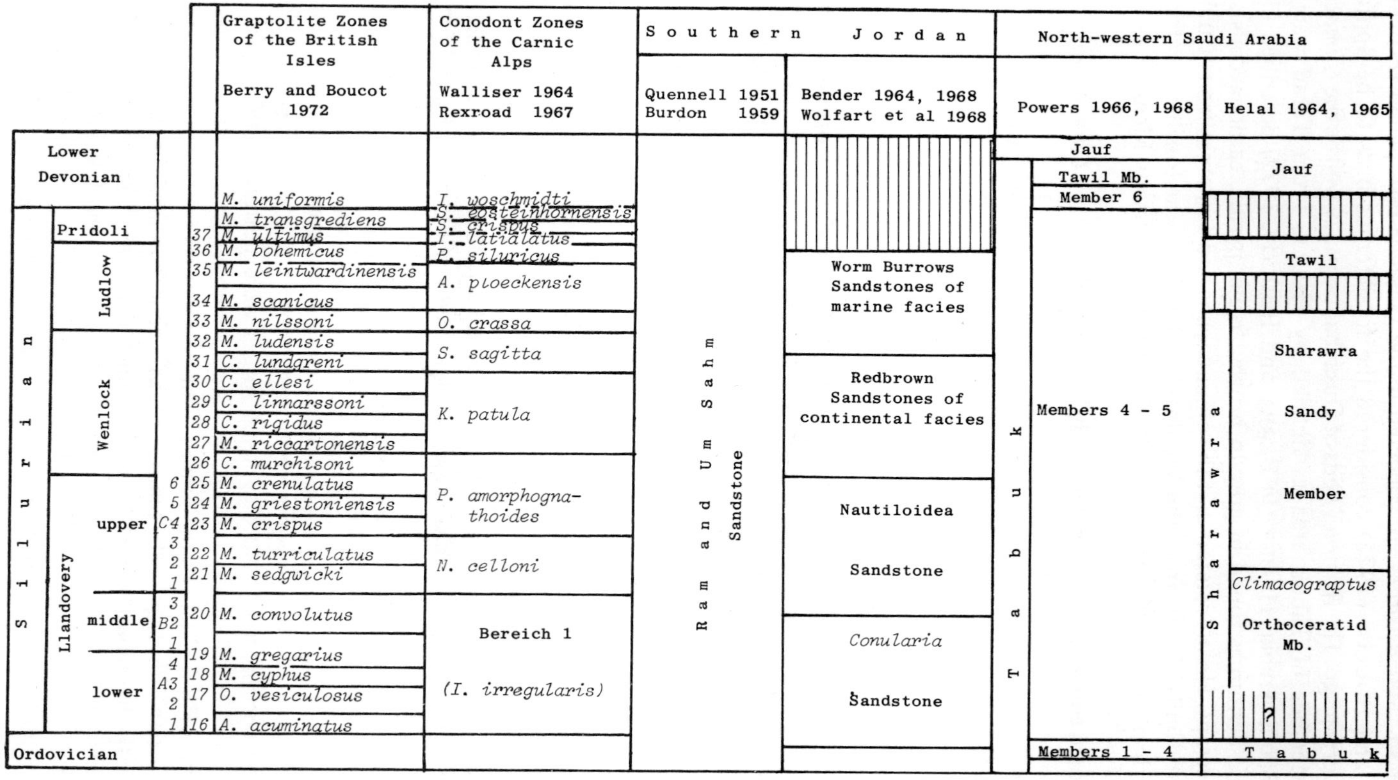

FIG. 27. Stratigraphical correlation of the Silurian rocks of southern Jordan and the Arabian Peninsula.

Sandstone and the Nautiloidea Sandstone, both of them nearly fully marine. The overlying Redbrown Sandstone BENDER ascribed to a continental origin, whereas the Worm Burrows Sandstone on top of the Silurian sequence is supposed to be marine. WOLFART and others (1968) placed the lower boundary of the Silurian somewhere in the upper part of the *Conularia* Sandstone (Section 3.2.1), below a faunal assemblage of definite late early to middle Llandovery age. Perhaps the lower Silurian boundary coincides with the rock salt horizon in the middle portion of the *Conularia* Sandstone. If so, a hiatus between the Ordovician and Silurian Systems might be present in southern Jordan, too, as well as in north-western Saudi Arabia (Section 3.2.1).

The clearly Silurian part of the *Conularia* Sandstone is composed of 30 metres of thin-banked, shaly micaceous fine sandstones which yielded the following faunal assemblage of late early to middle Llandovery age (*M. gregarius* to *M. convolutus* Zone), according to WOLFART and others (1968): '*Conularia*' sp., *Howellella* n. sp. A, Nautiloidea, *Climacograptus scalaris* cf. *scalaris* (closely related to *scalaris normalis* Lapworth, *Cl. innotatus jordaniensis* Stein, *Monograptus* (*Demirastrites*) sp. ex gr. *triangulatus* Harkness, *Distomodus kentuckyensis* Branson and Branson, and *Spathognathodus* sp. HELAL (1965) supposed the *Conularia* Sandstone to be a correlative unit of his Sharawra Sandy Member which seems rather to be an equivalent of the Nautiloidea Sandstone and the overlying Redbrown and Worm Burrows Sandstones (Figs. 26 and 27). Obviously, the upper graptolite-bearing part of the *Conularia* Sandstone is equivalent to part of the *Climacograptus*–Orthoceratid Member of the Tabuk area. BERRY and BOUCOT (1972) concluded from the ostensible occurrence of *Glyptograptus tamariscus* a middle–early late Llandovery age for the *Conularia* Sandstone. The identification of *G. tamariscus* turned out to be incorrect, and the upper part of the *Conularia* Sandstone is not younger than late early to middle Llandovery.

In southern Jordan, BENDER (1963) established the Nautiloidea Sandstone, exposures of which were noticed east of the Hejaz railway. The rocks dip slightly east-north-eastwards and form a cuesta landscape with chains of residual buttes. The lower boundary of the Nautiloidea Sandstone was placed approximately 100 metres above the salt-bearing horizon of the *Conularia* Sandstone, between beds bearing *Climacograptus* below and beds with numerous orthoceratids above. In the north, the Nautiloidea Sandstone is transgressively capped by Mesozoic sandstones; in the east, it is conformably overlain by the following unit of the Redbrown Sandstone. The total thickness of the Nautiloidea Sandstone is estimated to be about 130–160 metres. It is composed of an alternation of flaggy, manganese-dotted fine sandstones and limonitic, dark brown to red, micaceous, shaly fine sandstones. The monotonous sequence of thin-bedded sandstone containing horizons with ripple-marks and *Sabellarifex* is closely similar to the *Conularia* Sandstone. The lower 50 metres of the sequence yielded a faunal assemblage of late Ordovician type which must be attributed, however, to the middle to late Llandovery because it overlies the upper part of the *Conularia* Sandstone with its faunal assemblage of late early to middle Llandovery age. The assemblage comprises *Rostricellula*, *Bucanella*, Nautiloidea, numerous bivalves, three species of the trilobite *Brongniartella*, and numerous tracks. The upper 80 metres of the Nautiloidea Sandstone are unfossiliferous. As indicated by the facies and fossil content, the Nautiloidea Sandstone was deposited in a marine, littoral environment. HELAL (1965) proposed to correlate the Saudi Arabian Tawil Formation with the Nautiloidea Sandstone of southern Jordan. In the author's opinion, the Nautiloidea Sandstone is an equivalent of the upper part of the *Climacograptus*–Orthoceratid Member and of the lower part of the Sharawra Sandy Member in the Tabuk area.

In the eastern Tubeiq Mountains as far as the Saudi Arabian boundary, the fully marine Nautiloidea Sandstone is overlain by the Redbrown Shaly Sandstones which are 50–80 metres in thickness and unfossiliferous. BENDER (1968) interpreted the Redbrown Sandstones to be of continental origin. The top of the Silurian sequences is formed by a marine sequence of bedded, medium-grained sandstone—the Worm Burrows Sandstone—up to 200 metres thick, which contains numerous layers with vertically arranged worm burrows, besides *Arthrophycus*, *Rusophycus*, and *Onchus* (identified by HUCKRIEDE). Because of the presence of *Onchus*, the Worm Burrows Sandstone must be ascribed to a Ludlow age. It is transgressively overlain by chalk and chert of the Danian to Palaeocene. Considering the content of tracks and burrows and the succession of beds, the Worm Burrows Sandstone of southern Jordan must be correlated with the upper part of the Sharawra Sandy Member of north-western Saudi Arabia.

4.3. The Silurian rocks of Turkey and Syria

Silurian rocks are developed in two regions of Turkey and Syria, in the northern Anatolian geosyncline and in the mobile shelf region of southern Turkey and Syria surrounding the Arabian Shield on the north (Section 3.3). The two regions are separated by the intra-Mesozoic western and central Anatolian crystalline massifs (BRINKMANN, 1967). In the region of Bithynia, east of Istanbul, Silurian rocks and faunas are well known, especially by the works of PAECKELMANN (1925, 1938), PAECKELMANN and SIEVERTS (1932), WEISSERMEL (1939), and HAAS (1968). In southern Turkey and north-eastern Syria, Silurian rocks have been known since the work of FLÜGEL (1955), YALÇINLAR (1955–73), IMANDT (1962), and WEBER (1963).

In the Bosphorus region, it was PAECKELMANN (1938) who published the first extensive research work on Palaeozoic rocks. He attributed the lowermost conglomerates, arkoses, and shales to the upper Silurian. PAECKELMANN, himself, and SIEVERTS (1932) investigated the Silurian molluscs and brachiopods; WEISSERMEL (1939) studied the corals and stromatoporoids collected by PAECKELMANN. Much later, YALÇINLAR (1955a, 1956) reported the occurrence of *Monograptus*, but SAYAR (1962, 1964) considered YALÇINLAR's observation to be questionable concerning the locality and horizon of the fossils. Silurian–Devonian chitinozoans were described by TAUGOURDEAU and ABDUSSELAMOGLU (1962) from east of Istanbul. A first correlation of the Ordovician and Silurian formations in the surroundings of Istanbul was proposed by BAYKAL and KAYA (1965) on the basis of palaeontological and lithogical data. The entire sedimentary complex of the Palaeozoic was recently re-examined and subdivided by means of modern palaeontological knowledge by HAAS (1968b).

4.3.1. The north-western region of Turkey

HAAS (1968a) re-examined and named the various divisions of the Silurian rocks near the southern coast of the Bithynian Peninsula, comprising the upper part of the Bithynian Series and most of the overlying Akviran Series. The lower portion of the Bithynian Series (Kurtköy and part of the Ayazma Formation) has been assigned to the Ordovician, the upper portion (part of Ayazma and Yayalar Formation) has been attributed to the Llandovery. HAAS tentatively placed the Ordovician–Silurian boundary within the Ayazma Formation because the underlying Kurtköy yielded middle Ordovician and the overlying Yayalar upper Llandovery fossils. The Ayazma Formation is composed of unfossiliferous, thick-bedded, light red–brown to white quartzites with conglomeratic layers in the lower portion and of thin-bedded quartzites in the upper portion, showing ripple-marks and oblique bedding. Its thickness is estimated at 200 to 300 metres. All over Bithynia, the higher mountains are capped by the quartzites of the Ayazma Formation.

Most of the Llandovery is occupied by the Yayalar Formation which was introduced by HAAS (1968a). The Yayalar is composed of partly shaly greywackes, arkosic sandstone, and red shales. Regionally, the formation is capped by a chamosite horizon grading into the limestones of the

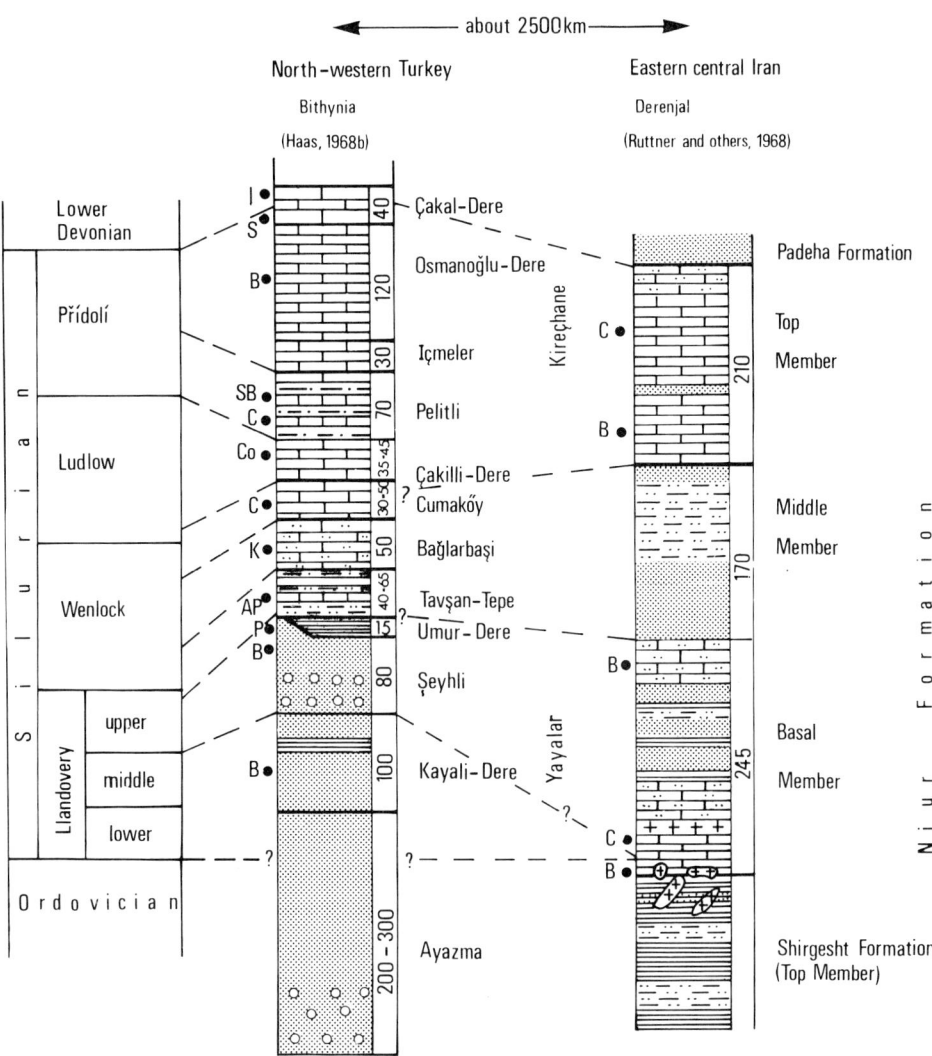

FIG. 28. Representative columnar sections and stratigraphical terminology for the Silurian
rocks of north-western Turkey and eastern central Iran. (A) *Amphistrophia*; (B)
brachiopods, corals; (C) corals; (Co) *Conchidium*, *Kockelella*; (I) *Icriodus woschmidti*
Ziegler and other conodonts; (K) *Kirkidium*; (P) *Pterospathodus amorphognathoides*
Walliser and other conodonts, brachiopods; (S) *Spathognathodus steinhornensis eosteinhornensis* Walliser. For lithological legend, see Fig. 3.

overlying Akviran Series. The Yayalar Formation is divided into three subdivisions, Kayali-
Dere Member, Şeyhli Member, and Umur-Dere Member.

The Ayazma Formation is either overlain or locally, in its upper part, interfingered with the
Kayali-Dere Member which is composed of very dense, greenish greywackes becoming grey–
brown if weathered. Large feldspar crystals and chamosite ooliths occur in single horizons. The
thickness of the Kayali-Dere is estimated at 100 metres. Locally, the greywackes are replaced by
black, sandy shales or by arkose and red shales. The fauna contains the brachiopods *Platystrophia*, *Plectatrypa*, *Cryptothyrella*, *Meristina*, *Leangella*, *Leptaena*, the ostracod *Craspedobolbina*,

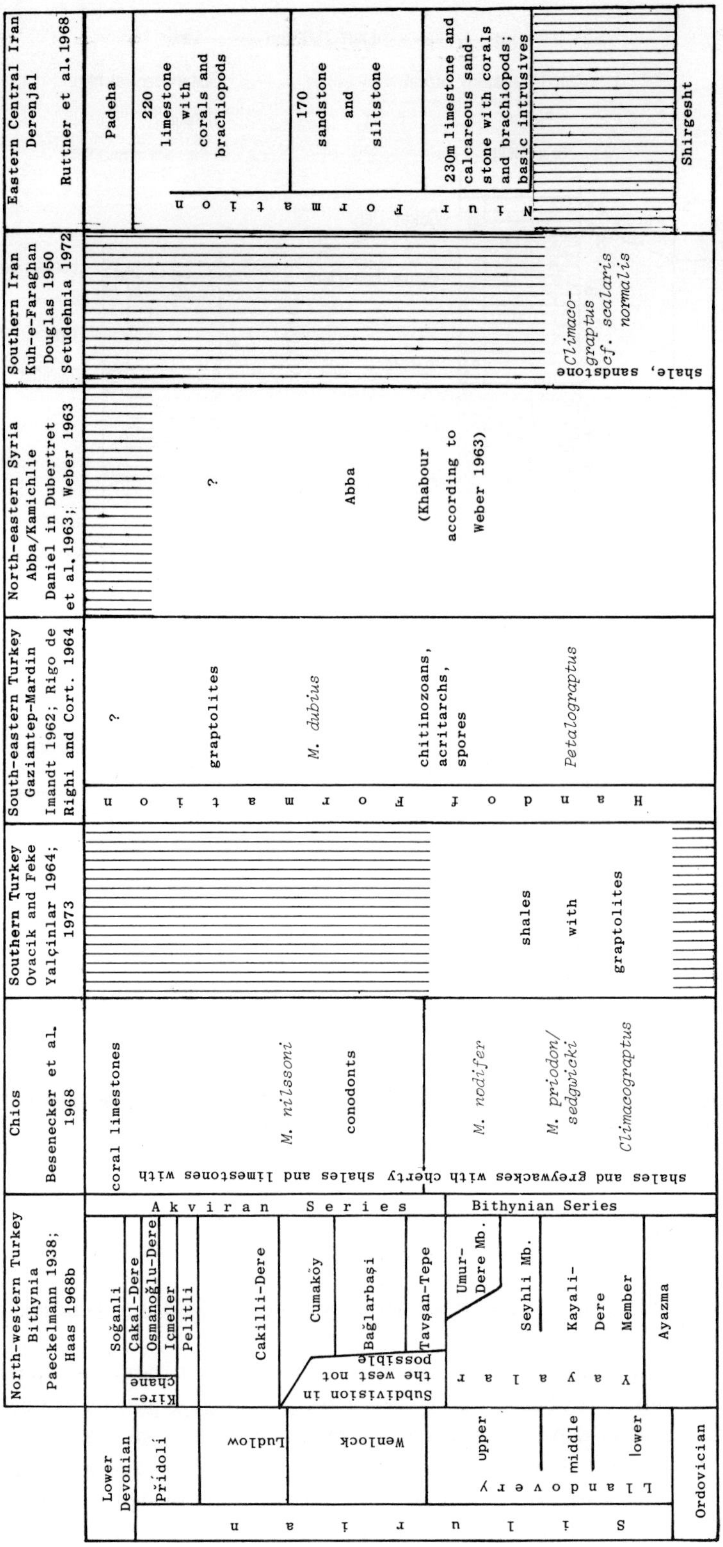

FIG. 29. Stratigraphical correlation of the Silurian rocks of Turkey and Iran.

and the tabulate coral *Halysites*, in addition to solitary rugose corals. The fauna indicates an upper Llandovery C2 age for the Kayali-Dere.

A clearly defined sequence about 80 metres in thickness was named the Şeyhli Member by HAAS. Lithologically sharply separated from the rocks of the Kayali-Dere Member, the Şeyhli Member begins with very coarse, light-coloured arkoses with well rounded quartz pebbles of 5 millimetres diameter and feldspar crystals. Upwards, the sequence passes into light-coloured, well bedded, cherty feldspathic sandstones, locally with minor content of feldspar. Beds filled with large shells of brachiopods and solitary corals are intercalated in the upper part of the member. A form of *Stricklandia lens* (Sowerby) intermediate between *S. lens progressa* Williams and *S. lens ultima* Williams, *Meristina* cf. *furcata* (Sowerby), and *Pentamerus* sp. aff. *oblongus* (Sowerby) indicate a late Llandovery C2–C3 age. In the eastern part of the region, where the Yayalar cannot be subdivided, intercalations of coarse feldspathic conglomerates were noticed, which are considered to be possible equivalents of the Şeyhli Member. This sequence does not bear any fossils.

Distribution of the Umur-Dere Member is restricted to the eastern part of Bithynia. The member is 15 metres thick and wedges out westwards. It consists of violet or rarely of green shales which are rich in chamosite ooliths. Upwards, the sequence grows increasingly calcareous. The fauna is very rich in brachiopods, predominantly consisting of small, roundish atrypids. *Cyrtia exporrecta* (Wahlenberg) and *Eospirifer* cf. *radiatus* (Sowerby) show that the age is not older than late Llandovery (C3) according to BERRY and BOUCOT (1972). A very rich conodont fauna belongs to the *amorphognathoides* Zone (*Carniodus carinthiacus* Walliser, *Hadrognathus staurognathoides* W., *Neoprioniodus costatus costatus* W., *Ozarkodina gaertneri* W., *Pterospathodus amorphognathoides* W., and *Roundya caudata* W.).

The chiefly coloured, clastic sequence of the Bithynian Series is overlain by about 400 metres of limestones and marls reaching from perhaps the higher upper Llandovery to the lower Gedinnian. In the eastern Bithynian region the lower parts of the Akviran Series can be well subdivided, but in the west a subdivision is not possible. In this region, the lower Akviran is marked by reduced thickness and by variable lithological development. The base is composed of 3–5 metres of greenish to yellowish brown, marly limestones rich in tabulate corals (*Halysites*). Locally, these rocks are interfingered with calcareous sandstones and intercalations of cross-bedded quartz conglomerates. The marly limestones are overlain by grey crystalline limestones growing coarsely crystalline towards the top. The lower Akviran yielded a conodont fauna of the upper part of the *amorphognathoides* Zone. Additionally, the age of the lower Akviran follows from its position between the Yayalar and the Çakilli-Dere Formation. The higher parts of the Akviran section have about the same appearance all over the area.

In the eastern Bithynian region, HAAS (1968a) established the Tavşan-Tepe Formation as the basal unit of the Akviran Series, the thickness of which ranges between 40 and 65 metres. Clearly separable from the Umur-Dere Member, the Tavşan-Tepe Formation begins with greenish–yellowish, marly limestones lacking chamositic ooliths and bearing only a few brachiopods similar to those of the Umur-Dere Member. Upwards the succession grows increasingly calcareous. In the upper part, grey limestones predominate locally containing *Stromatopora*. The upper boundary of the unit is sharply marked by the beginning of bedded, fine sandy limestones. *Cyrtia* and *Eospirifer* are lacking in the Tavşan-Tepe Formation. If there is really a true *Amphistrophia*, the age of at least part of the Tavşan-Tepe Formation would not be older than Wenlock. *Flexicalymene* (*F.*) sp. and *Encrinurus brevispinosus* Haas are widely distributed in the Umur-Dene Member as well as in the Tavşan-Tepe Formation. Conodonts belonging to the *amorphognathoides* Zone indicate that part of the unit is of late Llandovery age.

The Tavşan-Tepe Formation is conformably overlain by the Bağlarbaşi Fomation, which is about 80 metres in thickness. In an unweathered state, it consists of grey–green, fine sandy limestones locally grading into calcareous sandstones. If weathered, the rocks are red–brown. Additionally, impure crystalline limestones are intercalated. Thick beds are well stratified, but cross-bedding was also noticed. The upper boundary of the formation is marked by the occurrence of chiefly red, coarse crinoidal limestones. Some beds contain many shells of pentamerids including the genus *Kirkidium*. As the underlying beds belong to the *amorphognathoides* Zone, they cannot be younger than lowermost Wenlock. The Bağlarbaşi Formation must represent, therefore, a large part of the Wenlock, although *Kirkidium* does not appear in other parts of the world before late Wenlock times. In the western Bithynian region, the Bağlarbaşi Formation cannot be differentiated.

The Cumaköy Formation, between 30 and 50 metres in thickness, follows sharply above the Bağlarbaşi. It consists of light red, coarse, thick-bedded crinoidal limestones with the upper boundary placed at the base of the grey nodular limestones of the overlying unit. The Cumaköy is very rich in rugose and tabulate corals and in stromatoporoids. Locally, near the base, a bed is filled with many shells of *Chonetes* and *Howellella crispa* (Hisinger). Up to the present, however, no zonable fossils have been studied. From its position within the section, the age of the Cumaköy may be upper Wenlock or lower Ludlow.

The Cumaköy is conformably overlain by the Çakilli-Dere Formation represented by grey, chiefly irregularly bedded crystalline limestones between 35 and 45 metres in thickness. At the base, interbeds of whitish shales with chert nodules were noticed. The upper boundary was placed on top of the thickly bedded limestones containing the last specimens of *Conchidium* and *Halysites*. *Conchidium* cf. *pseudoknighti* (Černyčev) richly occurring everywhere and the conodont *Kockelella* from the base of the Çakilli-Dere indicate an early Ludlow age. *Polygnathoides siluricus* Branson and Mehl from the top of the Çakilli-Dere is indicative of the late Ludlow.

The Pelitli Formation emerges lithologically from the Çakilli-Dere Formation. It is composed of 70 metres of unevenly bedded limestones, nodular limestones, marly limestones, and marls. HAAS drew the upper boundary at the base of the black limestones of the overlying Içmeler Member and subdivided the Pelitli into a lower and an upper part without a very distinct boundary between them. The grey marly limestones of the Lower Pelitli are 30 metres thick and very rich in rugose and tabulate corals (*Favosites*, *Syringopora*, *Heliolites*, etc.). Brachiopods are very rare. There are no datable faunal elements, but the position of the Lower Pelitli above the *siluricus* Zone and below the *eosteinhornensis* Zone (=base of the Upper Pelitli) suggests that the Lower Pelitli represents more or less the *latialatus* and *crispus* Zones of the Přídolí. Without a sharp boundary, the marly limestones of the Upper Pelitli overlie the lower portion of the formation. Upwards, the rocks pass into nodular limestones locally terminated by grey–blue, well bedded, shaly marls. The thickness of the Upper Pelitli ranges between 30 and 40 metres. A rich brachiopod fauna contains *Dayia navicula* (Sowerby), *Platyorthis cimex* (Kozlowski), *Howellella* cf. *nucula* (Barrande), and *Quadrifarius* sp. indicating a Přídolí age. The conodont assemblage belongs to the *eosteinhornensis* Zone (*Spathognathodus steinhornensis eosteinhornensis* Walliser, *Sp. primus* (Branson and Mehl), *Sp. inclinatus* (Rhodes), *Ozarkodina denckmanni* Ziegler, and *Trichonodella* sp.). Near the base of the Upper Pelitli, *Spathognathodus crispus* Walliser was found besides the species cited above. The trilobites *Proetus* (*Lacunoporaspis*) *barrangus* (Haas), *Calymene arotia* Haas, and *Kosovopeltis crebristriata* (Lindström) seem to be restricted to the Upper Pelitli Formation having their closest relatives in the higher Silurian of the Baltic region.

The uppermost unit of the Bithynian Silurian sequence was named the Kireçhane Formation by HAAS, with the Silurian/Devonian boundary situated within its upper part. The unit comprises about 200 metres of mostly evenly bedded, black limestones beginning with laminated limestones and ending with thin platy flaser limestones. It can be subdivided into the three members: Içmeler, Osmanoğlu-Dere, and Çakal-Dere. All of the Içmeler Member consists of about 30 metres of laminated, black limestones with interbeds of light grey or reddish sheets of argillaceous material. There is no fauna permitting age determination. From the stratigraphical position between beds with *eosteinhornensis* Zone conodonts there follows a Přídolí age. With a sharp lithological boundary, the Içmeler Member is overlain by the well bedded, black, strongly bituminous limestone of the Osmanoğlu-Dere Member which is estimated to be 120 metres in thickness. The member yielded a poor fauna of brachiopods, tabulates, and stromatoporoids (*Howellella nucula* (Barrande), '*Rhynchonella*' *endrissi* Paeckelmann, *Syringopora* sp., and *Bollia* sp., large forms of *Leperditia*). The conodont faunas from the basal, middle, and upper part of the member contain *Spathognathodus steinhornensis eosteinhornensis* Walliser indicating the *eosteinhornensis* Zone of the late Přídolí. The Silurian/Devonian boundary lies approximately in the lower third of the Çakal-Dere Member about 40 metres in thickness. This member consists of black and grey flaser limestones the boundaries of which cannot be sharply defined. Besides single rugose and tabulate corals and *Atrypa reticularis* (Linné), the Çakal-Dere contains a rich conodont fauna by means of which the Silurian/Devonian boundary can be drawn fairly sharply. The lower beds yielded *Sp. steinhornensis eosteinhornenis*, the upper 28 metres containing conodonts of the *woschmidti* Zone: *Icriodus woschmidti* Ziegler, *Ozarkodina denckmanni* Ziegler, *Spathognathodus stein- hornensis remscheidensis* Ziegler, *Sp. wurmi* Bischoff and Sannemann, and *Trichonodella inconstans* Walliser.

From the Isle of Sedef Adasi (=Antirovitha) situated in the Marmora Sea neighbouring the Bithynian Peninsula, PAECKELMANN (1938) had already described Silurian rocks the lowest part of which were classified by HAAS (1968b) as Pelitli Formation by means of *Calymene arotia*. Earlier, ÜNSALANER-KIRAGLI (1958) found *Alveolites lemniscus* Smith in the same beds. The Silurian rocks of Antirovitha are exposed as far as the upper boundary of the Osmanoğlu-Dere Member.

A section similar to that described by HAAS east of Istanbul was noticed by HÖLL (1966) in the middle part of the Bithynian Peninsula. In that region, the Ordovician rocks consisting of conglomerates, greywackes, shales, etc., are unconformably overlain by limestones with inter- beds of conglomerates and sandstones about 135 metres in thickness. A rich faunal assemblage of rugose and tabulate corals, stromatoporoids, and some brachiopods (*Wilsonella wilsoni* (Sowerby), *Rhynchospirina baylei* (Davidson), *Eospirifer* cf. *insignis* Hedström, and *Howellella angustiplicatus* Kozlowski) indicates a late Silurian age. The Silurian rocks are conformably overlain by limestones of Gedinnian age. From south of Bilecik, south-east of the Gulf of Izmit, YALÇINLAR (1957b) described shales with intercalations of limestones containing 'des formes de *Monograptus*'.

EGEMEN (1947) published the only description of fossiliferous Silurian rocks from the Pontus region. Along the Güluc river, in the vicinity of Eregli, red–brown, sandy and marly shales are exposed, bearing the following faunal assemblage: *Orbiculoidea* sp aff. *circe* Billings, *Schell- wienella* cf. *pencki* Paeckelmann, *Stropheodonta* sp. aff. *ivanensis* Barrande, cf. *Trimerus*, cf. *Beyrichia*, *Monograptus* cf. *armoricanus* Philipot, *M.* cf. *miloni* Philipot, and *M.* cf. *dubius* Suess. EGEMEN compared the lithological facies with the Silurian rocks of Antirovitha. South of Eregli, TOKAY (1952) noticed unfossiliferous, coloured, quartzitic sandstones, ('Sandstones of

Hamzafakili'), below richly fossiliferous Lower Devonian shales. TOKAY assigned a Silurian age to the coloured rocks. According to the geological map (1 : 500,000, sheet Zonguldak) Silurian rocks are widely distributed in the western Pontus region, but HAUDE (1969) stated that there are no published descriptions. From the eastern Pontus region, no Lower Palaeozoic rocks are known.

From the Istranca Mountains, European part of Turkey, YALÇINLAR (1957a) described an alternation of well bedded, dark grey, cherty shales and quartzites with corals, brachiopods, and graptolites (*Monograptus*, ?*Climacograptus*). HAUDE (1969) held the identifications of YALÇINLAR to be doubtful because more recent investigations have not as yet confirmed his results.

Silurian rocks are also known from the closely neighbouring regions of the Peninsula of Karaburun, western Turkey (KTENAS, 1925, HÖLL, 1966) and the Greek Isle of Chios (KTENAS, 1925; PAECKELMANN, 1939; KAUFFMANN, 1965; BESENECKER and others, 1968; HERGET and ROTH, 1968). Both of the regions are, therefore, included in the present description. In the Aegaeis region, fossiliferous Silurian rocks are only rarely exposed. The Isle of Chios is of great importance to knowledge of the Silurian rocks, the comparatively large occurrences of which belong to an autochthonous complex which is locally overlain by allochthonous rocks. The earlier efforts by TELLER (1889), PHILIPPSON (1911), and MARAVELAKIS (1915–16) as well as KTENAS (1925) and PAECKELMANN (1939) to subdivide the Palaeozoic rocks were insufficiently based on fossils, as demonstrated by BESENECKER and others (1968). Lithologically, all of the Palaeozoic sequence of Chios is rather uniform and strongly folded and broken by several orogenies. Uninterrupted stratigraphical sections are not present, therefore, and the following descriptions are based on scattered localities.

The oldest rocks of Chios dated by fossils are Llandovery in age. They consist of bedded, red–brown or light grey weathering, green and dark cherty shales rarely bearing graptolites and conodonts. The thickness of the shales may reach more than 100 metres. According to BESENECKER and others (1968), the fossiliferous shales and limestones of the Silurian are intercalated in a sequence of greywackes, shales, and, occasionally, conglomerate (=Keramos–Aman Formation of HÖLL, 1966). The entire non-metamorphic complex rests with a tectonic contact on an epimetamorphic alternation of greywackes, sandstones, and shales, the age of which cannot be determined. The following graptolite zones (according to ELLES and WOOD, 1913) proved to be present at three localities in the north-eastern part of Chios (JAEGER *in* KAUFFMANN, 1965; HERGET and ROTH, 1968):

Zone 16–19 based upon *Climacograptus* sp. ex gr. *scalaris* (Hisinger); (probably it is zone 16–17 because monograptids are entirely lacking. Latest Ordovician cannot, however, be excluded).

Zone 19–22 based upon *Rastrites*, triangulate monograptids, and *Diplograptus* (*Petalograptus*) cf. *palmeus* (Barrande); (probably it is zone 21–22 if *Monograptus priodon* (Bronn) and *M. sedgwicki* Portlock are taken into consideration). Another exposure of contemporaneous beds yielded conodonts of the middle to late *celloni* Zone (*Neoprioniodus subcarnus* Walliser, *Carniodus carinthiacus* Walliser, *C.* cf. *carnulus* Walliser, and *Spathognathodus pennatus pennatus* Walliser).

Zone 22–25 probably 24, based upon *Monograptus nodifer* Törnquist, *M. galaensis* Lapworth, *Diplograptus* (*Petalograptus*) *palmeus tenuis* (Barrande), and other species such as *M. priodon* cf. *pandus* Lapworth, *M. veles* (Richter) ?, *M.* ex grege *spiralis* Geinitz, and *M.* e.g. *nudus* Lapworth.

In northern Chios, upper Silurian graptolitic shales of the same lithological nature as the Llandovery rocks have been noticed. *Monograptus nilssoni* Lapworth, *M. colonus* (Barrande),

M. roemeri (Barrande), *M. dubius frequens* (Jaekel), *M. bohemicus* (Barrande), *M.* cf. *micropomus* (Jaekel), and *M.* cf. *haupti* Kühne indicate zone 33 (early Ludlow). At another locality, the shales yielded *Monograptus* cf. *dubius* (Suess), considered to be Wenlock to late Ludlow in age. In addition to cherty shales, the first conodont-bearing limestones are already present in the Wenlock. The dark grey, sandy limestones alternate with black cherty shales containing conodonts of the *patula* Zone, middle Wenlock, such as *Neoprioniodus multiformis* Walliser and *Trichonodella excavata* (Branson and Mehl). In the northern Chios region, limestone lenses of the Ludlow and Lower Devonian are much more frequent than such of the Wenlock. According to KAUFFMANN (1965), light to blue–grey detrital lenses of limestones yielded conodonts of the: (1) *crassa* Zone, early Ludlow, with *Ozarkodina crassa* Walliser, *Spathognathodus inclinatus inclinatus* (Rhodes), etc.; (2) *siluricus* Zone, middle to early late Ludlow, with *Polygnathoides siluricus* Branson and Mehl and many other conodonts, in addition to orthocerids, *Cardiola*, echinoderms, and *Harpes* sp. aff. *crassifrons* Barrande; (3) *eosteinhornensis* Zone, latest late Ludlow or Přídolí, respectively.

In the northern Chios, the upper Silurian is also represented by the limestone of Agrelopos which is 0–100 metres in thickness and unconformably overlain by Carboniferous rocks. This consists of dense, chiefly massive, rarely thick-bedded, frequently coarsely crystalline carbonates with grey–brown to red–brown marly intercalations. Fossils are rare and badly preserved. The faunal assemblage is mostly composed of tabulate and rugose corals and some brachiopods (*Thamnopora*, *Favosites*, *Heliolites*, stromatoporoids, *Entelophyllum*, *Cyathophylloides*, *Arachnophyllum*, etc.). According to WEISSERMEL (1938), WELLNHÖFER *in* HÖLL (1966), and HERGET and ROTH (1968), the Agrelopos fauna must be Ludlow in age. Some of the coral species, for example *Thamnopora reticulata minor* Weissermel and *Cyathophylloides* sp. aff. *manipulatus* (Počta), indicate, however, a middle and early Devonian age respectively. It seems to be possible, therefore, that the Agrelopos limestone ranges from the Ludlow into the Lower Devonian.

The Lower Palaeozoic rocks of the Peninsula of Karaburun, east of Chios, have been studied by PHILIPPSON (1911), KTANAS (1925), and HÖLL (1966), who stated close relations between the Lower Palaeozoic rocks east of Istanbul and of Karaburun. HÖLL considered the Denizgiren Formation to comprise the oldest rocks of Karaburun, consisting of nearly 2000 metres of anchimetamorphic conglomerates, greywackes, sandstones, shales, and chert with rare interbeds of limestone lenses. Fossils are very rare in the Denizgiren; crinoids and the unconformably overlying upper Silurian limestone of Kaleçik indicate an Ordovician to early or middle Silurian age. KTENAS (1925) compared the limestone of Kaleçik with the limestone of Agrelopos, unconformably overlapping the folded rocks below. The limestone of Kaleçik is up to 80 metres in thickness, frequently beginning with a thin basal conglomerate. Concerning facies and fossil content, the Kaleçik Limestone is closely related to the Agrelopos Limestone. The corals are known from Chios or from Antirovitha, the brachiopods (*Platyorthis* cf. *cimex* (Kozlowski), *Camarotoechia* sp. aff. *bieniaszi* Kozlowski) from the Ludlow of Podolia. The Kaleçik Limestone is conformably overlain by 80 metres of shales, greywackes, and thin interbeds of carbonates, besides tuffites, lavas, and agglomerates. Several submarine effusions have been differentiated by HÖLL (1966) being connected with cinnabar mineralization (Section 4.8). These rocks, too, were considered to be late Silurian in age by HÖLL.

4.3.2. Southern Turkey and north-eastern Syria

From all over the Middle East, FLÜGEL (1955) and YALÇINLAR (1955b) were the first to describe Silurian rocks, with the exception of northern Turkey. Both of them studied Silurian rocks in the Antitaurus Mountains. According to YALÇINLAR (1955b, 1964), the Silurian succession of the region between Feke and Saimbeyli is composed of grey shales with graptolites at the base; quartzites, limestones, and shales with brachiopods, orthocerids, and trilobites in the middle part; and white crinoidal limestone and partly sandy limestone with brachiopods and corals in the top. According to the identifications of BERRY (*in* BERRY and BOUCOT, 1972), the basal shales yielded *Climacograptus medius* Törnquist, *C. scalaris* ? Hisinger, *Dimorphograptus* ? sp., *Diplograptus* cf. *thuringicus* Eisel, *Petalograptus* cf. *palmeus* Barrande, *Rastrites* cf. *approximatus* Perner, *Monograptus gregarius* Lapworth, and *M. millepeda* M'Coy. These species are indicative of a middle Llandovery zone 19 age. In addition to the graptolites, *Encrinurus* and *Plectodonta* were found in the Feke region.

From the Antitaurus region south of Pinarbaşi, east of Kayseri, FLÜGEL (1955) reported a Silurian fauna from black, platy, cherty shales locally full of ochreous particles. The faunal assemblage is rich in graptolite species considered to be late Llandovery in age (zone 22 of *M. turriculatus*). FLÜGEL identified the following species: *Petalolithus elongatus elongatus*? Bouček and Pribyl, *P. elongatus linearis* Bouček and Pribyl, *Monograptus lobiferus bulgaricus* Haberfelner, *M. halli* ? (Barrande), *M. (Streptograptus) admirandus* Bouček and Pribyl, *M. (S.)* cf. *nodifer* (Törnquist), *Spirograptus planus* (Barrande), *S. turriculatus minor* Bouček, *Rastrites linnaei* Barrande, *R. carnicus* Seelmeier, and *R. longispinus nafizi* Flügel. There are faunal relations with England and Bohemia. The graptolite-bearing shales are overlain by brown and ochreous, lumpy limestones with orthocerids, which were assigned a late Silurian age by FLÜGEL (1964) and VACHE (1966). The latter noticed conformably following Lower Devonian rocks of conglomeratic–sandy facies.

South of the Silurian exposures of the Antitaurus Mountains, on the eastern slope of the Amanos Mountains, LAHNER (1972) studied a sedimentary succession about 700–900 metres thick which has been named the Dedeler Formation. Its basal member overlies the upper Ordovician Kizlaç Formation and consists of 50 metres of quartzites and conglomerates. Thick-bedded, fine-graded, violet quartzites and shales form the basal part of the member which is 20 metres thick. The conglomerate part is of variable thickness. Its clasts consist chiefly of grey–whitish and violet quartzites up to 10 centimetres in diameter. Locally, a green, quartzitic sandstone with fossil plants is intercalated. From the basal portion of the Dedeler Formation, LAHNER obtained numerous brachiopods (*Eostrophonella, Resserella, Protomegastrophia, Eostropheodonta, and Mendacella*) attributed by DÜRKOOP to the Llandovery. The basal quartzites and conglomerates are overlain by a violet sequence 265 metres thick. It is composed of soft siltites and shales of intensive red–violet colours with interbeds of quartzitic and limonitic sandstone. Light, brownish weathering sandstones with occasional interbeds of grey–green siltites and shales 270 metres thick form the succeeding light sequence. The Dedeler Formation is terminated by a violet quartzitic sequence 120 metres thick, frequently capping the mountains and forming cuestas. From ripple-marks, fossil plants, and conglomerates, LAHNER concluded that the Silurian sediments were deposited in a shallow marine environment affected by fluviatile sedimentational factors. Except for the base, fossils are rare in the Dedeler Formation.

Southwards, the upper Ordovician Kizlaç and the Silurian Dedeler Formation pass laterally into the Fevzipaşa Formation (LAHNER, 1972), which consists of a uniform shale sequence with olive-grey and brown colours. Occasionally, bedded quartzites, sandstones, and siltites are intercalated as well as thin pebble layers which were locally noticed. The thickness of this formation has been estimated at 800 metres. Being unfossiliferous, the Fevzipaşa Formation could only be assigned to the upper Ordovician/Silurian on the basis of its lateral substitution by the Kizlaç and Dedeler Formations. The Fevzipaşa is overlain by the Kirtas Quartzite about 15 to 20 metres in thickness. It is thick-bedded, impure, and grey–green to brown in colour with interbeds of shales and siltites. Laterally it passes into the violet quartzites of the Dedeler

Formation and must therefore be attributed to the uppermost Silurian, possibly ranging into the Lower Devonian.

In the western part of southern Turkey, there are two more exposures of Silurian rocks in the region between the Antitaurus Mountains and Chios. In the Babadağ Mountain area, a few dendroid graptolites were collected from shales and quartzites by YALÇINLAR (1963). The dendroid fragments were indentified by BULMAN as species of *Dictyonema* and *Acanthograptus* which are suggestive of a Silurian age. On the basis of the results of NEBERT and RONNER (1956) and own field investigations, FLÜGEL (1964) and HAUDE (1969), however, considered it to be doubtful whether there are fossiliferous rocks in the Babadağ area at all. The rocks of this area are composed of an epizonally metamorphosed sequence of actinolitic chlorite schists and sericitic quartz-slates. In the Mediterranean Taurus region of Anamur–Ovacik–Silifke, YALÇINLAR (1964, 1969, 1973) studied a richly fossiliferous Lower Palaeozoic sequence comprising ?Upper Cambrian, Ordovician, and Silurian rocks. The Silurian succession is composed of four units overlying conglomerates and sandstones of the Ordovician. The lowest unit consists of shales and sandstones with brachiopods and graptolites whereas the next unit is formed of graptolitic shales only. Both of them are reported to be of early Silurian age. The succession continues with shales and sandstones bearing brachiopods and orthocerids; it terminates with cherty shales and interbeds of black limestones. Both of them are said to be Silurian in age. The Silurian rocks are overlain by Devonian limestones and shales. YALÇINLAR collected several graptolite assemblages without assigning them to the units referred to above. Assemblages from the vicinity of Ovacik contain *Monograptus*, *Climacograptus*, and *Rastrites* and are attributed to the pre-late Llandovery by BERRY (*in* BERRY and BOUCOUT, 1972). *Orthograptus vesiculosus* ? Nicholson and *Monograptus* sp. of the *M. cyphus* Lapworth type indicate an age in the span of zones 17 to 18. YALÇINLAR (1969) reported on two more graptolite collections from the Ovacik region identified by BERRY. The first one is composed of *Climacograptus* sp., cf. *Petalograptus palmeus tenuis* (Barrande), *Monograptus* cf. *decipiens* Törnquist, *M.* cf. *spiralis* (Geinitz), *M.* cf. *jaculum* (Lapworth), *M.* cf. *clingani* (Carruthers), and *M.* cf. *sedgwicki* (Portlock), which indicate a zone 19 to 21 range. The second collection contains *Climacograptus* cf. *scalaris* (Hisinger), *Petalograptus* sp., *Rastrites* sp., *Monograptus intermedius* (Carruthers), *M.* cf. *millepeda* (M'Coy), and *M. denticulatus* Törnquist. BERRY assigned these forms a zone 19 to 22 range.

Silurian rocks in south-eastern Turkey, east of the Antitaurus and Amanos Mountains, have been included in the Handof Formation by RIGO DE RIGHI and CORTESINI (1964), who considered this to range in age from the Ordovician into the Devonian. According to SHEPARD (*in* BERRY and BOUCOT, 1972), acritarchs, spores, and chitinozoans from shales and sandstones of the Hindof Formation encountered in wells near Mardin indicate a late Llandovery to Wenlock age. Shepard also found graptolites to be suggestive of a late Silurian age. JAEGER (*in* BERRY and BOUCOT, 1972) identified *Monograptus dubius* Suess and *M.* cf. *flemingii* Salter from the Handof Formation, which are indicative of a late Wenlock age. Shales encountered in the well Aril 1, south-east of Gaziantep, yielded *Petalograptus* sp., which is suggestive of a middle Llandovery age (IMANDT, 1962). TEMPLE and PERRY (1962) mentioned dark graptolite-bearing shales with siltstone interbeds of Silurian age from wells and surface sections in the vicinity of Diyarbakir.

In Syria, Silurian rocks are not exposed. They have only been encountered by some boreholes, such as Abba 1, Kamichlie, and El Bouab 1 (DANIEL, 1963; WEBER, 1963; WOLFART, 1967a). The Silurian rocks are part of the Abba Group (DANIEL, 1963), consisting of a shale succession with interbeds of fine sandstones ranging in age at least from the Ordovician into Early Devonian. In the borehole Abba 1, the entire Abba Group is more than 353 metres thick; in

Kamichlie, the thickness of the Silurian beds amounts only to nearly 500 metres. They yielded chitinozoans, hystrichosphaerids, and scolecodonts. MÄDLER (*in* WEBER, 1963) identified the genera *Veryhachium, Micrhistridium,* and *Hystrichosphaeridium* and considered the presence of highly developed chitinozoans to be indicative of a Silurian age of the shaly–sandy succession. According to WOLFART (1967b) and BERRY and BOUCOT (1972), equivalents of the Abba Group are widespread through much of Syria, Jordan, and Saudi Arabia. DANIEL reported rocks of the Abba Group type to be encountered also in a borehole in western Iraq.

4.4. The Silurian rocks of Iran

Outcrops of Silurian rocks are restricted to the north-eastern, east-central, and south-eastern part of Iran bordered eastwards by the Lut block and towards the north-west and south by possible Silurian highs. In the regions of the suspected highs, the Cambro–Ordovician beds are directly overlain by Devonian or younger rocks. South-westwards, a connection between the Iranian and the Arabian areas of Silurian sedimentation might have been possible. The thickest and most complete succession of Iranian Silurian rocks has been encountered in the northern Tabas area. In the eastern Alborz Mountains, middle Silurian deposits have been found in the Robat-e-Qarabil inlier described by BRICE and others (1973) and COCKS (1979). The third occurrence of rocks the Silurian age of which is proven by fossils is situated in southern Iran, north of Bandar Abbas. Rocks of suspected but uncertain Silurian age occur near Zarand, north-west of Kerman, and also near Anarak in central Iran and south of Kuh-e-Bazman in Baluchestan.

4.4.1. Derenjal and Ozbakh Kuh Mountains

FLÜGEL and RUTTNER (1962) were the first to publish studies on the Silurian rocks of the Ozbakh Kuh Mountains, eastern central Iran, and FLÜGEL investigated corals of Llandovery/Wenlock age. However, is was RUTTNER, NABAVI, and HAJIAN (1968) who carried out the most important research work on the Silurian of Iran. They introduced the name Niur Formation to distinguish a mainly carbonate unit in the Ozbakh Kuh Mountains, containing a rich Silurian coral and brachiopod fauna, from the underlying sandy–calcareous Shirgesht Formation and the overlying sandy–dolomitic Padeha Formation. The Niur Formation forms the lower part of the Gushkamar Group the upper part of which is represented by the Lower Devonian Padeha Formation.

About 60 kilometres south of the type area, RUTTNER and others (1968) studied a reference section of the Niur Formation in the Derenjal Mountains. Here, the unit is 628 metres thick and considerably more sandy than in the type section. Locally, the contact with the underlying Shirgesht Formation is disturbed by igneous intrusions, but in other places it is entirely conformable. The lower portion of the Niur, about 245 metres thick, is composed of an alternation of limestone, marlstone, shale, sandstone, and quartzite of white, grey, brown, and red colours. Upwards, the content in clastic components increases. Particularly in the lower part of the lower Niur, intrusive diabase is intercalated. In general, the calcareous, marly rocks bear brachiopods, corals, bryozoans, and some conodonts. According to BOUCOT (*in* BERRY and BOUCOT, 1972), the basal horizon of the Niur yielded late Ordovician gonambonitid brachiopods. The Ordovician/Silurian boundary must therefore be drawn somewhere above this basal horizon, which is overlain by coral-bearing, brownish grey limestones and marls considered to be identical in the Shirgesht and Ozbakh Kuh areas by RUTTNER and others. From the rich coral fauna discovered by RUTTNER in the type area, FLÜGEL (1962) identified, amongst others, the following genera: *Dinophyllum* ?, *Entelophyllum*, *Spongophyllum*, *Tryplasma*, *Cystiphyllum*, *Thecia*, *Palaeofavosites*, *Mesofavosites*, *Favosites*, *Striatopora*, *Coenites*, *Syringopora*, *Halysites*, *Heliolites*, and *Propora*, many of them with several species. FLÜGEL tentatively assigned the coral fauna a late Llandovery/early Wenlock age. He recognized relations with the English–Baltic faunal province on the one hand, and with Siberia, Japan, and Australia on the other. Sandy, brown limestones from the top of the lower member of the Niur Formation yielded, however, the following conodonts of the late Llandovery/early Wenlock, too: *Hadrognathus staurognathoides* Walliser, *Ozarkodina*, *Neoprioniodus*, and *Lonchodina* (WALLISER *in* RUTTNER and others, 1968).

From these results, two conclusions may tentatively be drawn, provided that the fossil identifications are correct: (1) most of the lower portion of the Niur Formation, which is 230 metres thick, is late Llandovery/Wenlock in age; and (2) sediments of early and middle Llandovery age have either never been deposited or have been eroded away. Basic igneous rocks intersperse the lower part of the Niur Formation and form finally a stratum considered to be a flow rather than a sill. No igneous rocks have been found above this level. RUTTNER and others, therefore, held the igneous rocks to be Silurian in age, which corresponds well to the conception of the early Llandovery as a tectonically particularly active time.

In the Shirgesht area, the lower marine member of the Niur is overlain by a sequence of sandstones, quartzites, and siltstones, about 170 metres in thickness, with white, grey, and brown colours. Upwards, the siltstone share increases considerably. Apart from some worm tracks, this middle member of the Niur is mostly unfossiliferous and was possibly deposited in a terrestrial or littoral environment. Near Shirgesht, the middle part of the Niur represents a conspicuous, mappable member, but in the type area of the Ozbakh Kuh Mountains it is completely missing. The age of the middle Niur member can only tentatively be concluded from the fossils below and above. As the fossils immediately below are of late Llandovery/early Wenlock age and fossils 150 metres above the upper boundary of the middle member are assigned a probable latest Ludlow age, the clastics of the middle member can be assumed to be chiefly Wenlock in age. The present author considers it to be possible that the middle Niur member is an equivalent unit of the Redbrown Sandstone of continental facies of southern Jordan and of part of the Sharawra Sandy Member of north-western Saudi Arabia.

In the Shirgesht region, the upper Niur member is composed of more than 210 metres of hard, massive, partly sandy limestone containing brachiopods, corals, bryozoans, and conodonts. Colours are mostly dark yellowish brown to reddish brown. SALEH (1969) studied the corals and determined the genera *Holmophyllum*, *Hedstroemophyllum*, *Cladopora*, *Parastriatopora*, *Thamnopora*, and *Palaeofavosites* which may be classified as Ludlow in age. The following conodonts from about 150 metres above the base of the upper member have been identified by WALLISER (*in* RUTTNER and others, (1968): *Spathognathodus steinhornensis* subsp. indet., *Sp. primus* (Branson and Mehl), *Ozarkodina denckmanni* Ziegler, *O. media* Walliser, *Lonchodina greilingi* Walliser, *Trichonodella inconstans* Walliser, and *Neoprioniodus primus* (Branson and Mehl). They indicate a late Ludlow to earliest Devonian, probably latest Ludlow (? Přídolían) age. The Niur Formation is overlain by the unfossiliferous sandstones of the Padeha Formation which has been estimated to exceed 700 metres in thickness. Presumably, the Padeha Formation is of Early Devonian age.

The lower and upper Niur members yielded many brachiopods the exact stratigraphical position of which has not been noted. STEPANOV (*in* RUTTNER and others, 1968) identified the following genera: *Fascicostella*, *Isorthis*, *Mendacella*, *Atrypina*, *Retziella*, *Protathyris*, *Eospirifer*, *Delthyris* (? = *Quadrifarius*), *Howellella*, and others. According to BERRY and BOUCOT (1972), the occurence of *Fascicostella* and *Quadrifarius* indicates the presence of Přídolí beds.

4.4.2. Eastern Alborz Mountains

The following succession is exposed in the Robat-e-Qarabil inlier of the eastern Alborz Mountains (BRICE and others, 1973). An unnamed and undated formation of lavas and tuffs is overlain by the Qarabil Limestone Formation, composed of marly–sandy limestone and interbedded marls with corals and brachiopods. The formation is 52 metres in thickness. A silicified brachiopod fauna has been found 33 metres above the base of the Qarabil Formation. A probable Wenlock age of that fauna is deduced by COCKS (1979). He determined the following genera: *Glyptorthis, Epitomyonia, Eoplectodonta, Shagamella, Pentamerus, Plico-plasia, Xerxespirifer*, and the trilobite *Diacalymene*. None of the bryozoans and corals listed by LAFUSTE (*in* BRICE and others, 1973) contradicts the Wenlock age deduced from the brachiopods. The question is unresolved whether the Wenlock faunas from the Robat-e-Qarabil inlier have more a European–Asian or a Gondwanan aspect. A stratigraphical break above the Qarabil Limestone Formation is followed by 90 metres of thin-banked limestones and marls with occasional intercalations of sandstones, succeeded by 200 metres of quartzites and shales, followed by a 1300 metres sequence of clastic and carbonate rocks of largely Devonian age.

4.4.3. Southern Iran and Kerman area

Occurrences of marine Silurian rocks in Iran other than Shirgesht, Ozbakh Kuh, and the eastern Alborz Mountains are known only from a few localities. At Kuh-e-Faraghun and Kuh-e-Gahkum, north of Bandar Abbas, dark fossiliferous shales form the top of a succession of sandstones and shales which is about 685 metres in thickness and of Ordovician age in the lower portion. According to DOUGLAS (1950), only the top shales have been attributed to the Silurian because they yielded graptolites of middle Llandovery age (*Climacograptus* cf. *scalaris normalis* Lapworth, *Monograptus* sp. aff. *incommodus* Törnquist, and some *Diplograptus* species). Small orthocerids, *Leptostrophia*, and *Dalmanella* were also found. LEES (1938) recorded the graptolite-bearing shales to be about 70 metres thick. The Silurian of southern Iran is disconformably overlain by sandstones of possible Carboniferous age. Near Zarand, north-west of Kerman, dark, fossiliferous limestones associated with dolomites, quartzites, and shales have been referred to as Ordovician–Silurian by HUCKRIEDE and others (1962). However, as the rocks are poorly fossiliferous (*Favosites, Monticuliporella, Platystrophia*, cf. *Rhynchotrema, Cyrtolites, Michelinoceras, Bollia, Acodus robustus, Distacodus, Icriodella, Roundya*), the Silurian at this place is palaeontologically not distinguished with certainty from the Ordovician. The Silurian rocks of southern Iran and Kerman have lithologically and faunistically nothing in common with the Niur Formation.

4.5. The Silurian rocks of Afghanistan

Marine sedimentary rocks of the Silurian have been deposited presumably only in the eastern part of Afghanistan (WEIPPERT and others, 1970); in the western part, igneous rocks and clastics of possible terrestrial origin have been noticed (MIRZAD and others, 1968). Generally, the Silurian regions of sedimentation were the same as in the Ordovician. Especially in the Logar trough, a richly fossiliferous, rather complete succession was deposited (FESEFELDT, 1964; DESPARMET, 1969; DÜRKOOP, 1970; KARAPETOV and others, 1971), whereas in the eastern Transafghan trough, region of Jalalabad/Nowshera, only Wenlock/Ludlow sediments were formed. From the Badakhshan trough, KOLČANOV and others (1971) described variously metamorphosed rocks of considerable thickness which are partly attributed to the Silurian, partly to the Devonian.

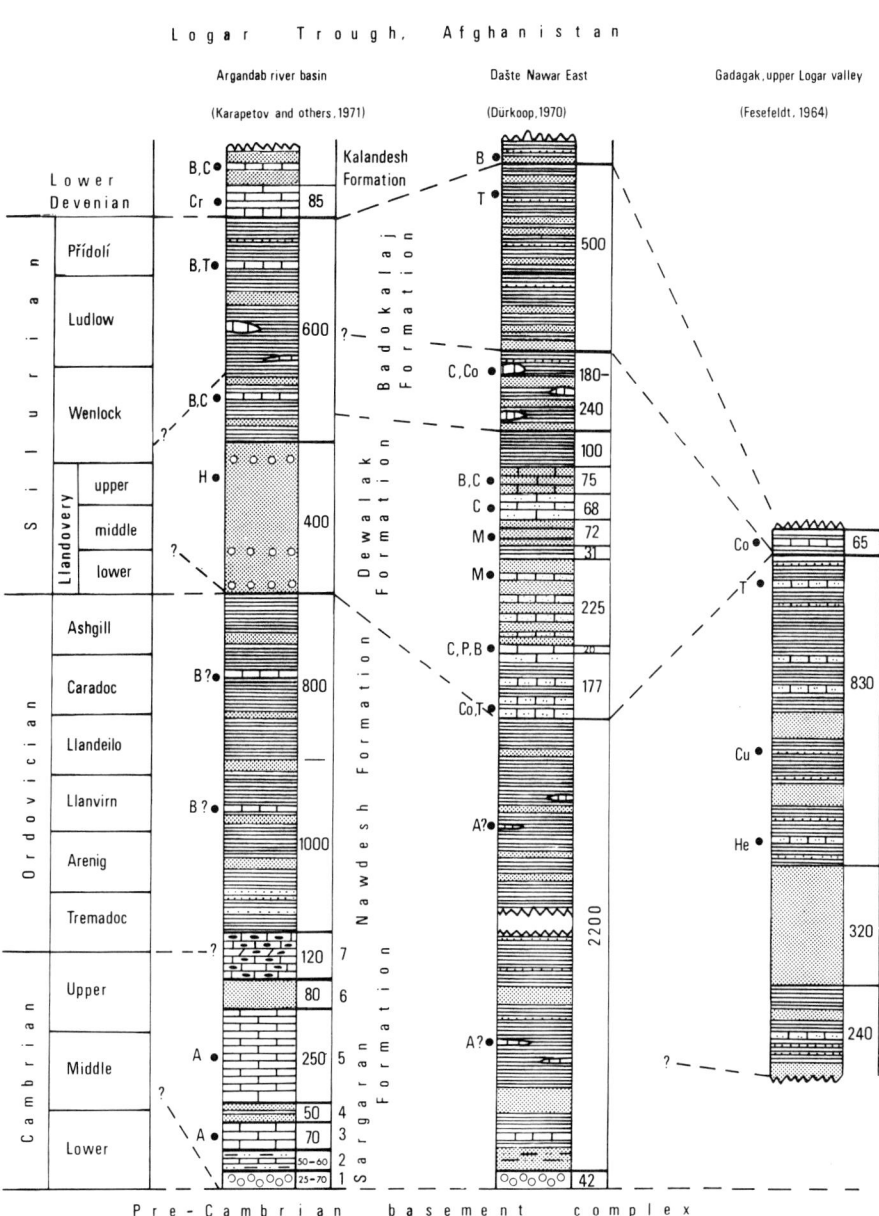

FIG. 30. Representative columnar sections and stratigraphical terminology for the Ordovician and Silurian rocks of the Logar trough, Afghanistan. (A) Algal limestones; (B) brachiopods; (C) corals; (Co) conodonts; (Cr) crinoids; (Cu) *Cruziana*; (H) *Howellella*; (He) *Hesperonomiella*; (M) *Metastromatoceras*; (P) *Pterognathodus amorphognathoides* Walliser; (T) trilobites; (?) position uncertain. For lithological legend, see Fig. 3.

4.5.1. Logar trough

Llandovery

In Afghanistan, Llandovery deposits have been palaeontologically proven only in the Logar trough. DÜRKOOP (1970) and PLODOWSKI (1970) were the first to report a chiefly clastic sequence up to 820 metres thick from the central part of the Logar trough which must be assigned a Llandovery age on the basis of its fossils. According to DÜRKOOP, the Silurian sequence of the eastern Dašte Nawar begins with an alternation of 177 metres of bedded, grey–black, sandy limestones and black shales overlying a monotonous shaly, sandy succession of the Ordovician. At the base of the Silurian, DÜRKOOP assumed a hiatus which corresponds to the conditions in the remaining part of Afghanistan. The basal alternation yielded the trilobites *Diacalymene* and *Dalmanites*, the brachiopods *Isorthis* and *Pentamerus*, and the conodonts *Ambalodus galerus* Walliser, *Hadrognathus staurognathoides* Walliser, *Ozarkodina* cf. *adiutricis* Walliser, and *Roundya detorta* Walliser. Because of the presence of *O.* cf. *adiutricis*, the sequence belongs presumably to the *celloni* Zone, i.e. the base of the late Llandovery.

The succeeding sequence is composed of 20 metres of grey and brown, partly sandy limestone containing tabulate corals (*Palaeofavosites, Halysites*) and the conodont *Pterospathodus amorphognathoides* Walliser, in addition to algae, bryozoans, brachiopods, and gastropods. The conodont is restricted to the *amorphognathoides* Zone which comprises the late late Llandovery and early Wenlock. As the overlying beds bear only fossils of the Llandovery, the present sequence must be considered to be late Llandovery in age.

The overlying succession is represented by grey–brown, shaly sandstone and brown calcareous sandstone 225 metres in thickness. About 175 metres above the base, the orthocerids *Metastromatoceras* and Proteoceratidae were found together with ?*Reocalymene*. The next sequence consisting of 31 metres of black shales and 72 metres of greenish brown, shaly sandstone contains *Metastromatoceras* abundantly. In the Soviet Union, *Metastromatoceras* is held to be a guide fossil of the Llandovery. The same applies to a trilobite of the family Illaenidae.

The overlying units consist of 68 metres of sandy limestone with lenticular intercalations of algal, coral, and bryozoan limestone bearing *Palaeofavosites* and *Halysites* and of 75 metres of bedded, grey–brown calcareous sandstones. In the lower portions of these sandstones, tabulate corals (*Favosites, Halysites*) are fairly frequent. In the upper portions, they are replaced by brachiopods which are abundant at the uppermost horizon (*Dolerorthis, Eoplectodonta, Isorthis, Pentamerus oblongus* (Sowerby), and *Stegocornu procerum* Dürkoop). During the late Llandovery, *P. oblongus* was distributed all over the world. In addition to the brachiopods, the trilobites *Diacalymene* and *Dalmanites* and the conodonts *Hindeodella, Lonchodina* cf. *fluegeli* Walliser, and *Spathognathodus* were found. The Llandovery sequence ends presumably with more than 100 metres of unfossiliferous, greenish black, sandy shales.

From the south-western Dašte Nawar, PLODOWSKI (1970) described Llandovery deposits approximately 700 metres thick, of which about 600 metres are of shales and sandstones. Calcareous sandstone yielded some unidentifiable orthocerids. PLODOWSKI attributed the beds to the Llandovery only to the grounds of their position within the succession and their resemblance to corresponding deposits of the eastern Dašte Nawar. The overlying 100 metres of grey to light brown calcareous sandstone, however, contain a rich faunal assemblage of brachiopods and some corals which is closely related to the late Llandovery faunas of the eastern Dašte Nawar (*Halysites, Dolerorthis*, cf. *Eoplectodonta, Pentamerus oblogus* (Sowerby), *Antirhynchonella*, and *Stegocornu procerum* Dürkoop). An overlying unfossiliferous alternation of dark, sandy shales and thin layers of sandstones was attributed to the Llandovery/Wenlock.

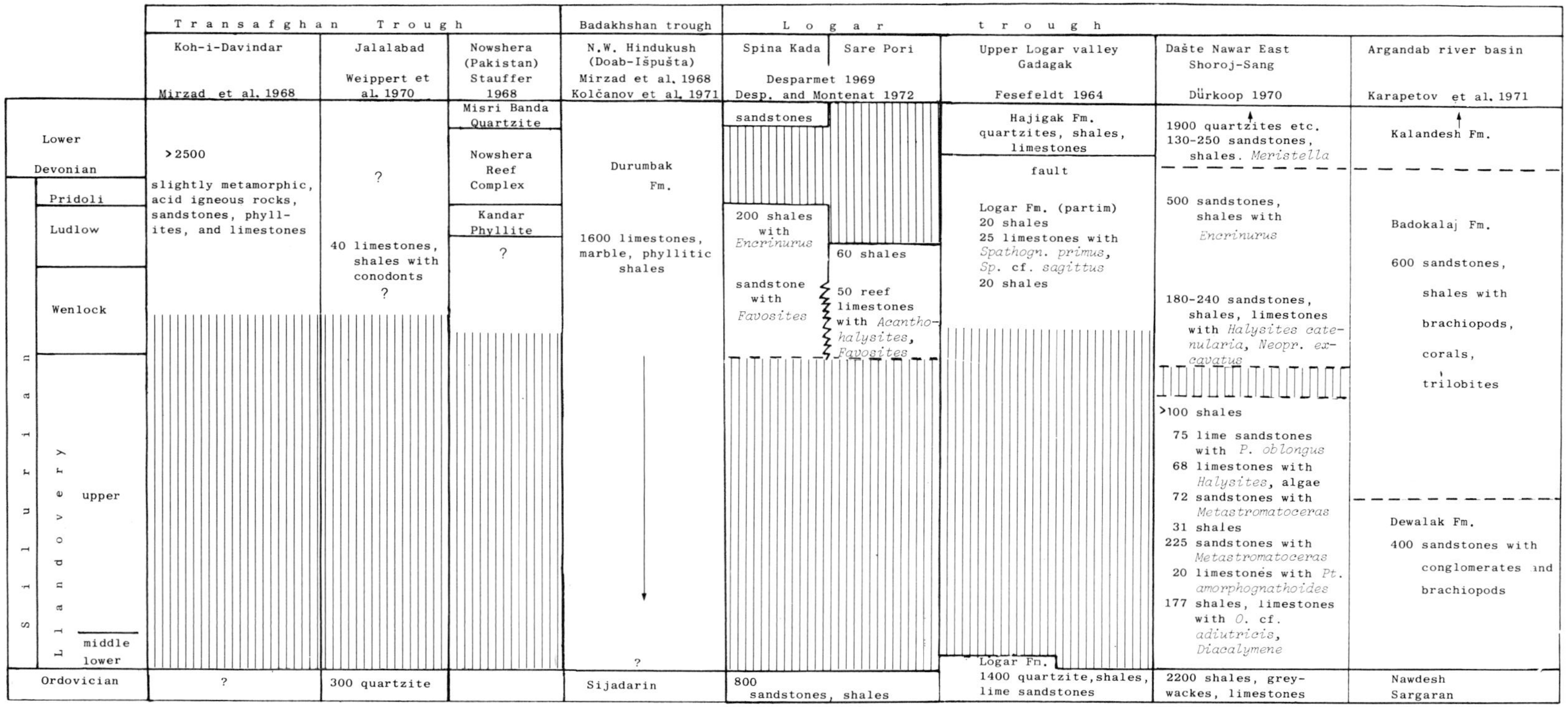

FIG. 31. Stratigraphical correlation of the Silurian rocks of Afghanistan.

	Transafghan Trough		Nowshera (Pakistan) Stauffer 1968	Badakhshan trough N.W. Hindukush (Doab-Ispušta) Mirzad et al.1968 Kolčanov et al.1971	Logar trough			Dašte Nawar East Shoroj-Sang Dürkoop 1970	Argandab river basin Karapetov et al.1971
	Koh-i-Davindar Mirzad et al.1968	Jalalabad Weippert et al.1970			Spina Kada Desparmet 1969	Sare Pori Desp. and Montenat 1972	Upper Logar valley Gadagak Fesefeldt 1964		
Devonian Lower	slightly metamorphic, acid igneous rocks, sandstones, phyllites, and limestones		Misri Banda Quartzite	Durumbak Fm.			Hajigak Fm. quartzites, shales, limestones	1900 quartzites etc. 130-250 sandstones, shales. *Meristella*	Kalandesh Fm.
Pridoli	>2500	?	Nowshera Reef Complex		sandstones			500 sandstones, shales with *Encrinurus*	Badokalaj Fm.
Ludlow		40 limestones, shales with conodonts ?	Kandar Phyllite ?	1600 limestones, marble, phyllitic shales	200 shales with *Encrinurus*	60 shales	Logar Fm. (partim) 20 shales 25 limestones with *Spathogr. prima, Sp. cf. sagittus* 20 shales	180-240 sandstones, shales, limestones with *Halysites catenularia, Neopr. eurcaptus*	600 sandstones, shales with brachiopods, corals, trilobites
Wenlock					sandstone with *Favosites*	50 reef limestones with *Acanthohalysites, Favosites*	fault	>100 shales	
Llandovery upper				?				75 lime sandstones with *P. oblongus* 68 limestones with *Halysites*, algae 72 sandstones with *Metastromatoceras*	Dewalak Fm. 400 sandstones with conglomerates and brachiopods
Llandovery middle								31 shales 225 sandstones with *Metastromatoceras* 20 limestones with *Pt. amorphognathoides*	
Llandovery lower							Logar Fm. 1400 quartzite, shales, lime sandstones	177 shales, limestones with *O. cf. adiutricis, Diacalymene*	
Ordovician	?	300 quartzite		Sijadarin	800 sandstones, shales			2200 shales, greywackes, limestones	Nawdesh Sargaran

Towards the south-western part of the Logar trough, in the Arghandab region, the thickness of the Silurian deposits decreases. KARAPETOV and others (1971) subdivided the Silurian into the Dewalak and the Badokalaj Formations. The Dewalak Formation comprises 400 metres of light coloured, fine- to medium-grained sandstones with lenticular intercalations of quartz conglomerates. On the basis of numerous brachiopods (*Howellella*, *Strophomena*), KARAPETOV and others assigned the Dewalak Formation an early Silurian age. The succeeding Badokalaj Formation was considered to range from the lower to the upper Silurian (Fig. 31). It consists of dark, greenish grey sandstones and aleurolites with a total thickness of 600 metres. Rare calcareous interbeds in the lower portion of the succession yielded the Llandovery brachiopods *Dalmanella neocrassa* (Nikiforova) and *Eocoelia*, together with tabulate corals (*Favosites*, *Mesofavosites*, and *Buranolites*) of late Llandovery/Wenlock age.

Wenlock

In Afghanistan, Wenlock deposits have been proved palaeontologically only in the Logar trough, and possibly also in the region of Jalalabad. In the remaining Lower Palaeozoic Afghan areas of sedimentation (Badakhshan trough, central and western part of the Transafghan trough), there is no palaeontological evidence for the presence of the Wenlock.

DESPARMET (1969) noticed that the Ordovician shales near Sare Pori, in the northernmost part of the Logar trough, are unconformably overlain by 50 metres of reef limestone with tabulate corals (*Palaeofavosites*, *Favosites*, *Mesofavosites*, *Acanthohalysites*). LAFUSTE (*in* DESPARMET, 1969) attributed them to the Wenlock because the species are elsewhere usually associated with fossils of this age. A few kilometres to the south-east, the reef limestone is replaced by micaceous, brown sandstone with *Favosites*. The hiatus between Silurian and Ordovician rocks comprises presumably more or less the entire Llandovery.

In the eastern Dašte Nawar, DÜRKOOP (1970) suggested a hiatus in the early Wenlock separating the shales of the latest Llandovery from greenish brown, shaly sandstones and grey–brown calcareous shales with interbeds of limestone rich in corals, bryozoans and crinoids (algae, *Palaeofavosites*, *Halysites catenularia* (Linné), *Hindeodella*, *Lonchodina*, *Neoprioniodus excavatus* (Branson and Mehl), *Spathognathodus inclinatus inclinatus* (Rhodes), *Trichonodella inconstans* Walliser). On the basis of *N. excavatus* and *H. catenularia*, DÜRKOOP considered the fossiliferous beds to be middle to late Wenlock in age.

In the south-western Dašte Nawar, the unfossiliferous shales of Llandovery/Wenlock age are overlain by about 200 metres of white, yellow, and reddish quartzites and sandstones with thin lenticular intercalations of conglomerates (PLODOWSKI, 1971). The latter contain well rounded quartz and lydite pebbles 2 to 3 centimetres in diameter. As this thick elastic sequence is thoroughly unfossiliferous, its age can only be concluded from the fossils beneath and above (Wenlock to Lower Devonian).

Part of the Badokalaj Formation may be Wenlock in age in the Arghandab region of the southern Logar trough, although specific fossils have not been recorded by KARAPETOV and others (1971). Some tabulate corals and some brachiopods were tentatively assigned a Llandovery/Wenlock and Wenlock/Ludlow age, respectively, but it was not possible to draw boundaries between the different Silurian series.

Ludlow

In all of the major Afghan regions of Lower Palaeozoic sedimentation, the presence of Ludlow rocks has been presumed, and in some of the regions even palaeontologically confirmed. Concerning the Ludlow succession of beds, facies, and fossil content, the Logar trough again must be considered to be the most informative region of Afghanistan. From its northernmost part, DESPARMET (1969) reported 200 metres of blue–black shales to overlie the Wenlock rocks. PILLET (*in* DESPARMET) identified the trilobite *Encrinurus konghsaensis* Reed from 150 metres above the base of the shales and considered it to be Ludlow in age. In Burma, *E. konghsaensis* occurs above *Monograptus priodon*; its age ranges, therefore, between early and late Ludlow. Locally, the thickness of the shales is reduced to 60 metres by the effects of late Silurian/Early Devonian erosion. They are overlain by conglomerates and sandstones which are presumably Lower Devonian in age.

Near Gadagak, in the upper Logar valley, WITTEKINDT (*in* FESEFELDT, 1964) identified the following conodonts from 25 metres of limestones with orthocerids intercalated between 20 metres of black shales beneath and 20 metres above: *Hindeodella equidentata* Rhodes, *Kockelella variabilis* Walliser, *Neoprioniodus multiformis* Walliser, *Ozarkodina media* Walliser, *Spathognathodus inclinatus* (Branson and Mehl), *Sp. primus* (Branson and Mehl), *Sp.* cf. *sagittus* Walliser, and *Trichonodella excavata* (Branson and Mehl). Owing to *Sp.* cf. *sagittus*, the limestones were assigned an early to middle Ludlow age by Wittekindt.

According to DÜRKOOP (1970), the Wenlock deposits of the eastern Dašte Nawar are overlain by a sequence of greenish brown to black shales and shaly greywackes which is poor in fossils and about 500 metres thick. Deposition in shallow water can be concluded from the coarse-grained nature, oblique bedding, fossil plants, and Lebensspuren. Numerous trilobites (*Encrinurus* n.sp., *Bohemoharpes*) have been collected in a shaly sandstone 0.5 metre thick, about 400 metres above the uppermost Wenlock fossils. In the opinion of HAAS (*in* DÜRKOOP, 1970), the *Encrinurus* n. sp. is indicative of a late Silurian age because of its evolutionary stage. DÜRKOOP placed the upper boundary of the Silurian between the beds with *Encrinurus* n. sp. and the overlying shales and sandstones with *Delthyris*, *Howellella*, and *Meristella*. In the southern Dašte Nawar, Ludlow deposits have not yet been identified palaeontologically. The upper portions of the 2500 metres of clastic rocks, the lower portion of which bears fossils of late Llandovery age, possibly must be attributed to the Ludlow and Lower Devonian. Late Silurian trilobites were collected from the upper part of the Badokalaj Formation in the Arghandab region together with some brachiopods (*Clorinda*, *Striispirifer*, *Spirigerina*, *Retziella*, *Cyrtia*, and *Lissatrypa*) considered to be indicative of the Wenlock (KARAPETOV and others, 1971). The Silurian Badokalaj Formation is conformably overlain by the Devonian Kalandesh Formation.

4.5.2. Jalalabad region

In the Jalalabad region, the Silurian System is presumably represented by Ludlow rocks only. From the region east of Jalalabad, WITTEKINDT (*in* WEIPPERT and others, 1970) reported 40 metres of black shales and crinoidal limestones with the conodonts *Kockelella* cf. *variabilis* Walliser and *Spathognathodus inclinatus inclinatus* Rhodes, which he considered to be early to middle Ludlow in age. The crinoidal limestone yielded the orthocerids *Ormoceras* ?, *Michelinoceras* and *Leurocycloceras* of the (Wenlock)/Ludlow (WOLFART *in* WEIPPERT and others, 1970).

STAUFFER (1968) described upper Silurian/Lower Devonian rocks from the northern part of Pakistan, north of Peshawar–Nowshera which might extend westwards into Afghanistan. The rocks exposed north of Nowshera comprise the Kandar Phyllite at the base, the Nowshera Formation in the middle, and the Misri Banda Quartzite at the top of the sequence. The Kandar Phyllite is greenish grey, rich in chlorite, in places silty, and contains, at irregular intervals, crinoidal limestones up to 3 metres thick. The exposed phyllites reach a thickness of about 70 metres. From the overlying Nowshera Formation, the phyllite is separated either by a fault or by a distinct angular discordance. The reef complex of the Nowshera Formation is characterized by a lower portion of fossil-poor carbonates overlain by abundantly fossiliferous carbonates. Where the limestones and dolomites have been well preserved, one can recognize an upward change in the type of fossils from delicate, quiet-water organisms to robust wave-resistant ones. Stromatoporoids are by far the most common fossils of the reef complex, in addition to tabulate and rugose corals, brachiopods, gastropods, pelecypods, and crinoids. An evaluation of the conodont fauna indicated a Ludlow to Gedinnian age (including *Plectospathodus extensus* Rhodes, *P. alternatus* Walliser, *Neoprioniodus bicurvatus* (Branson and Mehl), *Ozarkodina denckmanni* Ziegler, *Spathognathodus inclinatus* (Rhodes), and *Sp. remscheidensis* (Ziegler). Secondary alterations have changed the lithology of the original reef complex. Most of the reef outcrops have been partially or wholly dolomitized except for some of the reef cores. Metamorphism of the Nowshera reef has been fairly mild in the Nowshera region itself. Towards the Afghan–Pakistanian border range, north-west of Peshawar, however, the presumed equivalents of the Nowshera Reef are, according to STAUFFER (1968), obviously so strongly metamorphosed that no identifiable fossils are left. Further to the west, the crinoidal limestones in the area east of Jalalabad are less metamorphosed than in the Peshawar area. They are considered to be equivalents of the Nowshera Reef Complex. The latter is conformably overlain by the Misri Banda Quartzite, which must be younger than the Gedinnian and is up to 500 metres in thickness.

4.5.3. Badakhshan trough and western Afghanistan

Up to the present, Silurian rocks of the Badakhshan trough have been described only by MIRZAD and others (1968) and by KOLČANOV and others (1971). In the region of Doab-Išpušta, north-western Hindukush, the Durumbak Formation is composed of about 1000 metres of massive, black limestone and phyllitic intercalations. The lower part was attributed to the Silurian, the upper part to the Devonian. A similar sequence up to 1600 metres thick has been discovered in the north-eastern Hindukush, Iškamyš area. It consists of limestone, marmorized limestone, and marbles with thin interbeds (up to 100 metres) of aleurolites, greenish grey phyllites, and quartz albite mica schist. There are intercalations of conglomerates and effusive rocks of acid and intermediate composition. The upper portion of the Durumbak yielded corals and crinoids of the Emsian and Eifelian. According to MARKOVSKIJ (1959) and KRESTNIKOV (1962), during the Llandovery clastic sedimentation prevailed in Tadzhikistan and the border range between Afghanistan and the Tadzhik Pamir, whereas the Wenlock/Ludlow rocks are mainly calcareous. From these conditions, the conclusion is tentatively drawn that the lower part of the Durumbak chiefly consisting of calcareous rocks might be Wenlock/Ludlow in age.

There is no evidence for marine Silurian rocks in western Afghanistan. MIRZAD and others (1968) suggested that sedimentation and igneous activity in the Koh-i-Davindar area, western Afghanistan, possibly began during the course of the late Silurian and continued during the entire Devonian. The alternation of slightly metamorphosed acid effusive rocks, tuffs, sandstones, phyllites, and limestones reaches a thickness of up to 2500 metres.

4.6. Silurian stratigraphy and correlation in the Middle East

4.6.1. Correlations within the Middle East

In the western part of the Middle East—Turkey (except for the Bithynian Peninsula), Jordan, Saudi Arabia, and southern Iran—the biostratigraphy of the Llandovery rocks is chiefly based on graptolites, together with some brachiopods and trilobites; whereas in the eastern part—east central Iran, eastern Afghanistan (and the Bithynian Peninsula)—corals, brachiopods, and conodonts are the most important fossils. Llandovery graptolites have not yet been discovered in that part of the Middle East. The *Climacograptus scalaris* ? *normalis* group is widely distributed in southern Iran, Saudi Arabia, Jordan, southern Turkey.

Deposits of Wenlock age are presumably chiefly of continental origin in Saudi Arabia, Jordan, and Iran. In that region, the correlation of the Wenlock rocks, the age of which is palaeontologically not yet confirmed, can only be realized by means of the fossiliferous under- and overlying beds. The Wenlock faunal assemblages of the remaining areas are composed of brachiopods and corals in eastern Afghanistan and in Bithynia and of graptolites and some conodonts at Chios and in south-eastern Turkey.

Ludlovian rocks are very poor in fossils in the Saudi Arabian–Jordan region. They contain graptolites, brachiopods, and conodonts at Chios, questionable graptolites in south-eastern Turkey and the Pontus region, and corals, brachiopods, and conodonts in Bithynia. Corals are the characteristic component of the Ludlow faunal assemblage in eastern central Iran, in addition to brachiopods and conodonts, whereas corals are lacking in the Logar trough. In the latter region, conodonts, trilobites, and brachiopods are the biostratigraphically most indicative fossils. Stromatoporoids, corals, and conodonts are most common in the Nowshera–Jalalabad region. Many of the Silurian successions are poor in fossils or contain only few fossiliferous horizons such as the post-Llandovery rocks of Saudi Arabia and Jordan, Wenlock rocks of east central Iran, and Silurian rocks of north-eastern Syria and of the Badakhshan trough, north-eastern Afghanistan. For these regions, no definitive biostratigraphical conclusions can yet be presented.

4.6.2. The upper boundary of the Silurian

There are only very few sections in the Middle East—north-western Turkey and eastern Afghanistan, between Nowshera and Jalalabad—where the Silurian rocks pass without visible break into Devonian deposits. Probably only in Bithynia and near Nowshera, possibly also on Chios, the Silurian/Devonian boundary can be placed correctly on the base of conodonts which are present in limestones ranging in age from late Silurian into the Early Devonian. Nearly all of the other exposures in the Middle East show sudden changes in lithology between Silurian and Devonian rocks, sometimes distinctly indicating a hiatus between the two systems.

According to HAAS (1968), the Siluro–Devonian boundary must be placed within the Bithynian limestones of the Çakal-Dere Member, the lower portions of which yielded *Spathognathodus steinhornensis eosteinhornensis* (the latest Silurian zonal conodont), whereas the upper portions contain *Icriodus woschmidti* (the earliest Devonian zonal conodont). Based on the occurrence of Ludlow and Devonian species, BESENECKER and others (1968) assumed a Ludlow to Early Devonian age for the Agrelopos Limestone of Chios. The palaeontological evidence suggests a Ludlow to Gedinnian age for the entire reef complex in the Jalalabad–Nowshera region. BARNETT and others (1966) found no vertical or lateral variation in the conodonts and, therefore, treated all the conodonts as a simple assemblage. Most indicative conodonts, however, such as *S. eosteinhornensis* and *I. woschmidti* are missing up to the present. Consequently, the exact position of the Siluro–Devonian boundary within the Nowshera carbonates is still unknown.

All over southern Turkey as well as in southern Jordan and Saudi Arabia, there is no one section showing the transition from Silurian to Devonian rocks. In the Amanos Mountains, LAHNER (1972) presumed the Kirtas Quartzite to range from the uppermost Silurian into the Lower Devonian. The Kirtas is underlain by the Dedeler Formation containing Llandovery fossils and overlain by the Hasanbeyli Formation with Devonian and Lower Carboniferous fossils. BENDER (1964) reported the Ludlow Worm Burrows Sandstone of southern Jordan to be transgressively overlapped by Upper Cretaceous rocks. HELAL (1965) assumed a marked unconformity between the upper Silurian Tawil and the Lower Devonian Jauf Formation. There is no break in sedimentation between the Silurian Niur and the unfossiliferous Padeha Formation held to be Lower Devonian in age by RUTTNER and others (1968). In the central part of the Logar trough, eastern Afghanistan, DÜRKOOP (1970) noticed that the clastic deposits of the Ludlow continue into the Lower Devonian. The Devonian portion yielded a Gedinnian brachiopod assemblage (*Delthyris*, *Howellella*, *Meristella*); the Silurian portion contains *Encrinurus*. According to KOLČANOV and others (1971), in the Badakhshan trough the Siluro–Devonian boundary must be drawn somewhere in the lower part of the Durumbak Formation, the middle part of which yielded corals and crinoids of the Emsian and Eifelian.

4.7. Silurian history of the Middle East

4.7.1. Tectonics and sedimentation

In general, during the Silurian Period, as during the Ordovician, no major orogenic events happened in the Middle East. Marine sedimentation continued in many of the Ordovician depositional regions after a temporary regression during the early Silurian or part of it, affecting particularly the eastern part of the Middle East. In some areas of the Turkish–Afghan east–west belt of mobile shelf conditions, however, such as western Afghanistan, north-western Iran, and the south-eastern part of the Arabian Peninsula, being regions of marine sedimentation during the Ordovician, the marine environment obviously was transformed into a continental one from the beginning of the Silurian. These conditions may be taken as evidence for a differentiation of tectonic activity in the former Ordovician regions of subsidence which during the Silurian partly continued to subside and partly were uplifted. An indication of particular tectonic activity during the Llandovery is provided by effusive igneous rocks in the Badakhshan and Derenjal region. Stable shelf conditions with a more uniform geological development prevailed in the surroundings of the Arabian Shield during the Silurian as well as during the Ordovician and Cambrian.

In Afghanistan and eastern central Iran, distinct evidence can be found that the transition from Ordovician to Silurian time is generally marked by gaps in the sedimentary record of the latest Ordovician and early Silurian. The extent of these gaps depended on the palaeogeographical position of a locality. Gaps increased from the centre of a trough towards the adjoining swell. In the latter position, the gaps frequently comprise the time between latest Ordovician and early Wenlock. In southern Iran, no sedimentary record has been noticed from the post-middle Llandovery until the Carboniferous. In southern Jordan and in Turkey/north-eastern Syria, several sections are reported to show an uninterrupted sedimentary sequence of the Ordovician/Silurian transition. The transition from the Silurian to the Devonian seems to be marked by distinct gaps and alterations of environmental conditions over most of the Middle East (Section 4.6.2).

Concerning Caledonian orogenic activities, FLÜGEL (1964) compiled the observations published up to that time. He concluded that there is no evidence for alpinotype Caledonian movements in the region south of Anatolia. The same conclusion applies to Afghanistan (WEIPPERT and others, 1970). FLÜGEL (1971) emphasized that in numerous places in the Middle East paraconformable bedding can be found between the Cambrian–Silurian and the Devonian rocks. According to FLÜGEL, the hiatuses between the Cambro–Silurian and the Devonian can be attributed to an uplift of short duration without true orogenic movements. The various uplifts may be regarded as synorogenic movements related to Caledonian folding. All of the former major hiatuses regionally mainly to be encountered in the lower Silurian and in the upper Cambrian may be considered as synorogenic movements, too, which can be attributed to

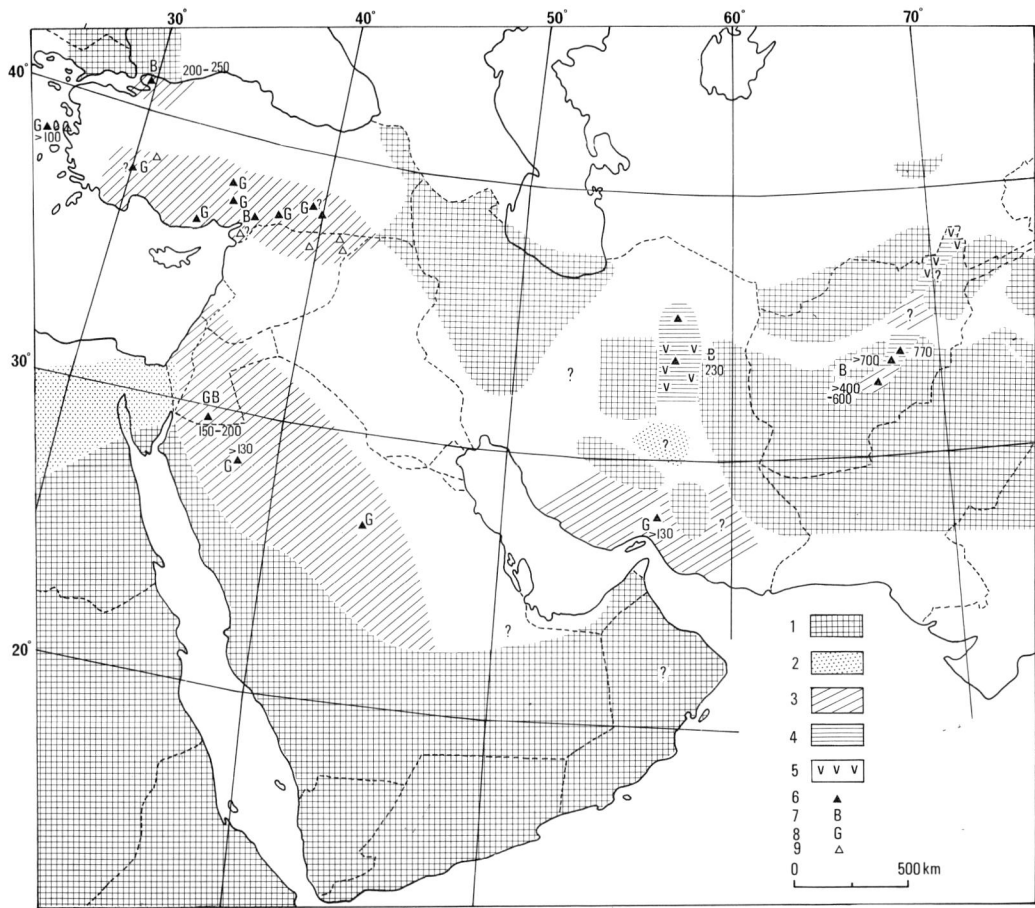

FIG. 32. *Palaeogeography*—Llandovery. (1) Potential source area; (2) continental to littoral environment (conglomerate, sandstone, shale); (3–4) marine environment; (3) sandstone, shale; (4)
quartzite, sandstone, shale, sandy limestone, limestone, and marl; (5) basic igneous rocks; (6)
Llandovery faunas; (7) brachiopods, corals, trilobites, orthocerids, etc.; (8) graptolites; (9)
supposed Llandovery rocks. Thicknesses in metres.

various widely distributed tectonic events of the Caledonian orogenic era. KAUFFMANN (1969),
who studied the geology of north-eastern Chios, believed in rather weak late Caledonian
movements of the epirogenic type on the basis of pre-upper Silurian unconformities and coarse
conglomerates of the upper Silurian.

In the Middle East, the palaeogeographical pattern of the Silurian land masses has been
considerably transformed in detail compared with that of the Ordovician. The dominating
palaeogeographical features of the Cambrian–Ordovician Systems, however, were still effective
during the Silurian. In the south-west, the Nubo–Arabian craton enlarged eastwards since the
Ordovician (Figs. 32–34). As in the former systems, the lithofacies in the stable shelf region
surrounding the Arabian craton is largely composed of quartz sandstones and shales predominantly deposited in marine environments during the Llandovery and part of the Ludlow,
and in continental environments during the rest of the Silurian. In the mobile shelf region,
the Silurian land masses covered the former regions of marine sedimentation in western

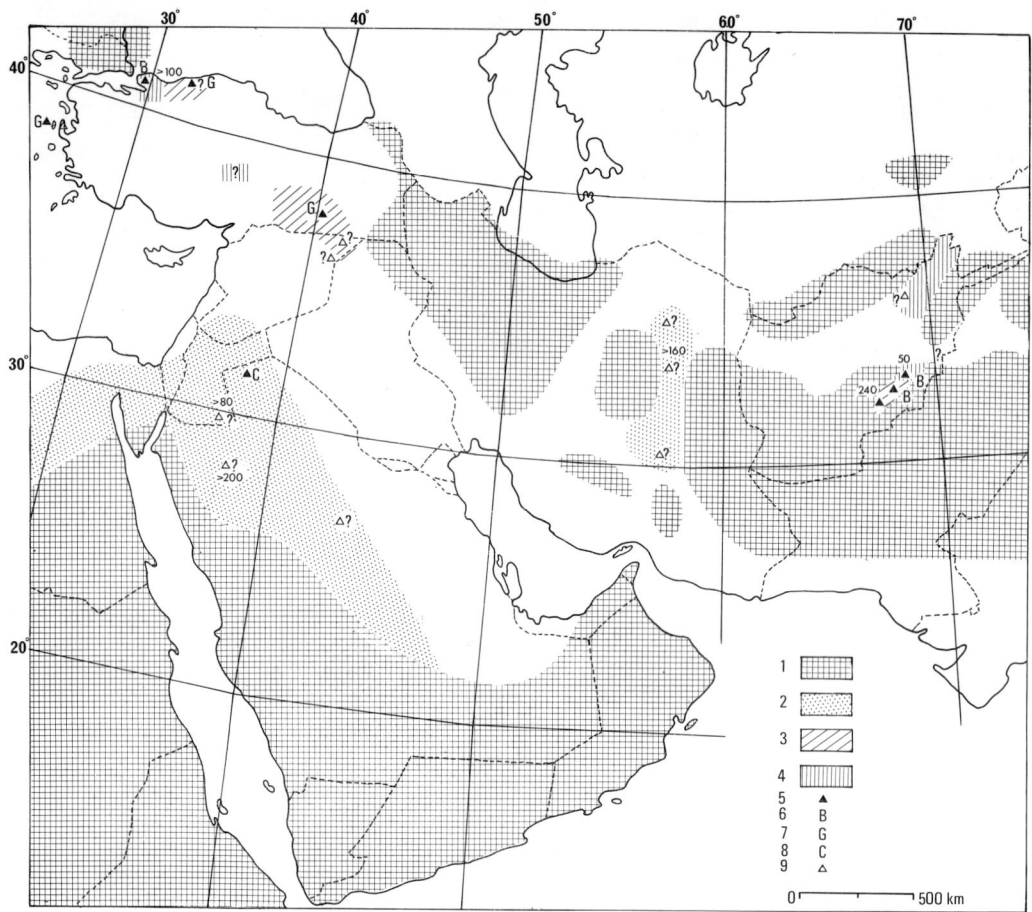

FIG. 33. *Palaeogeography*—Wenlockian. (1) Potential source area; (2) continental to littoral environment (conglomerate, sandstone, siltstone, shale, dolomite); (3–4) marine environment; (3) sandstone, shale, intercalations of limestone; (4) limestone with intercalations of aleurolite, shale, etc.; (5) Wenlock faunas; (6) brachiopods, corals, trilobites; (7) graptolites; (8) chitinozoans, acritarchs; (9) supposed Wenlock rocks. Thicknesses in metres.

Afghanistan and north-western Iran. The thickest Silurian sediments (up to 1600 metres) accumulated in the Logar and Badakhshan troughs, eastern Afghanistan. Clastic and calcareous marine sedimentation continued in these regions from about middle or late early Llandovery during the entire Silurian. Reef limestones of the Wenlock and of the Ludlow/Lower Devonian indicate shallow water conditions in the Logar trough and the Jalalabad–Nowshera region, respectively. Llandovery and Ludlow corals and benthonic faunas indicate shallow water conditions in eastern central Iran, too, which were interrupted only by an emergence during most of the Wenlock. FLÜGEL (1971) concluded deeper water conditions from the dark graptolitic shales of the Silurian distributed over large areas of Turkey. In the region of the Bithynian Peninsula, comparatively shallow water favoured the development of stromatoporoids and tabulate corals during most of the Silurian. The formation of reefs, however, was largely prevented by rapid, fine clastic sedimentation with the source area situated north of Bithynia.

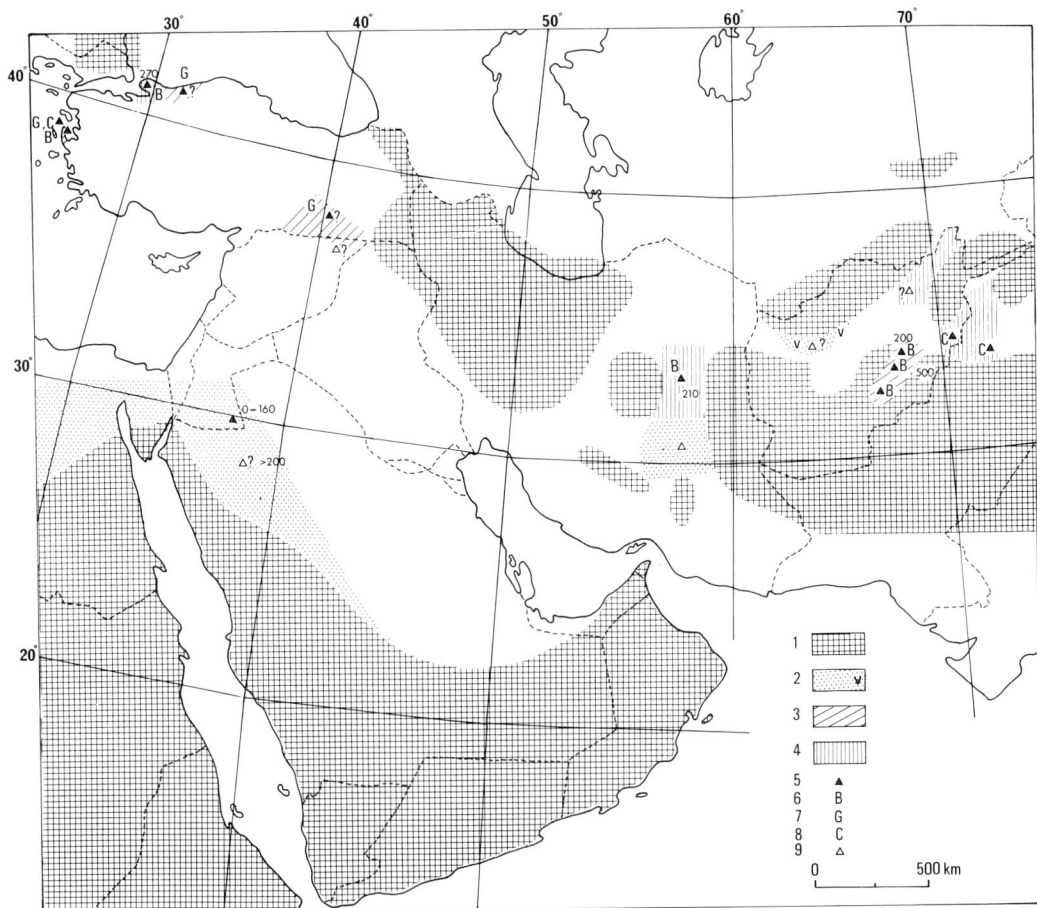

FIG. 34. *Palaeogeography*—Ludlow and Přidolí. (1) Potential source area; (2) continental to littoral environment (sandstone, shale, dolomite, igneous rocks); (3–4) marine environment; (3) sandstone, shale, intercalations of limestone; (4) limestone, intercalations of shale and sandstone; (5) Ludlow faunas; (6) brachiopods, corals, trilobites, conodonts; (7) graptolites; (8) conodonts; (9) supposed Ludlow rocks. Thicknesses in metres.

4.7.2. Zoogeographical features and climate

As has been outlined by HOLLAND (1971), worldwide studies on Silurian faunas yield the general impression that the Silurian appears to have been a period of uniformity of animal life. This impression HOLLAND added, may be due to the fact that outcrops of effectively observable Silurian rocks broadly follow the supposed Silurian equator. BOUCOT, JOHNSON, and TALENT (1967) stated that cosmopolitanism of Silurian brachiopod faunas went hand in hand with the widespread distribution of deposits belonging to that system, but brachiopod faunal provinces became marked during the Early Devonian when marine strata were more restricted. COCKS and MCKERROW 1973), however, demonstrated that there were two exceptions to the general assumption of cosmopolitan Silurian shelf faunas. The exceptions are South America and western Africa with the Silurian *Clarkeia* fauna and north-western America, northern Greenland, and the Soviet Union with the late Silurian *Atrypella* fauna. In the Middle East, the Silurian rocks have yielded only cosmopolitan faunas. There are two 'magnafacies' characterized by shelly faunas (corals, brachiopods, trilobites, etc.) and by graptolites, respectively.

The climatological background of the cosmopolitan distribution of the faunas possibly relates to the warmer climatic belt of the earth. It is evident, from the occurrence of corals and coral reefs in Afghanistan, Iran, and Bithynia, that the Middle East was part of this warmer belt during the Silurian, too. Additionally, HOLLAND (1971) hinted at the possiblity that the Silurian climate could have been in general relatively warm.

4.8. Economic geology

The Silurian rocks of the Middle East are only rarely mineralized, such as in western Turkey. From that region, HÖLL (1966) described some sedimentary ore deposits of antimony, tungsten, and mercury associated with fossiliferous Silurian rocks and epimetamorphic sediments of the Lower Palaeozoic. HÖLL stated that there were three metallogenetic epochs of the antimony–tungsten–mercury group in Turkey: the Silurian, Permian, and Lower Miocene. The Silurian metallogenetic epoch was by far the most important one, the effects of which were widely distributed in a Mediterranean and a Circumpacific belt (MAUCHER, 1965). In Turkey and on Chios it is represented by syngenetic sedimentary ores chiefly bound to a special graphitic facies. There is evidence for geographical and genetical relations between the origin of the ores and submarine volcanism of the same age. MAUCHER (1965) considered the ore deposits to be related to distinct fractures along the margins of continental plates.

HÖLL (1966) was not the first to describe the antimony deposits of the Keramos Formation near Keramos on Chios. Earlier, MARAVELAKIS (1915, 1916) interpreted the antimony ores to be the result of Tertiary volcanism, and MOUSOULOS (1962) considered them to be of hydrothermal origin. Near Keramos, several ore layers up to 1.5 metres thick are concordantly intercalated in a sequence of greywackes and shales below, tuffs between, and graphitic shales and chert up to 30 metres thick above the ore layers. These are represented by tuffaceous breccias consisting of a microcrystalline mica–quartz matrix with subordinate carbonaceous, carbonate, limonitic, and sericitic material. Fine-grained ores (antimony, pyrite, cinnabar, chalcopyrite) fill the cavities between the breccia fragments. Tungsten proved to be present in traces. The cherty beds contain fragments of graptolites identified as Monograptidae? indet. They are overlain by the upper Silurian Agrelopos Limestone. HÖLL, therefore, considered a middle Silurian age of the syngenetic ore layers to be probable.

In western Turkey, there are more ore deposits of the antimony–tungsten–mercury group which are also presumed to be middle Silurian in age by HÖLL, though some of them occur in the crystalline rocks of the Menderes Massif. These ore deposits also are of the syngenetic sedimentary type. Some of the more important are mercury deposits near Karareis on the Karaburun Peninsula opposite to Chios, mercury and antimony deposits near Alaşehir, mercury with traces of antimony and tungsten near Habibler, antimony and traces of mercury near Demirkapi and Sülüklü-Eymir, and antimony, some tungsten, and traces of mercury near Dudas.

The mercury deposit of Karareis on Karaburun is confined to the uppermost part of the Denizgiren Formation, consisting of graphitic and bituminous shales, carbonate beds, and quartz layers. The ore layer is up to 5 metres thick. Cinnabar is the most common ore mineral, in addition to pyrite, marcasite, arsenopyrite, and chalcopyrite. HÖLL interpreted the mercury deposit of Karareis as genetically connected with a submarine volcanism of the same age because, immediately below the ore layer, several tuffaceous layers are intercalated between obliquely bedded sandstones and shales of the uppermost Denizgiren Formation.

The mercury deposits of Kaleçik on the Karaburun Peninsula and Mudarli on the Bithynian Peninsula are related to rocks of the late upper Silurian. Near Kaleçik, HÖLL differentiated several submarine effusions which caused the deposition of seven layers of cinnabar ores in the Kaleçik Formation overlying the upper Silurian Kaleçik Limestone. The lowermost ore layer of Kaleçik, being the most important one, is underlain by greywackes and is up to 8 metres thick. Ore minerals are cinnabar, pyrite, marcasite, and arsenopyrite. The lowermost ore layer is overlain by tuffs 8 to 10 metres thick. The entire sequence is finally capped by massive cherts.

About 45 kilometres east of Istanbul, the mercury deposit of Mudarli is concordantly intercalated between richly fossiliferous limestones of the upper Silurian (HÖLL, 1966). The ore layer shows the characteristics of a sediment, such as fossil content, rhythmic sedimentary bedding, oblique bedding, and sedimentary breccias. It is chiefly confined to sandstone lenses whereas the limestones are only slightly ore bearing. The ore consists of cinnabar only.

All of the Turkish deposits of the antimony–tungsten–mercury group are comparatively small. Most of the ancient mines are, therefore, abandoned, and only a few of the mercury mines are still under exploitation.

References

ADIB, D. (1978). Geology of the metamorphic complex at the south-western margin of the Central-Eastern Iranian microplate (Neyriz area). *Abh. N. Jb. Geol. Paläont.*, **156**, 3, 393.

ALAVI, M. and AMIDI, M. (1968). Geology of western parts of Takab Quadrangle. *Geol. Surv. Iran., Geol. Note*, **49** (unpublished).

ALLENBACH, P. (1966). Geologie und Petrographie des Damavand und seiner Umgebung (Zentral-Elburz), Iran. *Mitt. Geol. Inst. E.T.H. Univ. Zürich, N.S.*, **63**, 144.

ALTINLI, J. E. (1966). Geology of eastern and southeastern Anatolia. *Bull. Min. Res. Explor. Inst. Turkey*, **66**, 35.

ARCHIPOV, I. V., LEONOV, JU. G., and NIKONOV, A. A. (1970). Grundzüge der Geologie des afghanischen Badakhshan. *Bjull. Mosk. Obsc. Isp. Pri., Geol.* **45**, 1, 46.

ASRARULLAH (1961). Cambrian formations in Pakistan, *in* El Sistema Cambrico, su paleogeografía y el problema de su base. *20th Congr. Geol. Internac., Mexico*, **3**, 414.

ASSERETO, R. (1963). The Paleozoic formations in Central Elburz (Iran). *Riv. Ital. Paleont. Strat.*, **69**, 503.

ASSERETO, R. (1966). Geological map of upper Djadjerud and Lar Valleys (Central Elburz, Iran), scale 1:50,000, with explanatory notes. *Inst. Geol. Univ. Milano*, G, 232.

ATHAVALE, R. N., BHALLA, M. S., and MITAL, G. S. (1970). Palaeomagnetism and drift of the Indian Peninsula. *Bull. Nat. Geophys. Res. Inst. India*, **8**, 3/4, 73.

ATHAVALE, R. N., RADHAKRISHNAMURTY, C., and SAHASRABUDHE, P. W. (1963). Palaeomagnetism of some Indian rocks. *Geophys. J. R. Astron. Soc.*, **7**, 304.

ATHAVALE, R. N., VERMA, R. K., BHALLA, M. S., and PULLAIAH, G. (1970). Drift of the Indian sub-continent since Pre-Cambrian times, *in* RUNCORN, S. K. (Ed.), *Palaeogeophysics*, 291, London, New York.

AULER PASCHA (1908). Die Hedschazbahn. 2. Ma'an bis El'Ula. *Petermanns Geog. Mitt.*, Ergänzungs-Heft, **161**.

BARKHATOV, B. P. (1963). *Die Tektonik des Pamirs*. Moscow.

BARNETT, S. G., KOHUT, J. J., RUST, C. C., and SWEET, W. C. (1966). Conodonts from Nowshera reef limestone (uppermost Silurian or lowermost Devonian), West Pakistan. *J. Paleont.*, **40**, 3, 435.

BAYKAL, F., and ATAN, O. (1965). Note préliminaire sur l'existence du Précambrien dans la montagne d'Amanos (Sud-Est de la Turquie). *Comptes Rendus Soc. Géol. Fr.*, 46.

BAYKAL, F., and KAYA, O. (1965). Note préliminaire sur le Silurien d'Istanbul. *Bull. Min. Res. Explor. Inst. Turkey*, **64**, 1.

BECKER, H., FÖRSTER, H., and SOFFEL, H. (1973). Central Iran, a former part of Gondwanaland? Palaeomagnetic evidence from Infra-Cambrian rocks and iron ores of the Bafq area, Central Iran. *Z. Geophys.* **39**, 6, 953.

BEER, H. (1966). Geologie der phosphatführenden Schichtfolge des Raumes Mardin-Derik-Mazidaği. *Bull. Min. Res. Explor. Inst. Turkey*, **66**, 106.

BEHAIN, C. (1970). Die Tektonik des Tschogart-Eisenerz-Massivs und seiner Umgebung bei Bafq im zentralen Iran. *Clausthaler Geol. Abh.*, **7**, 5.

BELLEN, R. C. VAN, DUNNINGTON, H. V., MORTON, D. M., and WETZEL, R. (1959). Iraq. *Lexique Strat. Internat.*, **3**, Asie, 10a.

BENDER, F. (1961). Stand der Exploration und Erdölaussichten in Jordanien. *Erdöl. Kohle*, **14**, 804.

BENDER, F. (1964). Jordanie (Extrême Sud de la Jordanie) *in Lexique Strat. Internat.*, **3** (Asie), 10c1 (Liban, Syrie, Jordanie).

BENDER, F. (1965). Zur Geologie der Kupfererzvorkommen am Ostrand des Wadi Araba, Jordanien. *Geol. Jb.*, **83**, 181.

BENDER, F. (1968). Geologie von Jordanien, *in* MARTINI, H.-J. (Ed.), *Beitr. Reg. Geol. Erde*, **7**, Borntraeger, Berlin.

BENDER, F. and HUCKRIEDE, R. (1964). Stratigraphie der 'Nubischen Sandsteine' in Süd-Jordanien. *Geol. Jb.*, **81**, 237.

BENDER, F. and others (1966). *Geologische Karte von Jordanien 1 : 250,000.* Geol. Survey F.R. Germany.

BENTOR, Y. K. (1956). The manganese occurrences at Timna (Southern Israel). A lagoonal deposit. *20th Internat. Geol. Congr.*, Symposium sobre Yacimientos de Manganeso, **4**, Asia y Oceania, 160.

BENTOR, Y. K. (1960). Israel. *Lexique Strat. Internat.*, **3** (Asie), 10c2.

BERRY, W. B. N. and BOUCOT, A. J. (1972). Correlation of the South-east Asian and Near Eastern Silurian rocks. *Spec. Pap. Geol. Soc. Am.*, **137**.

BESENECKER, H., DÜRR, S., HERGET, S., JACOBSHAGEN, V., KAUFFMANN, G., LÜDTKE, G., ROTH, W., and TIETZE, K. (1968). Geologie von Chios (Ägäis). *Geol. Paläont.*, **2**, 121.

BEYDOUN, Z. R. (1964). The stratigraphy and structure of the eastern Aden Protectorate. *Bull. Overseas Geol. Min. Resources*, Suppl. Ser., **5**.

BEYDOUN, Z. R. (1966). Contribution to 'Geology of the Arabian Peninsula: eastern Aden Protectorate and part of Dhufar'. *Prof. Pap. U.S. Geol. Surv.* **560-H.**

BEYDOUN, Z. R., and GREENWOOD, J. E. G. W. (1968). Protectorat d'Aden et Dhufar. *Lexique Strat. Internat.*, **3**, Asie, 10b2.

BLAISE, J. (1972). Études stratigraphiques, pétrographiques et tectoniques dans les montagnes de Maydan et de Wardak (Afghanistan central). *Rev. Géogr. Phys. Géol. Dyn.*, **2**, 14, 4, 357.

BLAISE, J., DESPARMET, R., and LAPPARENT, A. F. DE (1971). Stratigraphie et structure du Paléozoique de la région de Wardak, en Afghanistan. *Bull. Soc. Géol. Fr.* **7**, 13, 420.

BLAISE, J., DESPARMET, R., and PHILIPPOT, A. (1971). Découverte de graptolites ordoviciens dans la vallée de Sadmarda à l'Ouest de Kaboul (Afghanistan central). *Comptes Rendus Acad. Sci., Paris*, **272**, 5, 691.

BLAKE, G. S. (1935). *The stratigraphy of Palestine and its building stones.* Government of Palestine.

BLANCKENHORN, M. (1912). Kurzer Abriss der Geologie Palästinas. *Z. Deutsch. Palästina-Ver.*, **35**, 113.

BLANCKENHORN, M. (1914). Syrien, Arabien, Mesopotamien. *Handb. Region. Geol.*, **5**.

BLANFORD, W. T. (1872). Note on the geological formations seen along the coasts of Biluchistan and Persia from Karachi to the head of the Persian Gulf, and on some of the Gulf islands. *Rec. Geol. Surv. India*, **5**, 2, 41.

BLÜMEL, G. (1967) *Bericht über geologische Untersuchungen im Raum südlich und südöstlich Herat.* Geol. Survey F.R. Germany, unpublished report.

BLUMENTHAL, M. M. (1941). Un aperçu de la géologie du Taurus dans les vilâyets de Nigde et d'Adana. *Maden Tetkik Arama Enst.* Yayinlarindan, B, **6**.

BLUMENTHAL, M. M. (1947). Geologie der Taurusketten im Hinterland von Seydişehir und Beyşehir. Maden Tetkik Arama Enst. Yayinlarindan, D, *Beitr. Geol. Karte Türkei*, **2**.

BOBEK, H. (1934). Reise in Nordwestpersien 1934. *Z. Ges. Erdkunde*, **9/10**, 359.

BOUCOT, A. J., JOHNSON, J. G., and TALENT, J. A. (1967). Lower and Middle Devonian faunal provinces based on Brachiopoda, *in* OSWALD, D. H. (Ed.), *International Symposium on the Devonian System, Calgary, 1967*, Alberta Soc. Petrol. Geol., Calgary, 1239.

BRAMKAMP, R. A. and others (1963a). Geologic map of the Wadi as Sirhan quadrangle, Kingdom of Saudi Arabia, 1 : 500,000. *U.S. Geol. Surv. Misc. Geol. Inv.*, map 1–200 A.

BRAMKAMP, R. A. and others (1963b). Geologic map of the Wadi ar Rimah quadrangle, Kingdom of Saudi Arabia, 1 : 500,000. *U.S. Geol. Surv. Misc. Geol. Inv.*, map 1–206 A.

BRATASH, V. I., EGUPOV, S. V., PETSNIKOV, V. V., and ŠELOMENCEV, A. J. (1970). Geologie, Erdöl und Gas in Nord-Afghanistan. *Trudy VNIGNI*, **80**, 288 pp.

BRATASH, V. I., EGUPOV, S. V., PETSCHNIKOV, V. V., SCHELOMENTZEV, A. I., and PANTELEEV, F. P. (1964). *Geologische Karte des nördlichen Afghanistan* 1 : 500,000. Unpublished map.

BRICE, D., LAFUSTE, J., LAPPARENT, A. E. DE, PILLET, J., and YASSINI, I. (1973). Étude de deux gisements paléozoiques (Silurien et Dévonien) de l'Elbourz oriental (Iran). *Ann. Soc. Géol. N., Lille,* **93**, 177.

BRINKMANN, R. (1966). Geotektonische Gliederung von Westanatolien. Monatsh. *Neues Jb. Geol. Paläont.,* **1966**, 603.

BRINKMANN, R. (1967). Die Südflanke des Menderes-Massives bei Milâs, Bodrum und Oren. *Fac. Sci. Ege Univ., Sci. Rep.,* **43**.

BRINKMANN, R. (1971). *The geology of western Anatolia, in* CAMPBELL, A. S. (Ed.), Geology and history of Turkey. Petrol. Explor. Soc. Libya, 13th Ann. Field Conf., 171.

BROILI, F. (1911). *Geologische und paläontologische Resultate der Dr. Grothe'schen Vorderasienexpedition 1906 und* 1907, **1**, 1.

BROWN, G. F. and others (1963a). Geologic map of the northwestern Hijaz quadrangle, Kingdom of Suadi Arabia, 1 : 500,000. *U.S. Geol. Surv., Misc. Geol. Inv.,* map 1–204 A.

BROWN, G. F., and others (1963b). Geologic map of the north-eastern Hijaz quadrangle, Kingdom of Saudi Arabia, 1 : 500,000. *U.S. Geol. Surv., Misc. Geol. Inv.,* map 1–205 A.

BURDON, D. J. (1959). *Handbook of the geology of Jordan.* Amman.

BUREK, P. J. (1969). Device for chemical demagnetization of red beds. *J. Geophys. Res.* **74**, 27, 6710.

CARTER, H. J. (1852). Memoir on the geology of the south-east coast of Arabia. *J. R. Asiatic Soc., Bombay,* **4**, 21.

CHMYRIOV, V. M. and others (1973). Mineral resources of Afghanistan, *in* Geology and mineral resources of Afghanistan. *Ministry Mines Industries Rep. Afghanistan, Dept. Geol. Surv.,* **1**, 86.

CHMYRIOV, V. M. and others (1977). Grundzüge des geologischen Baues von Afghanistan. *Izv. Akad. Nauk SSSR, Ser. Geol.,* **1977**, 2, 29.

COCKS, L. R. M. (1979). A silicified brachiopod fauna from the Silurian of Iran. *Bull. Brit. Mus. Nat. Hist. (Geol.),* **32**, 1, 25.

COCKS, L. R. M. and MCKERROW, W. S. (1973). Brachiopod distributions and faunal provinces in the Silurian and Lower Devonian, *in* HUGHES, N. F. (Ed.), Organisms and continents through time. *Spec. Pap. Palaeont.,* **12**, 291.

COOPER, G. A. (1976). Lower Cambrian brachiopods from the Rift Valley (Israel and Jordan). *J. Paleontol.,* **50**, 2, 269.

COWIE, J. W. (1971). Lower Cambrian faunal provinces, *in* MIDDLEMISS, F. A., RAWSON, P. F., and NEWALL, G. (Eds.), *Faunal provinces in space and time,* Seel House Press, Liverpool, 31.

CRAWFORD, A. R. (1971). Góndwanaland and the growth of India. *J. Geol. Soc. India,* **12**, 3, 205.

DAVOUDZADEH, M., SEYED-EMAMI, K., and AMIDI, M. (1969). Preliminary note on a newly discovered Triassic section northeast of Anarak (central Iran), with some remarks on the age of the metamorphism in the Anarak region. *Geol. Surv. Iran, Geol. Note,* **52**, unpublished report.

DEAN, W. T. (1967a). The correlation and trilobite fauna of the Bedinan Formation (Ordovician) in south-eastern Turkey. *Bull. Brit. Mus. Nat. Hist. (Geol.),* **15**, 2, 81.

DEAN, W. T. (1967b). The distribution of Ordovician shelly faunas in the Tethyan region, *in* ADAMS, C. G. and AGER, D. V. (Eds.), *Aspects of Tethyan biogeography.* Systematics Association, Publication No. **7**, 11.

DEAN, W. T. (1971). The Lower Palaeozoic stratigraphy and faunas of the Taurus Mountains near Beyşehir, Turkey. 2. The trilobites of the Seydişehir Formation (Ordovician). *Bull. Brit. Mus. Nat. Hist. (Geol.),* **20**, 1.

DEAN, W. T. (1973). The Lower Palaeozoic stratigraphy and faunas of the Taurus Mountains near Beyşehir, Turkey. 3. The trilobites of the Sobova Formation (Lower Ordovician). *Bull. Brit. Mus. Nat. Hist. (Geol.),* **24**, 5.

DEAN, W. T. (1975). Cambrian and Ordovician correlation and trilobite distribution in Turkey, *in* BERGSTRÖM, J., *Evolution and morphology of the Trilobita, Trilobitoidea and Merostomata, Fossils and Strata,* No. 2, 353. Oslo.

DEAN, W. T. and KRUMMENACHER, R. (1961). Cambrian trilobites from the Amanos Mts., Turkey. *Palaeontology,* **4**, 71.

DEAN, W. T. and MONOD, O. (1970). The Lower Palaeozoic stratigraphy and faunas of the Taurus Mountains near Beyşehir, Turkey. 1. Stratigraphy. *Bull. Brit. Mus. Nat. Hist. (Geol.)*, **19**, 8, 411.

DEDUAL, E. (1967). Zur Geologie des mittleren und unteren Karaj-Tales, Zentral-Elburz (Iran). *Mitt. Geol. Inst. E.T.H. Univ. Zürich, N.S.*, **76**, 123.

DEMIRTAŞLI, E. (1975). Stratigraphic correlation of the Lower Palaeozoic rocks of Iran, Pakistan and Turkey. *Congr. Earth Sci., 1973, Ankara*, 210.

DESPARMET, R. (1969). Nouvelles données sur le Paléozoïque ancien d'Afghanistan central. *Comptes Rendus Acad. Sci., Paris*, **268**, 2389.

DESPARMET, R., and MONTENAT, CH. (1972). Les transgressions du Paléozoïque en Hazarajat. *Rev. Géog. Phys. Géol. Dyn.*, (2), **14**, 4, 397.

DESPARMET, R., TERMIER, G., and TERMIER, H. (1971). Sur un bivalve protobranche ante-arénigien trouvé au Nord du Wardak (Afghanistan). *Géobios*, **4**, 2, 143.

DIENEMANN, W. (1915). Älteres Paläozoikum von Südsyrien und Westarabien. *Centralbl. Mineralog. Geol. Palaeontol.*, 23.

DIETRICH, W. O. (1937). Ordoviz in Nordwest-Iran. *Centralbl. Mineralog. Geol. Palaeontol. B*, **10**, 401.

DIETZ, R. S. and HOLDEN, J. C. (1970). Reconstruction of Pangaea: Breakup and dispersion of continents, Permian to Present. *J. Geophys. Res.* **75**, 4939.

DOUGLAS, J. A. (1950). The Carboniferous and Permian faunas of south Iran and Iranian Baluchistan. *Mem. Geol. Surv. India, Palaeont. Indica, N.S.*, **22**, 7.

DRONOV, V. J. and others (1973). Scheme of stratigraphy of Afghanistan. A short explanatory note to the geological map of Afghanistan 1 : 1,000,000, *in* Geology and mineral resources of Afghanistan. *Min. Mines Industries Rep. Afghanistan, Dept. Geol. Surv.*, **1**, 86.

DUBERTRET, L. (1936). Stratigraphie des régions recouvertes par les roches vertes du nord-ouest de la Syrie. *Comptes Rendus Acad. Sci. Paris*, **203**, 1173.

DUBERTRET, L., DANIEL, E. J., and BENDER, F. (1963). Liban, Syrie, Jordanie. *Lexique Strat. Internat.*, **3**, Asie, 10cl.

DÜRKOOP, A. (1970). Brachiopoden aus dem Silur, Devon und Karbon in Afghanistan. Mit einer Stratigraphie des Paläozoikum der Dasht-e-Nawar/Ost und von Rukh. *Palaeontographica*, **A134**, 4–6, 153.

EGEMEN, M. R. (1947). A preliminary note on fossiliferous Upper Silurian beds near Eregli. *Bull. Geol. Soc. Turkey*, **1**, 1, 52.

ELLES, G. L. (1925). The characteristic assemblages of the graptolite zones of the British Isles. *Geol. Mag.*, **62**, 337.

ELLES, G. L. and WOOD, E. M. R. (1906–1913). A monograph of British graptolites. *Palaeontog. Soc. London*, **60–67**.

ERENTÖZ, C. (1966). Contribution á la stratigraphie de la Turquie. *Bull. Min. Res. Explor. Inst. Turkey*, **66**, 1.

ERKMEN, U. and BOZDOĞAN, N. (1979). Acritarchs from the Dadas Formation in Southeast Turkey, *Geobios*, **12**, 3, 445.

FALCON, N. L. (1967). The geology of the north-east margin of the Arabian basement shield. *Advancement Sci.*, **31**.

FESEFELDT, K. (1964). Das Paläozoikum im Gebiet der oberen Logar und im östlichen Hazarajat, südwestlich Kabul, Afghanistan. *Beih. Geol. Jb.*, **70**, 185.

FLÜGEL, H. (1955). Zur Paläontologie des anatolischen Paläozoikums. 5. Graptolithen aus dem Gotlandium des Antitaurus. *Monatsh. Neues Jb. Geol. Paläont.*, **1955**, 478.

FLÜGEL, H. (1962). Korallen aus dem Silur von Ozbakh-Kuh (NE-Iran). *Jb. Geol. Bundesanst.*, **105**, 287.

FLÜGEL, H. (1963). Graptolithen aus dem mittleren Ordovizium von Nord-Syrien. *Abh. Neues Jb. Geol. Paläont*, **118**, 21.

FLÜGEL, H. (1964). Die Entwicklung des vorderasiatischen Paläozoikums. *Geotekton. Forsch.*, **18**.

FLÜGEL, H. (1969). Stromatoporen aus dem Silur des östlichen Iran. *Monatsh. Neues Jb. Geol. Paläont.*, **1969**, 4, 209.

FLÜGEL, H. (1971). Paleozoic rocks of Turkey, *in* CAMPBELL, A. S. (Ed.), *Geology and history of Turkey*. Petrol. Explor. Soc. Libya, 13th Ann. Field Conf., 211.

FLÜGEL, H. and RUTTNER, A. (1962). Vorbericht über paläontologisch-stratigraphische Untersuchungen im Paläozoikum von Ozbakh Kuh (NE-Iran). *Verh. Geol. Bundesanst.*, **1962**, 146.

FORTEY, R. A. and RUSHTON, A. W. A. (1976). *Chelidonocephalus* trilobite fauna from the Cambrian of Iran. *Bull. Brit. Mus. Nat. Hist. (Geol.)*, **27**, 4.

FOX, C. S. (1947). *The geology and mineral resources of Dhofar Province, Muscat, and Oman.* Pub. Sultanate of Muscat and Oman.

FRECH, F. (1917). Geologie Kleinasiens im Bereich der Bagdadbahn. Ergebnisse eigener Reisen und paläontologischer Untersuchungen. *Z. Deutsch. Geol. Gesell.*, **68**, (1916), 1.

FUCHS, T. (1902). Über einige Hieroglyphen und Fucoiden aus den paläozoischen Schichten von Hadjin in Kleinasien. *Sitzungsber. Kaiser. Akad. Wiss., Math.-Naturw. Kl.*, 1, **111**, 327.

FURON, R. (1941). Géologie du plateau iranien (Perse–Afghanistan–Béloutchistan). *Mém. Mus. Nat. Hist. Natur., N.S.*, **7**, 2, 177.

GANSSER, A. (1955). New aspects of the geology in Central Iran. *Proc. 4th World Petrol. Congr. Rome*, **1A5**, 2, 280.

GANSSER, A. (1960). Über Schlammvulkane und Salzdome. *Vierteljahresschr. Naturf. Ges. Zürich*, **105**, 1.

GANSSER, A. and HUBER, H. (1962). Geological observations in the Central Elburz, Iran. *Schweizer. Min. Petrog. Mitt.*, **42**, 583.

GLAUS, M. (1965). Die Geologie des Gebietes nördlich des Kandevan-Passes (Zentral-Elburz), Iran. *Mitt. Geol. Inst. E.T.H. Univ. Zürich, N.S.*, **48**.

GREGORY, J. W. (Ed.) (1929). *The structure of Asia.* Methuen, London.

HAAS, W. (1968a). Trilobiten aus dem Silur und Devon von Bithynien (NW–Türkei). *Palaeontographica*, **A130**, 60.

HAAS, W. (1968b). Das Alt-Paläozoikum von Bithynien (Nordwest-Türkei). *Abh. Neues Jb. Geol. Paläont.*, **131**, 178.

HARRISON, J. V. (1930). The geology of some salt plugs in Laristan, south Persia. *Quart. J. geol. Soc. London*, **86**, 463.

HARRISON, J. V. (1931). Salt domes in Persia. Symposium on salt domes. *J. Inst. Petrol. Techn., London*, **17**, 300.

HAUDE, H. (1969). Das Alt-Paläozoikum—Präkambrium bis Silurium—in der Türkei. *Zentralbl. Geol. Paläont.*, 1, **4**, 702.

HAUDE, H. (1972). Stratigraphie und Tektonik des südlichen Sultan Dag (SW-Anatolien). *Z. Deutsch. Geol. Ges.*, **123**, 411.

HELAL, A. H. (1964). On the occurrence of Lower Paleozoic rocks in Tabuk area, Saudi Arabia. *Monatsh. Neues Jb. Geol. Paläont.*, **7**, 391.

HELAL, A. H. (1965). Stratigraphy of outcropping Paleozoic rocks around the northern edge of the Arabian Shield (within Saudi Arabia). *Z. Deutsch. Geol. Ges.* **117**, 506.

HEMER, D. O. (1965). Application of palynology in Saudi Arabia. *Fifth Arab. Petrol. Congr.*, Cairo.

HENSON, F. R. S. and ELLIOTT, G. F. (1958). Exhibition of specimens of *Collenia* ? from pre-Permian sediments in South Arabia. *Proc. geol. Soc. London*, **1561**, 89.

HERGET, G. and ROTH, W. (1968). Stratigraphie des Paläozoikums im Nordwest-Teil der Insel Chios (Ägäis). *Abh. Neues Jb. Geol. Palaeont.*, **131**, 46.

HILL, D. (1965). Archaeocyatha from Antarctic and a review of the phylum. *Sci. Rep. Trans-Antarctic Exped. 1955–1958*, **10**, geology, 3.

HIRSCHI, H. (1944). Über Persiens Salzstöcke. *Schweizer Min. Petrog. Mitt.*, **24**, 30.

HÖLL, R. (1966). Genese und Altersstellung der Sb-W–Hg-Formation in der Türkei und auf Chios/Griechenland. *Abh. Bayerische Akad. Wiss., Math.-Naturw. Kl., N.F.*, **127**.

HOLLAND, C. H. (1971). Silurian faunal provinces?, *in* MIDDLEMISS, F. A., RAWSON, P. F. and NEWALL, G. (Eds.). Faunal provinces in space and time. *Spec. Iss. Geol. J.* **4**, 61.

HUCKRIEDE, R. (1967). *Archaeonectris benderi* n. gen. n. sp., (Hydrozoa), eine Chondrophore von der Wende Ordovicium/Silurium aus Jordanien. *Geol. Palaeont.*, 1, 101.

HUCKRIEDE, R., KÜRSTEN, M., and VENZLAFF, H. (1962). Zur Geologie des Gebietes zwischen Kerman und Sagand (Iran). *Beih. Geol. Jb.*, **51**.

HULL, E. (1886). *Memoir on the physical geology and geography of Arabia Petraea, Palestine, and adjoining districts, with special reference to the mode of formation of the Jordan-Arabah depression and the Dead Sea.* The Survey of Western Palestine, Adelphi.

HUTCHINSON, R. D. (1956). Cambrian stratigraphy, correlation, and paleogeography of eastern Canada, *in* El Sistema Cambrico, paleogeografía y el problema de su base. 2. *20th Congr. Geol. Internac., Mexico*, 289.

IMANDT, J. (1962). Short paleontological note on some wells drilled in district VI (SE Turkey), *in* Petroleum activities in Turkey, 1962, *Bull. Petrol. Administration Pub.*, **7**, 29.

JONES, P. J., SHERGOLD, J. H., and DRUCE, E. C. (1971). Late Cambrian and Early Ordovician stages in western Queensland. *J. Geol. Soc. Aust.* **18**, 1.

KAADEN, G. V. D. (1959). Age relations of magmatic activity and of metamorphic processes in the north-western part of Anatolia. *Bull. Maden Tetkik Arama Enst., Ankara,* **52**, 15.

KALAFATÇIOĞLU, A. (1975). Distribution of Ordovician–Silurian formations in Turkey and in the neighbouring countries. *Bull. Min. Res. Explor. Inst. Turkey,* **84**, 1.

KAMEN-KAYE, M. (1971). A review of depositional history and geological structure in Turkey, *in* CAMPBELL, A. S. (Ed.), *Geology and history of Turkey.* Petrol. Explor. Soc. Libya, 111.

KARAPETOV, S. S., DOVGAL, JU. M., DEMIN, A. N., NAGALEV, V. S., MIRZAD, S. CH., and KOTOV, A. JA. (1971). The main stratigraphic features of Argandab river basin (Central Afghanistan). *Sov. Geol.,* **2**, 126.

KAUFFMAN, G. (1965). Fossilbelegtes Altpaläozoikum im NE-Teil der Insel Chios (Ägäis). *Monatsh. Neues Jb. Geol. Paläont.,* 647.

KELLOGG, H. E. (1960). *The geology of the Derik-Mardin area, south-eastern Turkey.* Rep. Explor. Div. Am. Overseas Petrol. Ltd., Ankara, unpublished.

KELLOGG, H. E. and KAYAR, M. (1960). *Report about oil exploration in the Mardin area.* Am. Overseas Petrol. Ltd., Ankara, unpublished.

KENT, P. E. (1958). Recent studies of south Persian saltplugs. *Bull. Am. Assoc. Petrol. Geol.,* **42**, 2951.

KENT, P. E. (1970). The salt plugs of the Persian Gulf region. *Trans. Leicester Lit. Philos. Soc.,* **64**, 56.

KETIN, I. (1963). *Explanatory text of the geological map of Turkey, 1 : 500,000, sheet Kayseri.* Maden Tetkik Arama Enst. Yayinlarindan.

KETIN, I. (1966). Cambrian outcrops in Southeastern Turkey and their comparison with the Cambrian of East Iran. *Bull. Min. Res. Explor. Inst. Turkey,* **66**, 77.

KING, W. B. R. (1923). Cambrian fossils of the Dead Sea. *Geol. Mag.,* **60**, 507.

KING, W. B. R. (1930). Notes on the Cambrian fauna of Persia. *Geol. Mag.,* **67**, 316.

KING, W. B. R. (1937). Cambrian trilobites from Iran (Persia). *Mem. Geol. Surv. India, Palaeont. Indica,* **22**, 5.

KING, W. B. R. (1941). The Cambrian fauna of the Salt Range of India. *Rec. Geol. Surv. India,* **75**, Prof. Pap., 9.

KOBAYASHI, T. (1936). The natural boundary between the Cambrian and Ordovician Systems discussed from the Asiatic standpoint. *16th Congr. Géol. Internat., Rep.,* Washington, Vol. 1.

KOBAYASHI, T. (1967). The Cambrian of Eastern Asia and other parts of the continent. *J. Fac. Sci. Univ. Tokyo,* 2, **16**, 3, 381.

KOBAYASHI, T. (1971a). Stratigraphy of the Chosen Group in Korea and South Manchuria and its relation to the Cambro–Ordovician formations and faunas of South Korea. 10, E. the Cambro–Ordovician faunal provinces and the interprovincial correlation discussed with special reference to the trilobites in Eastern Asia. *J. Fac. Sci. Univ. Tokyo,* 2, **18**, 1, 129.

KOBAYASHI, T. (1971b). The Eurasiatic faunal connection in the Ordovician period *in* Colloque Ordovicien–Silurien, Brest, Septembre 1971. *Mém. Bur. Rech. Géol. Min.,* **73**, 281.

KOBER, L. (1919). Geologische Forschungen in Vordersasien. 2C. Das nördliche Hegaz. *Denkschr. Akad. Wiss. Wien, Math.-Naturw. Kl.,* **96**, 779.

KOLČANOV, V. P., KULAKOV, V. V., MIKHAILOV, K. Y., and PASHKOV, B. R. (1971). New data on the stratigraphy of Precambrian and Paleozoic formations of the northern foothills of western Hindu–Kush. *Sov. Geol.,* **3**, 130.

KRESTNIKOV, V. N. (1962). History of oscillatory movements in the Pamirs and adjacent regions of Asia. *Izd. Akad. Nauk SSSR.,* 199.

KRYLOV, J. N. (1967). Riphean and Lower Cambrian stromatolites of Tien-Shan and Karatau. *Trans. Geol. Inst. Akad. Nauk USSR,* **171**, 1.

KTENAS, K. A. (1925). Contribution à l'étude géologique de la presque'ile d'Erythrée (Asie Mineure). *Ann. Sci. Fac. Sci.,* **A1**, 1.

KUSHAN, B. (1973). Stratigraphie und Trilobitenfauna in der Mila-Formation (Mittelkambrium bis Tremadoc) im Alborz-Gebirge (N-Iran). *Palaeontographica,* **144A**, 113.

LAFUSTE, J. and DESPARMET, R. (1969). Tabulés siluriens de Sar-e-Pori, Afghanistan. *Bull. Mus. Nat. Hist. Natur.,* 2, **41**, 5, 1299.

LAHNER, L. (1972). Geologische Untersuchungen an der Ostflanke des mittleren Amanos (SE-Türkei). *Geotekt. Forschungen,* **42**, 64.

LAPPARENT, A. F. DE (1962). Sur une nouvelle série antépaléozoïque en Afghanistan. *Comptes Rendus Somm. Séance Soc. Géol. Fr.* **1962**, 1, 15.

LAPPARENT, A. F. DE (1977). Sur l'âge Éocambrien et Cambrien de la "série de Zargaran" en Afghanistan central. *Mém Hors-sér. Soc. Géol. Fr.* **8**, 19.

LAPPARENT, A. F. DE, BLAISE, J., and DESPARMET, R. (1968). Découverte de faunes ordoviciennes (bilobites et trilobites) en Afghanistan central et sur ses conséquences. *Comptes Rendus Acad. Sci. Fr.* **266**, 7, 666.

LAPPARENT, A. F. DE and PILLET, J. (1969). Déscription de Trilobites ordoviciens, siluriens et dévoniens d'Afghanistan. *Ann. Soc. Géol. nord*, **89**, 323.

LEES, G. M. (1928). The geology and tectonics of Oman and of parts of southeastern Arabia. *Quart. J. geol. Soc.*, **84**, 585.

LEES, G. M. (1938). The geology of the oilfield belt of Iran and Iraq. *The science of petroleum*, **1**, 140.

LINDSTRÖM, M. (1972). Ice-marked sand grains in the Lower Ordovician of Sweden. *Geol. Palaeont.*, **6**, 25.

LOCHMAN-BALK, C. and WILSON, J. L. (1958). Cambrian biostratigraphy in North America. *J. Paleont.*, **32**, 312.

LOTZE, F. (1957). *Steinsalz und Kalisalze*. 1. Bornträger, Berlin.

LOTZE, F. (1969). *Die Salzlagerstätten in Zeit und Raum. Ein Beitrag zum Klima der Vorzeit.* Arbeitsgemeinschaft Forsch. Nordrhein-Westfalen, 195.

LOTZE, F. and SDUZY, K. (1970). El Cambrico de España. 1. F. Lotze: Estratigrafía. *Mem. Inst. Geol. Min. España*, **75**.

MARAVELAKIS, M. (1915, 1916). The eruptive rocks and metallogeny of Chios Island. *Archimidis*, **16**, 85; **17**, 18.

MARKOVSKIJ, A. P. (1959). Tadžikskaja SSR, *in Geologija SSR*, **24**, 1, Geologičeskoe opisanie.

MARTIN, H. (1969). *Die Kontinentaldrift-Hypothese aus heutiger Sicht.* Geowiss. Tagung, Berlin, 1967, 103.

MAUCHER, A. (1965). Die Antimon–Wolfram–Quecksilber-Formation und ihre Beziehungen zu Magmatismus und Geotektonik. *Freiberger Forschungshefte*, **C186**, 173.

McCLURE, H. A. (1978). Early Palaeozoic glaciation in Arabia. *Palaeogeogr., Palaeoclimatol., Palaeoecol.*, **25**, 4, 315.

MEYER, S. P. (1967). Die Geologie des Gebietes Velian-Kechiré (Zentral-Elburz, Iran). *Mitt. Geol. Inst. E.T.H. Univ. Zürich, N.S.*, **79**.

MILLER, J. F. (1969). Conodont fauna of the Notch Peak Limestone (Cambro–Ordovician), House Range, Utah. *J. Paleont.*, **43**, 2, 413.

MILLER, J. F. and ROBISON, R. A. (1974). Correlation of Tremadocian conodont and trilobite faunas, Europe and North America. *Abstr. 1974 Annual Meeting Geol. Soc. Am.*

MIRZAD, S. CH., KOLČANOV, V. P., and MANUČARYANC, O. A. (1968). Afghanistan, Kurze Angaben über den geologischen Bau und die Bodenschätze. *Bjull. Mosk. Obšč. Ispyt. Pir. Otdel. Geol., N.S.*, **43**, 31.

MONOD, O. (1967). Présence d'une faune ordovicienne dans les schistes de Seydişehir à la base des calcaires du Taurus occidental. *Bull. Min. Res. Explor. Inst. Turkey*, **69**, 79.

MORTON, D. (1959). The geology of Oman. *Proc. 5th World Petrol. Congr.*, **1**, 277.

MOSES, H. F. (1940). *Geological report on Mardin-Cizre.* Maden Tetkik Arama Enst. Yayinlarindan, Archive No. 212 (unpublished report).

MOUSOULOS, L. (1962). *Die Probleme des Untertageabbaues von Erzen in Griechenland.* 29. Antimonerze. 126, Athens.

NATIONAL IRANIAN OIL COMPANY (1959). *Geological map of Iran, scale 1 : 2,500,000, with explanatory notes.*

NEBERT, K. and RONNER, F. (1956). Alpidische Albitisationsvorgänge im Menderes Massiv und dessen Umrahmung. *Bull. Min. Res. Explor. Inst., Turkey*, **48**, 86.

NEUMAYR, M. (1885). Die geographische Verbreitung der Juraformation. *Denkschr. Akad. Wiss. Math.-Naturw. Kl., Wien*, **50**, 57.

NORTH, F. K. (1971). The Cambrian of Canada and Alaska, *in* HOLLAND, C. H. (Ed.), *Lower Palaeozoic rocks of the world.* 1. *Cambrian of the New World*, 219.

O'BRIAN, C. A. E. (1957). Salt diapirism in south Persia. *Geol. Mijnb., N.S.*, **19**, 9, 357.

OKAY, A. C. (1947). Geologische und petrographische Untersuchung des Gebietes zwischen Alemadağ, Karlidağ und Kayisdağ in Koçaeli (Bithynien, Türkei). *Rev. Fac. Sci. Univ. Istanbul, B*, **7**, 269.

OKULITCH, V. J. (1956). The Lower Cambrian of western Canada and Alaska, *in* El Sistema Cambrico, su paleogeografía y el problema de su base. 2. *20th Congr. Geol. Internat., Mexico,* 701.

OMARA, S. (1972). An Early Cambrian outcrop in Southwestern Sinai, Egypt. *Monatsh. Neues Jb. Geol. Palaeont.,* **5,** 257.

ÖPIK, A. A. (1956). Cambrian geology of Queensland, *in* El Sistema Cambrico, su paleogeografia y el problema de su base. 2. *20th Congr. Geol. Internac., Mexico,* 1.

ÖPIK, A. A. (1958). The Cambrian trilobite *Redlichia*: Organization and generic concept. *Bull. Bur. Min. Res. Geol., Geophys.,* **42,** 1.

PAECKELMANN, W. (1925). Beiträge zur Kenntnis des Devons am Bosporus, insbesondere in Bithynien. *Abh. Preussische Geol. Landesanst., N.F.,* **98,** 1.

PAECKELMANN, W. (1938). Neue Beiträge zur Kenntnis der Geologie, Paläontologie und Petrographie der Umgegend von Konstantinopel. 2. Geologie Thraziens, Bithyniens und der Prinzeninseln. *Abh. Preussische Geol. Landesanst., N.F.,* **142,** 1.

PAECKELMANN, W. (1939). Ergebnisse einer Reise nach der Insel Chios. *Z. Deutsch. Geol. Ges.,* **91,** 341.

PAECKELMANN, W. and SIEVERTS, H. (1932). Neue Beiträge zur Kenntnis der Geologie, Paläontologie und Petrographie der Umgegend von Konstantinopel. 1. Obersilurische und devonische Faunen der Prinzeninseln, Bithyniens und Thraziens. *Abh. Preussische Geol. Landesanst., N.F.,* **142.**

PAFFENGOLZ, K. N. (1963). Geologischer Abriss des Kaukasus. *Fortschr. Sowjet. Geol.,* **5/6.**

PALMER, A. R. (1968). Cambrian trilobites of East-Central Alaska. *Prof. Pap. U.S. Geol. Surv.,* **559 B.**

PALMER, A. R. (1969). Cambrian trilobite distributions in North America and their bearing on Cambrian paleogeography of Newfoundland, *in* KAY, H. (Ed.), North-Atlantic—Geology and Continental Drift. *Mem. Am. Assoc. Petrol. Geol.,* **12,** 139.

PALMER, A. R. (1972). Problems of Cambrian biogeography. *24th Internat. Geol. Congr., Canada, 1972,* **7,** 310.

PAMIR, N. H. and CHAPUT, J. E. (1960). Turquie. *Lexique Strat. Internat.,* **3,** Asie, 9c.

PARNES, A. (1971). Late Lower Cambrian trilobites from the Timna area and Har 'Amram (southern Negev, Israel). *Israel J. Earth Sci.,* **20,** 179.

PENCK, W. (1919). Grundzüge der Geologie des Bosporus. *Veröffentl. Inst. Meereskunde, N.F.,* **A4,** 1.

PETRUSHEVSKY, B. A. (1971). On the problem of the horizontal heterogeneity of the earth's crust and uppermost mantle in southern Eurasia. *Tectonophysics,* **11,** 1, 29.

PHILIPPSON, A. (1911). Reisen und Forschungen im westlichen Kleinasien. *Petermanns Geog. Mitt.,* Ergänzungsheft, **172.**

PICARD, L. (1942). New Cambrian fossils and Paleozoic problematica from the Dead Sea and Arabia. *Bull. Geol. Dept. Hebrew Univ. Jersualem,* **4,** 1.

PICHON, X. LE (1968). Sea-floor spreading and continental drift. *J. Geogphys. Res.* **73,** 12, 3661.

PILGER, A. (1971). Die zeitlich-tektonische Entwicklung der iranischen Gebirge. *Clausthaler Geol. Abh.,* **8.**

PILGER, A. and RÖSLER, A. (1977). Gondwana, Indik and Tethys in ihren zeitlich-tektonischen Zusammenhängen. *Z. Deutsch. Geol. Ges.,* **128,** 153.

PILGRIM, G. E. (1908). The geology of the Persian Gulf and the adjoining portions of Persia and Arabia. *Mem. Geol. Surv. India,* **34,** 4.

PILGRIM, G. E. (1922). The sulphur deposits of southern Persia. *Rec. Geol. Surv. India,* **53,** 4, 343.

PILGRIM, G. E. (1924). The geology of parts of the Persian provinces of Fars, Kirman, and Laristan. *Mem. Geol. Surv. India,* **48,** 2.

PILLET, J. (1973). Sur quelques trilobites ordoviciens d'Iran oriental. *Ann. Soc. Géol. Nord.* **93,** 33.

POJETA, J., JR., KŘÍŽ, J., and BERDAN, J. M. (1976). Silurian–Devonian pelecypods and Paleozoic stratigraphy of subsurface rocks in Florida and Georgia and related Silurian pelecypods from Bolivia and Turkey. *Prof. Pap. U.S. Geol. Surv.* **879,** 32 pp.

POWERS, R. W. (1968). Saudi Arabia (excluding Arabian Shield). *Lexique Strat. Internat.,* **3,** Asie, 10bl.

POWERS, R. W., RAMIREZ, L. F., REDMOND, C. D., and ELBERG, E. L., JR. (1966). Geology of the Arabian Peninsula. Sedimentary geology of Saudi Arabia. *Prof. Paper U.S. Geol. Surv.* **560D.**

QUENNELL, A. M. (1951). The geology and mineral resources of (former) Transjordan. *Colon. Geol. Min. Res.* **2,** 85.

REPINA, L. N. (1969). Triloblity nižnego i srednego kembrija juga Sibiri (nadsemijstvo Redlichioidea). 2. *Akad. Nauk SSSR, Sibirskoe Otdel., Trud. Inst. Geol. Geofiz.,* **67,** 108.

REXROAD, C. B. (1967). Stratigraphy and conodont paleontology of the Brassfield (Silurian) in the Cincinnati Arch area. *Bull. Geol. Surv. Indiana,* **36,** 64 pp.

RICHARDSON, R. K. (1926). Die Geologie und die Salzdome im südwestlichen Teile des Persischen Golfes. *Verh. Naturh.-Med. Ver. Heidelberg. N.S.*, **15**, 51 pp.

RICHARDSON, R. K. (1928). Weitere Bemerkungen zu der Geologie und den Salzaufbrüchen am Persischen Golf. *Zentralbl. Mineralog. Geol. Palaeontol. B*, **43**.

RICHTER, R. and RICHTER, E. (1941). Das Kambrium am Toten Meer und die älteste Tethys. *Abh. Senckenberg. Naturforsch. Ges.*, **460**, 50 pp.

RIGO DE RIGHI, M. and CORTESINI, A. (1964). Gravity tectonics in foothill structure belt of southeast Turkey. *Bull. Am. Assoc. Petrol. Geol.*, **48**, 1911.

ROBISON, R. A. and PANTOJA-ALOR, J. (1968). Tremadocian trilobites from the Nochixtlán region, Oaxaca, Mexico. *J. Paleont.*, **42**, 3, 767.

ROGNON, P., BIJU-DUVAL, B., and CHARPAL, O. (1972). Modelés glaciaires dans l'Ordovicien supérieur saharien: phases d'érosion et glaciotectonique sur la bordure Nord des Eglab. *Rev. Géog. Phys. Géol. Dyn.*, (2), **14**, 5, 507.

ROZANOV, A. YU. (1967). The Cambrian lower boundary problem. *Geol. Mag.*, **104**, 415.

RUNNEGAR, B. (1977). Marine fossil invertebrates of Gondwanaland: palaeogeographic implications. *4th Internat. Gondwana Sympos., Calcutta, India*, **3**, 1.

RUSSEGGER, J. (1847). *Reisen in Europa, Asien und Afrika. 3. Reisen in Unter-Ägypten, auf den Halbinsel des Sinai und im gelobten Lande.* Schweizerbart, Stuttgart.

RUTTNER, A., NABAVI, M. H., and ALAVI, M. C. Geology of the Ozbakh Kuh Mountains (Tabas area, East Iran). *Rep. Geol. Surv. Iran*, **5**.

RUTTNER, A., NABAVI, M. H., HAJIAN, J., and others (1968). Geology of the Shirgesht Area (Tabas Area, East Iran). *Rep. Geol. Surv. Iran*, **4**.

SADEK, A. (1977). Early Paleozoic sediments of the Zagros-Taurus Ranges. *Geol. Rdsch.*, **66**, 1, 263.

SAHASRABUDHE, P. W. and MISHRA, D. C. (1966). Palaeomagnetism of Vindhyan rocks of India. *Bull. Nat. Geophys. Res. Inst.*, **4**, 2, 49.

SALEH, H. (1969). A new coral fauna from the Niur Formation (Silurian) of East Iran (preliminary note). *Verh. Geol. Bundesanst.*, **1**, 33.

SAYER, C. (1962). New observations in the Paleozoic sequence of the Bosphorus and adjoining areas, Istanbul, Turkey. *Symp. Silur/Devon-Grenze*, 1960, 222.

SAYAR, C. (1964). Ordovician conulariids from the Bosphorus area, Turkey. *Geol. Mag.*, **101**, 3, 193.

SCHINDEWOLF, O. H. and SEILACHER, A. (1955). Beiträge zur Kenntnis des Kambriums in der Salt Range (Pakistan). *Abh. Akad. Wiss. Mainz, Math.-Naturwiss. Kl.*, 1955, **10**, 257.

SCHMIDT, G. (1965). *Proposed rock unit nomenclature, Petroleum District V, South-east-Turkey.* Turkish Assoc. Petrol. Geol.

SCHMIDT, G. C. (1966). Stratigraphy of the Lower Paleozoic rock units of petroleum district V—Turkey, *in* Petroleum activities in Turkey. *Bull. Petrol. Adm. Pub.*, **11**, 73.

SCHRÖDER, J. W. (1946). Géologie di l'île de Larak. Contribution à l'étude des dômes de sel du Golfe Persique (Comparaison avec la Salt Range). *Arch. Sci. Phys. Nat.*, **28**, 5.

SCHWARZBACH, M. (1974). *Das Klima der Vorzeit. Eine Einführung in die Paläoklimatologie.* 3. Aufl., 380 pp. Stuttgart.

SDZUY, K. (1967). The Tethys in Cambrian time, *in* ADAMS, C. G. and AGER, D. V. (Eds.), *Aspects of Tethyan biogeography.* Systematics Assoc., Publication No. **7**, 5.

SDZUY, K. (1971). Acerca de la correlacion del Cambrico Inferior en la Peninsula Iberica. *1st Congr. Hispano-Luso-Americano Geol. Econ.*, **A1**, 1-1, 753.

SEILACHER, A. (1961). The oldest fossils of Iraq. *Al-Geology*, **2**, 17.

SEILACHER, A. (1963). Kaledonischer Unterbau der Irakiden. *Monatsh. Neues Jb. Geol. Paläont.*, **1963**, 527.

SELLEY, R. C. (1972). Diagnosis of marine and non-marine environments from the Cambro–Ordovician sandstone of Jordan. *J. geol. Soc. London*, **128**, 135.

SETUDEHNIA, A. O. (1972). Iran du sudouest. In Iran. *Lexique Strat. Internat.*, **3**, Asie, 9b, 285.

SLAVIN, V. J. (1976). *Tektonik Afghanistans.* 205 pp. Moscow.

SLAVIN, V. J. and MIRZAD, S. CH. (1969). Tektoničeskoe rajonirovanie Afganistana. *Sov. Geol.*, **1969**, 4, 68.

SMITH, A. G. (1971). Continental drift, *in* GASS, J. G., SMITH, P. J., and WILSON, R. C. L. (Eds.), *Understanding the earth*, Open University, Milton Keynes, 213.

SMITH, A. G., BRIDEN, J. C., and DREWRY, G. E. (1973). Phanerozoic world maps, *in* HUGHES, N. F. (Ed.), Organisms and continents through time. *Spec. Pap. Palaeont.*, **12**, 1.

SMITH, A. G. and HALLAM, A. (1970). The fit of the southern continents. *Nature, London,* **225,** 139.

SPJELDNAES, N. (1961). Ordovician climatic zones. *Norsk Geol. Tidsskr.,* **41,** 45.

STAHL, A. F. (1911). Persien, *in Handbuch der regionalen Geologie,* **8,** 5, Winter, Heidelberg.

STAUFFER, K. W. (1968). Silurian-Devonian reef complex near Nowshera, West Pakistan. *Bull. Geol. Soc. Am.,* **79,** 1331.

STEIGER, R. (1966). Die Geologie der West-Firuzkuh-Area (Zentral-Elburz/Iran). *Mitt. Geol. Inst. E.T.H. Univ. Zürich, N.S.* **68,** 145 pp.

STEINEKE, M., BRAMKAMP, R. A., and SANDER, N. J. (1958). Stratigraphic relations of Arabian Jurassic oil, *in* WEEKS, L. G. (Ed.), Habitat of oil. *Spec. Pub. Am. Assoc. Petrol. Geol.,* 1294.

STÖCKLIN, J. (1960). Ein Querschnitt durch den Ost-Elburz. *Eclogae Geol. Helvet.,* **52,** 2, 681.

STÖCKLIN, J. (1961). Lagunäre Formationen und Salzdome in Ostiran. *Eclogae. Geol. Helvet.,* **54,** 1.

STÖCKLIN, J. (1968a). Salt deposits of the Middle East. In International Conference on Saline Deposits, Houston, Texas, 1962. *Spec. Pap. Geol. Soc. Am.,* **88,** 157.

STÖCKLIN, J. (1968b). Structural history and tectonics of Iran. A review. *Bull. Am. Assoc. Petrol. Geol.,* **52,** 1229.

STÖCKLIN, J. (1972). Iran central, septentrional et oriental. In Iran. *Lexique Strat. Internat.,* **3,** Asie, 9b5.

STÖCKLIN, J. (1977). Structural correlation of the alpine ranges between Iran and Central Asia. *Mém. Hors-Sér. Soc. Géol. Fr.* **8,** 333.

STÖCKLIN, J., NABAVI, M., SAMIMI, M., and others. (1965). Geology and mineral resources of the Soltanieh Mountains (Northwest Iran). *Rep. Geol. Surv. Iran,* **2.**

STÖCKLIN, J., RUTTNER, A., NABAVI, M., and others. (1964). New data on the Lower Paleozoic and Pre-Cambrian of North Iran. *Rep. Geol. Surv. Iran,* **1.**

STOPPEL, D. (1968–70). *Conodonten von Afghanistan.* Geol. Survey F.R. Germany, unpublished repts.

SUDBURY, M. (1957). *Diplograptus spinulosus* sp. nov., from the Ordovician of Syria. *Geol. Mag.,* **94,** 6, 503.

SUESS, E. (1893). Are great ocean depths permanent? *Nat. Sci.,* **2,** 180.

SYLVESTER-BRADLEY, P. C. (1967). The concept of Tethys, *in* ADAMS, C. G. and AGER, D. V. (Eds.), *Aspects of Tethyan biogeography,* Systematics Association Publication No. 7, 1.

TAKIN, M. (1972). Iranian geology and continental drift in the Middle East. *Nature, London,* **235,** 147.

TASMAN, C. E. (1949). Stratigraphy of Southeastern Turkey, *Bull. Am. Assoc. Petrol. Geol.,* **33,** 22.

TAUGOURDEAU, P. and ABDUSSELAMOGLU, S. (1962). Présence de Chitinozoaires dans le Siluro-Dévonien turc des environs d'Istanbul. *Comptes Rendus Soc. Géol. Fr.* **1962,** 238.

TAVERNIER, J. B. (1642). *Les six voyages de Jean Baptiste Tavernier en Turquie, en Perse, et aux Indes,* **1,** 5.

TEICHERT, C. and STAUFFER, K. W. (1965). Paleozoic reef in Pakistan. *Science,* **150,** 3701, 1287.

TELLER, F. (1880). Geologische Beobachtungen auf der Insel Chios. *Denkschr. Math.-Naturw. Kl. Kaiser. Akad. Wiss.,* **11,** 340.

TERMIER, H. and others (1975). Cambrian stromatolites from Lakkarkuh, east-central Iran. *Rep. Geol. Surv. Iran,* **32,** 35.

TERMIER, H. and TERMIER, G. (1977). Étude bio-sédimentaire de Stromatolithes recueillis dans la sérlie de Zargaran (Afghanistan central). *Mém. Hors-Sér. Soc. Géol. Fr.* **8,** 263.

THOMAS, A. T. (1977). Classification and phylogeny of homalonotid trilobites. *Palaeontology,* **20.** 1, 159.

THRALLS, H. W. and HASSON, R. C. (1956). Geology and oil resources of eastern Saudi Arabia. *20th Internat. Geol. Cong., Mexico, Symp. Yacimientos Petrol. Gas,* **2,** 9.

TOKAY, M. (1952). Contribution à l'étude géologique de la région comprise entre Ereğli, Alapi, Kiziltepe et Alacaağzi. *Bull. Min. Res. Explor. Inst. Turkey,* **42/43,** 37.

TOLUN, N. (1960). Stratigraphy and tectonics of southwestern Anatolia. *Rev. Fac. Sci. Univ. Istanbul,* **B25,** 201.

TOLUN, N. and TERNEK, Z. (1952). Notes géologiques sur la région de Mardin. *Bull. Geol. Soc. Turkey,* **3,** 15.

TROMP, S. W. (1941). Preliminary compilation of the stratigraphy, structural features and oil possibilities of South Eastern Turkey. "Meteae", *Pub. Min. Res. Inst. Turkey,* **A4,** 1.

TSCHOPP, R. H. (1967). The general geology of Oman. *Proc. 7th World Petrol Congr. Mexico,* **2,** 231.

TUCKER, M. E. and REID, P. C. (1973). The sedimentology and context of late Ordovician glacial marine sediments from Sierra Leone, West Africa. *Palaeogeogr., Palaeoclimatol., Palaeoecol.,* **13,** 289.

TÜRKÜNAL, S. (1953). Géologie de la région de Hakkari et de Başkale (Turquie). *Maden Tetkik Arama Enst. Yayinlarindan,* **B,** 18.

ÜNSALANER-KIRAGLI, C. (1958). *Alveolites lemniscus* Smith from the Upper Silurian of Sedef Adasi (Antirovitha) with remarks on the genera *Roxoporella* and *Kitakamiia*. *Bull. Min. Res. Explor. Inst. Turkey*, **50**, 83.

U.S. GEOLOGICAL SURVEY und ARABIAN–AMERICAN OIL COMPANY (1963). *Geologic map of the Arabian Peninsula, 1 : 2,000,000*.

VACHÉ, R. (1966). Zur Geologie der Varisziden und ihrer Lagerstätten im südanatolischen Taurus. *Min. Depos.*, **1**, 1, 30.

WALLISER, O. H. (1964). Conodonten des Silurs. *Abh. Hessisches Landesamt Bodenforschung*, **41**.

WALTHER, H. W. (1968). On the genesis of the iron ore of the Hormuz-Series near Bandar Abbas (Cambrian, SE-Iran). *Abstr. 23rd Internat. Geol. Congr.*, 224.

WALTHER, H. W. (1972). Über Salzdiapire in Südost-Iran. *Geol. Jb.*, **90**, 359.

WEBER, H. (1963). Ergebnisse erdölgeologischer Aufschlußarbeiten der DEA in Nordost-Syrien. 1. Schichtenfolge, Fazies und Tektonik in der Haute Djésireh. *Erdöl Kohle*, **16**, 6, 669.

WEIPPERT, D. (1967). *Bericht über geologische Untersuchungen im Gebiet Panjaw-Germao-Deh Kundi (Zentral Afghanistan)*. Geol. Survey F.R. Germany, unpublished report.

WEIPPERT, D., WITTEKINDT, H., and WOLFART, R. (1970). Zur geologischen Entwicklung von Zentral- und Südafghanistan. *Beih. Geol. Jb.*, **92**.

WEIPPERT, D. and WOLFART, R. (1968). Altordovizische Trilobiten von Kirman Panjao im östlichen Zentral-Afghanistan. *Monatsh. Neues Jb. Geol. Palaeont.*, **1968**, 11, 699.

WEISSERMEL, W. (1939). Neue Beiträge zur Kenntnis der Geologie, Paläontologie und Petrographie der Umgegend von Konstantinopel. 3. Obersilurische und devonische Korallen, Stromatoporiden und Trepostome von der Prinzeninsel Antirovitha und aus Bithynien. *Abh. Preussische Geol. Landesanst., NF.*, **190**, 1.

WETZEL, R. and MORTON, D. M. (1959). Contribution a la géologie de la Transjordanie. *Notes Mém. Moyen-Orient*, **7**, 95.

WHITTINGTON, H. B. (1966). Phylogeny and distribution of Ordovician trilobites. *J. Paleont.*, **40**, 3, 696.

WHITTINGTON, H. B. and HUGHES, C. F. (1972). Ordovician geography and faunal provinces deduced from trilobite distribution. *Phil. Trans. R. Soc. London*, **B263**, 235.

WHITTINGTON, H. B. and HUGHES, C. P. (1973). Ordovician trilobite distribution and geography, *in* HUGHES, N. F. (Ed.), Organisms and continents through time. *Spec. Pap. Palaeont.*, **12**, 235.

WILLIAMS, A. (1973). Distribution of brachiopod assemblages in relation to Ordovician palaeogeography, *in* HUGHES, N. F. (Ed.), Organisms and continents through time. *Spec. Pap. Palaeont.*, **12**, 241.

WITTEKINDT, H. (1967). *Bericht über geologische Untersuchungen im Gebiet Jalalabad (Ost-Afghanistan)*. Geol. Survey F.R. Germany, unpublished report.

WOLFART, R. (1967a). Geologie von Syrien und dem Libanon, *in* MARTINI, H.-J. (Ed.), *Beitr. Region. Geol. Erde*, **6**, Born-träger, Berlin.

WOLFART, R. (1967b). Zur Entwicklung der paläozoischen Tethys in Vorderasien. *Erdöl Kohle*, **20**, 168.

WOLFART, R. (1969). Die kambro-ordovizische Schichtenfolge von Surkh Bum bei Panjaw im östlichen Zentralafghanistan. *Geol. Jb.*, **87**, 545.

WOLFART, R. (1970a). Fauna, Stratigraphie und Paläogeographie des Ordoviziums in Afghanistan. *Beih. Geol. Jb.*, **89**.

WOLFART, R. (1970b). The age of the Early Tremadocian and of the *Saukia* Zone and the boundary between Cambrian and Ordovician. *Newsl. Strat.*, **1**, 1, 10.

WOLFART, R. (1973). Das Kambrium im mittleren Südasien (Iran bis Nordindien). *Zentralbl. Geol. Paläont.*, **1**, **1972**, 5/6, 227.

WOLFART, R., BENDER, F., and STEIN, V. (1968). Stratigraphie und Fauna des Ober-Ordoviziums (Caradoc–Ashgill) und Unter-Silurs (Unter-Llandovery) von Südjordanien. *Geol. Jb.*, **85**, 517.

WOLFART, R., and KÜRSTEN, M. (1974). Fauna, Stratigraphie und Paläogeographie des Kambriums in Südostiran und Afghanistan. *Geol. Jb.*, **B8**, 234 pp.

WOLFART, R. and WITTEKINDT, H. (1980). Geologie von Afghanistan, *in* BENDER, F. *Beitr. Region. Geol. Erde*, **14**, Bornträger/Schweizerbart, Berlin/Stuttgart.

YALÇINLAR, I. (1955a). Note préliminaire sur les schistes à graptolithes du Silurien découverts près d'Istanbul. *Publ. Inst. Géog. Univ. Istanbul*, **18**, 167.

YALÇINLAR, I. (1955b). Note préliminaire sur les schistes à graptolithes du Silurien de Feke au nord d'Adana (Turquie). *Publ. Inst. Géog. Univ. Istanbul*, **18**.

YALÇINLAR, I. (1956). Istanbulda bulunan graptolithi Silur sistleri hakkinda. *Istanbul Univ. Geog. Enst. Dergisi*, **4**, 157.

YALÇINLAR, I. (1957a). Une faune graptolithique dans la chaîne de Stranca (Turquie). *Rev. Georg. Inst. Istanbul*, **3**, 13.

YALÇINLAR, I. (1957b). Présence de schistes à graptolithes au S de Bileçik (Turquie). *Comptes Rendus Soc. Géol. Fr.* **1957**, 283.

YALÇINLAR, I. (1959a). Note sur des schistes à graptolithes découvertes au NW d'Izmit (Turquie). *Comptes Rendus Soc. Géol. Fr.* **1959**, 82.

YALÇINLAR, I. (1959b). Sur le terrain du Primaire ancien au Sud d'Akşehir (Turquie). *Comptes Rendus Soc. Géol. Fr.* **1959**, 215.

YALÇINLAR, I. (1959c). Decouvertes de séries à Stromatolithes anciens en Turquie. *Comptes Rendus Soc. Géol. Fr.* **1959**, 215.

YALÇINLAR, I. (1963). Graptolite series belonging to Silurian found in the Mediterranean region of Turkey. *Pub. Inst. Géog. Univ. Istanbul*, **36**.

YALÇINLAR, I. (1964). Les couches du Paléozoique inférieure dans la Turquie meridionale. *Pub. Inst. Géog. Univ. Istanbul*, **39**.

YALÇINLAR, I. (1969). Le Massif d'Anamuret ses caracteres géomorphologiques (Turquie). *Rev. Geog. Inst. Univ. Istanbul, Internat. Ed.*, **12**.

YALÇINLAR, I. (1973). Observations sur la faune du primaire ancien trouvée dans la région méditerranéenne de la Turquie. *Bull. Geol. Soc. Turkey*, **16**, 1, 101.

Lower Palaeozoic of the Middle East, Eastern and Southern Africa, and Antarctica
Edited by C. H. Holland
© 1981 John Wiley & Sons Ltd.

LOWER PALAEOZOIC ROCKS OF LIBYA, EGYPT, AND SUDAN*

Eberhard Klitzsch

Institut für Geologie und Paläontologie, Technische Universität, Berlin, Germany

Contents

* The author notes that for Libya his publication reflects knowledge of the early 1970s. He has, however, incorporated his recent research in south-west Egypt and north-west Sudan. (C.H.H.)

1. Introduction

More than two thirds of Libya are underlain or covered by rocks of Lower Palaeozoic age. In Egypt, part of western Egypt and Sinai are underlain by Lower Palaeozoic rocks which are only locally at the surface. In Sudan, Lower Palaeozoic strata are present only in the north-western corner of the country. Consequently, this description deals mainly with the Lower Palaeozoic rocks of Libya. It has to be stated from the beginning that our knowledge of the Cambrian and Ordovician strata is still very incomplete.

Palaeozoic strata were first described here by OVERWEG (1851), a young German geologist, who accompanied BARTH and RICHARDSON on their way to Bornu. The first systematic reconnaissance was carried out mainly by Italian and French geologists and resulted in the identification of Silurian shale together with underlying and overlying sandstone formations in western Libya and bordering areas of Algeria. The main explorers of this period between the early 1920s and the early 1950s were C. KILIAN, A. DESIO, M. LELUBRE, and J. M. FREULON (*see* References). Strata of possible Lower Palaeozoic age at Jebel Auenat and surrounding areas at the Egyptian, Libyan, and Sudanese borders were first reported by K. S. SANDFORD.

When oil exploration began during the 1950s, much of the Palaeozoic strata at the borders of the Homra, Murzuk, and Kufra Basins were not subdivided and the real extent of the Silurian transgression was not known. It was mainly because of the Silurian shale, which is considered to be a potential source rock for oil, that most parts of south-western and south-eastern Libya and bordering areas in Sudan and Egypt were investigated by geological teams of various oil companies. Unfortunately, much of their work was not published. Under the leadership of P. F. BUROLLET, the Petroleum Exploration Society of Libya soon began to encourage its members to discuss—among others—problems of Palaeozoic strata. As a result, the 'Lexique Strati-graphique International' on Libya was published in 1960 and a symposium on Saharan geology was organized in 1963. A second symposium followed in 1969, organized by the Faculty of Science of the Libyan University under C. GRAY. In 1964, L. C. CONANT and G. H. GOUDARZI published their map (1:2 million) of the geology of Libya, which in large parts is based on contributions of several oil companies. Present day knowledge of Lower Palaeozoic sediments in Libya and bordering areas in Egypt and Sudan results mainly from the publications of P. F. BUROLLET, R. CARDELLO, G. R. COLLOMB, M. FUERST, V. HAVLÍČEK, H. JAEGER, M. JACQUÉ, E. KLITZSCH, D. MASSA, J. J. MENNING, B. PLAUCHUT, L. K. RATSCHILLER and P. VITTIMBERGA. R. SAID published results of subsurface exploration in north-western Egypt. A. J. WHITEMAN summarized the scanty knowledge of Lower Palaeozoic rocks of Sudan and T. WEISSBROD described the Lower Palaeozoics of Sinai.

During a 9-year stay in Libya the present author visited most of the areas where Lower Palaeozoic rocks are exposed in Libya, Niger, Chad, and along the Sudanese and Egyptian border. The remote position and the difficult accessibility of several of these areas were the major difficulties, but excellent exposures made up for this. Since 1976 the author has investigated also the geology of south-western Egypt and north-western Sudan.

2. Regional distribution

Rocks of Lower Palaeozoic age cover large areas in central and southern Libya (Figs. 1 and 2). They are exposed at the southern edge of the Homra Basin in the area of Jebel Gargaf, along the western and eastern flanks of the Murzuk Basin (extending into Algeria, Niger, and Chad), and in some areas in the surroundings of the Kufra Basin. The latter marginally extends into south-western Egypt and—across north-eastern Chad—into north-western Sudan. In Egypt and Sudan the exposures of Lower Palaeozoic rocks cover relatively small areas. While in Sudan the subsurface extension of these strata is most likely limited to the north-western corner of the country, in Egypt sediments of mainly Silurian age underlie younger strata in a larger area along the Libyan border and in north-western Egypt. Sandstone of Cambrian age is also present in a small area in Sinai, forming part of the 'Lower Paleozoic Sandstone' (see Fig. 1).

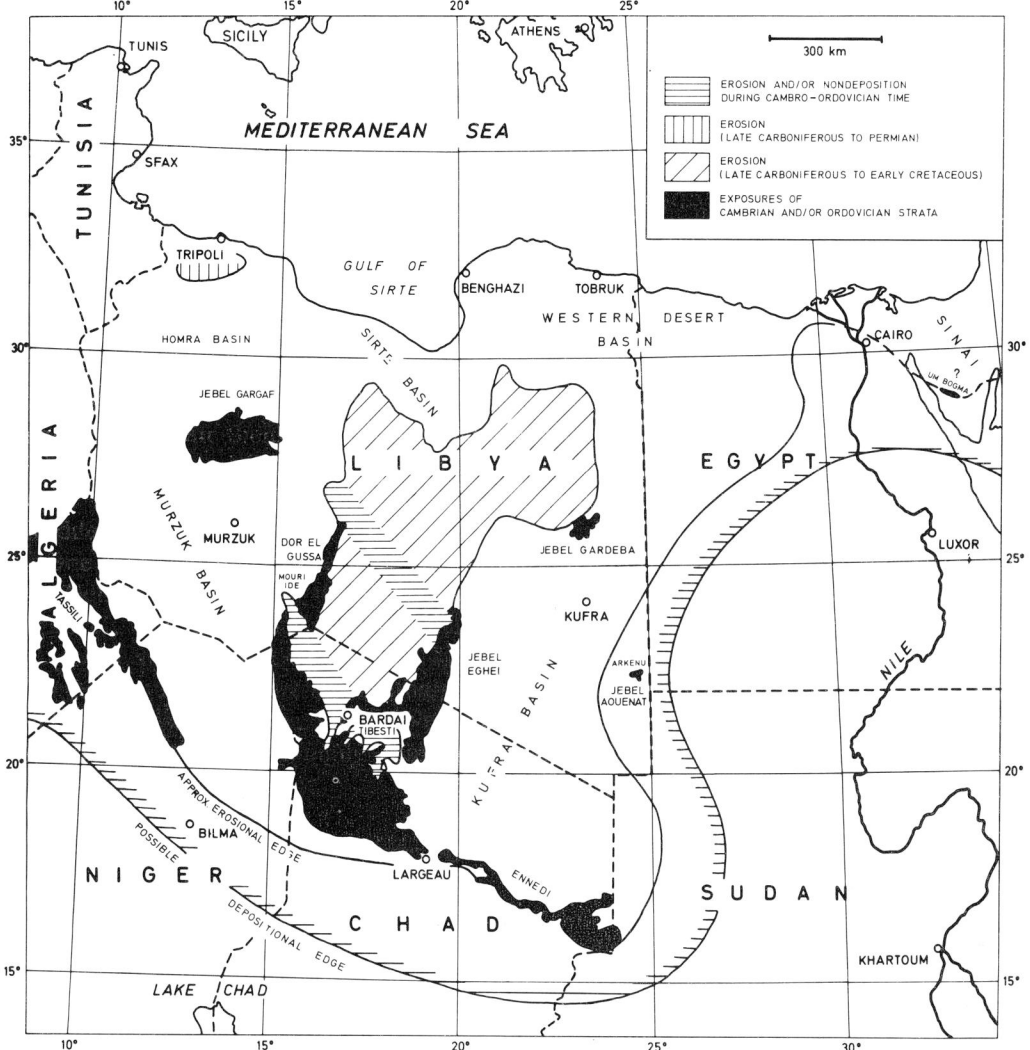

FIG. 1. Distribution of Cambrian and Ordovician strata (after KLITZSCH, 1970).

While the subsurface distribution of Silurian strata (Fig. 2) is limited to the Homra Basin in north-western Libya, the major part of the Murzuk Basin in south-western Libya, the Kufra Basin in south-eastern Libya (and bordering areas of Sudan and Egypt), and the Western Desert Basin in north-eastern Libya and north-western Egypt, Cambrian and Ordovician strata in addition occur in the subsurface of most of the Sirte Basin in north-central Libya (Fig. 1). Here

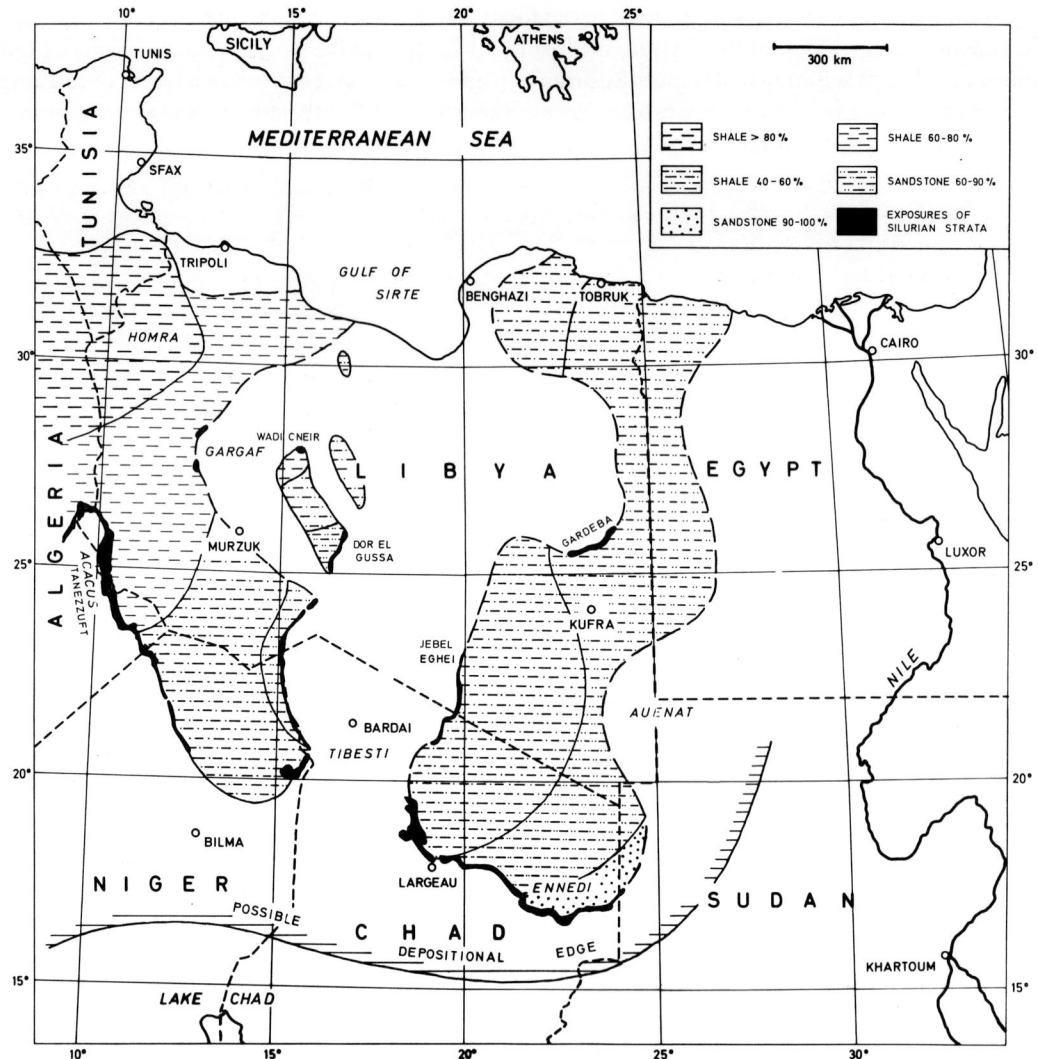

FIG. 2. Distribution and facies of Silurian strata (after KLITZSCH, 1970).

Cambrian and Ordovician rocks directly underlie sediments of Mesozoic age, while in the other basins Cambrian to Silurian sediments from the lower part of more or less complete Palaeozoic successions.

For the stratigraphical, palaeogeographical, and tectonic understanding of Lower Palaeozoic rocks in the countries concerned, the most important areas of exposure are the following: for stratigraphy of Cambrian and Ordovician strata, Jebel Gargaf in northern Fezzan; for stratigraphy of Silurian strata, Wadi Tanezzuft and Jebel Acacus in south-western Fezzan; and for tectonic and palaeogeographical development, Dor el Gussa in eastern Fezzan.

3. Tectonic framework and palaeogeographical development

Consolidation of central and eastern North Africa took place in Pre-Cambrian times. Since then, most of the area has formed a marginal northern part of the African craton, which by turns was transgressed by Palaeozoic to recent seas or exposed to erosion and continental deposition. Most Palaeozoic transgressions originated from western and north-western directions, while Mesozoic and Tertiary transgressions came from the north (present day Mediterranean Sea, Tethys geosyncline).

The last Pre-Cambrian orogeny was followed by a long period of erosion and penelplanation. In Libya, Cambrian sediments overlie the Pre-Cambrian rocks with a strong unconformity. While Pre-Cambrian strata are intensively folded, Cambrian and younger sediments are almost flat lying and dip only with a very slight angle towards the centre of the different basins.

The sediments of Cambrian age are exclusively clastic; they consist of fine- to coarse-grained and partly conglomeratic sandstone and arkose. Most of the strata are cross-bedded and seem to represent a fluvio-continental environment with intercalations of windblown deposits. Near the top of the Cambrian strata, marine influence is evident (*Tigillites* beds). Probably the paleogeographical situation is to be characterized as follows. During Cambrian time erosion in more central parts of the African craton (present day republics of Niger, Chad, and Sudan) continued and most of the material was redeposited at the northern margin of the craton under continental depositional conditions. Some of the material originated from areas of erosion within this predominantly depositional northern foreland of the craton.

In two areas at the eastern edge of the Murzuk Basin it is evident that during Cambrian time a N.N.W.–S.S.E. striking structural relief developed (KLITZSCH, 1970). At Mouri Ide (around 24° N and 15°45′ E) a 40 kilometres wide horst striking N.W. to N.N.W. was formed. At its core Silurian strata directly overlie Pre-Cambrian rocks; north-east and south-west of the horst several hundred metres of Cambrian clastics are present.

At northern Dor el Gussa (around 25°55′ N and 16°50′ E) the south-western foreland of a N.W. to N.N.W. striking horst is exposed. In this area more than 1700 metres of Cambrian clastics, dipping up to 35 degrees south-west, are unconformably overlain by 5 degree dipping strata of Ordovician age. According to direction of transport and sediment characteristics (high feldspar content, coarse-grained and conglomeratic sandstone and arkose) the thick Cambrian sequence originates from the horst farther north-east, which evidently began to be formed while the clastics were deposited, with major movements towards the end of Cambrian times (strong unconformity between Cambrian and Ordovician strata).

During Ordovician and Silurian times, the Cambrian horsts became the core of wide uplifts striking north-north-westwards (*see* Fig. 3 and KLITZSCH, 1970). These uplifts are characterized by reduced deposition and by temporary erosion. In palaeogeographical respect they are important because their direction of strike is more or less the same as the direction of the Palaeozoic transgressions. It was probably because of this N.N.W.–S.S.E. striking structural relief that the Palaeozoic seas were able several times to transgress large parts of the African craton (unlike the younger transgressions, most of which were limited to the northern edge of the continent, because direction of transgressions and structural relief were approximately perpendicular).

After early Caledonian movements most of Libya and bordering areas of north-western Egypt were temporarily transgressed by the Ordovician sea, leaving fine-grained sandstone with trace fossils (Haouaz) and locally silty shale (Melez Chograne) behind. At the eastern edge of the Murzuk Basin, I found glacial deposits (tillite) below the Silurian shale. Similar sediments have been reported from Jebel Gargaf (COLLOMB, 1962; HAVLÍČEK and MASSA, 1973). During late Ordovician or early Silurian times Libya was partly included in the area of glaciation, which is known already from southern Algeria (GARIEL and others, 1968). It is likely that the late Ordovician sandstone which unconformably covers a pre-existing Ordovician relief is partly a glacial deposit.

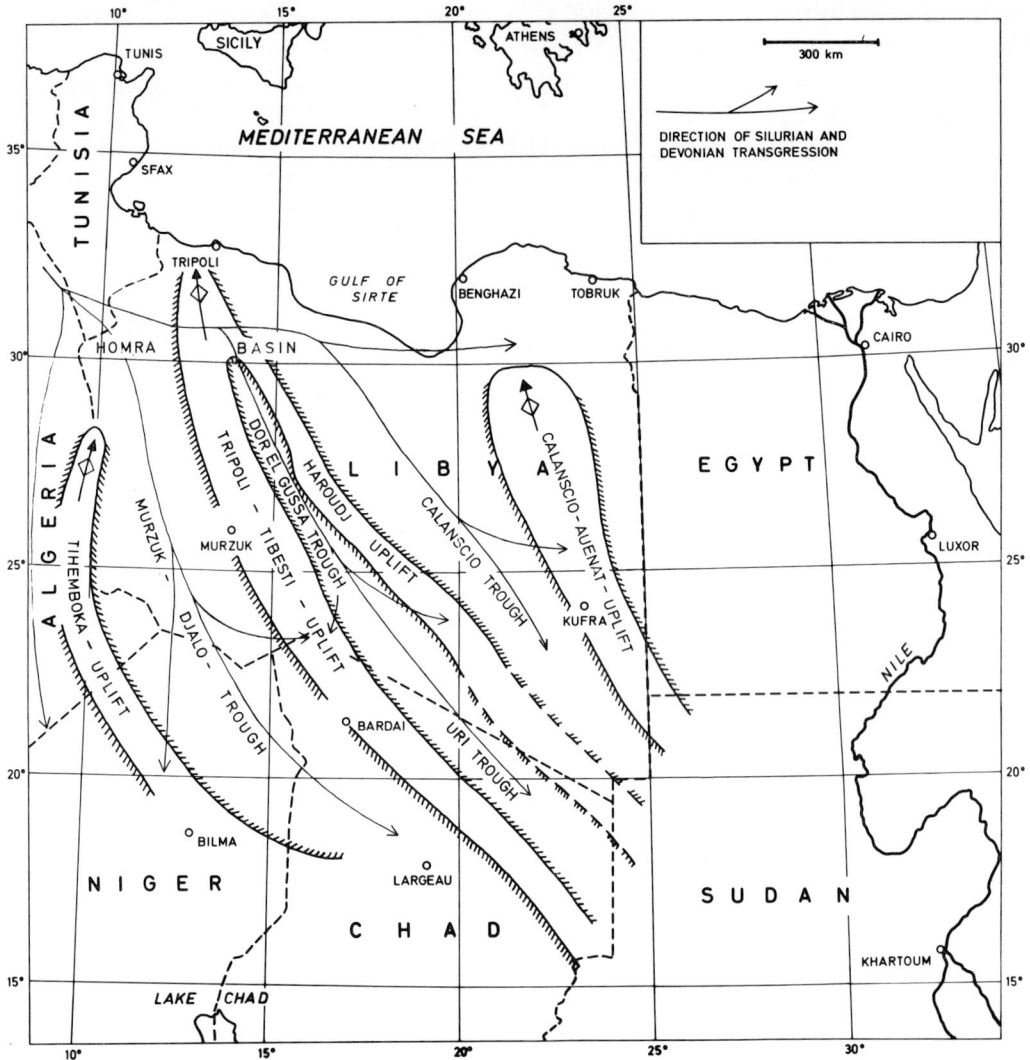

FIG. 3. Structural relief during Lower Palaeozoic and Devonian times (after KLITZSCH, 1970).

In Silurian times, the subsidence of the north-north-westerly striking troughs between the different uplifts continued and the sea returned. Within the troughs the Silurian sea left several hundred metres of sediment behind, while at the uplifts only a thin and incomplete sequence was deposited, which in some areas was eroded in late Silurian time. Locally the tectonic movements were accompanied by basic volcanism. In the middle Llandovery (zone 19) the Silurian sea reached its greatest extension (Fig. 2), covering most of Libya, western Egypt, north-western Sudan, and the northern parts of Chad and Niger.

In the south, regression began during Llandovery time, characterized by sediments typical of a shallow water environment. This regression facies (Acacus Sandstone) reached the Homra Basin in north-western Libya not before late Silurian time (Ludlow; see Fig. 4).

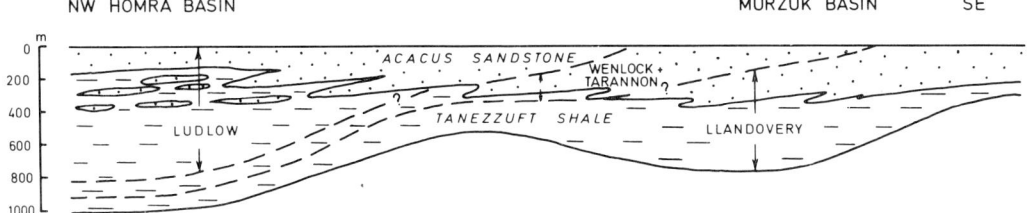

FIG. 4. Schematical time and facies section N.W.–S.E. through Silurian strata.

4. General description of Cambrian and Ordovician strata

In large parts of the area concerned subdivision of Cambrian and Ordovician strata has not so far been carried out.

In the surroundings of the Kufra Basin and in most areas around the Murzuk Basin, sediments of the two systems have been mapped as undifferentiated Cambro–Ordovician. The few areas where clear breakdown of the different Cambrian and Ordovician rock units is possible are not yet sufficient to separate the different formations from each other over larger distances. The only two regions where separation is relatively clear and easy are Jebel Gargaf, between the Homra and Murzuk Basins, and northern Dor el Gussa, at the eastern edge of the Murzuk Basin (*see* Sections 6.2 and 6.3.1).

According to SAID (1962) and WEISSBROD (1969), Cambrian sediments found in the subsurface of northern Egypt (Baharia) and in Sinai are clastic, reach a thickness of several hundred metres, and contain marine fossils (trilobites, or trilobite tracks, and brachiopods).

Most of the strata thought to be of Cambrian age in Libya nevertheless seem to be of continental origin. They consist of the exclusively clastic Hassaouna Sandstone Formation (type area at Jebel Gargaf; *see* Section 6.2.2), which is made of sandstone and arkose, conglomeratic at the base, mainly cross-bedded, azoic with the exception of the upper part, where beds containing trace fossils (*Tigillites*) occur locally. The thickness in most areas is of the order of 250 to 500 metres; locally it exceeds 1000 metres. The same or similar sandstone is known from the subsurface of northern Libya.

The Hassaouna Formation seems to be a fluvio-continental sediment with aeolian intercalations, which was deposited along the northern edge of the African craton. Marine intercalations near the top of the formation and the presence of marine fossils of Cambrian age in northern and north-eastern Egypt indicate that the Cambrian sea was situated in the eastern part of North Africa during part of Cambrian time.

Where clear separation of Cambrian and Ordovician formations is possible—at Jebel Gargaf and northern Dor el Gussa—the Ordovician strata are subdivided into three formations. The basal Ordovician Haouaz Formation unconformably overlies the Cambrian Hassaouna Formation and is conformably overlain by the Melez Chograne Shale which, in turn, is unconformably overlain by the late Ordovician Memouniat Sandstone. The total thickness of all three Ordovician formations is between 100 and 350 metres. Equivalents of the three Ordovician rock units can be found at several places in the southern surroundings of the Murzuk Basin and locally also at the edge of the Kufra Basin.

The basal Haouaz Formation consists of well bedded fine- to coarse-grained sandstone containing abundant trace fossils (mainly *Tigillites*, rare *Cruziana* sp., *Harlania* sp.). Subsurface correlations with wells in the Homra Basin indicate a lower to middle Ordovician age (Tremadoc to Llandeilo: COLLOMB, 1962; HAVLÍČEK and MASSA, 1973). *Tigillites* beds also occur near the top of the Cambrian Hassaouna Formation and locally within the Memouniat Formation; consequently, identification of the Haouaz Formation is difficult where the Ordovician succession is incomplete. This is the case over large distances in the surroundings of the Murzuk and Kufra Basins.

The Melez Chograne Formation consists mainly of silty multi-coloured shale and siltstone, interbedded with fine-grained sandstone. This relatively thin unit (60 metres maximum) represents the culmination of the Ordovician transgression. According to its fossil content (mainly brachiopods) it is of Caradoc to Ashgill age (HAVLÍČEK and MASSA, 1973). West of Jebel Gargaf in the type area the formation contains sediments of glacial origin. Along the eastern and western edge of the Murzuk Basin and around the Kufra Basin, the Melez Chograne Shale is only locally present. Probably much of the formation was eroded after late Ordovician movements and possibly also by glacial events. At southern Dor el Gussa (eastern Murzuk Basin) the upper part of the Cambro–Ordovician strata consists of tillite (Fig. 8).

The Memouniat Formation is at the top of the Ordovician sequence and consists of fine- to coarse-grained, partly ferruginous sandstone. At the type area it occurs in different facies laterally grading into each other (COLLOMB, 1962). While much of the formation is cross-bedded, barren sandstone of fluvio-continental (possibly partly fluvio-glacial) origin, some of it is well bedded and contains marine fossils (mainly brachiopods) of uppermost Ordovician age (HAVLÍČEK and MASSA, 1973). Apart from the type area and northern Dor el Gussa, typical Memouniat Sandstone is present west of northern Jebel Acacus (western edge of the Murzuk Basin) and at several other localities in the surroundings of the Kufra and Murzuk Basins.

5. General descriptions of Silurian strata

Sediments of Silurian age cover large areas of north and north-east Africa (Fig. 2). They are characterized by two typical types of facies: the lower part of the Silurian sediments generally consists of fine-clastic material (Tanezzuft Shale), while the upper part is dominated by more or less fine-grained sandstone and siltstone (Acacus Sandstone). The Tanezzuft Shale contains mainly planktonic and nektonic fauna (graptolites, orthoceratids, small bivalves). Trace fossils typical of a shallow water environment are frequent in the Acacus Sandstone (*Tigillites* sp., *Cruziana* sp., and especially *Harlania* sp.). Thus the Tanezzuft Shale represents the sediments deposited during the transgression and in areas far from the coast, while the Acacus Sandstone generally was deposited during the regression of the Silurian sea and near its shoreline.

In general, the sand content in both formations increases towards the southern edge of the Silurian transgression and the total content of shale decreases. Both formations are typical rock-stratigraphical units and not time units: the Silurian transgression began in early Llandovery times (zone 16) and reached its maximum in the middle Llandovery (zone 19). At this time the Silurian sea covered more or less all of Libya, northern Chad and Niger, north-western Sudan, and western Egypt. Within the outer areas of the Silurian transgression the regression facies (Acacus Sandstone) is of Llandovery age (zone 19–20 at Jebel Acacus) and it reached the Homra Basin in northern Libya not before late Silurian time (Ludlow; *see* Fig. 4). In accordance with the palaeogeographical situation sediments of lower Silurian age are very thick within the southern areas of transgression (up to 650 metres in Jebel Acacus and up to 625 metres in Dor el Gussa) and very thin in the northern Libyan Homra Basin, which was far from any source area and possibly also cut off from major sediment supplies by submarine relief. Sediments of late Silurian age are very thick in the north (up to 800 metres in the Homra Basin), because at that time the Silurian sea had left southern Libya and the areas south and east of it and the source areas were close to the north Libyan Homra Basin.

Unlike the different Ordovician formations, the two rock-stratigraphical units of Silurian age are easy to differentiate from each other and from other formations, owing to their typical fossil assemblage. In general graptolites become rare towards the edge of the transgression, but they do not disappear. With the exception of Jebel Auenat and some areas of late Silurian erosion at the northern and eastern edge of Murzuk Basin, sediments of Silurian age are well exposed around the Murzuk and Kufra Basins.

6. Regional description

6.1. Homra Basin

The Homra Basin—also called the Ghadames Basin in some of the literature—is situated in north-western Libya and extends into Tunisia and Algeria. This Palaeozoic basin is covered by Mesozoic and Palaeocene sediments. A north–south striking zone of relatively thin Tanezzuft Shale and thin or missing Acacus Sandstone (Tripoli–Tibesti–Uplift, *see* Fig. 3) subdivides the basin into an eastern and a western part. Only at the southern edge of Homra Basin at Jebel Gargaf are Lower Palaeozoic strata exposed (*see* Section 6.2). According to unpublished oil company reports, the Homra Basin contains several thousand metres of Palaeozoic strata, the lower part of which is of Cambrian to Silurian age. The little information available on Cambrian and Ordovician strata indicates similarity with the Cambrian and Ordovician formations of Jebel Gargaf, with a tendency for greater thickness and more evident marine influence in the subsurface of the Homra Basin. BONNEFOUS (1963) has published details of some exploration wells in southern Tunisia close to the Libyan border, which partly or totally penetrated the Silurian sequence. The publication also includes long lists of fauna (mainly graptolites), which indicate that over a relatively thin shale of Llandovery and Wenlock age several hundred metres of Ludlow age sediments (including sandstone in the upper part) are present. BONNEFOUS also reports radioactive tracer beds within the Llandovery and at the top of the Wenlock shale. MASSA and JAEGER (1971) are the only authors to have published in some detail on the stratigraphy of Silurian sediments of the Libyan part of the Homra Basin, including a well-defined fauna. They reached conclusions similar to those of BONNEFOUS: above relatively thin strata of Llandovery and Wenlock age (mainly shale, less than 100 metres), several hundred metres of Ludlow age shale occur, which are intercalated with sandstone and siltstone within the upper part of the sequence. The uppermost 200 metres of the Silurian strata are characterized by increasing amounts of spores and plants (*Cooksoniaceae*) and by the absence of graptolites, while fragments of other marine fossils still continue to be present (tentaculites, ostracodes, crinoids, orthoceratids, and others). Thus increasing continental influence is documented towards the end of Ludlovian sedimentation, which was the result of Caledonian movements. Early Devonian continental sandstone overlies unconformably.

MASSA and JAEGER propose to replace the term Tanezzuft Shale by 'Principal Shale Formation' and the term Acacus Sandstone by 'Interbedded Sandy-Shale Formation'. Because the older terms are commonly used in Libya, it is not likely that the later ones will replace them.

6.2. Jebel Gargaf

6.2.1. General description and exploration history

At or near Jebel Gargaf are the type areas of Libyan strata of Cambrian and Ordovician age. Formation names and brief descriptions were given by the Names and Nomenclature Committee of the Petroleum Exploration Society of Libya and published in 1960 in the 'International Stratigraphic Lexicon'. Unfortunately, no exact type section or detailed description was published. Cambrian and Ordovician strata of the Jebel Gargaf area together were called the Gargaf Group. It does, however, not seem advisable to place the formations of the Libyan Cambrian and Ordovician in one group, because they were deposited in very different environments and are partly separated by strong unconformities.

The Jebel Gargaf area is an east–west striking morphological high in northern Fezzan. It is part of the late Palaeozoic to Mesozoic uplift (Gargaf uplift; KLITZSCH, 1970), which separates the Murzuk Basin in south-western Libya from the Homra Basin in north-western Libya. The central part of Jebel Gargaf (also called Jebel Fezzan and Jebel Hassaouna) is formed of mountains up to approximately 650 metres high. They consist of Cambrian Hassaouna Sandstone. At several places in this area, Pre-Cambrian basement is exposed below the base of the Cambrian strata. The mountainous central part of Jebel Gargaf is surrounded by rocky plains which are 350–550 metres high and are also made of Cambrian Hassaouna Sandstone, which is covered in the north and east by marine sediments of Cretaceous age. Southward towards Wadi Chatti, the Cambrian sandstone is directly overlain by sandstone and shale of Middle to Upper Devonian age. This unconformity is the result of uplift and erosion during the Caledonian episode, when part of Jebel Gargaf was included in a structural high striking N.N.W.–S.S.E. (Klitzsch, 1970, and Fig. 3), perpendicular to the present striking direction of Jebel Gargaf and the Gargaf uplift. At a small exposure east of Jebel Gargaf and a few kilometres north of Wadi Cneir the presence of Silurian Tanezzuft Shale and underlying Ordovician Haouaz Sandstone indicate that the Lower Palaeozoic strata become complete towards the eastern flank of the Caledonian uplift (northern part of the Dor el Gussa Trough). At the western edge of Jebel Gargaf in the areas of Jebel Haouaz, Melez Chograne, and Memouniat (type areas of the Ordovician strata of Libya) a complete sequence of the sediments of the Libyan Ordovician as well as some Silurian (and some early Devonian) strata are present, indicating the western edge of the Caledonian uplift (Fig. 5). In this area the stratigraphical subdivision of Cambrian and Ordovician strata was carried out first in Libya, which remains best investigated within the countries concerned.

Although OVERWEG (1851) had discovered marine Devonian sediments near Auenat Ouenine, Italian authors thought that most of the Jebel Gargaf strata are part of the so called 'Nubian Sandstone'. Others believed that they were all of Devonian age. The modern interpretation of the area began when LELUBRE (1946) discovered Silurian shale near Bir el Gasr. MASSA and COLLOMB (1960) published the first detailed subdivision of the Early Palaeozoic formations, using stratigraphical names which the Petroleum Exploration Society of Libya (PESL, 1960) suggested at the same time. COLLOMB (1962) completed the description and published a detailed map (Fig. 5). Finally, HAVLÍČEK and MASSA (1973) added new information on fauna, age, and environment of part of the Ordovician strata.

6.2.2. Hassaouna Formation

The type area is Jebel Hassaouna (Fig. 5), according to MASSA and COLLOMB (1960) and COLLOMB (1962). The basal part of the formation consists of about 10 metres of conglomerate including boulders and pebbles of the underlying Pre-Cambrian basement (Pharusien), such as lydite and quartz. It is interbedded with arkosic sandstone and partly has a ferruginous and clayey matrix. This basal conglomerate is exposed in the central part of Jebel Hassaouna in the surrounding of exposures of Pre-Cambrian rocks, for example near 14° East and 28° North.

It is overlain by 350–400 metres of homogeneous and monotonous sandstone, coarse-grained in the lower part and medium-grained in the upper 250–300 metres; quartz grains are well rounded; the matrix is partly kaolinitic and partly siliceous, rarely carbonate; most beds are cross-stratified. The uppermost 30 metres are coarse-grained again and contain abundant

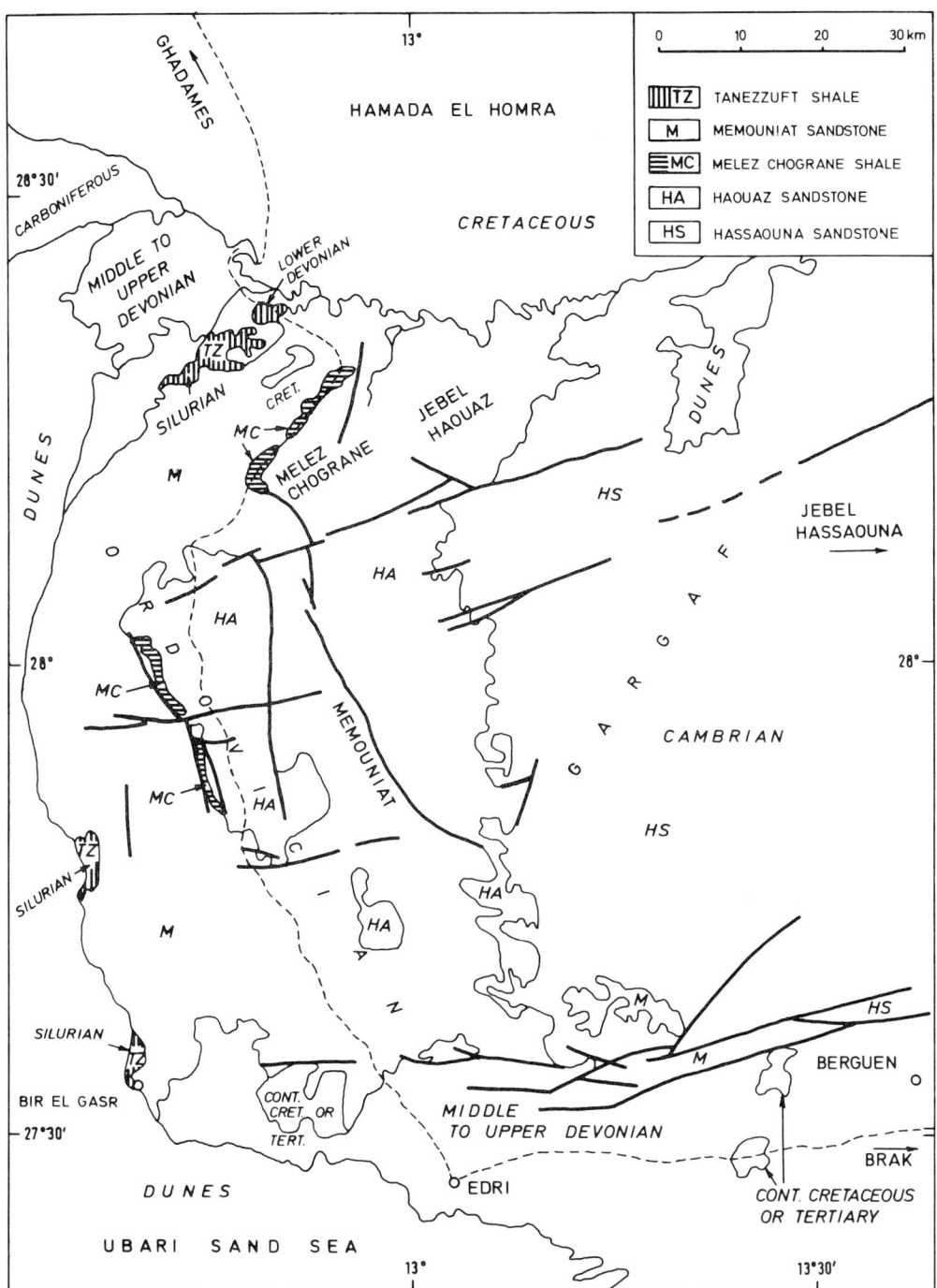

FIG. 5. Geological map showing type areas of Cambrian and Ordovician strata at Jebel Gargaf (after COLLOMB, 1962).

vertical worm borings (*Tigillites*). The top is silicified. The presence of *Tigillites* beds probably expresses a marine influence in parts of the strata. The silicified top possibly indicates an old land surface and a disconformity or unconformity.

The Cambrian age of the unfossiliferous Hassaouna Formation cannot be proved. It is nevertheless very likely. The formation is overlain by sediments of early Ordovician age and it unconformably overlies metamorphic Pre-Cambrian basement. Sediments similar to the Infracambrian rocks of the eastern edge of the southern Murzuk Basin are not present at Jebel Gargaf.

6.2.3. Haouaz Formation

The type area is Jebel Haouaz (Fig. 5). COLLOMB (1962) subdivides the Haouaz Sandstone Formation into three parts: a basal sandstone containing *Tigillites*, an unfossiliferous intermediate sandstone, and an upper sandstone with *Tigillites*. The total thickness is between 120 and 190 metres.

The basal sandstone is 15–20 metres thick; it is well bedded and most beds contain *Tigillites* and other trace fossils. The sandstone is fine-grained with partly angular quartz fragments and has a silty to clayey matrix. It is interbedded with clayey siltstone.

The intermediate sandstone is 10–50 metres thick and consists of massive beds of medium to coarse-grained sandstone. The upper sandstone is 90–120 metres thick, fine to very fine-grained, and well bedded. It contains abundant *Tigillites* and other trace fossils such as *Cruziana* sp. and *Harlania* sp.

MASSA (*in* MASSA and COLLOMB, 1960) and COLLOMB (1962) disagree in their judgement concerning the contact relationships: COLLOMB states that there is conformity between the Haouaz and the underlying Hassaouna Formation and that the overlying Melez Chograne Shale 'transgresses' over the Haouaz Sandstone. In MASSA's opinion there is a disconformity or unconformity between the Hassaouna and Haouaz Sandstones while the Melez Chograne Shale overlies conformably.

My own visits to the area concerned do not allow a judgement. In other areas of Libya, however, Ordovician Haouaz Sandstone rests definitely unconformably upon Cambrian Hassaouna Sandstone (for example, northern Dor el Gussa; *see* Fig. 7). The contact relationship between the Melez Chograne Shale and the Haouaz Sandstone seems to be conformable in several areas and unconformable in others.

HAVLÍČEK and MASSA (1973) separate the sandstone directly overlying the Hassaouna Formation from the Haouaz Formation and call it the 'Achebyat Formation'. They refer to DEUNFF and MOREAU-BENOIT (publication in preparation), who are establishing chronostratigraphical correlations of the Cambrian and Ordovician strata of the Homra Basin (called Ghadames Basin in the French literature). According to these investigations the Achebyat Formation of HAVLÍČEK and MASSA (1973) is possibly of early Ordovician age (Tremadoc to Arenig), while the overlying major part of the Haouaz Formation is probably of middle Ordovician age (Llanvirn to Llandeilo).

6.2.4. Melez Chograne Shale

The type area is Melez Chograne south of Auenat Ouenine (Fig. 5). This formation is up to 60 metres thick and consists of interbedded shale and silty shale, siltstone, and fine-grained sandstone. The colour is grey to multi-coloured. The basal 15 metres contain ferruginous and oolitic beds and conglomerate and breccia.

Some shale and sandstone beds and phosphatic nodules contain trilobites, brachiopods, and other marine fossils. The middle and upper parts of the strata locally consist of sediments of a glacial environment: within green shale boulders of granite, gneiss, schist, lydite, quartz, and quartzite appear (COLLOMB, 1962), as well as redeposited boulders and pebbles of shale and sandstone partly containing fauna. These glacial sediments are probably equivalent to the glaciation of southern Algeria (GARIEL and others, 1968; BEUF and others, 1969).

HAVLÍČEK and MASSA (1973) describe the following fauna of upper Ordovician (Caradoc) age: *Fezzanoglozza fezzanica* nov. sp., *Libyaeglossa colombi* nov. sp., *Orbiculoides massai* nov. sp., *Orbiculothyris castellata* Wolfart, *Drabodiscina* cf. *grandis* (Barrande), *Drabovinella* sp., *Rhynchotrema* cf. *clariondi* (Termier and Termier), and *Rafinesquina* cf. *pseudoloricata* (Barrande). COLLOMB (1962) reported among others *Synhomalonotus tristani* (Brongniart) and *Kloucekia* sp.

The Melez Chograne Shale is unconformably overlain by the late Ordovician Memouniat Sandstone and it rests on Haouaz Sandstone of middle Ordovician age. The existing basal conglomerate indicates unconformable or disconformable contact relationship in the type area. The upper Ordovician, Caradoc age of the formation is proved.

6.2.5. Memouniat Sandstone

The type area is the landscape of Memouniat between Edri and Auenat Ouenine (Fig. 5). According to COLLOMB (1962), this formation is between 90 and 130 metres thick and consists of three different facies types: in the eastern part of the area concerned it is made of well sorted, medium-grained, thick-bedded sandstone and characterized by *Tigillites* (lower 80 metres). Here, the upper 15 metres are very thick-bedded and conglomeratic at their base.

The middle part in the Memouniat area consists of 90–130 metres of monotonous and homogeneous sandstone, cross-bedded, unfossiliferous and with ferruginous crusts near the top.

Within the western area the Memouniat Sandstone is marine. Here it consists of fine-grained to silty sandstone with clayey and ferruginous matrix and with shale beds. In these beds bivalves and brachiopods were found, for example *Orthis (Orthis) calligramma* (Dalman) (COLLOMB, 1962). Within the whole area the Memouniat Formation rests unconformably on formations of different age, which at least locally seems to be the result of glacial erosion. According to HAVLÍČEK and MASSA (1973), the upper part of the Memouniat Sandstone in the areas north of Ghat and in the type area contain a fauna of uppermost Ordovician age (Ashgill), for example *Plectothyrella libyca* nov. sp. and *Hirnantia* aff. *sagittifera* (M'Coy). The Memouniat Sandstone is conformably overlain by Tanezzuft Shale of Silurian (Llandovery) age.

6.2.6. Tanezzuft Shale

The type locality is at the southern part of Wadi Tanezzuft in south-western Fezzan (*see* Section 6.3.3.2 and KLITZSCH, 1965a). Owing to Caledonian age movements and following erosion, in the Jebel Gargaf area the Silurian Tanezzuft Shale is present only in the extreme west and at its eastern edge. The Silurian Acacus Sandstone is completely missing. Exposures in the west are near Bir et Gasr and near Auenat Ouenine (Fig. 5) and in the east they lie north of Wadi Cneir.

At Auenat Ouenine near the road to Edri the formation begins with 3–5 metres of quartzitic sandstone, which rests conformably on the underlying Ordovician Haouaz Sandstone. *Climacograptus* sp. and other Diplograptidae are frequent. The major part of the formation consists of 30–35 metres of grey to yellow shale containing abundant Monograptidae and Diplograptidae (after MASSA and JAEGER, 1971) such as *Climacograptus* cf. *scalaris* (Hisinger), *Retiolites* sp., *Monograptus* cf. *sedgwicki* (Portland), and others in the lower part of the section, and among others in the upper part *Retiolites (Pseudoretiolites) perlatus* (Nicolson), *Diplograptus*

(*Petalograptus*) cf. *intermedius* Bouček & Pribyl, *Monograptus* cf. *regularis* (Tornquist). Most of the section also contains *Orthoceras* sp. and undetermined small bivalves.

According to its fauna the Tanezzuft Shale of Auenat Ouenine correlates with zones 19–22 (middle Llandovery). Facies and fauna at Bir el Gasr are more or less identical with those of Auenat Ouenine. The formation is unconformably overlain by Tadrart Sandstone of Lower Devonian age.

East of Jebel Gargaf and 12 kilometres north of Wadi Cneir, the Tanezzuft Shale is exposed at the inner escarpment of a 4 by 2 kilometres wide and 20–40 metres deep morphological basin formed partly by wind erosion (KLITZSCH, 1974). These exposures can easily be reached by the road to Brak; the northern edge is 1–2 kilometres south of 'kilometre 16' on this road.

Here 40–50 metres of grey to yellow shale crop out, containing *Rastrites peregrinus* (Barrande), *Glyptograptus tamariscus* (Nicholson) and *Orthoceras* sp. These strata correlate with zones 19–20 (middle Llandovery). They are overlain by Cretaceous Ben Afen Beds and are without direct contact with under- and overlying Palaeozoic strata. North of the basin poorly exposed Cambrian and/or Ordovician sandstone is present.

6.3. Murzuk Basin

The western, eastern, and southern edges of the Murzuk Basin are framed by a system of cuesta-type escarpments exposing the Palaeozoic strata of this basin, which during early Palaeozoic times consisted of several N.N.W.–S.S.E. striking uplifts and troughs (Fig. 3). In Mesozoic times these different tectonic units were included in a basin developing more or less perpendicular to the Caledonian structural elements. Because of this tectonic development, parts of early Palaeozoic troughs as well as parts of uplifts are exposed at the eastern edge of Murzuk Basin, where the structural development of Lower Palaeozoic time can best be studied.

The western edge of Murzuk Basin was more or less parallel to the Caledonian tectonic structures and therefore shows only little differentiation. Within this section two well exposed areas of Murzuk Basin will be described in some detail, each typical for western and eastern Murzuk Basin.

6.3.1. Dor el Gussa

6.3.1.1. *General description and exploration history.* Dor el Gussa is situated at the eastern flank of the Murzuk Basin. It is the most important key area for the understanding of Lower Palaeozoic tectonic and palaeogeographical development in Libya (KLITZSCH, 1963, 1970). Dor el Gussa forms a north-north-easterly striking mountain range made of Palaeozoic sandstone and shale. It reaches an altitude of almost 1000 metres. At its sudden northern end the strike of the Palaeozoic sediments changes within a short distance from a north-north-easterly into a north-westerly direction (Fig. 6) parallel to the Haroudj uplift, indicating the original tectonic trend of this area. The southern end is less distinct. As the thickness of Lower Palaeozoic and Devonian sediments decreases towards the Tibesti–Tripoli–Uplift (Fig. 3), the altitude of the mountains decreases also and the landscape loses its mountainous character. The present day Dor el Gussa mountains are made of that part of the basin filling of the formerly north-north-westerly striking Dor el Gussa Trough, which was uplifted during Mesozoic movements and included in the eastern edge of the Murzuk Basin. The connection between the mountainous character of the landscape and the tectonic situation (thick basin filling) is evident and can be noticed even on Gemini photographs (PESCE, 1968; Fig. 7).

The most complete stratigraphical section is to be found within the mountains of northern Dor ei Gussa, which can be entered only through a narrow valley at approximately 25°58′ N and

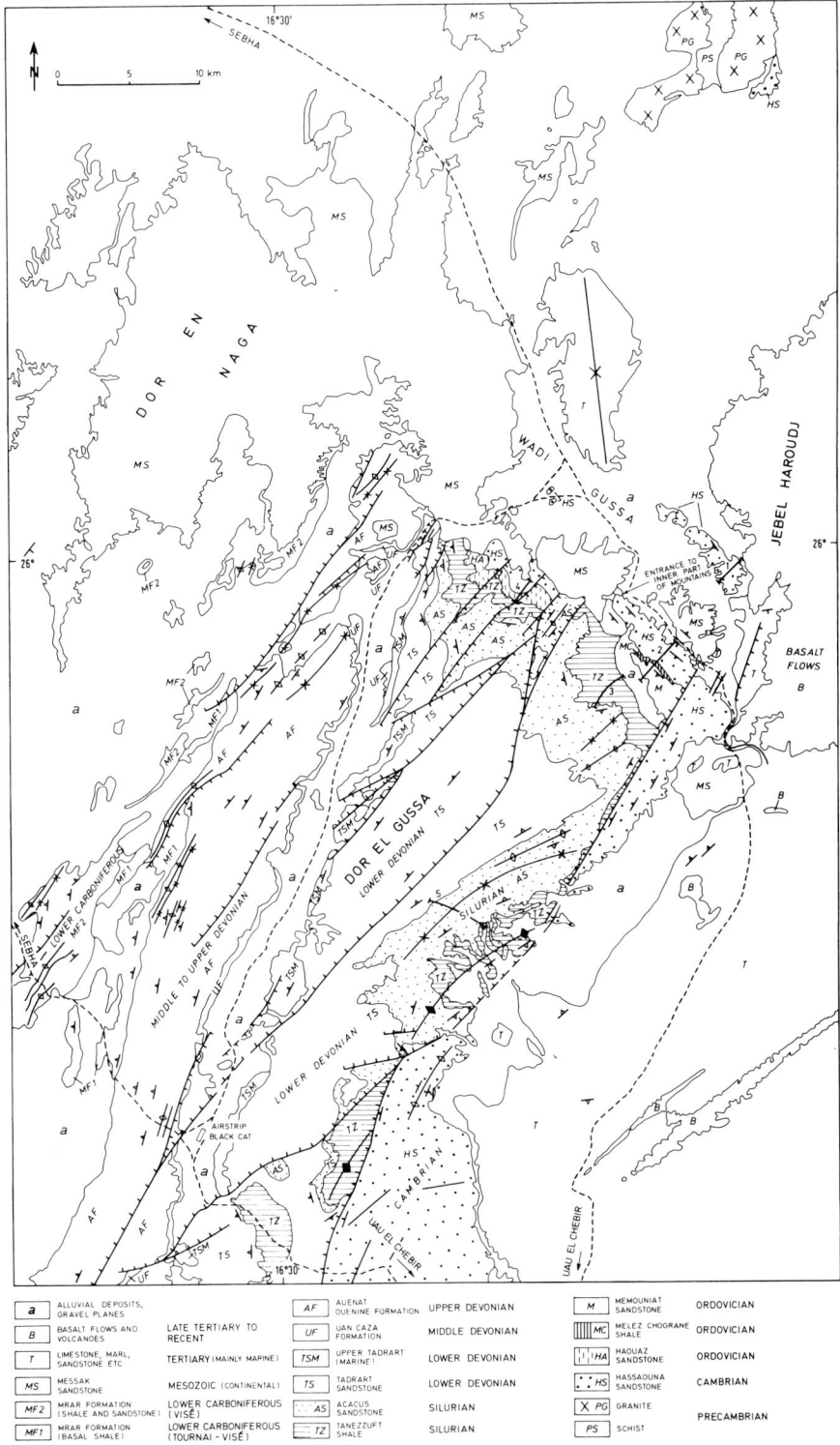

FIG. 6. Geological map of Dor el Gussa, eastern Murzuk Basin, (for sections see Fig. 8).
(1) Basal Hassaouna; (2) Hassaouna, Melez Chograne, Memouniat; (3) Tanezzuft;
(4) Haouaz; (5) Acacus.

16°45' E (Fig. 6). Here also the strong unconformity between Cambrian and Ordovician strata can be seen.

The thick sequence of Silurian strata typical of the central part of the Dor el Gussa Trough is exposed at the eastern edge of the mountains along approximately 25°45' N latitude and from there westward towards the contact of Silurian and Devonian strata.

Reduced thickness of Silurian sediments and unconformity between Silurian and Devonian strata, both typical of Caledonian uplifts, can be studied along the Sebha–Uan el Chebir track at approximately 25°33' N and 16°29' E and at 25°11' N and 16°24' E. At the latter locality glacial sediments (tillite) underlie the Silurian Tanezzuft Shale.

The first descriptions and sections of Dor el Gussa sediments were published by DESIO (1935), who thought that all of Dor el Gussa was of Carboniferous age. In the course of oil exploration the area was mapped by, among others, French and German geologists, who also published first details on the geology of Dor el Gussa and bordering areas (JACQUÉ, 1963; KLITZSCH, 1963, 1966). More recent field investigations by the present author resulted in the discovery of glacial sediments of probably Ordovician age south of Dor el Gussa and of plants within the upper part of the Silurian Acacus Sandstone (KLITZSCH, LÉJAL-NICOL, and MASSA, 1973).

6.3.1.2. *Hassaouna Sandstone.* The Hassaouna Sandstone of probable Cambrian age (type locality at Jebel Gargaf) is predominantly a continental, unfossiliferous sandstone sequence unconformably overlying folded Pre-Cambrian and overlain unconformably by Ordovician strata (Plate 1).

North of Dor el Gussa are scattered outcrops of Hassaouna Sandstone directly on metamorphic or granitic rocks of Pre-Cambrian age. The Hassaouna Sandstone also crops out at the northern edge and along the whole east flank of Dor el Gussa. In an isolated crest approximately 5 kilometres north-east of the mountain chain, 600 metres of sandstone crop out. In the lower half of the outcrop the sandstone is very conglomeratic and includes numerous clasts of the metamorphic basement (quartzite, quartz, siliceous slate). The sandstone is thick-bedded to massive, cross-bedded, and frequently cemented by kaolin, probably owing to the decay of the abundant feldspar in the rock. The upper beds of the Hassaouna Sandstone at this locality dip towards the south-west under the alluvium of Wadi Gussa (Fig. 6).

South of Wadi Gussa at the northern edge of Dor el Gussa, another 1100 metres of Hassaouna Sandstone crop out in the escarpment. It is definitely less conglomeratic than the sandstone of the isolated crest north-east of Wadi Gussa and is probably younger. The sandstone (and arkose) is fine to coarse-grained throughout, in parts weakly conglomeratic,

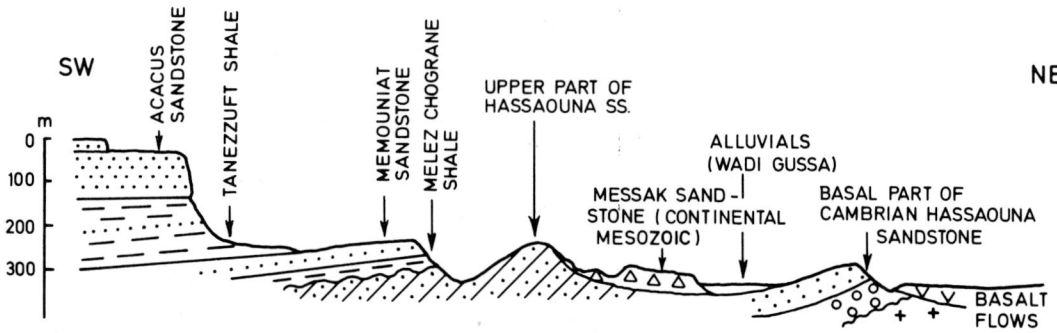

FIG. 7. Schematic cross-section northern edge of Dor el Gussa, eastern Murzuk Basin.

normally massive, cross-bedded, generally with kaolin cement, and has a high feldspar content. The beds dip 20–45 degrees south-west. In the mountains Ordovician age strata dipping only 5 degrees south-westerwards unconformably overlie the Hassaouna Sandstone (Fig. 7).

At the northern edge of Dor el Gussa, the Hassaouna Sandstone has a minimum thickness of 1700 metres. JACQUÉ (1963) believed that the upper part of this section is equivalent to the Haouaz Sandstone. Evidently JACQUÉ did not see the exposures of the real Haouaz equivalents at 16°43′ E and 25°58′ N as described below under Haouaz Sandstone.

Along the eastern flank of the mountain chain, a distinct angular unconformity between Cambrian and Ordovician strata has not been observed. In this area the Ordovician Melez Chograne Shale, a distinct marker, is usually missing; consequently, it is difficult to differentiate between rocks of Cambrian and Ordovician age. Therefore, south of latitude 25°48′ N these rock sequences were mainly mapped as undifferentiated Cambro–Ordovician.

6.3.1.3. *Haouaz Sandstone.* The type localities of the three formations of Ordovician age are situated in the western part of the Jebel Gargaf (*see* Section 6.2). At Dor el Gussa the three Ordovician formations can be distinguished in a narrow stretch south of, and parallel to, the Hassaouna Sandstone at the northern end of Dor el Gussa. Their thicknesses, however, are less than at the type localities (*see also* Fig. 8).

In the north-western part of the outcrop of Ordovician age rocks, approximately 16°43′ E and 25°58′ N, Haouaz Sandstone overlies the Hassaouna Sandstone with an unconformity of 20 degrees and more. The Haouaz Sandstone is a sequence of thin- to thick-bedded, fine-grained quartzitic sandstone containing abundant *Tigillites* sp. and other trace fossils (*Harlania* sp.). Badly preserved casts of brachiopods and bivalves were found. The Haouaz Sandstone is here approximately 50 metres thick and wedges out in a south-eastern direction between the Hassaouna Sandstone and the Melez Chograne Shale. Apparently, the Haouaz Sandstone is missing in the major part of the eastern Dor el Gussa.

6.3.1.4. *Melez Chograne Shale.* The Haouaz Sandstone is overlain by the Melez Chograne Shale, which is up to 25 metres thick. In the eastern part of northern Dor el Gussa, the Melez Chograne Shale overlies the Hassaouna Sandstone with an unconformity of up to 30 degrees (Fig. 7). The Melez Chograne Shale is silty to sandy, grey and red, and contains many intercalated fine-grained sandstone beds. No fossils were found, except for doubtful casts. The Melez Chograne Shale, like the Haouaz Sandstone, is missing in most parts of the eastern flank of Dor el Gussa. JACQUÉ (1963, p. 21) introduced the term 'Emi Daoun Formation' for these strata.

6.3.1.5. *Memouniat Sandstone.* A highly ferruginous sandstone bed approximately 2 metres thick which passes laterally into nearly pure haematite indicates a stratigraphical break between the Melez Chograne Shale and the overlying Memouniat Sandstone. South-eastwards, the Memouniat Sandstone lies directly on the Hassaouna Sandstone with an angular unconformity of 25–30 degrees. The Memouniat consists of fine- to coarse-grained, in part conglomeratic, often cross-bedded sandstone partly containing *Tigillites* sp. and reaches a thickness of about 30 metres. It is possible that the Memouniat Sandstone extends into the lower Silurian (COLLOMB, 1962). JACQUÉ (1963, p. 21) introduced the term 'Dor el Fatta Formation' for this sandstone. At Dor el Fatta scarp, however, only Hassaouna Sandstone is exposed.

At approximately 25°11′ N and 16°24′ E the uppermost 28 metres of undifferentiated Cambrian–Ordovician strata directly underlying Silurian Tanezzuft Shale consist of tillite (Fig. 7). It is poorly sorted material comprising all grades from clay to fine-grained gravel. This glacial sediment probably results from the late Ordovician or early Silurian glaciation, which is well known from southern Algeria and which must have affected central parts of Libya.

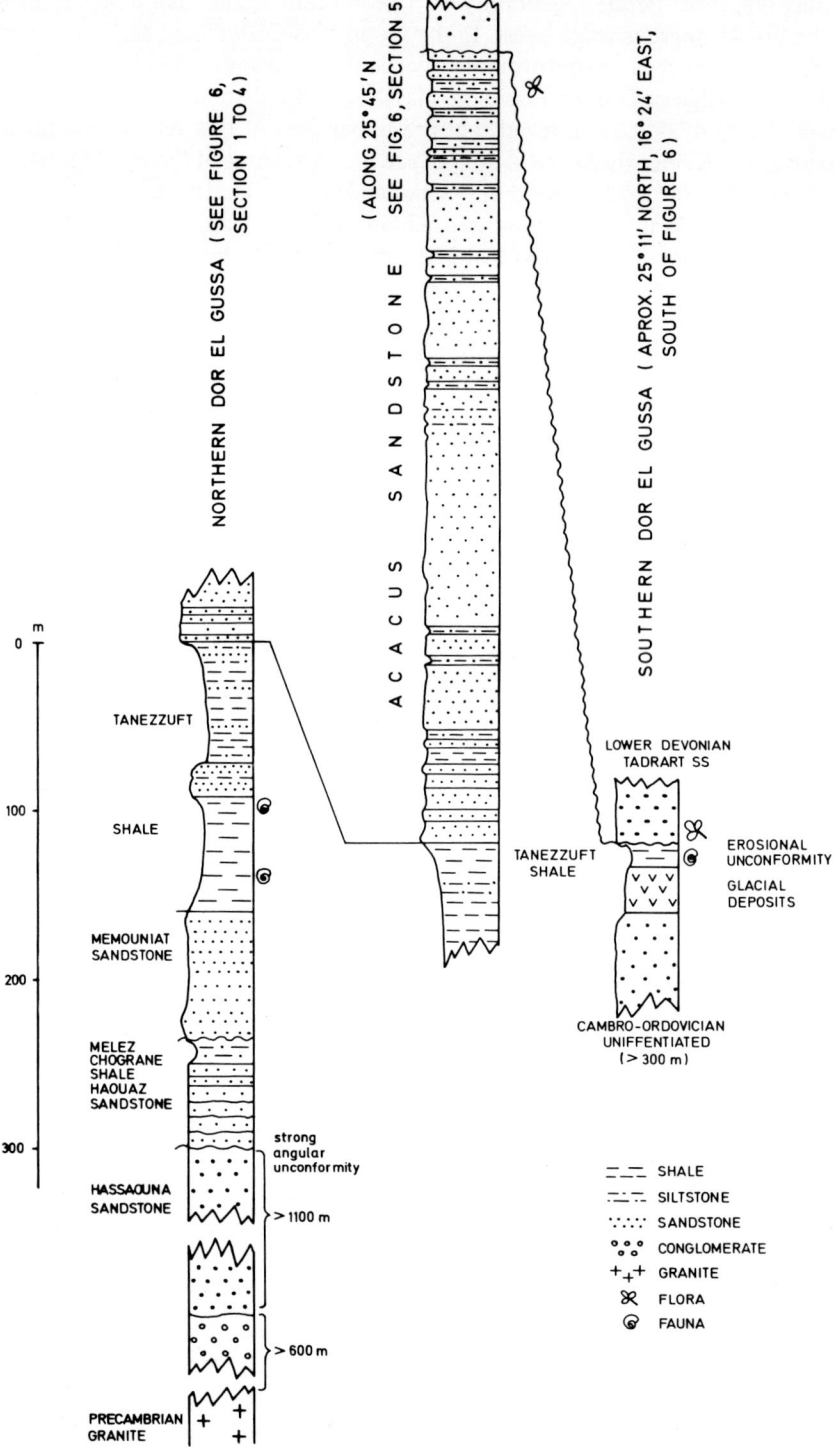

FIG. 8. Representative sections of Cambrian, Ordovician, and Silurian strata from Dor el Gussa.

6.3.1.6. *Tanezzuft Shale.* The type localities for the Tanezzuft Shale and Acacus Sandstone, both of Silurian age, are situated at the western flank of the Murzuk Basin at Wadi Tanezzuft (Plate 2) and Jebel Acacus (*see* Section 6.3.3).

In northern Dor el Gussa, Memouniat Sandstone is possibly conformably overlain by Tanezzuft Shale. In this region the Tanezzuft Shale is grey, about 160 metres thick, and intercalated in the lower and upper thirds with numerous beds of fine-grained sandstone. The Tanezzuft Shale is especially well exposed in a large steep-walled valley at approximately 25°55′ N and 16°45′ E. At this locality, well preserved in all parts of the section, graptolites are associated, especially in the middle third of the section, with numerous orthocones and some bivalves. The following graptolites were found approximately 100 metres above the bottom of the Tanezzuft Shale: *Climacograptus scalaris scalaris* (Hisinger), *C. scalaris normalis* (Lapworth), and *Cryptograptus* sp. aff. *tamariscus* (Nicholson). Based on the graptolites, the locality belongs in zone 19 of the English subdivision (Llandovery). Silurian faunas younger than Llandovery are not known from the Dor el Gussa.

Along the eastern flank of the mountain chain, the Tanezzuft Shale consists of a rhythmical alternation of pure and silty shale with clayey siltstone and silty sandstone. There trace fossils like *Bifungites* sp. occur together with graptolites.

Near the track from El Gaf to Uau el Chebir at approximately 25°33′ N and 16°29′ E, the Tanezzuft is predominantly shale with relatively little silt and sand, but owing to erosion during late Silurian time its thickness is very much reduced. At this locality Lower Devonian Tadrart Sandstone rests above an unconformity of 3–5 degrees on the middle part of the Tanezzuft Shale. The upper part of the Tanezzuft Shale and the entire Acacus Sandstone are missing. The thickness of Tanezzuft Shale is progressively reduced southward. At about 25°11′ N and 16°24′ E, in a narrow and very steep ravine at the bottom of an escarpment, the thickness is approximately 12 metres of slightly silty shale. At this shale section the following graptolites were found (KLITZSCH, 1963): *Climacograptus medius* (Tornquist), *C. rectangularis* (M'Coy), *C. scalaris* Hisinger, and *Glyptograptus sinuatus* (Nicholson). This fauna is also within zone 19 of the English subdivision (middle Llandovery). There are also numerous ostracodes in some beds. The Tanezzuft Shale is unconformably overlain by Tadrart Sandstone of Lower Devonian age. The lowermost sandstone bed contains plant imprints characteristic of the Tadrart Sandstone. The Tanezzuft Shale overlies glacial sediments (*see* Section 6.3.1.5).

Some distance southward, north of Mouri Ide, are the next localities of Silurian shale which can be clearly recognized as such. Probably between approximately 25° N and 24° N most of the Silurian strata were eroded before or during the early Devonian.

6.3.1.7. *Acacus Sandstone.* Within the area concerned, the Acacus Sandstone occurs only in the Palaeozoic trough of the northern Dor el Gussa. Its outcrop extends from the northern flank of the mountain chain southward to about 25°38′ N. From the centre of the basin outward, the Tadrart Sandstone of Lower Devonian age lies with a clear unconformity on progressively older parts of the Acacus sandstone, and finally at approximately 25°38′ N, directly on the Tanezzuft Shale.

The Acacus Sandstone is distinguished from the Tanezzuft Shale by gradual change of the predominant shale lithology into the predominantly sandy–silty strata of the lower Acacus Sandstone. An unconformity or indication of a stratigraphical break was not observed in the Dor el Gussa. Since the Tanezzuft Shale belongs to the Llandovery, the overlying Acacus Sandstone may be placed higher in the Silurian. From the Acacus Sandstone of the Dor el Gussa we know only numerous trace fossils such as *Harlania harlani* (Conrad) and *Cruziana* sp., which are also known in the Acacus Sandstone from all marginal zones of the Murzuk Basin. When occurring in abundance, these forms are used as a correlation criterion within the Murzuk Basin. They have, however, no value as stratigraphical key fossils as they also occur in Ordovician and in Devonian age strata, but with less frequency.

In the northern Dor el Gussa, the lower Acacus consists of approximately 65 metres of a rhythmical alternation of clayey siltstone with predominantly silty, thin-bedded, fine-grained sandstone and some shale beds. Trace fossils are abundant. In the upper part of this basal 65 metres are some thick sandstone beds, fine- to medium-grained, in part cross-bedded, and grading upwards into a massive sandstone, fine- to medium-grained but in part slightly conglomeratic. This upper massive sandstone is very light-coloured, nearly always cross-bedded, and interbedded with some thin, shaly siltstone layers partly containing *Harlania* sp. The clean cross-bedded sandstone unit is probably fluviatile–continental in origin. In the centre of the northern Dor el Gussa trough, at about 25°45′ N and 16°35′ E, this massive sandstone is approximately 400 metres thick, decreasing north and south as the unit is unconformably overlain by the Tadrart Sandstone.

In the uppermost part of this section of central Dor el Gussa, I recently found, within clayey siltstone, the following flora which was identified by Mrs A. Léjal-Nicol (KLITZSCH, LÉJAL-NICOL, and MASSA, 1973): *Taeniocrada* sp., *Psilophyton princeps* Dawson, *P. goldsmithii* Halle, *Dawsonites* sp., *Drepanophycus eximius* Frenguelli, *Protolepidodendron helleri* Léjal-Nicol, *P.* sp., *Archaeosigillaria kidstoni* Kräusel and Weyland, *Precyclostigma tardradense* Léjal-Nicol, and *P.* sp. Some of the psilophytes are known from the early Devonian of Europe, Siberia, and North America. Since the strata are unconformably overlain by Tadrart Sandstone of early Devonian age, it is likely that the above flora is of Silurian age, indicating an area of early development of lycophytes and psilophytes. It is, however, not to be excluded that within the Dor el Gussa Basin sedimentation of Acacus Sandstone continued until early Devonian time and that the above flora is of early Devonian age.

The maximum thickness of the Acacus Sandstone is 465 metres in the northern Dor el Gussa. The maximum thickness of the entire Silurian in this area is approximately 625 metres.

6.3.2. Infracambrian of Mourizidié

South of 24° N the Cambrian Hassaouna Formation of the eastern Murzuk Basin is unconformably underlain in some areas by a quartzitic and arkosic sandstone series up to 200 metres thick, which in turn unconformably overlies metamorphic rocks of Pre-Cambrian age. JACQUÉ (1963) called this sequence the Mourizidié Formation and attributed it to the Infracambrian.

It is separated from the Hassaouna Sandstone by an unconformity and the basal conglomerate of the Hassaouna contains boulders of rhyolite, which probably has the same age as the Mourizidié Formation. Such rhyolite flows are typical of strata underlying Cambrian sediments in other parts of North Africa, for example the Nigritien of Adrar de Iforas. It is possible that the Mourizidié Formation is an equivalent of the Nigritien, the age of which is also still uncertain (late Algonkian or Cambrian).

6.3.3. Jebel Acacus and Wadi Tanezzuft

6.3.3.1. *General description and exploration history.* Jebel Acacus is an almost north–south striking mountainous plateau facing Wadi Tanezzuft towards the west with an escarpment up to about 600 metres high. The top region of Jebel Acacus is above 1400 metres. Both landscapes are situated along the western edge of the Murzuk Basin. The western Acacus escarpment is the result of regressive erosion moving eastward, which took advantage of the relatively soft and thick Tanezzuft Shale forming the lower part of the escarpment.

Plate 1. Unconformity between Ordovician and Cambrian strata at Dor el Gussa (eastern Murzuk Basin).

Plate 2. Base of Silurian Tanezzuft Shale near Ghat (Wadi Tanezzuft, western Murzuk Basin).

Plate 3. Silurian Tanezzuft Shale and Acacus Sandstone at Jebel Idinene (Wadi Tanezzuft, western Murzuk Basin).

Plate 4. Acacus Sandstone at Jebel Acacus, taken approximately 1500 metres above ground (western Murzuk Basin).

Plate 5. Takarkhouri Pass at the Algeria–Libya border (Jebel Acacus).

Plate 6. Cruziana II (Seilacher, 1969), Trilobite burrow from Silurian Acacus Sandstone (Wadi Takarkhouri, Jebel Acacus, western Murzuk Basin).

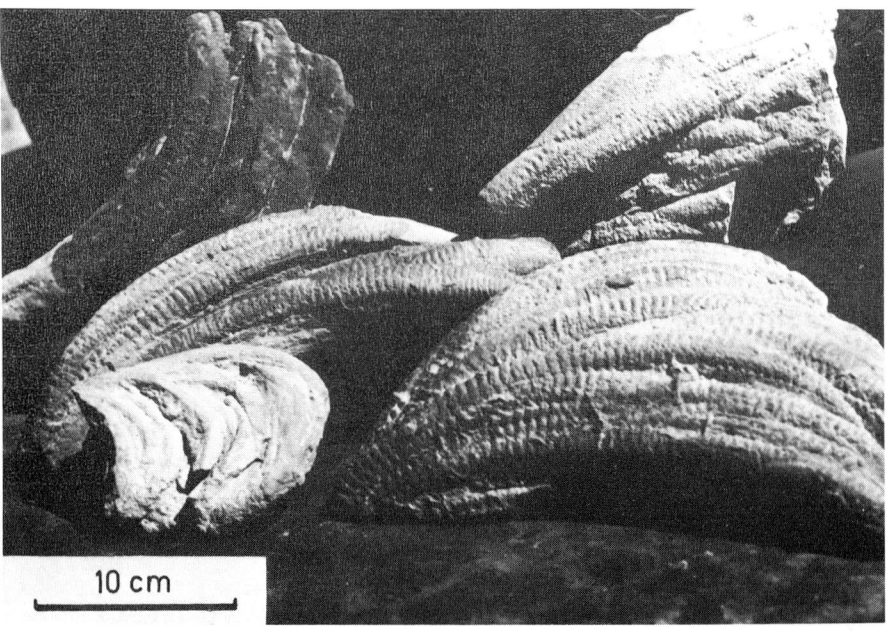

Plate 7. Harlania (= *Arthrophycus*), trace fossil representing sediment eaters, Acacus
Sandstone, Jebel Acacus (western Murzuk Basin).

Plate 8. Cross-bedded sandstone of Cambrian age (Jebel Eghei, western Kufra Basin).

Plate 9. Ordovician Haouaz Sandstone with vertical tubes (*Tigillites* or *Skolithus*) made by worm-like suspension feeders. Jebel Eghei (western Kufra Basin).

Jebel Acacus (Plate 4) and Wadi Tanezzuft (Plate 3) are the type areas of the Silurian strata of the Central Sahara. West of Wadi Tanezzuft, Cambrian and Ordovician rocks cover a large region which at the foothills of Tassili in south-eastern Algeria is bordered by Pre-Cambrian basement. The Cambrian and Ordovician age sediments, however, are not well investigated at the western edge of the Murzuk Basin and will therefore not be described in detail.

At scattered localities equivalents of the different Ordovician formations are known (see, for example, HAVLÍČEK and MASSA, 1973), but neither a summarizing description is given nor do individual publications or any personal knowledge of the area allow one to give a satisfactory account of the Ordovician strata of the western Murzuk Basin. Most of the sediments forming the large Tassili Plateau dipping eastward towards Wadi Tanezzuft probably consist of equivalents of the Cambrian Hassaouna Sandstone Formation.

The oldest subdivision of early Palaeozoic strata in south-eastern Algeria and south-western Libya has been established by French geologists (KILIAN, 1931; LELUBRE, 1952; and several others). Lower Sandstone ('grès inférieurs') comprising Cambrian and Ordovician strata was separated from an Upper Sandstone ('grès supérieurs', Silurian and early Devonian sandstone) by the shale sediments of Wadi Tanezzuft. DESIO (1936) introduced the terms Tanezzuft Shale and Acacus Sandstone, without publishing a type section. Both terms were taken over by the Names and Nomenclature Committee of the Petroleum Exploration Society of Libya (1960). In 1965, I published a type section to fill the existing gap. It was revised and completed in a later publication (KLITZSCH, 1969). Finally, MASSA and JÄGER (1971) published new detailed stratigraphical information on the Silurian strata. The following description is mainly after KLITZSCH (1969).

6.3.3.2. *Tanezzuft Shale.* Approximately 400 metres of marine shale, grey, partly silty, with sandstone lenses and some thin sandstone beds in the lower part. Abundant graptolites are present, especially in the lower part. In some beds there are abundant shells including *Orthoceras*. The age is lower Silurian, zones 16–18 (Llandovery).

Description of the Takarkhouri Section (*Figs. 9 and 10, Plate 5*):

(1) *Underlying strata: Cambrian to Ordovician beds, not subdivided. Sandstone, fine- to medium-grained, partly coarse-grained to conglomeratic, quartzitic near the top.

——unconformity——

(2) 28 metres shale, grey to grey–yellow, some fine-grained sandstone beds up to 10 centimetres thick. *Diplograptus* and *Orthoceras* are common. The following graptolites are from 12 metres above the base:

Climacograptus rectangularis (M'Coy)	zones 16–19
Glyptograptus persculptus (Salter)	zones 15–17
Diplograptus modestus Lapworth	zones 16–18
D. modestus parvulus Lapworth	zones 16–(?)

The same beds, 40 kilometres north of the Takarkhouri section, contain in addition:

Climacograptus scalaris miserabilis Elles & Wood	zones 14–17
Diplograptus (*Diplogr.*) *modestus fezzanensis* Desio	zones 16–18
D. (*Diplogr.*) *modestus modestus* Lapworth	zones 16–18

(3) 33 metres shale, grey, with thin layers and lenses of fine grained sandstone, rare graptolites.

---

* The marginal numbers refer to unit numbers in Fig. 10.

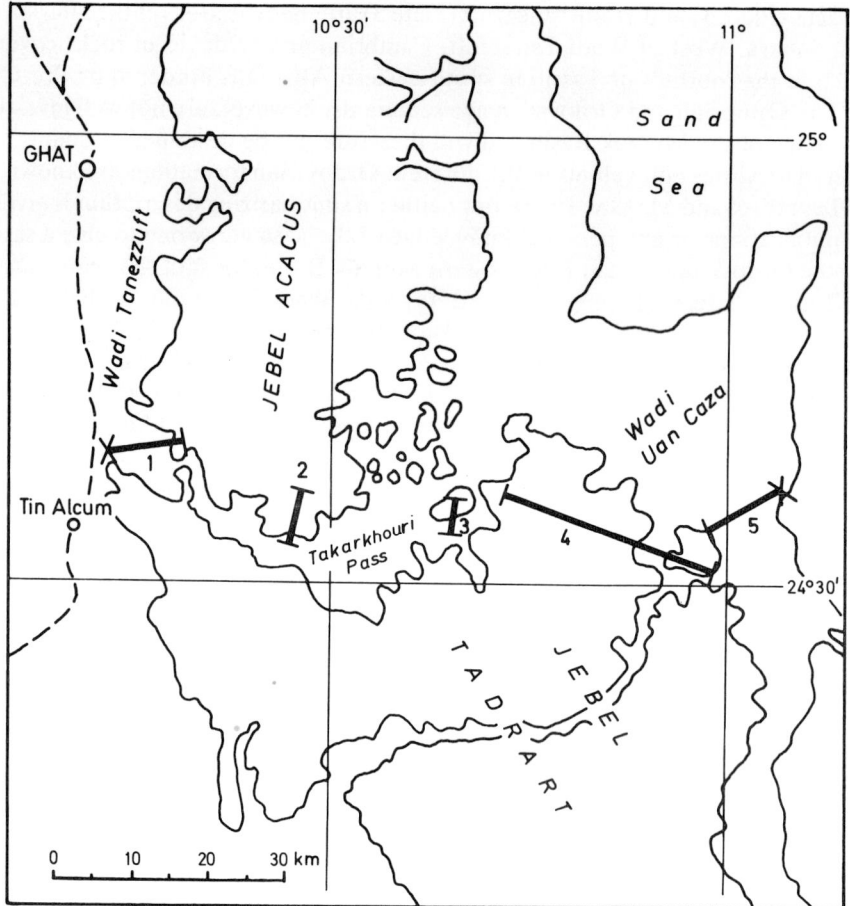

FIG. 9. Situation map of Takarkhouri section near Libyan–Algerian border.
(1) Tanezzuft Shale; (2) lower Acacus; (3) upper Acacus Ss; (4) top Acacus Ss
and Tadrart Ss; (5) middle Devonian strata (after KLITZSCH, 1969).

(4) 21 metres shale, grey, rare graptolites.

(5) 20 metres shale, grey, with many sand lenses and some fine-grained sandstone beds up to 1.5 metres thick. The basal part contains:

Diplograptus (Diplogr.) modestus fezzanensis Desio	zones 16–18
D. (Diplogr.) modestus modestus Lapworth	zones 16–18
D. (Diplogr.) modestus parvulus Lapworth	zones 16–(?)

Because *Climacograptus scalaris miserabilis* is missing, unit (5) possibly belongs above the basal part of zone 17.

(6) 130 metres shale, grey to violet–grey, in parts silty and with frequent sand lenses and thin, fine-grained sandstone beds. Some beds contain numerous shells and '*Orthoceras*'. *Diplograptus (Diplograptus) modestus* (Lapw.) zones 16–18 and *Climacograptus scalaris normalis* Lapworth zone 16 to basal part of 19 are common.

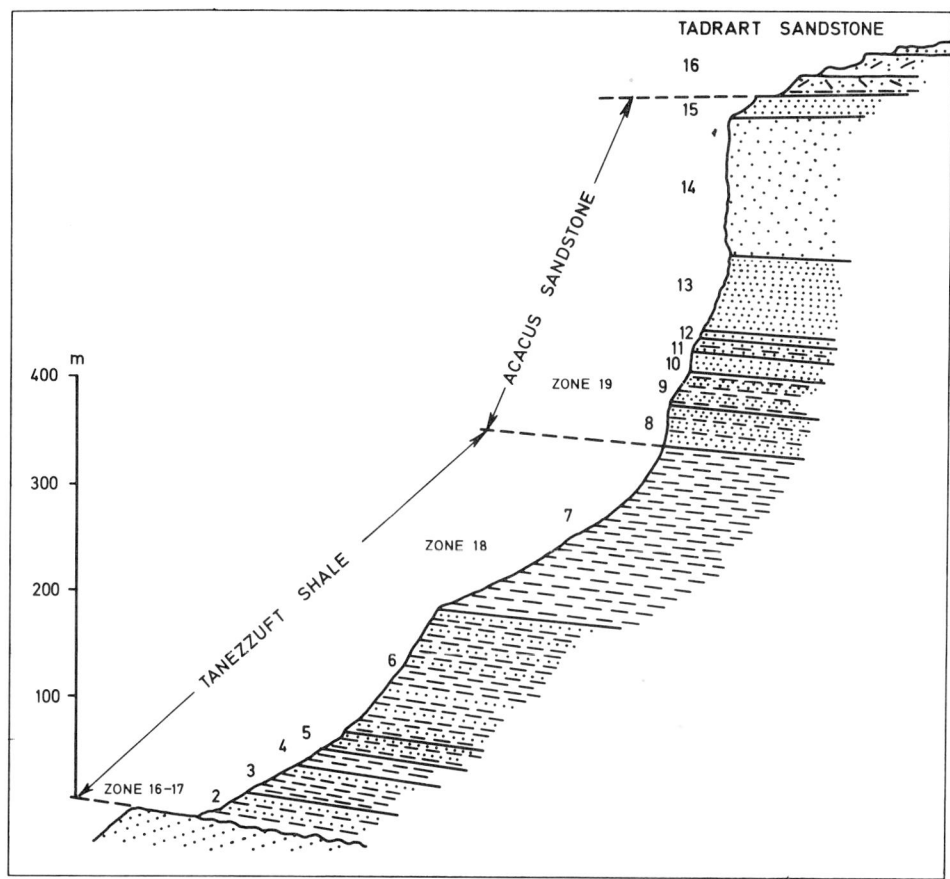

FIG. 10. Type section of Silurian strata at Wadi Takarkhouri, western Murzuk Basin (after KLITZSCH, 1969).

(7) 170 metres shale, greenish grey to black, partly bituminous, numerous small shells as well as *Diplograptus (Diplogr.) modestus fezzanensis* Desio zones 16–18 and *Monograptus revolutus* (Kurck)? zones 18–19.

6.3.3.3. *Acacus Sandstone*. Approximately 345 metres of marine to subcontinental sandstone, white, grey and brown, mainly fine-grained, thin- to thick-bedded, quartzitic near the top, shale beds in the lower part. Frequent trace fossils (Plates 6 and 7) occur on thin-bedded sandstone indicating shallow water deposition. The Acacus Sandstone overlies the Tanezzuft Shale conformably.

The lower part probably belongs, according to the graptolites collected in unit (9), to zone 19 of the middle Llandovery. Because of the lack of diagnostic fossils, the top of the Acacus Sandstone cannot be dated at the Takarkhouri section. Taking into consideration that most sandstone beds above unit (9) were deposited much faster than the conformably underlying strata, it seems very likely that the top of the formation is not younger than middle Llandovery or upper Llandovery.

Continuation of Takarkhouri Section:

(8)   40 metres sandstone, white to grey, fine-grained, thin to medium-thick beds, interbedded with some silty, grey shale.

(9)   35 metres shale, grey, partly silty, sand lenses and some fine-grained sandstone beds. The following graptolites were collected after the original publication of the type section (KLITZSCH, 1965a):

Climacograptus hughesi (Nicholson)	zones 16–21
Monograptus atavus Jones	zones 17–19
M. cf. *concinnus* Lapworth	
M. n. sp. aff. *Monograptus variabilis* Perner	zones 18–(?)
M. *lobiferus* M'Coy?	zones 19–20

(10) 19 metres sandstone, brown to grey, fine-grained, partly clayey, thick-bedded, laminated, some beds with *Harlania harlani* (Conrad).

(11) 13 metres sandstone, light brown, fine-grained, well bedded, some shale and siltstone layers, abundant trace fossils, e.g. *Cruziana* sp.

(12) 7 metres sandstone, brown to grey, fine-grained, thick-bedded.

(13) 77 metres sandstone, grey to partly brown, some white sandstone, fine-grained, thin- to mainly thick-bedded, with thin shale layers near the base, abundant trace fossils: *Harlania harlani* (Conrad), *Cruziana* sp., *Tigillites* sp., and others.

(14) 135 metres sandstone, white to dark brown, fine-grained to medium-grained, some beds coarse-grained to conglomeratic, mainly thick-bedded and cross-stratified, some beds with strong ferruginous matrix, but most beds with no matrix, very friable and very porous, near the top some beds with *Harlania harlani* (Conrad).

(15) 19 metres sandstone, grey, fine-grained, thin-bedded, partly with strong clayey matrix, several beds contain *Harlania harlani* (Conrad) and *Tigillites* sp., near the top very ferruginous, at the top 1–2 metres quartzite.

——unconformity——

(16) Basal part of Lower Devonian Tadrart Sandstone: 54 metres sandstone, light to dark brown, fine-grained to slightly conglomeratic, thick-bedded, cross-stratified, very ferruginous in some beds, but most beds without matrix and very friable, very porous, at the base 1.2 metres white silty shale, at the top 0.5 metres red quartzitic siltstone.

MASSA and JAEGER (1971) added new information on stratigraphical knowledge of the Silurian strata of the western Murzuk Basin by publishing the fauna of several sections taken between the Takarkhouri area and Serdeles. Their conclusions, however, are partly not convincing. Because the Silurian shale begins with zone 16 near Takarkhouri (south of Ghat) and possibly not before zone 19 near Serdeles, MASSA and JAEGER separate a lower Silurian shale formation (D'Iyadhar Shale) from the Tanezzuft Shale, which they relate to zone 19 (middle Llandovery). The changing age of the beginning of the Silurian transgression and changing thickness of different time units reflect differences in the palaeogeographical situation. The Serdeles section is close to the Tihemboka uplift (Fig. 3) and consequently sedimentation might have started later than in the Takarkhouri area. Moreover, the Tanezzuft Shale at Takarkhouri does not, or not much, exclude zone 18 (KLITZSCH, 1969, after graptolite identifications of JAEGER) and is not mainly of zone 19 as at Serdeles, as indicated by MASSA and JAEGER. This only proves that the Tanezzuft Shale is no time unit and that subdivisions of this formation by age criteria are questionable. The same can be stated about the age of the Acacus Sandstone. It is zone 19 (middle Llandovery) 50 metres *above* the base at Takarkhouri (not at the base or at the top of Tanezzuft Shale, as MASSA and JAEGER interpret my section) and, according to the two authors, the base is of lower Ludlow age in the Homra Basin. This reflects different stages of regression of the Silurian sea and does not imply a different geological history of the two basins.

6.4. Kufra Basin

The Kufra Basin forms a large area in south-eastern Libya, extending into Chad (the southern part of the Kufra Basin there is called the Erdis Basin) and marginally into north-western Sudan and south-western Egypt. North-eastwards, the Kufra Basin is in connection with the Western Desert Basin of Egypt and of north-eastern Libya. Most of the basin is covered by continental strata of Carboniferous and Mesozoic age (mainly Nubian facies) and by sand. Good exposures of Lower Palaeozoic sediments are at Jebel Gardeba (northern edge of the basin), at Jebel Eghei and along the eastern edge of the Tibesti Mountains (western edge of the basin), and in the southern Ennedi Mountains in the Republic of Chad.

Along the eastern frame of the basin are relatively small exposures of the Lower Palaeozoic strata 'at and near Jebel Arkenu, and also at the north-western part of Jebel Auenat, at the western edge of Gilf Kebir, and south-east of Jebel Kissu. Much of the basin's edge there is covered by younger sediments, probably owing to the absence of several of the Palaeozoic formations and to the very reduced thickness of others. Distinct escarpments of early Palaeozoic strata only developed west of Wadi Sura at both sides of the Egyptian–Libyan border near the western part of Gilf Kebir. Around Jebel Arkenu and Jebel Auenat at the borders between Libya, Egypt, and Sudan Pre-Cambrian basement is exposed in a 200 by 120 kilometres wide area. In some areas it is directly overlain by Carboniferous (locally Devonian) and younger strata.

The exploration history of the early Palaeozoic strata of the Kufra Basin is recent and the results are not complete. None of the expeditions carried out before oil exploration in Libya started in the 1950s proved the existence of Lower Palaeozoic strata, but SANDFORD (1935) had already reported Lower Palaeozoic sandstone from Jebel Auenat and from the Ennedi Mountains which partly is of Carboniferous age. Geologists of various oil companies discovered equivalents of the Silurian Tanezzuft Shale at the northern, western, and southern edges of the basin, which contain very rare graptolites (KLITZSCH, 1966, 1968a; DE LESTANG, 1968; and several unpublished oil company reports) and which are probably of Llandovery age. The underlying sandstone, partly containing trace fossils, consequently should be of Cambrian to Ordovician age. Cambrian and Ordovician sandstone was more or less proved with granulometric and palaeobathymetric methods and correlations by VITTIMBERGA and CARDELLO (1963) and by BUROLLET (1963). HOTTINGER (1959) and PESCE (1968) interpreted the geology in the border zones of Libya, Egypt, and Sudan. The latest investigations by the author (KLITZSCH, 1979, 1980) proved the presence of Silurian strata in north-western Sudan and south-western Libya.

In general, the early Palaeozoic formations in the Kufra Basin area are less complete and shale formations are more silty and sandy than farther north and west, because the Kufra Basin was close to the stable African craton and not all transgressions covered the entire area. The rare published knowledge about the Lower Palaeozoics of the Kufra Basin allows only a brief description.

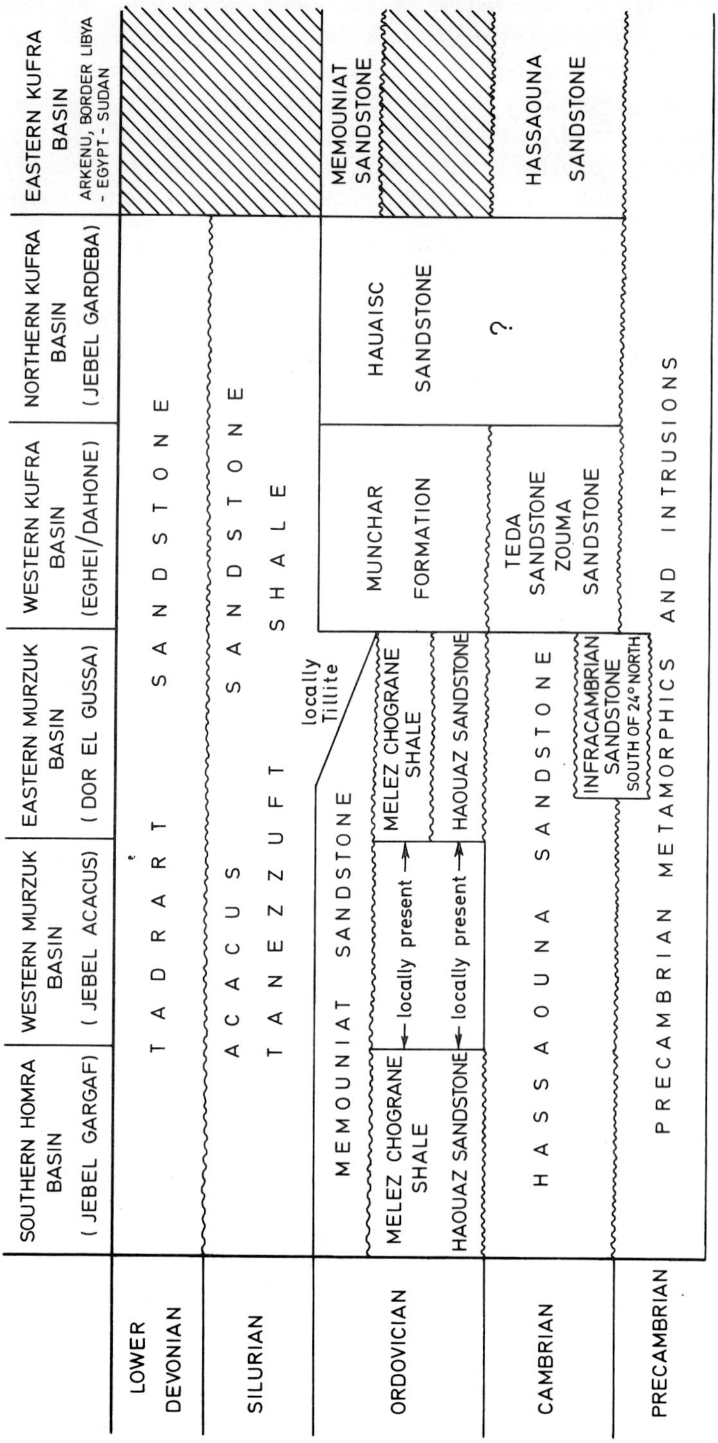

FIG. 11. Correlation chart of Lower Palaeozoic formations in Homra, Murzuk, and Kufra Basins.

6.4.1. Jebel Eghei

Jebel Eghei is an eastward-dipping mountain range with a very high scarp towards the west, where it is underlain by Pre-Cambrian basement. It is mainly made of Cambrian (Plate 8) and Ordovician sandstone. Some areas are covered by young basalt flows.

The lowermost part of the Lower Palaeozoics unconformably overlies Pre-Cambrian rocks and consists of sandstone which is more than 150 metres thick. This Zouma Sandstone is fine-grained to conglomeratic, normally cross-bedded and has an argillaceous and kaolinitic matrix. The overlying Teda Sandstone is up to 100 metres thick and is similar to the Zouma Sandstone, but it contains little or no kaolin (VITTIMBERGA and CARDELLO, 1963). Both formations are probably equivalent to the Cambrian Hassaouna Formation of the Murzuk and of the Homra Basin. They are unconformably overlain by the Munchar Sandstone Formation, which is approximately 140 metres thick at Jebel Eghei. It consists of fine- to coarse-grained, partly cross-bedded sandstone with an argillaceous matrix. The basal part is conglomeratic. Several levels contain trace fossils (Plate 9) such as *Tigillites* sp., *Cruziana* sp., and *Harlania* sp.

The basal part of the upper third of the formation, at least locally, consists of a 10–20 metres thick shale and silty shale sequence, which probably is an equivalent of the Melez Chograne Shale Formation of the Murzuk and Homra Basins. Thus the Munchar Sandstone Formation of Jebel Eghei probably represents the Haouaz, Melez Chograne, and Memouniat Formations of western Libya.

The overlying Tanezzuft Shale (KLITZSCH 1965b, Fig. 4, and KLITZSCH, 1968a, Fig. 2), which is up to 120 metres thick, consists of greenish to grey shale interbedded with many siltstone and some sandstone beds especially within the upper two thirds. Graptolites (*Climacograptus* sp.) are rare; trace fossils occur on silt and sandstone beds (for example *Bifungites* sp.). The conformably overlying Acacus Sandstone reaches a thickness of 40–50 metres (locally more than 100 metres) and consists of fine- to medium-grained sandstone. Some trace fossils occur locally (*Harlania* sp., *Tigillites* sp., and *Cruziana* sp.). Both Silurian formations wedge out towards the north mainly because of late Caledonian erosion. They are unconformably overlain by Tadrart Sandstone of Lower Devonian age and north of 22°15′ N by continental Mesozoic sediments.

6.4.2. Jebel Gardeba

At Jebel Gardeba and Jebel Hauaisc, an isolated area of escarpments formed of Palaeozoic strata north of Kufra, sediments of Lower Palaeozoic age are exposed. The basal part consists of the more than 100 metres thick Hauaisc Sandstone of Cambrian and/or Ordovician age. It is cross-bedded, fine- to coarse-grained sandstone with kaolinitic and ferruginous cement. Towards the base, this formation is unconformably truncated by sediments of early Tertiary age and by recent dunefields. The Palaeozoic–Pre-Cambrian contact is not exposed. The Hauaisc Sandstone is overlain by Tanezzuft Shale, up to 120 metres thick and interbedded with many siltstone and sandstone layers and beds. Graptolites are rare; they indicate a Llandovery age. The conformably overlying Acacus Sandstone reaches a thickness of more than 120 metres. It is mainly fine-grained and locally contains trace fossils such as *Harlania* sp. and *Tigillites* sp. It is unconformably overlain by the Lower Devonian Tadrart Sandstone.

6.4.3. Gilf Kebir, Jebel Arkenu, Jebel Auenat, Jebel Kissu (S.W. Egypt, N.W. Sudan, S.E. Libya)

Jebel Arkenu and Jebel Auenat are large post-Carboniferous magmatic intrusions in southwestern Libya near the Egyptian and Sudanese border. According to BUROLLET (1963), sandstone of Cambrian and Ordovician age covers part of the Jebel Arkenu intrusion and the

Pre-Cambrian metamorphics directly north-east of Jebel Arkenu. The early Palaeozoic sediments consist of a lower formation, which is a 200 metres thick sandstone, fine- to coarse-grained, with cement of illite and silica. The lower 100 metres contain beds with *Tigillites* sp., the upper 100 metres are mainly cross-bedded and conglomeratic. This formation is unconformably overlain by a 150 metres thick sequence of cross-bedded conglomeratic sandstone with intercalations of well bedded, white, fine-grained sandstone. The base is marked by a very distinct conglomerate.

BUROLLET (1963) compares the lower formation with the Cambrian Hassaouna Sandstone and the upper formation with the Ordovician Memouniat Sandstone of western Libya. According to BUROLLET (1963), Silurian strata are completely missing at Jebel Arkenu and in the Jebel Auenat area. Recent fieldwork by the author in Egypt and Sudan resulted in a different interpretation (see below).

According to the interpretations of HOTTINGER (1959), PESCE (1968), and SAID (1971), it is likely that no rocks of Lower Palaeozoic age are exposed within the Egyptian and Sudanese part of the Jebel Auenat area and its surroundings. This interpretation, however, has recently been proved to be wrong (KLITZSCH, 1980). At Karkur Talh (north-eastern part of Jebel Auenat) Pre-Cambrian basement is overlain by 70–80 metres of sandstone with abundant *Tigillites* sp. and which also contains *Harlania* sp.

Farther north, near the Umm Ras passage of the western Gilf Kebir plateau, similar sandstone at several levels contains *Cruziana acacensis* Seilacher, *Cruziana* sp., and other trace fossils, apart from *Harlania* sp. and *Tigillites*. This sequence reaches a thickness of almost 400 metres. According to SEILACHER (personal communication) the trilobite track *Cruziana acacensis* is of Silurian, possibly Llandovery, age. As at Karkur Taln of Jebel Auenat, this sequence rests more or less directly on Pre-Cambrian rocks and is overlain by sediments of Devonian to Carboniferous age. A similar but thinner succession is present east of Jebel Auenat in isolated hills in the Ras el Abd area directly south of the Egyptian–Sudanese border.

Furthermore, 30–50 metres of Silurian sandstone containing *Cruziana acacensis* and *Harlania* sp. is present between 40 kilometres south and 90 kilometres east-south-east of Jebel Kissu in north-western Sudan. In both cases the Silurian strata consist of fluvio-continental to deltaic and shallow marine sandstone and they rest directly on Pre-Cambrian strata. The Silurian beds are disconformably overlain by a fluvio-continental sandstone and, above, by plant-bearing sandstone, siltstone, and shale of Devonian to Carboniferous age.

The different exposures of Silurian strata along the Libyan, Egyptian, and Sudanese borders certainly represent Silurian sedimentation at the south-eastern edge of the Kufra Basin. Shale layers are present but very thin (2 metres maximum); sandstone dominates and represents environments from both sides of the shore line.

After having found Silurian strata more or less directly over Pre-Cambrian rocks south of Jebel Kissu in Sudan and west of Gilf Kebir in Egypt, as well as in between at the Egyptian part of Jebel Auenat and east of it, I doubt that the strata described by BUROLLET (1963) from Jebel Arkenu in Libya are older than Silurian.

A first geological map of the Sudanese and Egyptian part of this area has recently been published (KLITZSCH and LIST, 1979). It seems that the Libyan part needs some new interpretation.

6.5. Western Desert Basin

The marine Tertiary and Mesozoic strata covering north-western Egypt and north-eastern Libya are underlain by Palaeozoic sediments of Cambrian to Carboniferous age. None of the Lower Palaeozoic formations is exposed at the surface in this area or close to it. The only

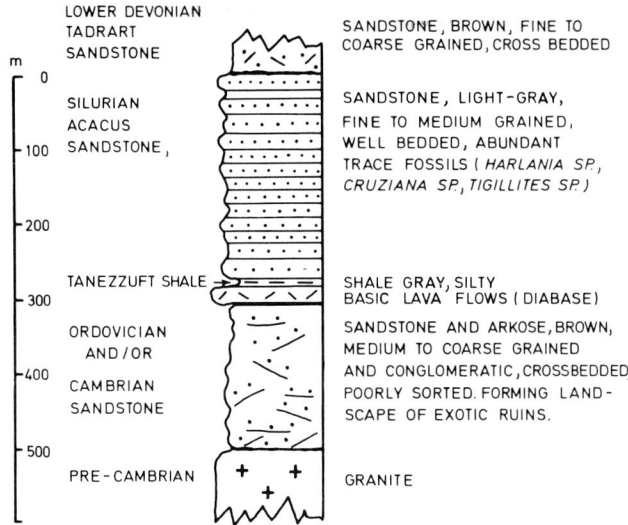

FIG. 12. Lower Palaeozoic strata of Wadi Sini (approximately
16°50′ N, 22°30′ E), north-eastern Chad near Sudanese
border, southern Kufra Basin.

information officially available is in scarce publications about exploration wells of oil companies
(for example SAID, 1962; BAER and KLITZSCH, 1964). According to these, and to unofficial
information, the northern part of Cyrenaica and the north-western part of Egypt west of
Alexandria are part of a Palaeozoic basin, which to the south is in connection with the Kufra
Basin (Figs. 1 and 2). At Baharia (SAID, 1962) approximately 470 metres of Cambrian
sandstone containing trilobites and Lingulae was penetrated. Silurian strata have been found in
several wells. Within the Silurian sediments the content of sandstone is considerably higher and
that of shale is lower than in north-western Libya and graptolites seem to be less frequent. Thus
it is very likely that the Western Desert Basin was closer to the edge of the Silurian transgression
than the Homra Basin (Fig. 2). The thickness of Silurian sediments is less than in the Murzuk and
Homra Basin.

6.6. Sirte Basin

The Sirte Basin is the only oil-producing province in Libya. Several thousand wells have been
drilled there, some of which reached the base of the Mesozoic and others even reached the
Pre-Cambrian basement. From unofficial oil company information it has long been known that
the marine and, in their lower parts, locally continental strata of the Sirte Basin are underlain by
sandstone and quartzitic sandstone of Cambrian and/or Ordovician age, which by some oil
companies is called the Sirte Quartzite and by others the Hofra or Amal Formation or Gargaf
Group (BARR and WEEGAR, 1972).

BONNEFOUS (1972), in a very detailed publication, differentiates Cambrian to Ordovician age
strata from continental Mesozoic strata of the Sirte Basin and he proves, from microfossils, the
Cambrian to Ordovician age and the Cretaceous age of the two stratigraphical units, which are
similar in lithology and consequently may be confused. Within the Cambrian to Ordovician
section forms such as *Cymatiogalea cuvillieri* and *Attritasporites messaoudi* were found as well as
fragments of graptolite siculae (upper strata).

The tectonic and palaeogeographical conclusions of BONNEFOUS, however, are not necessarily acceptable. Because BONNEFOUS proved the presence of continental strata of Cretaceous age he concluded that the Sirte Basin had the same post-Palaeozoic history as the Homra ('Ghadames') Basin and the Western Desert Basin ('Cyrenaica Shelf'). According to KLITZSCH (1970), the present day Sirte Basin was the northern part of a late Palaeozoic to Mesozoic uplift (Tibesti–Sirte–Uplift), from which during Permian to Jurassic times Palaeozoic strata were eroded. The fact that continental sediments of early Cretaceous age are present proves only that development of the Sirte Basin locally started in early Cretaceous time. The Cambrian to Ordovician age strata still present in the subsurface of the Sirte Basin are probably part of the early Palaeozoic Calanscio Trough (Fig. 3). The post-Ordovician Palaeozoic sediments of this trough were eroded in Permian to Jurassic times, after the area of the present day Sirte Basin was uplifted by Hercynian movements.

The Cambrian to Ordovician sandstone and quartzitic sandstone are preserved in thicknesses of more than 500 metres locally. These strata are the reservoir of several oil fields in the Sirte Basin, the source rock of which is late Cretaceous shale (submigration from graben floors into sediments of uplifted horst-blocks).

6.7. Cambrian of Sinai

Palaeozoic rocks have been known from Sinai in eastern Egypt for more than 100 years. Recent investigations by WEISSBROD (1969) and correlations with Lower Palaeozoic strata farther east suggest a Cambrian age for the Lower Sandstone 'Series' of the Um Bogma–Jebel Dhalal area in south-western Sinai. Most previous authors had tentatively placed the Lower Sandstone in the Carboniferous.

The Lower Sandstone Series unconformably overlies peneplained Pre-Cambrian basement and is unconformably overlain by dolomite of the Carboniferous Um Bogma Formation. Its thickness increases from 20 metres in the north-western part of the Um Bogma area to 150 metres in the south-east. The sequence consists of reddish sandstone at the base overlain by green, red, and brown shale and by reddish to brown sandstone. The middle part of the sequence includes a 6 metres thick sandy dolomite and dolomitic sandstone unit. WEISSBROD (1969) correlates this unit with the Nimra Formation, which contains trilobites of Lower Cambrian (Georgian) age. Trilobite traces and other trace and problematic fossils are known from several parts of the Lower Sandstone Series.

7. Volcanism

Volcanic rocks are known from the Silurian of the Homra Basin (BONNEFOUS, 1963) in bordering areas of Tunisia and from the southern parts of the Murzuk and Kufra Basins in southern Libya, northern Niger, and northern Chad (PLAUCHUT, 1960; KLITZSCH, 1968a).

South-west of Mourizidié at the eastern edge of the Murzuk Basin sills, flows, and tuffs of a basic magma (diabase) occur within the Tanezzuft Shale Formation. They range in thickness from a few decimetres to more than 10 metres and have large horizontal extension. In Chad, Niger, and Tunisia these magmatic rocks occur in the same stratigraphical position. In the Ennedi Mountains they can be used to identify the basal part of the Silurian sequence, where the Tanezzuft Shale is more or less completely replaced by near shore sandstone (Acacus, see Fig. 12). This basic volcanism is possibly linked to tectonic movements of the Caledonian episode.

8. Economic value

The Silurian Tanezzuft Shale is one of the major source rocks of oil in North Africa. In Algeria the oil of the largest oil field (Hassi Messaoud) originates from Silurian shale. When oil exploration started in Libya, Palaeozoic prospects therefore came within the first aims of exploration. The first oil strike in Libya, made by Esso in December 1957, was from Devonian sandstone near the Algerian border. Since then, oil from Silurian source rocks was discovered in more than 100 wells in different parts of the Homra Basin and in bordering areas near the Murzuk Basin. In some wells gas was found. Reservoir rocks are mainly Silurian and Devonian sandstone, and at some locations also Ordovician sandstone. The total test rate of oil and gas related to Lower Palaeozoic sources is of the order of at least 35,000 BOPD and 3500 million cubic feet of gas per day. Unlike those of some of the fields in Algeria most structures in north-western Libya are small and the individual fields contain only relatively small quantities of recoverable oil. As a result, none of the oil and gas fields of north-western Libya has so far been developed. The Murzuk Basin of south-western Libya probably also contains oil fields originating from Silurian source rocks as must be expected because of oil shows tested in several of the few wells drilled in this area. The remote position of this basin (far from the coast) has so far been a hindering factor against intensitive oil exploration.

No oil which can be related to Silurian source rocks has been found in the Western Desert Basin of Egypt and north-eastern Libya. Probably the Silurian shale there is too silty and sandy to be a source rock for oil. The same has to be expected for the entire Kufra Basin.

From the tectonic point of view, it may be important in an economic sense that the formation of north-north-westerly striking horsts and uplifts during early Palaeozoic times produced structural situations which might have been very favourable for the accumulation of Palaeozoic oil (KLITZSCH, 1970; Figs. 5 and 10), especially in the Murzuk Basin.

References

BAER, C. B. and KLITZSCH, E. (1964). Introduction to the Geology of Egypt, *in Guidebook to the Geology and Archaeology of Egypt*. Petroleum Exploration Society of Libya, Tripoli, 71–98.

BAIRD, D. W. (1969). Geological Bibliography of the Murzuk Basin Region, *in Geology, Archaeology and Prehistory of the Southwestern Fezzan, Libya*. Petroleum Exploration Society of Libya, Tripoli, 139–150.

BARR, F. T. and WEEGAR, A. A. (1972). *Stratigraphic Nomenclature of the Sirte Basin, Libya*. Petroleum Exploration Society of Libya, Tripoli.

BENDER, F. (1968). Geologie von Jordanien, *in* MARTINI, H.-J. (Ed.), *Beitr. Reg. Geol. Erde*, **7**, Borntreger, Berlin.

BEUF, S., BIJU-DUVAL, B., STEVAUX, J., and KULBICKI, G. (1969). Extent of 'Silurian' Glaciation in the Sahara, its Influences and Consequences upon Sedimentation, *in Geology, Archaeology and Prehistory of the Southwestern Fezzan, Libya*. Petroleum Exploration Society of Libya, Tripoli, 103–116.

BONNEFOUS, J. (1963). Synthèse stratigraphique sur le Gothlandien des Sondages du Sud Tunisien. *Rev. Inst. Fr. Pétrole*, **18/10–11**, 123–133.

BONNEFOUS, J. (1972). Geology of the quartzitic 'Gargaf Formation' in the Sirte Basin Libya. *Bull. Centre Rech. Pau–SNPA*, **6/2**, 225–261.

BUROLLET, P. F. (1963). Reconnaissance geologique dans le sud-est du bassin de Kufra, *Rev. Inst. Fr. Pétrole*, **18/10–11**, 219–227.

BUROLLET, P. F. and BYRAMJEE, R. (1969). Sedimentological Remarks on Lower Paleozoic Sandstones of South Libya, *in Geology, Archaeology and Prehistory of the Southwestern Fezzan, Libya*. Petroleum Exploration Society of Libya, 91–101.

COLLOMB, G. R. (1962). Etude Géologique du Jebel Fezzan et sa Bordure Paléozoique. *Compagnie Francaise des Pétroles, Notes et Memoires*, **1**.

CONANT, L. C. and GOUDARZI, G. H. (1964). Geologic Map of the Kingdom of Libya, 1:2 Million. *Misc. Geol. Invest*, Map 1–350A, U.S. Geological Survey, Washington.

CONANT, L. C. and GOUDARZI, G. H. (1967). Stratigraphic and Tectonic Framework of Libya. *Bull. Am. Assoc. Petrol Geol.*, **51/5**, 719–730.

DE LESTANG, J. (1968). Das Paläozoikum am Rande des Afro–Arabischen Gondwanakontinents. *Z. Deut. Geol. Ges.*, **117/2–3**, 479–488.

DESIO, A. (1935). Studi geologici sulla Cirenaica, sul Deserto libico, sulla Tripolitania e sul Fezzan Orientali. *Missione Sci. Reale Accad. Ital. Cufra* (1931), **1**.

DESIO, A. (1936). Brevi notizie sulla presenza del Silurico fossilifero nel Fezzan. *Boll. Soc. Geol. Ital.*, **55**, 116–120.

DESIO, A. (1936). Riassunto sulla costizione geologice del Fezzan. *Boll. Soc. Geol. Ital.* **55**, 319–356.

DESIO, A. (1939). Le nostre conoscenze geologiche sulla Libia sino al 1938. *Ann. Mus. Libico Storia Naturale*, **1**, 13–54.

DESIO, A. (1940). Fossili neosilurici del Fezzan occidentale. *Ann. Mus. Libico Storia Naturale*, **2**, 13–45.

FREULON, J. M. (1951). Sur la série primaire de Fezzan nord-occidental. *C.R. Soc. Géol. Fr.*, **12**, 216–218.

FREULON, J. M. (1964). Etude géologique des séries Primaires du Sahara Central. *C. Nat. Rech. Sci.*, *Ser. Géol.*, **3**.

FUERST, M. and KLITZSCH, E. (1963). Late Caledonian Paleogeography of the Murzuk Basin. *Rev. Inst. Fr. Pétrole*, **18/10–11**, 158–170.

GARIEL, O., DE CHARPAL, O., and BENNACEF, A. (1968). Sur la sedimentation des grès du Cambro–Ordovicien (Unité II) dans l'Ahnet et le Mouydir (Sahara Central). *Bull. Serv. Géol. Algerie*, **38**, 7–37.

HAVLÍČEK, V. and MASSA, D. (1973). Brachiopodes de l'Ordovicien Supérieur de Libye Occidentale. *Geobios*, **6/4**, 267–290.

HECHT, F., FUERST, M., and KLITZSCH, E. (1963). Zur Geologie von Libyen. *Geol. Rdsch.*, **53**, 413–470.

HOFFMANN-ROTHE, J. (1966). Zur Stratigraphie und Tektonik des Paläozoikums der Algerischen Ostsahara. *Geol. Rdsch.*, **55/3**, 736–774.

HOTTINGER, A. F. (1959). *Geological Reconnaissance of the North-Western Sudan by Royal Dutch Shell and British Petroleum Survey Party.* M.S. The Hague, Open File Report, Sudan. Geological Survey Library.

JACQUÉ, M. (1963). Reconnaissance Géologique du Fezzan Oriental. *Compagnie Francaise de Pétroles*, *Notes et Memoires* **5**.

KILIAN, C. (1928). Sur la Présence du Silurien à L'est et au Sud de L'Ahaggar. *C.R. Acad. Sci., Paris*, **186/8**, 508.

KILIAN, C. (1931). Sur l'âge des grès à Harlania et sur l'extension du Silurien dans le Sahara oriental. *C.R. Acad. Sci., Paris*, **192/26**, 1742–1743.

KLITZSCH, E. (1963). Geology of the North-East Flank of the Murzuk Basin (Djebel Ben Ghnema–Dor el Gussa Area). *Rev. Inst. Fr. Pétrole*, **18/10–11**, 97–113.

KLITZSCH, E. (1965a). Ein Profil aus dem Typusgebiet gotlandischer und devonischer Schichten der Zentralsahara (Westrand Murzuk-becken, Libyen). *Erdöl Kohle*, **18/8**, 605–607.

KLITZSCH, E. (1965b). Zur regionalgeologischen Position des Tibesti-Massivs. *Max Richter Festschrift Clausthal-Zellerfeld*, 111–125.

KLITZSCH, E. (1966). Comments on the Geology of Central Parts of Southern Libya and Northern Chad (etc.). South Central Libya and Northern Chad. Petroleum Exploration Society of Libya, Tripoli, 1–17, 19–32, 75–87.

KLITZSCH, E. (1968a). Die Gotlandium-Transgression in der Zentral-sahara. *Z. Deut. Geol. Ges.*, **117/2–3**, 492–501.

KLITZSCH, E. (1968b). Outline of the Geology of Libya, *in Geology and Archaeology of Northern Cyrenaica*. Petroleum Exploration Society of Libya, Tripoli, 71–78.

KLITZSCH, E. (1969). Stratigraphic Section from the Type Areas of Silurian and Devonian Strata at Western Murzuk Basin (Libya), *in Geology, Archaeology and Prehistory of the southwestern Fezzan*, Libya, Petroleum Exploration Society of Libya, Tripoli, 83–90.

KLITZSCH, E. (1970). Die Strukturgeschichte der Zentralsahara. Neue Erkenntnisse zum Bau und zur Paläogeographie eines Tafellandes. *Geol. Rdsch.*, **59/2**, 459–527.

KLITZSCH, E. (1974). Bau und Genese der Grarets und Alter des Grossreliefs im Nordostfezzan (Südlibyen). *Z. Geomorph.*, *N.F.*, **18/1**, 99–116.

KLITZSCII, E. (1978). Geologische Bearbeitung Südwest Ägyptens. *Gcol. Rdsch.*, **67**, 2, 509 520.

KLITZSCH, E. (1979). Zur Geologie des Gilf Kebir Gebietes in der Ostsahara. *Clausth. Geol. Abh.*, **30**, 113–132.

KLITZSCH, E. (1980). Neue stratigraphische und paläogeographische Ergebnisse aus dem Nordwest-Sudan. *Berliner Geowiss. Abh., A*, **20**, 217–222.

KLITZSCH, E. and LIST, F. (1979). *Geological Interpretation Map 1:500,000*, Sheet 2521, *Gebel Uweinat*, and Sheet 2523, *Gilf Kebir*. Techn. Fachhochsch. Berlin.

KLITZSCH, E., LÉJAL-NICOL, A., and MASSA, D. (1973). Le Siluro-Devonien à Psilophytes et Lycophytes du bassin de Mourzouk (Libye). *C.R. Acad. Sci., Paris Sér. D*, **277**, 2465–2467.

LELUBRE, M. (1946). Sur le Paléozoique du Fezzan. *C.R. Acad. Sci., Paris*, **222**, 1403–1404.

LELUBRE, M. (1952). Apercu sur la Géologie du Fezzan. *Bull. Serr. Carte Géol. Algerie, Trav. Recents Collab.*, **3**, 109–148.

MASSA, D. and COLLOMB, G. R. (1960). Observations Nouvelles sur la Région d'Aouinet Ouenine et du Djebel Fezzan (Libye). *XXI Internat. Geol. Congress, Norden*, Section 12, 65–73.

MASSA, D. and JAEGER, H. (1971). Données Stratigraphiques sur le Silurien de L'Ouest de la Libye. *Mem. B.R.G.M.*, **73**, 313–321.

MENNING, J. J. and VITTIMBERGA, P. (1962). Application des méthodes petrographiques à l'étude du Paleozoique ancien du Fezzan. *Companie Francaise des Pétroles, Notes et Memoires*, **2**.

OMARA, S. and CONIL, R. (1965). Lower Carboniferous foraminifera from south-western Sinai, Egypt. *Ann. Soc. Geol. Belg.*, **88/5**, 222–242.

OVERWEG, A. (1851). Geognostische Bemerkungen auf der Reise von Phillippeville über Tunis nach Tripoli und von hier nach Murzuc im Fezzan. *Z. Deutsch. Geol. Ges.*, **3**, 93–102.

PESCE, A. (1968). Gemini space photographs of Libya and Tibesti. Petroleum Exploration Society of Libya, Tripoli.

PETROLEUM EXPLORATION SOCIETY OF LIBYA, NAMES AND NOMENCLATURE COMMITTEE (1960). Libye, *in Lexique Stratigraphique International*, **4**, Africa, IVa, C. Nat. Rech. Scientifique, Paris.

PETROLEUM EXPLORATION SOCIETY OF LIBYA (1963). *Field Trip Guidebook of the Excursion to Aouinat Ouenine.*

PICARD, L. (1953). Silurian in Negev (Israel). *C.R. Intern. Geol. Congr. Algiers*, Sect. 2, fasc. 2, 87–92.

PLAUCHUT, B. (1960). Notice explicative sur la carte géologique du Djado 1:500,000, Bureau de Recherches Géologiques et Minières, Dakar.

RATSCHILLER, L. K. (1967). Sahara, Correlazioni geologico–lithostratigrafiche fra Sahara centrale et occidental. *Mem. Mus. Tridentino Sci. Nat.*, **15/1**.

ROGNON, P., DE CHARPAL, O., BIJU-DUVAL, B., and LEGRAND, PH. (1968). Les glaciations "siluriennes" dans Ahnet et le Mouydir (Sahara central). *Bull. Serr. Géol. Algerie*, **38**, 53–81.

SAID, R. (1962). *The Geology of Egypt*. Elsevier, Amsterdam.

SAID, R. (1971). Explanatory Notes to accompany the geological map of Egypt. *Geol. Surv. Egypt*, **56**.

SAID, R. and SHUKRI, N. M. (1955). Ancient Shore lines of Egypt, part 1: The Paleozoic. *Bull. Geogr. Egypte*, **28**, 41–49.

SANDFORD, K. S. (1935). Geological observations on the north-west frontiers of the Anglo–Egyptian Sudan and the adjoining part of the Southern Libyan Desert. *Q. J. geol. Soc.*, **91**, 323–381.

SEILACHER, A. (1969). Sedimentary Rhythms and Trace Fossils in Paleozoic Sandstones of Libya, in *Geology, Archaeology and Prehistory of the southwestern Fezzan, Libya*. Petroleum Exploration Society of Libya, Tripoli, 117–123.

VITTIMBERGA, P. and CARDELLO, R. (1963). Sédimentologie et pétrographie du Paléozoique du bassin de Kufra. *Rev. Inst. Fr. Pétrole*, **18/10–11**, 228–240.

WEISSBROD, T. (1969). The Paleozoic of Israel and adjacent countries part II, the Paleozoic outcrops in South-Western Sinai and their correlation with those of Southern Israel. *Bull. Geol. Surv. Israel*, **48**.

WHITEMAN, A. J. (1971). *The Geology of the Sudan Republic*. Clarendon Press, Oxford.

Lower Palaeozoic of the Middle East, Eastern and Southern Africa, and Antarctica
Edited by C. H. Holland
© 1981 John Wiley & Sons Ltd.

LOWER PALAEOZOIC ROCKS OF SOUTHERN AFRICA

I. C. Rust

Department of Geology, University of Port Elizabeth, South Africa

Contents

1. Introduction

Sediments of proven Lower Palaeozoic age are scarce in southern Africa south of latitude 15° S, but several formations are considered to belong in this age group. The sedimentary formations are the Mulden Group, the Fish River Formation, and the Table Mountain Group (Fig. 1). The latest authoritative publications on these formations are by FRETS (1969) and GUJ (1970) on the Mulden, GERMS (1972) on the Fish River, and RUST (1967) and VISSER (1974) on the Table Mountain.*

Igneous suites belonging wholly or partly to the Lower Palaeozoic are the Cape Granite Complex (SCHOLTZ, 1946; SCHOCH, 1972), the Kuboos Granite Complex (DE VILLIERS and SÖHNGE, 1958), the Salem Granite Complex (MILLER, 1973), and widely distributed pegmatites belonging to the Mozambique and Zambezi belts (CLIFFORD, 1967). Reviews of all these formations have been published by DU TOIT (1954), HAUGHTON (1969), and TRUSWELL (1970).

2. Stratigraphical boundaries of the Lower Palaeozoic in southern Africa

A primary problem in southern Africa is the lack of definite information on the exact position in the sedimentary record of both the lower and the upper boundaries of the Lower Palaeozoic. Nowhere in southern Africa is the base of the Cambrian known even approximately. If this time-plane is preserved at all within the local sedimentary record, it must be located somewhere in the Damara/Nama sequence of the Pan-African Geosyncline (KENNEDY, 1964) in South West Africa.

---

* The manuscript of this contribution was completed June 1975 and did not contain reference to applicable papers published after January 1975.

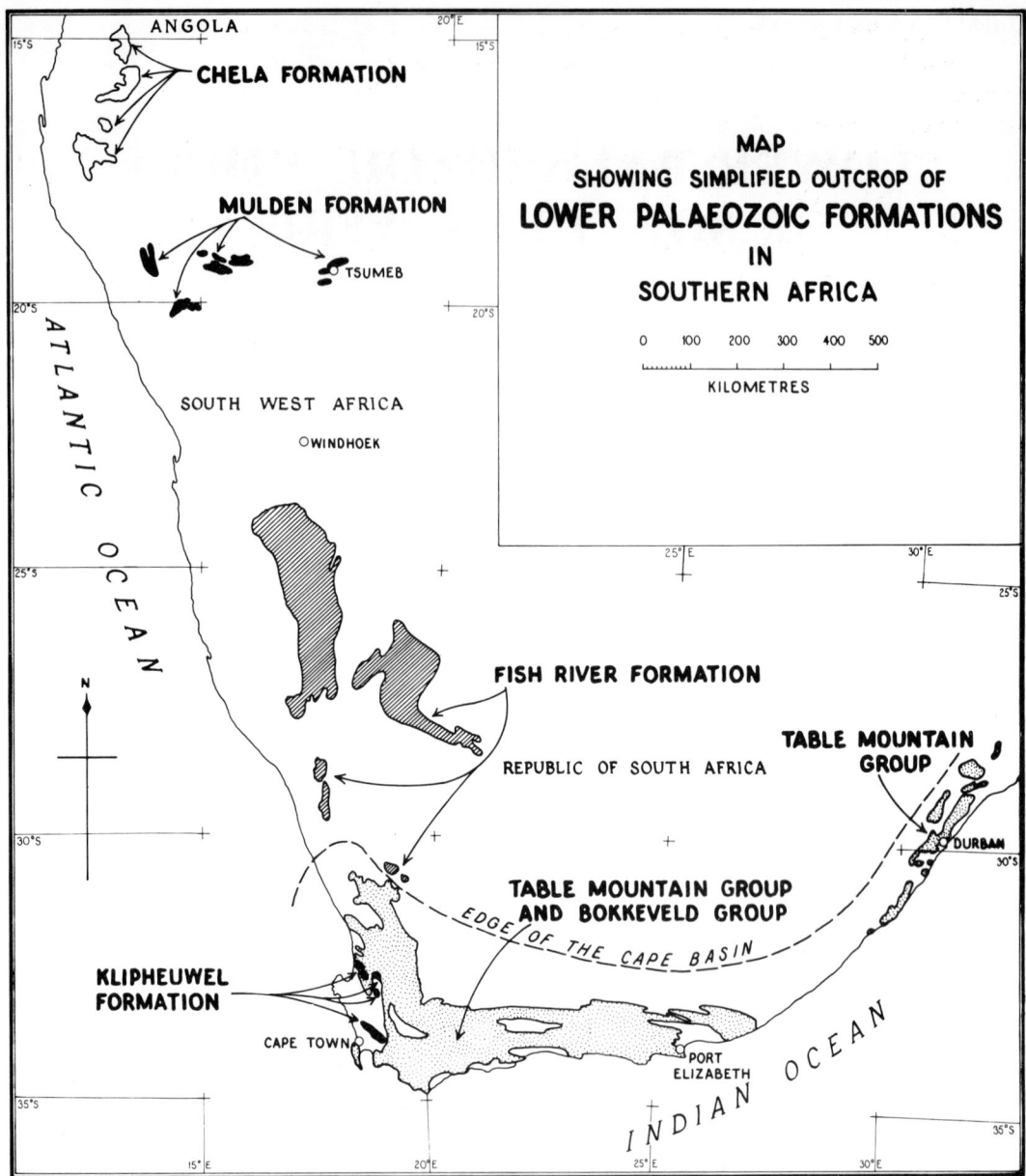

FIG. 1. Map showing simplified outcrop of Lower Palaeozoic formations in southern Africa.

The Nama sediments below the Fish River Formation are characterized by a late Pre-Cambrian Ediacara fauna (GLAESSNER, 1971), whereas *Phycodes podem* (Cambrian?) occurs in the Fish River Formation. This relationship as well as certain radiometric data have led GERMS (1972a, p. 210) to speculate that the age of the sediments above the prominent unconformity in the upper Schwarzrand Formation is '...most probably Cambrian'. This age estimate is accepted in the present review. Future work may well amend the view put forward in this paper that it is convenient to regard the base of the Fish River Formation as the base of the Cambrian in southern Africa.

The Cape Granite Complex was probably intruded during the late Pre-Cambrian and Early Cambrian (SCHOCH, 1972), but some of the radiometric dates conflict with field evidence; work continues to resolve these problems. The more reliable U–Pb analyses indicate an age range between 610 ± 20 and 530 m.y.

Radiometric data relating to the age of pegmatite emplacement associated with the Damaran orogeny in South West Africa suggest a Cambrian age for this event (CLIFFORD, 1967). The Khomas shales which were metamorphosed during the orogeny may in part also be of Cambrian age, but all evidence relating to such a possible date seems to have been destroyed. Ultrametamorphism led to extensive melting and production of the magmatic Salem Granite Suite (MILLER, 1973).

The Klipheuwel Formation in the southern Cape Province could possibly be of Cambrian age. As far as is known the sequence is devoid of fossils, nor is it intruded by datable igneous rocks. The Klipheuwel clastics unconformably overlie the Cape Granite, and are in turn overlain by the Table Mountain Group.

The age of the oldest formation in the Table Mountain Group, the Piekenier Conglomerate Formation, is also unknown, but from indirect evidence (RUST, 1967), it seems reasonable to guess that its base falls within the early Ordovician.

This information, meagre as it is, represents the full evidence relating to the base of the Palaeozoic in southern Africa.

Data on the location of the upper boundary of the Lower Palaeozoic in southern Africa are almost non-existent. This is particularly frustrating because the stratigraphical succession included in the Table Mountain and Bokkeveld Groups seems to be essentially unbroken, spanning the Ashgill (or Llandovery) and the Emsian without any conspicuous break in the sedimentation, but between these two widely spaced palaeontological markers no datable fossils or other datable rock horizons are known.

Brachiopods and other marine invertebrates of the Cedarberg Formation (Table Mountain Group) suggest an Ashgill age according to COCKS and others (1969), but BERRY and BOUCOT (1973) prefer a Llandovery (or possibly an early Llandovery) age for the fossiliferous member of the Cedarberg Formation. BERRY and BOUCOT consider the presence of the Pakhuis Tillite (about 30 metres stratigraphically below the fossil horizon in the Cedarberg Formation) an added reason that, in the light of other glaciations in North Africa and South America, the Cedarberg should be dated as early Silurian rather than late Ordovician.

The precise age is actually of little significance in our present quest for the upper boundary of the South African Silurian. However, as BERRY and BOUCOT (1973) noted, the stratigraphical position of the Nardouw Formation (between the Cedarberg Formation and the Bokkeveld Group) does suggest that much of the Nardouw may be of Silurian age.

The only other relevant datable horizon is the lowermost shale formation of the marine Bokkeveld Group. The brachiopods in this formation indicate an Emsian age (THERON, 1972). The associated ophiuroids are not particularly useful (Early Devonian to Early Carboniferous), nor are the trilobites (Early Devonian). So, on this account, too, it is not possible to locate with any measure of certainty the position of the lower boundary of the Devonian, and thereby locate the upper limit of the Lower Palaeozoic.

However, as a practical measure, and certainly as one of convenience, it may have merit to select the boundary between the dominantly super-mature Table Mountain Group and the conspicuously immature Bokkeveld Group as the upper boundary of the Lower Palaeozoic in South Africa. Even if the time value of this contact (the upper boundary of the Nardouw Sandstone) is in some doubt, it may still be useful on at least three accounts: firstly, in time value the contact is probably not too far removed from the desired time plane; secondly, the contact is located in a sequence which seems to display no evidence of a major time hiatus over the larger part of the basin; and thirdly, the contact is reasonably easily located in the field by virtue of the distinct lithological change across it.

To summarize, in this paper the stratigraphical limits for the Lower Palaeozoic in southern Africa are considered to be the base of the Cambrian? Fish River Formation and the top of the Silurian? Nardouw Formation. It is clear that the problem of precisely locating and extending time markers in the Lower Palaeozoic sequence of southern Africa is one that urgently needs to be investigated.

3. Orogenic and thermal events

The regional structural setting of South West Africa in terms of Proterozoic to Lower Palaeozoic rocks has been discussed by MARTIN (1965), CLIFFORD (1967), KRÖNER (1971), and GERMS (1974).

MARTIN (1965) put forward useful models for the Damaran orogeny and its associated sedimentation and proposed three possible correlations of the various sedimentary formations in the major tectono-sedimentary units, namely:

(1) The Nama Group is younger than the entire Damara sequence;
(2) The Nama Group is a correlate of the upper part of the Damara sequence;
(3) The Nama Group and the now severely metamorphosed Damara sequence are coeval, representing respectively shelf and geosynclinical facies of the Upper Proterozoic/Lower Palaeozoic Pan-African Geosyncline. GERMS (1974) lists several lines of evidence which indicate that this model can be discounted.

The problem of the intercorrelation of the various units in this Damara–Nama complex has not been solved, and several schemes, differing more or less radically, have been proposed (GERMS, 1974a). MARTIN (1974) now thinks that the Nama Group is younger than the metamorphosed Swakop Group of the Damara sequence on the basis of new data relating to the nappe area in the Naukluft Mountains (KORN and MARTIN, 1959). GERMS (1974a) supports Martin's recent view and in addition concludes that the base of the Mulden Group is slightly older than the Nama Group; again, however, such correlation is based on circumstantial evidence, namely supposed progressive tectonic movements in the Damara geosyncline.

The major tectono-thermal event of the Damara orogeny involved regional metamorphism of the sediments deposited in the central part of the Damara geosyncline, as well as emplacement of granite and pegmatite bodies in the same area.

The Salem Granite Complex was produced during ultrametamorphic transformation of the Khomas schist (SMITH, 1965), mainly by melting of greywacke, crystal settling, and progressive *in situ* crystallization (MILLER, 1973). The generation of granitic magma occurred at a very late stage in the Damaran tectonic cycle and the actual intrusive phase post-dates the tectonism. The suite consists of a differentiated sequence grading from diorite along the periphery of the intrusive complex through monzonite into the main and central Salem adamellite ('granite').

CLIFFORD (1967) indicates that seven K–Ar age determinations of biotite from a wide variety of Khomas schists, youngest of the central Damara sediments [MARTIN's (1965) Swakop Facies], span the interval from about 452 to 550 m.y.; 14 analyses of other Damaran rocks give similar results. BURGER and COERTZE (1973) list a further nine U–Pb dates ranging between 430 ± 15 and 565 ± 50 m.y. for rocks belonging to this tectono-thermal event. They also list a Rb–Sr age of 506 ± 10 m.y. for a syenite from the Bremen pluton in South West Africa.

It is not possible to determine the sedimentary age of the Khomas schist, except to note that the regional setting of the Damara orogen suggests that Khomas sedimentation probably took place not long before the metamorphic episode. In view of this uncertainty the geology of the Khomas schist will not be discussed here; an excellent review is given by MARTIN (1965).

KRÖNER and WELIN (1973) indicate that a 500 m.y. thermal episode can be recognized in the Bogenfels shale (whole-rock Rb–Sr) in the area south of the central Damara complex. KRÖNER (1969) reports an age of 504 ± 10 m.y. in the Vanrhynsdorp area where the southernmost outcrops of the Nama Group are known to occur. These dates suggest that the Damaran thermal event operated noticeably far to the south of the type area (KRÖNER and WELIN, 1973).

The Cape Granite Complex (SCHOLTZ, 1946; SCHOCH, 1972; VISSER and SCHOCH, 1973) consists of several phases occurring in a series of plutons distributed from George in the east to Saldanha Bay in the north. At least seven intrusive phases can be recognized in the west, ranging from oldest to youngest in the following order: Ysterfontein diorite and gabbro; Hoedjiespunt and Darling granites; Saldanha quartz porphyry; Saldanha granite; Cape Columbine and Contreberg granites; aplogranite; quartz porphyry and granite porphyry dykes. The complex is intrusive into already deformed Malmesbury sediments and, as the radiometric ages suggest, the Malmesbury Group may be much older than the Nama Group, with which it has been correlated (VAN EEDEN, 1972). Of this correlation HARTNADY and others (1973) cautioned that it may be '... a great oversimplification'.

Radiometric age values for the Cape Granite Complex are 630 m.y. (HOLMES and CAHEN, 1957), 610 m.y. (Cape Peninsula granite, SCHOCH, 1972), 550 ± 50 m.y. (ALDRICH and others, 1958), and 550 and 496 m.y. (ALLSOPP and KOLBE, 1965). In the Saldanha area SCHOCH and others (1975) measured U–Pb ages of 600 ± 20 m.y. for the Hoedjiespunt granite and 530 ± 15 m.y. for the Vredenburg granite. They suggest that the Hoedjiespunt granite can be correlated with the Darling granite and the coarsely porphyritic Cape Peninsula granite.

VAN EEDEN (1972) suggests that the Kuboos Granite Complex near the mouth of the Orange River should be correlated with the Cape Granite thermal event on the basis of a radiometric age of 550 ± 20 m.y. (DE VILLIERS and BURGER, 1967) for the Kuboos granite. Like the Cape Complex, the Kuboos granite plutons intrude Pre-Cambrian rocks; essentially undeformed and unmetamorphosed Nama sediments occur in the vicinity but are nowhere known to be in contact with the Kuboos intrusives.

The only palaeomagnetic measurements on Lower Palaeozoic rocks in southern Africa relate to the middle Ordovician Graafwater Formation (GRAHAM and HALES, 1961; note the 'Silurian' age allocated to these sediments). The more reliable results place the Cape Town area near the Ordovician equator with the magnetic pole near present day Morocco. CREER (1970) and McELHINNY and others (1968) point out certain inconsistencies and suggest that more work be done.

4. The Mulden Group

The Mulden Group (formerly known as the Mulden Formation but here referred to as the Mulden Group in accordance with KRÖNER's (1974b) suggestion) is a sequence of clastics which forms the third and upper part of MARTIN's (1965) Outjo Facies. The following discussion on the geology of the Mulden Group is based extensively on the work of FRETS (1969) and GUJ (1970).

The Mulden Group has nowhere been dated directly. Its estimated Cambrian age is based entirely on the correlation of the Mulden Group with the Cambrian Fish River Formation and their supposed relationship to the tectonic event of the Damaran orogeny.

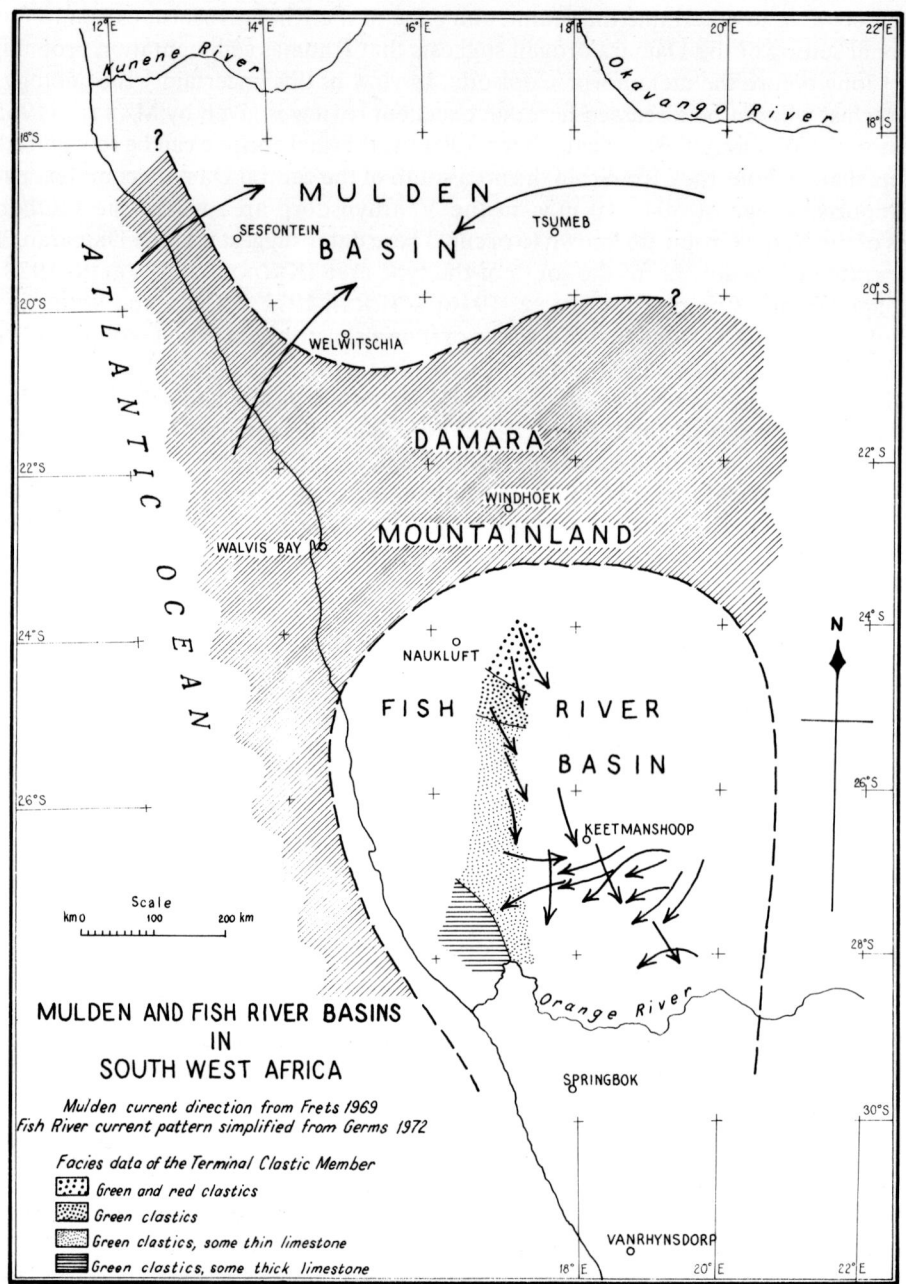

FIG. 2. Mulden and Fish River basins in South West Africa.

A mirror-image relationship relates the Mulden Group to the Fish River Formation (Fig. 2). Towards the south of the central Swakop Facies the relationship of the Fish River Formation to the underlying Nama rocks is almost exactly paralleled by the relationship of the Mulden Group to the underlying Otavi sequence north of the Swakop core. Considering the evidence, it is a reasonable postulate to regard these two formations as having formed essentially simultaneously and representing syntectonic deposition related to the Damara tectonic episode.

The Chela Formation in southern Angola (BEETZ, 1933; VALE, 1968) is a mixed sequence of sediments and volcaniclastics. It has been correlated with the Nama Group by VALE and regarded as of Cambrian age. DE VILLIERS (1972) suggested a possible correlation with the Mulden Group. KRÖNER (1974a), assisted by H. CORREIA, shows that the Chela Formation is appreciably older than either the Nama or the Mulden.

The Mulden Group overlies rocks of Otavi age and older. The contact between the Mulden and the Otavi is a disconformity near Tsumeb and eastward, and becomes an angular unconformity towards the west and north. GUJ (1974) supplies proof that Mulden sediments overlie Khomas schist with a marked structural unconformity and a rather thick basal conglomerate on the farm Austerlitz 515 west of Welwitschia.

Mulden and Fish River sediments are not in contact with one another, nor are the Fish River sediments in contact with any part of the central Damara complex, although the two sequences approach one another to within a few tens of kilometres (KRÖNER, 1971).

The Mulden Group has a preserved thickness of 2800 to 3000 metres, and can be subdivided into three major units. The basal unit contains some conglomerate but consists mainly of greywacke and shale, and represents a topographic fill-in. Some of these filled-in valleys near Welwitschia are up to 150 metres deep (Fig. 3). The Otavi phyllite is probably an eastern correlate of the Welwitschia valley-fill unit. In the west it seems as if some of the basal conglomerate represents near-shore coastal cliff breccias. The middle part of the Mulden Group in the west is a quartzite unit, about 1600 metres thick but diminishing in thickness southwards. It consists of a monotonous sequence of feldspathic current-bedded quartzite which grades eastwards into greywacke, arkose, and shale.

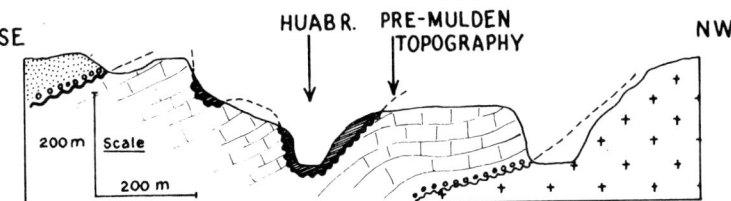

FIG. 3. Pre-Mulden topography developed on Otavi dolomite near Welwitschia (after FRETS, 1969).

The upper unit of the group is a thinly laminated shale and greywacke, occasionally ripple-marked. Its maximum preserved thickness in the west is less than 1000 metres.

In the Welwitschia area FRETS (1969) has suggested that the immature Mulden clastics indicate a high, rugged, uplifted provenance. This tectonism was related to the folding and thrusting movements of the Damara orogeny which preceded and accompanied the deposition of Mulden sediments in a rapidly subsiding intra-montane basin; GUJ (1974) supports this view. Lateral facies changes indicate a sediment source south-west and north-west of Welwitschia.

In the south-western Kaokoveld the unconformity-bounded immature Mulden clastics exhibit significant facies changes (Fig. 4). East and north of this area the Sesfontein phyllite, which is interbedded with calcarenite and the Warmquelle monogenetic dolomite conglomerate, represents the Mulden Group. Towards the south-west this sequence grades into and is progressively overlain by greywacke and chlorite–feldspar–quartz sandstone with rare

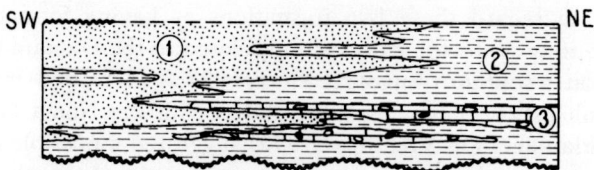

Reconstructed facies relationships, vertical section
Mulden, 700 m, greywacke sequence.

FIG. 4. Facies relationships in the Mulden Formation,
showing (1) non-marine bedded feldspathic quartzite
and greywacke; (2) marine, flysch-type phyllite;
(3) carbonate conglomerate and other clastics derived
from an Otavi provenance (after GUJ, 1970).

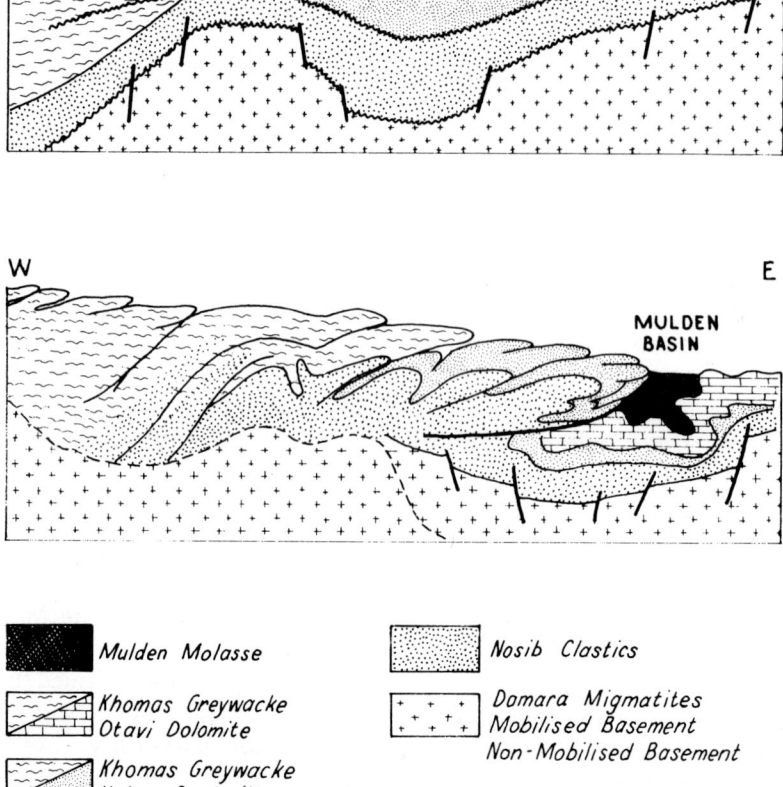

FIG. 5. Reconstruction of the pre-Mulden basin (above) showing stable shelf
deposition of clastics and carbonates, and the syntectonic Mulden basin
(below) in the vicinity of Welwitschia (after GUJ, 1970).

cross-bedding. The Warmquelle conglomerate was probably derived from elevated Otavi carbonates and clastics in the east and north. However, the sandstones towards the south-western part of the basin must have been derived from metasediments of the Swakop facies of the Damara complex farther to the south, and not from the Otavi sequence.

GUJ suggested that the shallow-burial, non-marine greywacke fill of the Mulden basin displays a discontinuous syn- or post-orogenic character typical of exogeosynclines or molasse foredeeps (Fig. 5). Much of the original basin could now be buried below the Sesfontein thrust sheet.

In the Tsumeb area the Mulden Group follows upon the Otavi dolomites with apparent conformity, and the entire sequence is gently folded along an east–west strike (SÖHNGE, 1964). The Mulden sediments are quartzite, arkose greywacke, and slate, in total about 700 metres thick.

In the Etosha Pan deep drilling has indicated a sequence of quartzite and shale, informally referred to as the Ovamboland formation, which overlies the Otavi dolomite; this is probably a correlate of the Mulden Formation (VAN EEDEN, 1972; BRAND, personal communication, 1973).

5. The Fish River Formation

The Fish River Formation is known only in part, the more detailed information about it being provided in passing by GERMS (1972a), whose work forms the basis of the discussion which follows.

The Fish River Formation is normally regarded as an integral part of the Nama Group (HAUGHTON, 1969), of which it forms the uppermost section. GERMS (1972a) has subdivided the Nama Group in the southern part of South West Africa into three formations (Fig. 6). The basal unit is the Kuibis Formation, up to 300 metres thick, which contains, amongst others, the Schwarzkalk limestone. The middle unit is the Schwarzrand Formation, consisting of a sequence of clastics near its base and limestone towards the top, with a thickness ranging between 20 and 1200 metres. The Fish River Formation overlies the Schwarzrand Formation with an uncon-formity.

The fossil assemblage of the Nama Group is an Ediacara type (Fig. 6) and suggests a Late Pre-Cambrian age for the sequence below the Fish River beds. *Cloudina* (GERMS, 1972b), which occurs in limestone beds below the Fish River Formation, might indicate a Cambrian age (following Soviet ideas), but radiometric data make it unlikely that the Kuibis and Schwarzrand Formations are in fact Cambrian; they are probably slightly older.

Phycodes pedum in the Fish River Formation suggests a Cambrian age for these sediments; the same can be said of the so-called Terminal Clastic Member which immediately underlies the Fish River Formation in places.

The Fish River Formation is a clastic sequence up to about 750 metres thick, consisting of mainly reddish to greenish sandstone, siltstone, shale, shale pebble conglomerate, some interbedded black sandstone and coarse feldspathic sandstone with basal conglomerate in places (Plate 1). Cross-bedding is conspicuous and ripple-marks are common. Carbonates and thick shale units are absent in the Fish River Formation. It has been reported (Société Nationale des Pétroles d'Aquitaine, *in* GERMS, 1972) that five members can be recognized in the Fish River Formation, each based on a distinctive sandstone to shale ratio.

The Fish River Formation overlies the older Nama rocks with an unconformity which is developed with various degrees of clarity depending on tectonic movements in various parts of the Fish River basin. It seems as if the so-called Terminal Clastic Member (uppermost unit of the

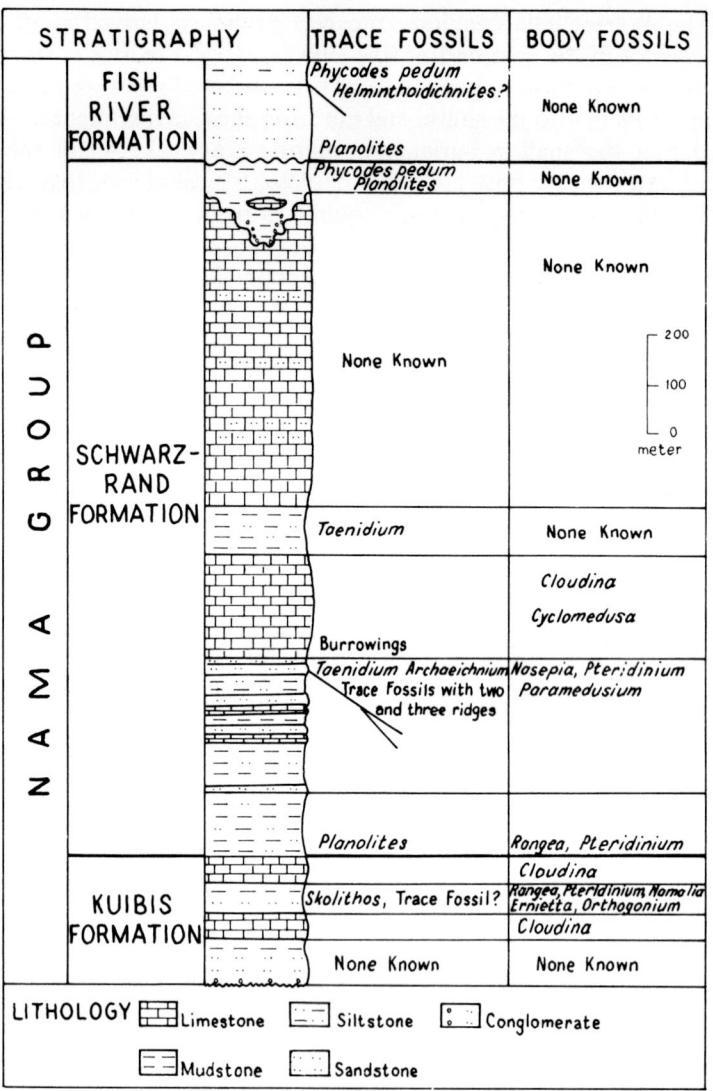

FIG. 6. Stratigraphic column of the Nama Group in southern South
West Africa (after GERMS, 1972b).

Schwarzrand Formation) was a precursor of the subsequent development of the Fish River
Formation. GERMS suggested that the basal unconformities of both the Terminal Clastic
Member and the Fish River Formation can be related to tectonic movements (uplifts) which also
produced the two phases of gravity sliding in the Naukluft Mountains (KORN and MARTIN,
1959). Future work may indicate that the Terminal Clastic Member should be included in the
Fish River Formation.

Basal conglomerates of the Fish River Formation are sporadically developed more or less
west and north of the Klein Karas Mountains, and are absent to the south-east. The pebbles are
mainly black chert, red jasper, silicified oolite, and stromatolitic cherts of unknown origin, with

some Tsumis quartzite pebbles to the north. Variations in pebble size seem to indicate source areas to the north and west. The palaeocurrent map of the Fish River basin (Fig. 2) supports this view. No facies data for the Fish River Formation are available. Data relating to the Terminal Clastic Member may be relevant and have been superimposed upon the palaeocurrent map.

The areas of uplift were generated by tectonic processes associated with the deformation of the core areas of the Damara geosynclinal complex, and in this respect the Fish River Formation should be regarded as a syn- or post-orogenic deposit. The sequential development of the Nama basin clearly shows how the initial basin irregularities, notably the early Osis ridge, which formed a conspicuous salient into the basin from the north-east, as well as the later 'islands', were flooded out by the Fish River sediment influx coupled with the sustained downwards movement in the basin itself. The general immaturity of the Fish River sandstones indicates considerable tectonic activity in the basin, and a molasse-type sedimentary environment is envisaged.

The exact depositional milieu is in doubt. Detailed investigation into the nature and possible causes of the red colouration of the sediments seems to indicate that it is due to diagenetic processes in an oxidizing environment, the climate being unknown. The absence of evaporites may be significant. One specimen of Fish River shale analysed by Guj contained 83 p.p.m. of boron and 249 p.p.m. of vanadium; both values are high compared with the average values for the rest of the Nama sequence, and suggest a marine environment. HÄLBICH (1964) has described primary structures in the Fish River Formation which could indicate a cold climate at times.

The fossils of the Fish River Formation are the traces *Planolites*? in some lower beds, and *Phycodes pedum*, mainly in the upper beds, and some spiral *Helminthoidichnites*. A marine environment seems to be indicated.

The cross-bedding data indicate a consistency ratio of 70 and an average variance of $s^2 = 110$, which could be indicative of non-fluviatile currents, but GERMS thinks that fluviatile dispersion should not be discounted.

The Fish River Formation can be recognized as far south as Brandkop near Vanrhynsdorp, where it is slightly thicker than 100 metres. It follows concordantly on older Nama sediments, and consists of typical shallow-water, reddish, fine-grained sandstone with shale and mudstone (VON BACKSTRÖM and others, 1960). The provenance lay to the west. The trace fossil *Phycodes pedum* also occurs in this area (GERMS, 1974b).

Present knowledge favours the interpretation that the Mulden Group and the Fish River Formation represent all or part of the Cambrian System in southern Africa.

It seems proved that both of these sedimentary sequences are closely linked with the main orogenic phase of the Damara geosynclinal complex. In regional context they represent the final stage in the evolution of the Pan-African Geosyncline, a structural feature which was in operation for 200 to 300 million years around the end of the Proterozoic and the beginning of the Palaeozoic, and which occupied much of the present west coast of southern Africa as well as a broad belt striking north-east through South West Africa and Zambia (DE VILLIERS and SIMPSON, 1974; CLIFFORD, 1967).

Following VALENTINE and MOORES (1970, 1972), GERMS (1974a) has interpreted the geology of the Pan-African Geosyncline, of which the Damara formed part, in terms of plate tectonic theory. The sediments of the Gariep Group (DE VILLIERS, 1961) could be interpreted as being deposited during the early phase of a continental disruption which started about 800 to 900 m.y. The lower transgressive Nama sequence could indicate the commencement of plate convergence. The Fish River Formation and presumably the Mulden Group were deposited on marginal troughs which developed during the continuing assemblage of two continental fragments. The geotectonical reconstruction is supported by the palaeographical reconstruction of the Nama Group as based on its faunal assemblage.

6. The Klipheuwel Formation

The Klipheuwel Formation is a sequence of unmetamorphosed reddish clastics with a maximum known thickness of about 2100 metres (VISSER, 1967; VISSER and TOERIEN, 1971). It overlies the Pre-Cambrian Malmesbury Group and its associated intrusive Cape Granite unconformably in the south-western Cape Province. In places the Klipheuwel Formation is conformable below the Table Mountain Group but there is no doubt that its upper boundary is an erosional surface. The main outcrop area is between Elands Bay and Klapmuts, but it has tentatively been identified far to the east (ROSSOUW and others, 1964). Its degree of deformation is low. The attitude of the beds varies from subhorizontal with almost no tectonic structures, to mild folding with steeply dipping cleavage (DE VILLIERS and others, 1964). The style of deformation and the character of the sediments clearly separate the Klipheuwel Formation from the older Malmesbury Group and the younger Table Mountain Group, but its lateral correlation is in doubt. VAN EEDEN (1972) tentatively suggests, without arguing the case, that the Klipheuwel Formation could be correlated with the Mulden/Ovamboland Formations, but on the other hand he indicates the possibility that the Fish River Formation, in this paper considered to be a correlate of the Mulden, is older than the Mulden. It is probably more meaningful, pending further information to the contrary, to consider the Klipheuwel as a terrestrial phase of the Fish River Formation rather than to correlate it directly with the Mulden.

In the type area at Klipheuwel the lower 900 to 1000 metres consists of medium- to coarse-grained cross-bedded quartz sandstone and feldspathic sandstone. Conglomerate and grit beds are conspicuous and numerous small lenticular and thin shale beds occur. It is a feature of this zone that lateral continuity is lacking, and the sediments are coloured mainly brown, purple, and reddish. The middle zone, also about 900 to 1000 metres thick, consists mainly of pale red to white sandstone, with some grey to yellowish shale. The uppermost zone in the type sequence consists of a few hundred metres of reddish brown and purple shale with interbedded sandy shale. Mudcracks, clay pellet zones, and oscillation ripple-marks indicate extremely shallow water deposition.

North of the type area near Heuningberg is exposed between 1000 and 1300 metres of predominantly very coarse-grained petromict conglomerate, interbedded with shale, clay pellet conglomerate, and sandstone (Plate 2). The colours are conspicuously brown, red and purple, with some green and yellow tints. Two types of conglomerate are developed: coarse conglomerate, with clasts up to 30 centimetres in diameter, containing mainly angular clasts of Malmesbury slate, greywacke, and vein quartz; and the other type consisting of small vein quartz pebbles and grading into a shale pellet conglomerate. In both types of conglomerate the clasts are set in a clayey matrix which determines the predominantly purple, red, and brown colours. In many instances it appears as if the shale slabs incorporated into the conglomerate are similar to Klipheuwel shale. Other clasts are granite types derived from the Cape Granite plutons, and a variety of rock types obviously derived locally from the Malmesbury Group.

No trace fossils or body fossils of any sort have been reported from the Klipheuwel Formation. The dominantly coarse-grained clastics indicate a vigorous depositional environment. The high degree of oxidation points towards deposition on land; this inference is supported by the shallow-water structures in the shales. The fairly limited distribution of the Klipheuwel Formation and its special character seem to suggest that the formation represents a terrestrial clastic wedge, or a series of wedges, deposited against rising fault blocks. It is not yet possible to

Plate 1. Typical sequence of reddish shale and sandstone in the Fish River Formation on the Farm Nordeck in the Klein Karas Mountains between Karasburg and Keetmanshoop, South West Africa. (*Photograph by G. J. B. Germs*).

Plate 2. Immature conglomerate at the base of the ?Cambrian Klipheuwel Formation at Klapmuts Hill, near Cape Town. Similar conglomerate occurs at Heuningberg. (*Photograph by H. Pienaar*).

Plate 3. The folded Lower Palaeozoics of South Africa in the syntaxis area near Ceres where the east-trending ranges of the Cape Fold Range meet the north-trending mountains of the Cedarberg Range. In the foreground the breached anticline in the Hex River Mountains shows the lower cliffs of Peninsula Sandstone, capped by soft slopes of Cedarberg Shale, followed by the precipitous cliffs of the Nardouw Sandstone. The broad synclinal valley in the middle ground is underlain by the Devonian Bokkeveld Group and the low plateau to the right is the Witteberg Sandstones. The range in the background is part of the regional Cedarberg range. The far background is mainly Pre-Cambrian formations. *(Infrared photograph by S. A. Hiemstra).*

Plate 4. Typical arthropod track in Graafwater Sandstone. Named *Petalichnus capensis* sp. nov. by ANDERSON (1974).

Plate 5. Cliff exposure at Donkerkloof in the Cedarberg Range, showing some 850 metres of the Table Mountain Group. Note the prominent and rather massive bedding in the Peninsula Formation (lower cliff) in contrast to the thinner bedding of the Nardouw Formation (upper cliffs). The soft slope is underlain by the Pakhuis Tillite and the Cedarberg Shale. Below the slope the fold zone can be seen.

Plate 6. Pakhuis Tillite with a faceted and striated quartzite erratic, Cedarberg Range.

Plate 7. Typical planar cross-bedding and intraformational conglomerate in the Nardouw Formation, Clanwilliam.

be sure about the relationships of this presumed fault system, except to note that the Klipheuwel Formation is traversed by a prominent set of north-west trending faults which obviously had been active for a long time. This set of faults is recorded in the Cape Granite Complex and was active during emplacement of the granite (SCHOCH, 1972). Faults belonging to the same set have displaced the Klipheuwel (HAUGHTON, 1933; VISSER, 1967), and later movement along this set has displaced the Table Mountain sandstone (DE VILLIERS and others, 1964). However, the sedimentology and basin structure of the Klipheuwel Formation need to be determined before this clastic wedge hypothesis can be tested.

7. The Table Mountain Group

The Table Mountain Group (Fig. 7) is a sequence of supermature quartz arenite in which other sediments, mainly conglomerate, siltstone, shale, and some tillite with glaciogene sediments, are distinctly subordinate.

The main outcrop forms the magnificent Cape fold ranges (Plate 3), which curve in an L-shape round the southern margin of the African continent, running out to sea at Port Elizabeth. Peak elevations range between 1300 and 2000 metres. In Natal, tectonic deformation is limited to faulting and minor tilting and the topographic expression of the sediments is subdued.

The northern edge of the Table Mountain basin is buried by Karoo sediments, but its position is known from deep drilling (VISSER, 1972, 1974; WINTER and VENTER, 1970). The other depositional boundaries have either been destroyed by tectonic processes, or reside on other fragments in Gondwanaland.

The volume of clean quartz sand included in the Table Mountain Group is staggering. Known distribution and thickness indicate a minimum volume of about $5 \times 10^5 - 10 \times 10^5$ km^3 of clean quartz sand (RUST, 1967; *see also* VISSER, 1974). The implication of this in terms of provenance composition, and probably more so in terms of the sorting ability of the Table Mountain depositional agents, is impressive. VISSER (1974) points out that the source rocks of the Table Mountain Group were the granite gneisses of the Pre-Cambrian basement complex in the Namaqualand area, as well as a variety of sediments and lavas in the northern part of the Cape Province. In terms of outcrop area the provenance of the Table Mountain Group might have consisted of 60% gneissose rocks, 25% medium- to fine-grained clastic rocks, 5–10% non-clastics, and 5–10% volcanic rocks. VISSER calculates that about 1.35×10^6 km^3 of source rock must have been weathered during production of the observed Table Mountain sediments. Approximately 0.75×10^6 km^3 of fine-grained material were produced during the sedimentary cycle, but essentially all of this was moved through the observed basin and deposited elsewhere. This lost part of the Table Mountain basin may now be represented by some metamorphic terrain in Antarctica.

Facies changes seem to be limited to the minimum in the Table Mountain formations. The major formations, such as the Peninsula Sandstone, the Cedarberg Shale, and the Nardouw Sandstone, persist essentially unchanged for hundreds of kilometres along strike. VISSER (1974) recognizes three main facies in the Table Mountain basin. In the south and far west the so-called Cape Facies is represented by supermature quartz sandstone deposited in an open beach, wave-dominated environment. This facies grades gradually north-eastwards into the so-called Pondoland facies, where channel sandstones and other features suggest deposition of sand on beaches and in submerged bars in a shallow neritic environment. The Natal facies along the north-eastern edge of the Table Mountain basin is represented by a variety of sediments, presumed to have been deposited under varied conditions, mainly deltaic, but ranging from alluvial fans and braided flood-plain deposits to marine beach deposits. The Natal facies is

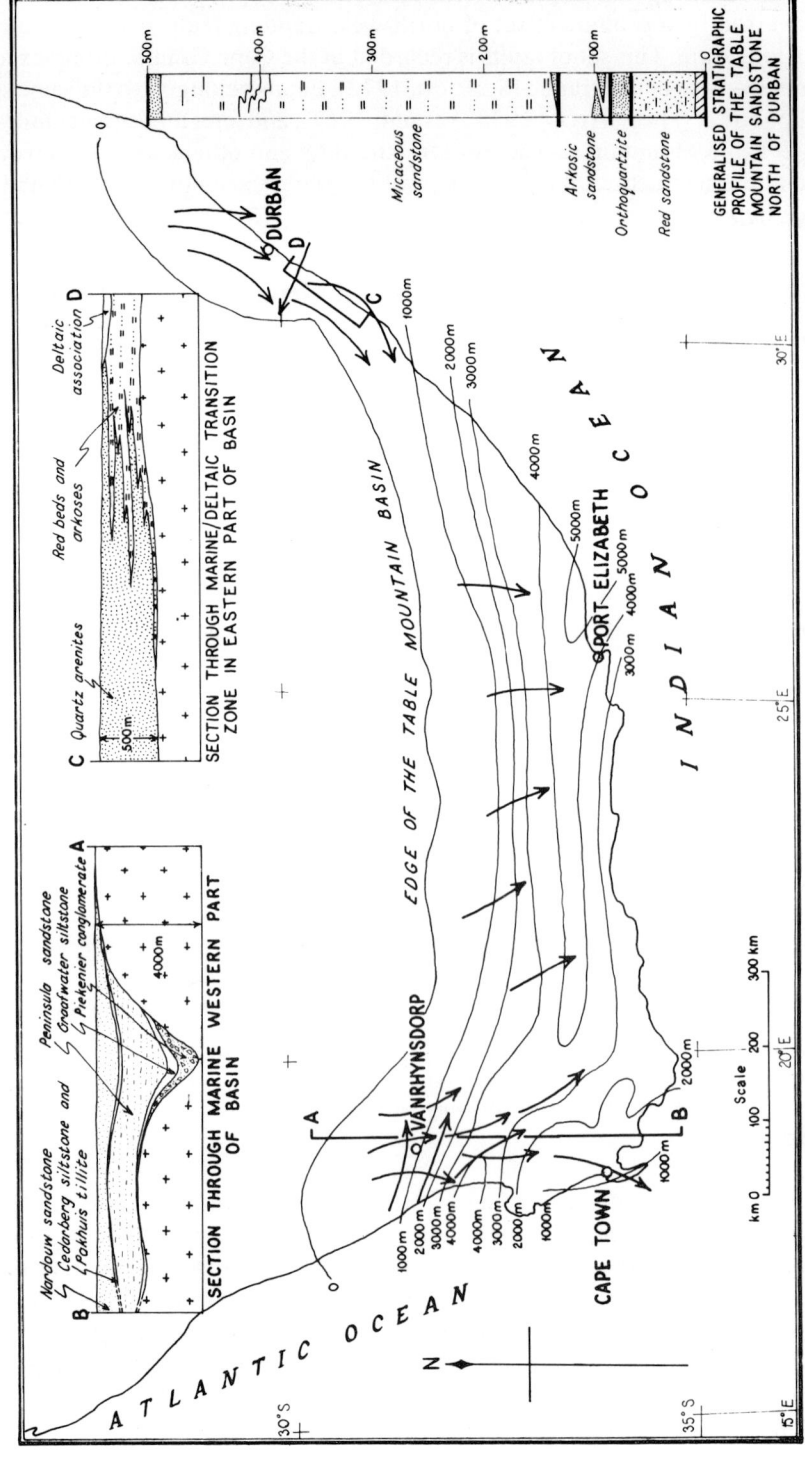

Fig. 7. The Table Mountain basin showing basin structure and sediment dispersal pattern (after RUST, 1973; KINGSLEY, 1975; and RHODES and LEITH, 1967).

distinctly unlike the rest of the Table Mountain basin, so much so that these sediments cannot be dove-tailed with stratigraphical subdivisions towards the west (HOBDAY and MATHEW, 1974).

So far only one regional inter-formational unconformity has been identified, namely, at the base of the Pakhuis Formation. Detailed stratigraphical studies over the entire outcrop area might point out the existence of more such intrabasinal breaks. However, the lack of useful marker beds and suitable fossils complicate the matter appreciably.

Tectonically the Table Mountain basin was extremely stable, certainly the main basin between Cape Town and Port Elizabeth. The trough axis remained practically stationary during the life of the basin (RUST, 1974), but the northern basin edge is distinguished by its marked transgression (RUST, 1967; VISSER, 1972). In the west the rate of sedimentation is estimated to have been about 30 metres/m.y. (RUST, 1967). VISSER (1974) stresses the implications of the tectonic stability in the basin, specifically that the supermature quartz-rich Table Mountain sediments are dominantly first cycle products.

In the discussion that follows the type area of the Table Mountain Group, namely, the south-western Cape Province, will be dealt with first, followed by a discussion of the more easterly exposures.

Several regional formations can be recognized in the Table Mountain Group in the west (Table 1); the names are those proposed by RUST (1967).

TABLE 1

Summary of the Stratigraphy of the Table Mountain Group in the Western Cape Province

Formation	Maximum thickness (m)	Fossils	Age
Nardouw Sandstone	1300	*Skolithos, Helminthoida*	Silurian?
Cedarberg Shale	160	Brachiopods, architrachs	Ashgill
Pakhuis Tillite	2–150	None	Late? Ordovician
Peninsula Sandstone	2000	*Skolithos, Helminthoida,* arthropod tracks	Middle to late? Ordovician
Graafwater Siltstone	300	Arthropod tracks, *Skolithos,* *Helminthoida*	Mid? Ordovician
Piekenier Conglomerate	0–1000	None	?Early Ordovician to Late Cambrian??

Regional Unconformity

In terms of the entire Table Mountain basin, the Piekenier Formation occurs only in the extreme western part. Much interfingering occurs between quartz sandstone (which is actually the dominant constituent) and oligomict conglomerate, the formation being distinctly more conglomeratic towards the north-west. Cross-bedding data, facies information, and bedding structure suggest a beach-type environment. The topography of the plane of unconformity is remarkably subdued and over many tens of kilometres practically no buried topography can be recognized.

It seems reasonable to assume that the Piekenier Formation represents a tongue-shaped clastic apron which invaded the newly formed Table Mountain Basin from the north-west. For various reasons this initial depositional phase did not persist far to the east, nor was any shaly sediment deposited in the part of the basin preserved in southern Africa. The provenance area has been termed the Atlantic highland and it was intermittently active as a source of sediment throughout the Palaeozoic.

The precise age of the Piekenier Formation is unknown. By inference, and by making use of some debatable sedimentation rates, RUST (1967) estimated that Piekenier sedimentation could have started during the very latest Cambrian. Most South African geologists think this puts too old an age on the Piekenier Formation but no other age estimate is available.

'The redbed Graafwater Formation marks a dramatic change from a high-energy environment in the basin to a low-energy tidal flat situation. The basin increased appreciably in size but the eastern boundary was probably not much beyond 21° E.

Five stratigraphical members can be recognized in the north-west, but their extension elsewhere in the Graafwater basin is uncertain or unknown. At the northern edge of the basin a small pebble vein quartz conglomerate is present but elsewhere no basal conglomerate is developed, even in places where the Graafwater Formation directly overlies the Pre-Cambrian basement. In a more central part of the basin the lower half of the Graafwater Formation consists of purple shale, fine-grained orthosandstone, and shale-pellet conglomerate, but laterally the upper sandy members overstep this lower shaly unit. The Loop Sandstone, a member of the Graafwater Formation, contains important arthropod (trilobite?) tracks and burrows (Plate 4). ANDERSON (1974) has described these and named the trackways *Petalichnus capensis* sp. nov. and the burrows *Metaichna rustica* gen. et sp. nov. *Skolithos* and *Helminthoidichnites* are very common in this thinly-bedded reddish fine-grained sandstone, as well as in the upper three members of the Graafwater Formation.

The age of the Graafwater Formation cannot be fixed accurately by means of the available trace fossils except to note that an Ordovician age is probable.

The Peninsula Sandstone Formation (Plate 5) consists essentially of pure quartz sandstone, now recrystallized to a metaquartzite due to burial metamorphism and dynamic metamorphism, especially in the Cape fold belt (DE SWARDT and ROWSELL, 1974). In the west the formation reaches a maximum thickness of about 2000 metres near Worcester. The Peninsula Sandstone decreases in thickness towards the north until it pinches out near Vanrhynsdorp. Here it seems as if the depositional environment was a beach situated in a very even topography underlain by Pre-Cambrian Malmesbury greywackes and marbles. Locally, sink hole formation preceded the deposition of the Peninsula sand. Away from the actual edge of the basin the formation is monotonous and it is hardly possible to identify marker horizons.

Vein quartz pebbles, all invariably very well rounded and rarely more than about 25 millimetres in diameter, occur scattered throughout the Peninsula Formation in thin stringers, but the concentration of pebbles increases appreciably towards the top of the formation in many places. This probably represents a precursor of the glaciation which shortly followed.

Towards the east, the Peninsula Formation persists as a white to blueish grey coarse-grained metaquartzite. In the Tsitsikamma area near Port Elizabeth the formation is about 1480 metres thick, increasing to about 2700 metres in the Baviaanskloof area to the north (TOERIEN, personal communication, 1973). Deep drilling has indicated that the Peninsula Sandstone is probably thin, or even absent, near Graaff-Reinet (VISSER, 1972). WINTER and VENTER (1970) indicate that the deepest part of the Peninsula trough trended eastwards more or less through present Port Elizabeth.

Cross-bedding is the most obvious primary structure of the Peninsula Formation. In the west the sediment transport direction is consistently to the south, and this direction remains constant up to at least Oudtshoorn. Towards Port Elizabeth the trend is directed more towards the south-east and, in addition, a north-easterly source becomes noticeable. Throughout the cross-beds are mainly the planar type; trough cross-beds are relatively rare. The average consistency ratio is 65%.

Skolithos burrows are common in places. Indistinct arthropod-like tracks have been found in several places (ANDERSON, 1974; RUST, 1967).

RUST (1967) has postulated that the Peninsula Formation in the west formed a shallow marine shelf deposit. A marine environment is generally accepted for the Table Mountain sediments (VISSER, 1974). The sand was probably supplied by numerous rivers, dune fields, and delta plains, and was sorted by surf and swept out to sea by marine currents which also removed the finer clastics.

Following on the Peninsula Formation in the west and south-west is the glaciogene Pakhuis Formation. This lithologically complex formation consists predominantly of glaciogenic sediments which range from true tillite through diamictite of variable origin, sandstone, laminated shale with dropstones, varvite, and shale. Faceted erratics are exceedingly common in the tillite and diamictite (Plate 6). At the edge of the basin polished and striated glacial pavements occur. Soft-sediment glacial pavements occur within the basin (VISSER, 1962).

The formation is thin, being usually only a few metres thick, but at the edge of the basin, near Vanrhynsdorp, the tillite facies is up to 80 metres thick locally. Several well defined members can be recognized and traced over a distance of about 300 kilometres from north to south. In addition, the formation is known to crop out for at least 300 kilometres east of Cape Town.

Basically, the formation can be subdivided into three units. The lower unit consists of the Sneeukop member, a remarkable structureless glaciogene sandstone which is intimately associated with large scale soft-sediment deformation folds (Plate 5). There seems to be accord that glacial processes were responsible for the association between the fold zone and the Sneeukop sediment, but interpretation varies. HAUGHTON (1929) and BLIGNAULT (1969) suggest that ice sheets moving eastward produced the fold structures. RUST (1967) and VISSER (1962) think that the main glacial centre was to the north. The central unit in the Pakhuis Formation is a thin, unconformable orthosandstone which represents an intraglacial phase. It is overlain by the glaciogene upper unit, consisting of the Kobe and Steenbras tillitic members. The bulk of the upper glaciogene unit consists of water-reworked sediment.

Little is known of the Pakhuis Formation east of Worcester.

ROSSOUW and others (1964) report a tillite thickness of less than a metre in the Swartberg Pass between Oudtshoorn and Prince Albert.

The Cedarberg Shale Formation, which overlies the Pakhuis Formation conformably, is a conspicuous marker (Plate 5) which occurs almost unchanged from the north-west near Vanrhynsdorp, to the Cape Town area in the south, and all the way east to Port Elizabeth. In the east dynamic metamorphism of Permo-Triassic age has transformed the shale and siltstone into a sheared slate.

The main importance of the Cedarberg Shale lies in its value as a structural and stratigraphical marker, and in the presence of a datable marine fauna which occurs in the extreme west. The fossiliferous zone is about 10 metres thick and occupies roughly the middle of the 90-metre thick sequence. The fauna includes the brachiopods *Plectothyrella*, *Eostropheodonta*, *Plectoglossa*?, *Trematis*, *Orbiculoidea* sp., *Marklandella*, and an unidentified homalonotid trilobite, crinoid fragments, and bryozoa. The estimated age of this assemblage is Ashgill (COCKS and others, 1969) or Llandovery (BERRY and BOUCOT, 1973). The chitinozoans in the lower member of the Cedarberg Formation are *Ancyrochitina*, *Conochitina*, *Cyatochitina*, *Desmochitina*, and *Hoegisphaera*; this assemblage suggests an upper Ordovician age (CRAMER and others, 1974).

The Cedarberg Shale almost certainly represents the regressional phase of the Pakhuis glaciation but some additional and unknown factors operating in the basin must have been responsible for the preservation of this shale accumulation, a sediment type otherwise quite rare in the Table Mountain basin.

The sediments in the Transkei and Natal which have been correlated with the Table Mountain Group form a rather distinctive unit, and direct correlation of formations across the erosion window between Port Elizabeth and Port St. Johns is not possible. VISSER (1974) suggests correlating the upper part of the Table Mountain Group with the sediments in the Natal area, but his evidence (the presumed correspondence of marine incursions) is not convincing. Several workers in Natal have indicated that the situation should be recognized by formally naming the sequence the Natal Group; however, the matter has not been finalized and the name Table Mountain Group is still being used.

The discussion that follows is based mainly on the work of HOBDAY and MATHEW (1974), VISSER (1974), KINGSLEY (1975), DU TOIT (1931), RHODES and LEITH (1967), and MATTHEWS (1961).

Facies and environmental data show that the Natal basin operated in a different way from the main basin to the west. North of Durban the environment in the Natal basin was mainly fluvial. About 100 kilometres south of Durban a shoreline separated the dominantly fluvial area from the marine environment to the south. The stratigraphy is therefore more complex and the facies changes are much more marked than in the western part of the Table Mountain basin.

Along the northern edge of the basin the floor topography is as much as 260 metres (DU TOIT, 1931). Red and chocolate-coloured conglomerate, grit, and breccia, greatly variable in thickness and character, fill the palaeovalleys in between granite and quartzite hills. The clasts in the conglomerate are mainly locally derived Insuzi quartzite and are well rounded, up to 1 metre in length, and set in very little matrix in some places. The reddish breccia contains angular fragments of vein quartz, quartzite, banded iron formation, and schist. The indistinct nature of the plane of unconformity suggests deep weathering of the source rocks before deposition of the Table Mountain sediments.

Away from the edge of the basin the floor topography evens out and the place of the very coarse-grained basal sediments is taken by reddish sandstone, feldspathic sandstone, and gritty sandstone, all containing few pebbles. No basal conglomerate rests upon the gneiss surface. The maximum thickness of the sediments in northern Natal is about 650 metres.

RHODES and LEITH (1967) were the first to analyse the stratigraphy and palaeoenvironment of the Table Mountain sediments in the central part of the Natal basin. They recognized four units.

The basal zone is generally thinner than 100 metres and, owing to variations in the basin floor topography, its thickness varies appreciably. It consists mainly of reddish maroon sandstone, grit, and shale with minor basal conglomerate. Cross-bedded layers up to 1 metre thick indicate sediment transport towards the south. Upward-fining sequences are common.

The orthoquartzite marker is a remarkably homogeneous, vitreous quartz–arenite about 20 metres thick. It is well bedded, with a bed thickness exceeding 1 metre in places, but is seldom visibly cross-bedded. Ripple-marks and rill-marks occur on the bedding planes. BRAUTESETH (1970) found that the orthoquartzite marker loses its identity south of Scottburgh because practically the entire Table Mountain sequence grades into a super-mature orthosandstone. HOBDAY and MATHEW (1974) and KINGSLEY (1975) consider this transition as indicative of the change in environment from fluvial in the north to marine in the south.

The arkosic zone consists of about 50 metres of mauve to purple coarse-grained feldspathic sandstone, grit, and pebbly grit. Some grit beds are as much as 7 metres thick. Cross-bedding and grading are present. As a whole, the arkosic zone becomes distinctly more gritty towards the top.

The micaceous zone is more than 300 metres thick. It consists of muscovite-bearing sandstone, flagstone, and shale, but interbedded gritty layers are usually not micaceous. Soft-sediment slump structures (MATTHEWS, 1961) are confined to this unit. These structures consist of 'overturned folds, overthrusts and sharp discordant contacts between superimposed sheets of contorted strata' (MATTHEWS, 1961, p. 59). The deformed sheets are up to 15 metres thick and extend laterally for 1500 metres and more. The stratigraphical thickness affected by this intraformational deformation is at least 70 metres. The eastern areas are more severely affected. Contemporaneous tectonic tilting of the eastern basin edge probably triggered the deformation.

In the Transkei, HOBDAY and MATHEW (1974) recognized three types of palaeoenvironments in the up to 1000 metres thick Table Mountain sandstone sequence. KINGSLEY (1975) named several formations in addition to recognizing a red bed and arkose association as well as a deltaic association towards the north-east, and a marine quartz–arenite association towards the south-west. The sheet-sandstone facies of HOBDAY and MATHEW (1974) contains structures and fossil traces which suggest accumulation in a sheltered shallow marine environment, possibly a nearshore back-bar area. Their lenticular sandstone facies contains sandstone beds with a thickness to length ratio of less than 1:50. Trough cross-bedding cosets are distinctive, and a complex interplay of erosional and depositional processes is indicated. The cross-bedding pattern is strongly unimodal, with sediment movement to the south-west. Off-shore sand bars modified by tidal effects are proposed as the palaeoenvironment. The channel sandstone facies is characterized by erosion channels, all aligned north-east, and regarded as scour fill produced by tidal currents, rip currents, or estuarine currents.

The trace fossils found in the Table Mountain sequence are *Scolicia* gastropod? traces, *Planolites* burrows, ramifying tubular burrows, U-shaped spreiten burrows, and ovoid surface impressions. *Rusophycus* also has been identified (HOBDAY and others, 1971).

KINGSLEY (1975) joins MATTHEWS (1970) in pointing out that the palaeocurrent system was mainly down the axis of the Natal Trough and confirms that the sediment source was in the north and north-east. The supply of sediment slightly exceeded the rate of downwarp. The tectonic framework produced several transgression–regression cycles but Kingsley questions the interpretation by RHODES and LEITH (1967) that the orthoquartzite marker (KINGSLEY's Mkunya Formation) represents a major shallow marine transgression, flooding the basin for about 250 kilometres beyond the normal shoreline. Kingsley suggests deposition of windblown sand in a shallow lake. In the southern marine facies cross-bedding patterns are indicative of both wave and longshore transport mechanisms.

To summarize, there is a noticeable difference in the character of the Table Mountain basin in Natal and towards the west, and correlation of events is not possible at this stage. Overall, however, the basin was a marine environment, important facies changes being restricted to the basin edges. The sedimentary fill represents an impressive volume of supermature quartz sandstone, making it one of the world's largest sandstone accumulations.

8. Subsequent history

In terms of its supermature sedimentary fill, the Table Mountain basin was terminated rather abruptly, but in space deposition continued in that basin up to at least the end of the Devonian Period.

The Bokkeveld basin (Fig. 8) occupied essentially the same place but it was characterized by a notably immature fill. It seems that a change in tectonic tempo was the main reason for this change in sediment type. Otherwise, the environment remained marine, now with a much

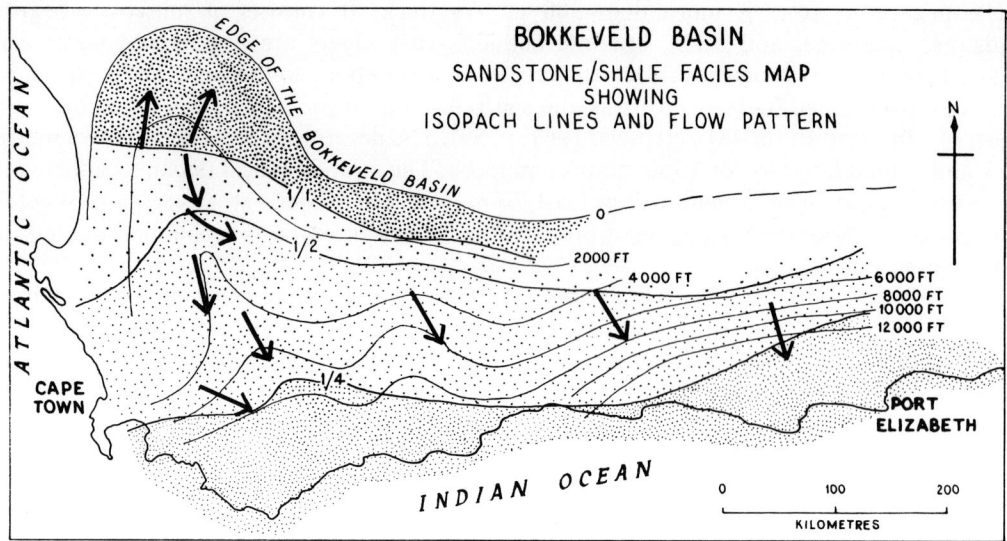

FIG. 8. The Bokkeveld basin showing sandstone/shale ratios and sediment dispersal pattern superimposed on isopach contours (after THERON, 1972).

expanded fossil community, and the sediment source was still to the north. A marked regression was set in motion, however, and during the final stages of the basin it seems that the Witteberg sandstones and siltstones were laid down in a narrow and elongated trough. The Natal embayment was inactive after the deposition of the Table Mountain Sandstone.

The onset of Carboniferous Dwyka glaciation, and the associated tectonism in the far south, combined to finally destroy the distinctive character of the Cape Basin.

9. Acknowledgements

Without generous assistance from many colleagues I would not have been able to compile this review. I particularly thank Dr Gerhard Germs, Dr S. A. Hiemstra, and Mr Herbert Pienaar for supplying some photographs. Mrs D. Read drafted the diagrams. Mrs Shelagh Matthews assisted in compiling the reference list.

References

ALDRICH, L. T., WETHERILL, G. W., DAVIS, G. L., and TILTON, G. R. (1958). Radioactive ages of micas from granitic rocks by Rb–Sr and K–Ar methods. *Trans. Am. Geophys. Union*, **39**, 1124.

ALLSOPP, H. L. and and KOLBE, P. (1965). Isotopic age determination on the Cape granite and intruded Malmesbury sediments, Cape Peninsula, South Africa. *Geochim. Cosmochim. Acta*, **29**, 1115.

ANDERSON, H. M. (1974). Arthropod trackways in the Ordovician Table Mountain Group. *Palaeont. Afr.*, **17**, 1.

BEETZ, P. F. W. (1933). Geology of South West Angola, between Cunene and Luanda axis. *Trans. Geol. Soc. S. Afr.*, **36**, 137.

BERRY, W. B. N. and BOUCOT, A. J. (1973). Correlation of the African Silurian rocks. *Spec. Pap. Geol. Soc. Am.*, **147**.

BLIGNAULT, H. J. (1970). *Contemporaneous Glacial Folding in the Table Mountain Group, Western Cape.* Unpublished Masters Thesis, University of Stellenbosch.

BRAUTESETH, S. V. (1970). The nature of the Table Mountain Series in southern Natal with special reference to the Distribution of the Orthoquartzite Marker. *Petros*, **2**, 4.

BURGER, A. J. and COERTZE, F. J. (1973). Radiometric age measurements on rocks from southern Africa to the end of 1971. *S. Afr. Geol. Surv. Bull.*, **058**.

CLIFFORD, T. N. (1967). The Damaran episode of the Upper Proterozoic–Lower Paleozoic structural history of southern Africa. *Spec. Pap. Geol. Soc. Am.*, **92**, 100.

COCKS, L. R. M. (1972). The origin of the Silurian *Clarkeia* shelly fauna of South America, and its extension to West Africa. *Palaeontology*, **15**, 623.

COCKS, L. R. M., BRUNTON, C. H. C., ROWELL, A. J., and RUST, I. C. (1969). The first Lower Paleozoic fauna proved from South Africa. *J. geol. Soc. Lond.*, **125**, 583.

CRAMER, F. H., RUST, I. C., and DIEZ DE CRAMER, M. (1974). Upper Ordovician chitinozoans from the Cedarberg Formation of South Africa. *Geol. Rdsch.*, **63**, 340.

CREER, K. M. (1970). A review of palaeomagnetism. *Earth Sci. Rev.*, **6**, 369.

DE SWARDT, A. J. M. and ROWSELL, D. M. (1974). Note on the relationship between Diagenesis and Deformation in the Cape Fold-belt. *Trans. Geol. Soc. S. Afr.*, **77**, 239.

DE VILLIERS, J. (1961). The Gariep system. *4th meeting, Sth. Reg. Comm. Geol.*, C.C.T.A., Pub. 80, Pretoria.

DE VILLIERS, J. (1972). Results of research work. 7th, 8th and 9th Annual Reports, Precambrian Research Unit, University of Cape Town, Cape Town.

DE VILLIERS, J. and BURGER, A. J. (1967). Notes on the minimum age of certain granites from the Richtersveld area. *Ann. Geol. Surv. S. Afr.*, **6**, 83.

DE VILLIERS, J., JANSEN, H., and MULDER, M. P. (1964). Die geologie van die gebied tussen Worcester en Hermanus. *S. Afr. Geol. Surv. Publ.*, Expl. Sheets 3319C, 3419A, 3318D, 3418B.

DE VILLIERS, J. and SIMPSON, E. S. W. (1974). Late-Precambrian tectonic patterns in south-western Africa. *Precambrian Res. Unit Bull., Univ. Cape Town*, No. 15.

DE VILLIERS, J. and SÖHNGE, P. G. (1959). The geology of the Richtersveld, *S. Afr. Geol. Surv. Mem.*, No. 48.

DU TOIT, A. L. (1931). The geology of the country surrounding Nkandhla, Natal. *S. Afr. Geol. Surv. Pub.*, Expl. Sheet 109.

DU TOIT, A. L. (1954). *The Geology of South Africa*, Oliver & Boyd, Edinburgh, 611.

FRETS, D. C. (1969). Geology and structure of the Haub-Welwitschia area, South West Africa, *Precambrian Res. Unit. Bull., Univ. Cape Town*, No. 5.

GERMS, G. J. B. (1972a). The stratigraphy and paleontology of the lower Nama group, South West Africa. *Precambrian Res. Unit. Bull., Univ. Cape Town*, No. 12.

GERMS, G. J. B. (1972b). New shelly fossils from Nama Group, South West Africa. *Am. J. Sci.*, **272**, 752.

GERMS, G. J. B. (1974a). The Nama Group in South West Africa and its relationship to the Pan African Geosyncline. *J. Geol.*, **82**, 301.

GERMS, G. J. B. (1974b). Observations on the Nama of the Brandkop Basin, Vanrhynsdorp District, *in Tenth and Eleventh Annual Reports, 1972 and 1973*, Precambrian Research Unit, University of Cape Town, Cape Town, 89.

GLAESSNER, M. F. (1971). Geographic distribution and time range of the Edicara Precambrian fauna, *Bull. Geol. Soc. Am.*, **82**, 509.

GRAHAM, K. W. T. and HALES, A. L. (1961). Preliminary palaeomagnetic measurements on Silurian sediments from South Africa. *Geophys. J.*, **5**, 318.

GUJ, P. (1970). The Damara mobile belt in the South Western Kaokoveld, South West Africa. *Precambrian Res. Unit Bull., Univ. Cape Town*, No. 8.

GUJ, P. (1974). A revision of the Damara stratigraphy along the southern margin of the Kamanjab Inlier, South West Africa. *Precambrian Res. Unit Bull., Univ. Cape Town*, No. 15.

HÄLBICH, I. W. (1964). Observations on primary features in the Fish River Series and the Dwyka Series in South West Africa. *Trans. Geol. Soc. S. Afr.*, **67**, 95.

HARTNADY, C. J., NEWTON, A. R., and THERON, J. N. (1974). The stratigraphy and structure of the Malmesbury Group on the southwestern Cape. *Precambrian Res. Unit Bull., Univ. Cape Town*, No. 15.

HAUGHTON, S. H. (1933). The geology of Cape Town and adjoining country. *S. Afr. Geol. Surv. Publ.*, Expl. Sheet 247, Pretoria.

HAUGHTON, S. H. (1969). *Geological History of Southern Africa*. Geological Society of South Africa, Pretoria, 528 pp.

HAUGHTON, S. H., KRIGE, L. J., and KRIGE, A. V. (1925). On intraformational folding connected with the glacial bed in the Table Mountain sandstone. *Trans. Geol. Soc. S. Afr.*, **28**, 19.

HOBDAY, D. K., BRAUTESETH, S. V., and MATHEW, D. (1971). The Table Mountain Series between the Mtentu River mouth and Waterfall Bluff, Pondoland. *Petros*, **3**, 51.

HOBDAY, D. K. and MATHEW, D. (1974). Depositional environment of the Cape Supergroup in the Transkei. *Trans. Geol. Soc. S. Afr.*, **77**, 223.

HOLMES, A. and CAHEN, L. (1957). Géochronologie Africaine, 1956. *Acad. Roy. Soc. Col. (Bruxelles) Cl. Sc. Nat. Mem. B, New Ser.*, **5**, 169.

KENNEDY, W. Q. (1964). The structural differentiation of Africa in the Pan African (500 My) tectonic episode. *Ann. Rep. Res. Inst. Afr. Geol. Univ. Leeds, 1962–63*, **8**, 48.

KINGSLEY, C. S. (1975). A new stratigraphic classification implying a facies change in the Table Mountain Sandstone in southern Natal. *Trans. Geol. Soc. S. Afr.*, **78**, 43.

KORN, H. and MARTIN, H. (1959). Gravity tectonics in the Naukluft Mountains, South West Africa. *Bull. Geol. Soc. Am.*, **70**, 1047.

KRÖNER, A. (1969). The correlation of the Pre-Cape sediments in the Vanrhynsdorp region, Cape Province. *Trans. Geol. Soc. S. Afr.*, **72**, 127.

KRÖNER, A. (1971). Late Precambrian correlation and the relationship between the Damara and Nama systems of South West Africa. *Geol. Rdsch.*, **60**, 1513.

KRÖNER, A. (1974a). Report on field work in Angola, *in Tenth and Eleventh Annual Reports, 1972 and 1973*. Precambrian Research Unit, University of Cape Town, Cape Town, 74.

KRÖNER, A. (1974b). Editorial comments concerning the present confusion over lithostratigraphic nomenclature and a proposal for a new Damaran terminology. *Precambrian Res. Unit Bull., Univ. Cape Town*, No. 15.

KRÖNER, A. and WELIN, E. (1973). Evidence for a ±500 m.y. old thermal episode in southern South-West Africa. *Earth Planet. Sci. Lett.*, **21**, 149.

MARTIN, H. (1965). *The Precambrian geology of South West Africa and Namaqualand*. Precambrian Research Unit, University of Cape Town, Cape Town, 159 pp.

MARTIN, H. (1974). Damara rocks as nappes on the Naukluft Mountains, South West Africa. *Precambrian Res. Unit Bull., Univ. Cape Town*, No. 15.

McELHINNY, M. W., BRIDEN, J. C., JONES, D. L., and BROCK, A. (1968). Geological and geophysical implications of palaeomagnetic results from Africa. *Rev. Geophys.*, **6**, 201.

MATTHEWS, P. E. (1961). Slump structures in the Table Mountain series of Natal. *Trans. Geol. Soc. S. Afr.*, **64**, 55.

MATTHEWS, P. E. (1970). Paleorelief and the Dwyka glaciation in the eastern region of South Africa. *Proc. Pap. 2nd Gond. Symp., S. Afr.*, C.S.I.R., Pretoria, 491.

MILLER, R. McG. (1973). The Salem Granite Suite, South West Africa: genesis by partial melting of the Khomas schist. *Geol. Surv. S. Afr. Mem.*, No. 64, 97 pp.

RHODES, R. C. and LEITH, M. J. (1967). Lithostratigraphic zones in the Table Mountain Series of Natal. *Trans. Geol. Soc. S. Afr.*, **70**, 15.

ROSSOUW, P. J., MEYER, E. I., MULDER, M. P., and STOCKEN, C. G. (1964). Die geologie van die Swartberge, die Kangovallei en die omgewing van Prins Albert, K.P. *S. Afr. Geol. Surv. Publ.*, Expl. Sheets 3321B, 3322A.

RUST, I. C. (1967). *On the sedimentation of the Table Mountain Group in the western Cape Province.* Unpublished doctoral dissertation, University of Stellenbosch.

RUST, I. C. (1974). The evolution of the Paleozoic Cape Basin, Southern margin of Africa, *in Ocean Basins and Margin*, Vol. 1, Plenum, New York, 247.

SCHOCH, A. E. (1972). *The Darling granite batholith.* Unpublished D.Sc. Thesis, University of Stellenbosch.

SCHOCH, A. E., LEYGONIE, F. E., and BURGER, A. J. (1975). U–Pb ages for Cape granites from the Saldanha Batholith: a preliminary report. *Trans. Geol. Soc. S. Afr.*, **78**, 97.

SCHOLTZ, D. L. (1946). On the younger Precambrian granite plutons of the Cape Province. *Proc. Geol. Soc. S. Afr. C.S.I.R.*, **49**, 25.

SMITH, D. A. M. (1965). The geology of the area around the Khan and Swakop Rivers in South West Africa. *Geol. Surv. S. Afr. Mem.*, No. 3 (SWA Series), 113 pp.

SÖHNGE, P. G. (1964). The geology of the Tsumeb Mine, *in The Geology of some Ore Deposits in Southern Africa*, Vol. II, Geological Society of South Africa, Pretoria, 367.

THERON, J. N. (1972). *The stratigraphy and sedimentation of the Bokkeveld Group.* Unpublished doctoral dissertation, University of Stellenbosch.

TRUSWELL, J. F. (1970). *An introduction to the historical geology of South Africa.* Purnell, Johannesburg, 167 pp.

VALE, F. SOUSA DO (1968). Carta geologica de Angola Escale 1/100,000, Noticia explicativa de folha 355 (Humpata-Cainde), Luanda.

VALENTINE, J. W. and MOORES, E. W. (1970). Plate tectonic regulation of faunal diversity and sea level: a model. *Nature (London)*, **228**, 657.

VALENTINE, J. W. and MOORES, E. W. (1972). Global tectonics and the fossil record. *J. Geol.*, **80**, 167.

VAN EEDEN, O. R. (1972). The geology of the Republic of South Africa; an explanation of the 1:1,000,000 map, 1970 edition. *S. Afr. Geol. Spec. Publ.*, No. 18.

VISSER, H. N. (1967). Distribution and correlation of the Klipheuwel Formation in the Swartland and the Sandveld (in Afrikaans with English summary). *Ann. Geol. Surv. S. Afr.*, **6**, 31.

VISSER, H. N. and SCHOCH, A. E. (1973). The geology and mineral resources of the Saldanha Bay area. *Geol. Surv. S. Afr. Mem.*, No. 63.

VISSER, H. N. and TOERIEN, D. K. (1971). Geology of the area between Vredendal and Elands Bay (in Afrikaans with English summary). *S. Afr. Geol. Surv. Publ.*, Expl. Sheets 3118C and 3218A.

VISSER, J. N. J. (1962). *Die voorkoms en oorsprong van die tillietband in die serie Tafelberg*, M.Sc. Thesis, University of the Orange Free State, Bloemfontein.

VISSER, J. N. J. (1972). Die Kaapsupergroep in 'n diep boorgat wes van Graaff-Reinet. *Trans. Geol. Soc. S. Afr.*, **75**, 111.

VISSER, J. N. J. (1974). The Table Mountain Group: a study in the deposition of quartz arenites on a stable shelf. *Trans. Geol. Soc. S. Afr.*, **77**, 229.

VON BACKSTRÖM, J. W., JANSEN, H., COETZEE, C. B., SNYMAN, A. A., and SPIES, J. (1960). Die geologie van die gebied om Nieuwoudtville, Kaapprovinsie. *S. Afr. Geol. Surv. Pub.*, Expl. Sheet 241.

WINTER, H. DE LA R. and VENTER, J. J. (1970). Lithostratigraphic correlation of recent deep boreholes in the Karroo–Cape Sequence. *Proc. Pap. 2nd Gond. Symp.*, *S. Afr.*, C.S.I.R., Pretoria, 395.

Lower Palaeozoic of the Middle East, Eastern and Southern Africa, and Antarctica
Edited by C. H. Holland
© 1981 John Wiley & Sons Ltd.

LOWER PALAEOZOIC TRACE FOSSILS OF AFRICA

T. P. Crimes*

Department of Geology, University of Liverpool, England

Contents

1. Introduction

The Lower Palaeozoic sequences of Africa contain an unparalleled wealth of trace fossils, the full stratigraphical potential of which has yet to be realized. They occur at many horizons and are often common in otherwise unfossiliferous strata, such as shallow-water sandstones and shales.

Recent investigations in accurately dated sequences have shown that some ichnospecies have restricted time ranges. Their occurrence in otherwise unfossiliferous rocks can therefore be useful in correlation: this appears to have wide significance for many of the North African Lower Palaeozoic successions.

Many trace fossils are also markedly facies controlled. They can therefore assist in environmental interpretation and in the construction of both local and regional palaeogeographical syntheses.

The purpose of this part of the present volume is both to review ichnological investigations in Africa and also to provide a framework to stimulate further research. Most of the discussion is concerned with Africa itself but important studies on markedly similar successions in the adjacent territories of south Jordan and North-west Saudi Arabia have also been included. The Lower Palaeozoic sediments which yield the trace fossils pass upwards, in some sections without a break, into Devonian strata. Since these younger rocks are also rich in trace fossils they have been included in this account for completeness.

2. History of research

The abundance of trace fossils at many levels in North Africa led to their recognition in even the earliest surveys. Most authors were content, however, simply to record their presence. References to many papers listing trace fossils from specific localities or horizons can be found in

---

* The author notes that his manuscript was first submitted 1973, revised 1974, and brought up to date in June 1975.

FURON (1957, 1963). The first attempt at more detailed work was by FRITEL (1925), who recorded trace fossils from the L'Ouadau area of Chad, North Africa, and briefly described some of them. The most significant early study was that of DESIO (1940), who recorded, described, and figured a wide variety of trace fossils from Libya. He also made the first systematic attempt to present their occurrence stratigraphically (DESIO, 1940, p. 61). More recently, extensive investigations in North Africa have been undertaken by L'Institut Français du Petrole and were summarized by BEUF and others (1971), who recorded and described many trace fossils and discussed their environmental significance. The facies control shown by trace fossils in Ordovician and Devonian sections in the Fezzan, Libya, and south Jordan, respectively, has been briefly discussed by SEILACHER (1969). The same author recorded and described 16 species of *Cruziana* from North Africa and evaluated their stratigraphical significance (SEILACHER, 1970).

Trace fossils have also been recorded at many levels within the Ordovician, Silurian, and Devonian rocks of north-west Saudi Arabia by HELAL (1965) and south Jordan by BENDER (1963). The detailed relationship between trace fossils and facies has been discussed with particular reference to the last named area by SELLEY (1970).

To the south of the Sahara, records of trace fossils are few, although a paper by COCKS and others (1970) mentions trace fossils, including trilobite tracks, at several levels from the lower Ordovician to the upper Silurian in western Cape Province, South Africa. Also, GERMS (1972) recorded a variety of traces from the late Pre-Cambrian to Lower Cambrian Nama Group in south-west Africa.

3. The trace fossils

Despite the abundance of trace fossils in the Lower Palaeozoic successions of North Africa, relatively few ichnogenera have been recorded. This probably reflects a lack of detailed investigations rather than low trace fossil diversity. Only those forms occurring abundantly seem to have been mentioned. Observation also seems to have centred more on forms occurring within beds than on those only apparent after careful inspection of lower bedding surfaces.

The common trace fossils are shown in Fig. 1. The most frequently recorded ichnogenus is the vertical pipe-like structure *Skolithos* (= *Sabellarifex* = *Scolites* = *Tigillites*). A similar form (*Monocraterion*), showing a more complex structure resembling a series of stacked filter-funnels, also occurs. Other vertical burrows are mostly U-shaped and amongst these *Arenicolites*, *Glossifungites*, *Corophioides*, and *Diplocraterion* have been recorded. These last four ichnogenera are all similar. *Arenicolites* can be distinguished by an absence of spreite (laminae between the limbs of the U). *Glossifungites* is normally associated with erosion surfaces while *Diplocraterion* and *Corophioides* are distinguished by the presence of funnels in the former (*see* OSGOOD, 1970, p. 317).

Bivalve burrows (*Pelecypodichnus*) have also been recognized, but only in the Tassili des Ajjers region of the Sahara (BEUF and others, 1971, p. 239). The dumbell-shaped burrow *Bifungites* occurs widely but is particularly common in the Devonian of the Fezzan, Libya (DESIO, 1940). According to DUBOIS and LESSERTISSEUR (1964), it may represent a horizontal bedding plane section through a U-tube. The meandering sole traces *Nereites*, *Neonereites*, and *Scolicia* have been recorded from the Devonian, the first named from the Fezzan (DESIO, 1940) and last two from south Jordan (SEILACHER, 1969). Simple unbranched burrows (*Planolites* = *Pholeus*) are also widely distributed.

More complex burrow systems are also well distributed, both geographically and stratigraphically. *Arthrophycus* (= *Harlania*), a transversely ribbed, palm-like form, is perhaps the most abundant but the spiral *Spirophyton* is common in the Devonian right across North Africa.

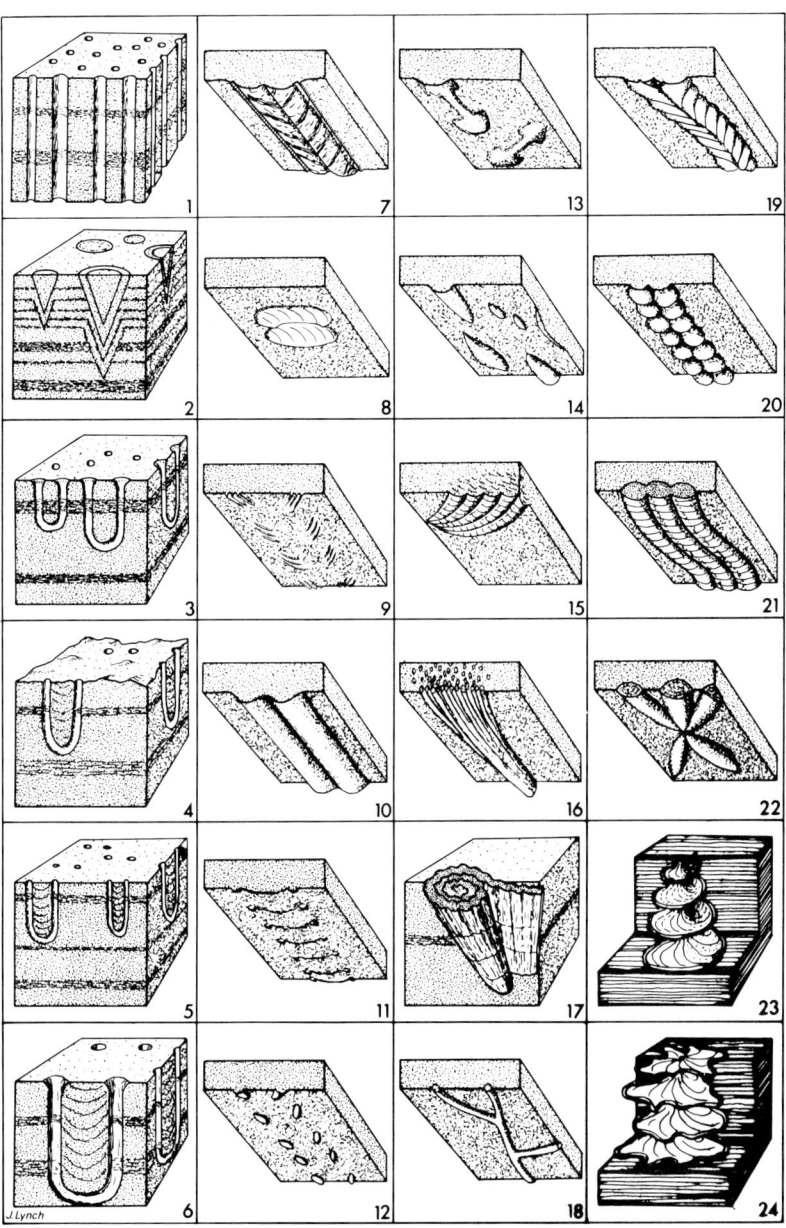

FIG. 1. Common trace fossils in the Lower Palaeozoic rocks of North Africa. 1, *Skolithos*; 2, *Monocraterion*; 3, *Arenicolites*; 4, *Glossifungites*; 5, *Corophioides*; 6, *Diplocraterion*; 7, *Cruziana*; 8, *Rusophycus*; 9, *Diplichnites*; 10, *Rouaultia*; 11, *Myrianites*; 12, *Merostomichnites*; 13, *Bifungites*; 14, *Pelecypodichnus*; 15, *Arthrophycus*; 16, *Phycodes*; 17, *Daedalus*; 18, *Paleophycus*; 19, *Nereites*; 20, *Neonereites*; 21, *Scolicia*; 22, *Asterosoma*; 23, *Spirophyton*; 24, *Zoophycos*.

The related *Zoophycos* (= *Spirophyton canda galli*, = *Taonurus*) also occurs. Other less common branched burrows and burrow systems include *Phycodes, Paleophycus, Daedalus* (= *Vexillum*), and *Asterosoma*. The detailed morphology and environmental interpretation of specimens of *Daedalus* from North African localities is discussed by LESSERTISSEUR (1971).

Perhaps the most intrinsically interesting of the trace fossils are the trilobite traces. So far, records have been limited to furrows (*Cruziana*), resting excavations (*Rusophycus* = *Rhyzophycus*), and walking traces (*Diplichnites*), although *Rouaultia* (= *Fraena*) might also represent trilobite activity. There are also less diagnostic traces, probably referrable to arthropods, if not trilobites, including *Myrianites* and *Merostomichnites*.

A few other forms given ichnogeneric names by DESIO (1940) appear from examination of his figures to be of inorganic origin. *Kinneyia labyrynthica* appears to be interference ripple-marking. *Eolanthus* may be of similar origin, while *Interconulites* and *Polygonolites* were names suggested by Desio (1940) for cone in cone and mud cracks, respectively.

4. Use of trace fossils in defining marker horizons

Trace fossils have long been used to define marker horizons in North Africa and many units have been described or defined by reference to them, for example the sandstones with *Tigillites* of SOUGI (1956) or the *Sabellarifex* Sandstone of BENDER (1963). Similarly, HELAL (1965) differentiated a lower group of quartzites in Saudi Arabia by reference to their contained trilobite traces and named them the *Cruziana* series. Fritel (1925) appreciated that, in the Ouadau area (Chad), upper Silurian and Lower Devonian could be differentiated by an abundance of *Arthrophycus* (= *Harlania*) and *Spirophyton*, respectively. This differentiation was subsequently recognized in widespread areas including the Tibesti, L'Ennedi, and Le Borkou districts of Chad (FURON, 1957), Fezzan, Libya (DESIO, 1940; JACQUE, 1963), and Ahnet (BEUF and others, 1971).

In south-west Africa, GERMS (1972) has recorded the following trace fossils from the Nama Group: *Archaeichnium, Phycodes pedum, Planolites, Skolithos*, and *Taenidium*. The lowest beds of this group have yielded the body fossils *Rangea, Pteridinium*, and *Cloudinia*, which have been recorded elsewhere in late Pre-Cambrian rocks. However, *Phycodes pedum*, found low in the succession, indicates a Cambrian age for some of the group.

The use of trace fossils to define marker horizons is, however, open to pitfalls. Many of the traces used are reliably identified only to ichnogeneric level and in all cases their stratigraphical range is considerably in excess of the time interval represented by the marker horizon. For example, the range of *Skolithos* is Pre-Cambrian to Recent and that of *Arthrophycus* Ordovician to Silurian, yet both have been used to trace individual units laterally over wide areas. *Spirophyton*, also used extensively to differentiate upper Silurian from Lower Devonian, is, however, more restricted and probably occurs only in the Devonian (SIMPSON, 1970).

Thus, while the use of these trace fossils to follow specific beds or units laterally in near-continuous exposure is valid, regional correlations based on the occurrence of horizons rich in these traces could be hazardous. Detailed regional correlation becomes possible only with accurate identification to ichnospecific level of forms of known and restricted time range.

5. Correlation using trace fossils

The possibility of using trace fossils stratigraphically has long been recognized. In the 19th century the Geological Survey of Ireland (KINAHAN, 1878) used *Oldhamia* to date the Bray Series (now Bray Group) of south-eastern Ireland as Cambrian and this has been confirmed for the *Oldhamia*-bearing beds (CRIMES and CROSSLEY, 1968). Later, MÄGDEFRAU (1934) showed that the feeding burrow *Phycodes circinatum* was of cosmopolitan distribution in the Arenig and restricted to the lower Ordovician. SEILACHER (1960) discussed the use of *Cruziana* to differentiate Cambrian and Ordovician strata. CRIMES (1968) demonstrated that Upper Cambrian and Arenig strata could be distinguished by their contained *Cruziana* and later used trace fossils to date otherwise unfossiliferous successions in North Wales (CRIMES, 1969, 1970b). SEILACHER (1970) described 30 species of stratigraphically restricted but widely distributed *Cruziana* and *Rusophycus* and listed many from localities within the Lower Palaeozoic sequences of North Africa.

TABLE 1
Known Ranges of *Cruziana* Species Recorded from Africa.

	ORDOVICIAN						SILURIAN		DEVONIAN		
	TREMADOC	ARENIG	LLANVIRN	LLANDEILO	CARADOC	ASHGILL	LOWER	UPPER	LOWER	MIDDLE	UPPER
C. imbricata		—									
C. furcifera	—				—						
C. rugosa		—			—						
C. goldfussi		—			—						
C. petraea					—						
C. flammosa					—						
C. perucca					—						
C. lineata					—						
C. almadenensis					—						
C. pudica					—	—	—				
C. ancora						- - - -	- - -				
C. acacensis								—			
C. pedroana								—			
C. quadrata								—			
C. uniloba									—		
C. lobosa										—	
C. rhenana									—		

The importance of *Cruziana* and *Rusophycus* in correlation stems from their common occurrence throughout the Lower Palaeozoic and their varied, distinctive, easily recognized, and short-ranging species. More than 30 species have been recognized as stratigraphically significant (see Table 1), although further work will doubtless extend the ranges of some of these. Detailed descriptions of the traces can be found in CRIMES (1968, 1970a), OSGOOD (1970), and, more particularly, SEILACHER (1970).

Although attention has focused on trilobite traces, some other fossils occurring in Africa appear to have restricted time ranges. For example, *Spirophyton* probably occurs only in the Devonian (SIMPSON, 1970) and *Phycodes circinatum* is restricted to the lower Ordovician (MÄGDEFRAU, 1934).

The possibilities of using trace fossils to date and correlate within the Lower Palaeozoic successions in Africa are considerable. For example, during an Algerian Petroleum Institute's field excursion to the Vire de Mouflon in the Tassili d'Ajjer (Algeria), SEILACHER (1970) dated Unit II of that area as lower Ordovician on the occurrence of *C. goldfussi* and the upper part of Unit III as Caradoc on the presence of *C. almadenensis*.

6. Palaeogeographical significance of trace fossils

Trace fossils reflect the behavioural responses of the animals which produced them. Thus, since the animals are normally very sensitive to their environment, trace fossils can be expected to be, at the very least, as good environmental indicators as inorganic sedimentary structures.

The distribution of trace fossils will therefore have environmental significance and may allow the determination of depositional energy state and palaeobathymetry. This has been treated extensively in a series of papers, including SEILACHER (1964, 1967), CRIMES (1970b), and FREY and HOWARD (1970). Many of the North African trace fossils have demonstrable environmental significance, as was appreciated by DESIO (1940). This has also been discussed by BEUF and others (1971). Trace fossils with palaeobathymetric significance recorded from Africa are listed in Table 2.

Perhaps the most significant contribution which trace fossils can make to environmental interpretation in Africa is in distinguishing between fresh water and marine sediments. Reliable sedimentological criteria are tantalisingly elusive. It is accepted, however, that many trace

TABLE 2
Bathymetric Zonation of African Lower Palaeozoic Trace Fossils
Few of these forms are strictly limited to any one zone but they normally occur
most abundantly within the zone indicated.

Freshwater	Intertidal	Neritic (0–200 m)	Bathyal (upper part)
Scoyenia	*Arenicolites*	*Arthrophycus*	*Asterosoma*
	Bifungites	*Cruziana*	*Neonereites*
	Corophioides	*Daedalus*	*Nereites*
	Diplocraterion	*Diplichnites*	*Spirophyton*
	Glossifungites	*Merostomichnites*	*Zoophycos*
	Monocraterion	*Myrianites*	
	Skolithos	*Pelecypodichnus*	
		Phycodes	
		Rouaultia	
		Rusophycus	

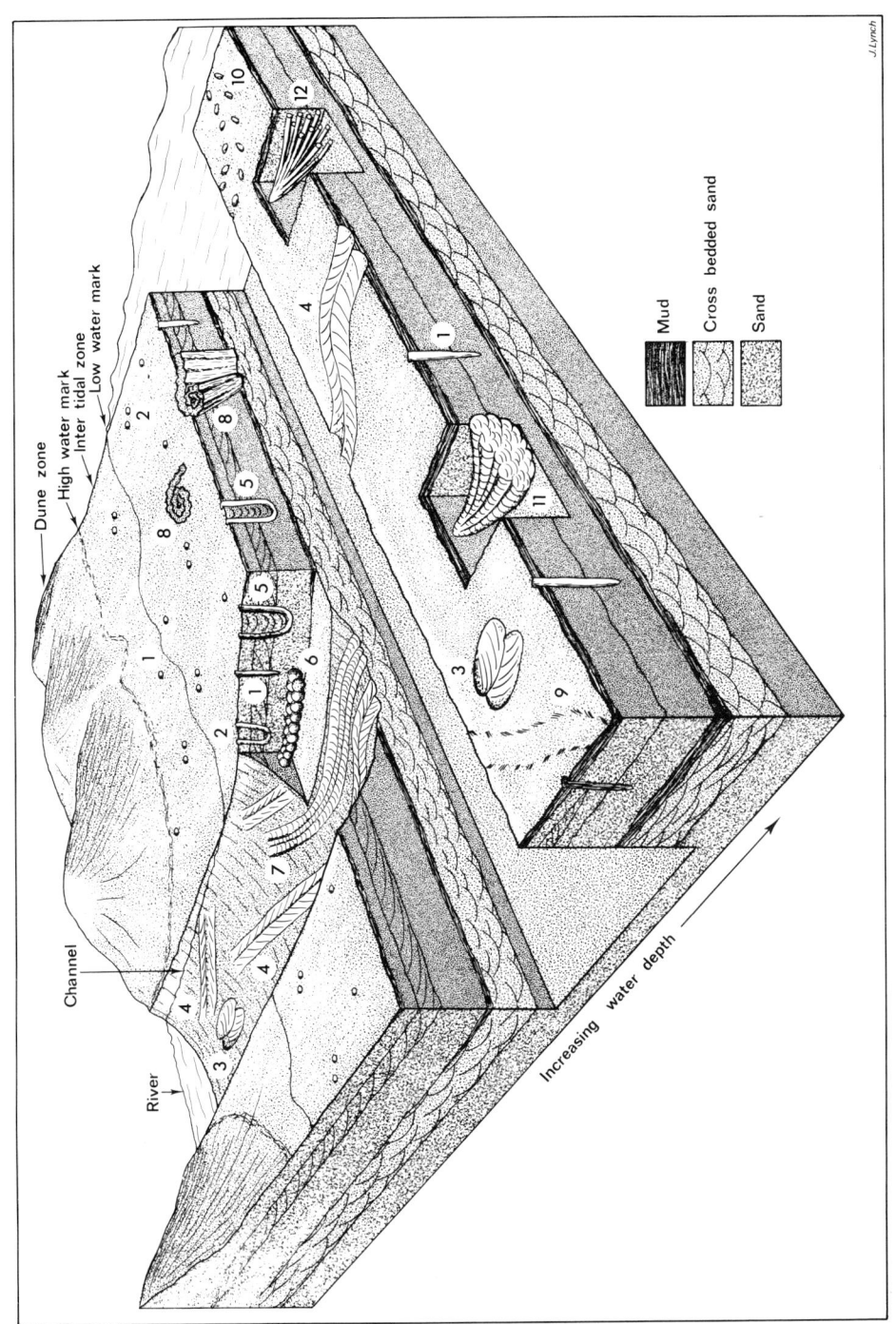

FIG. 2. Schematic representation of trace fossil distribution on the sea floor and in abandoned fresh water channels in North Africa during the Ordovician. 1, *Skolithos*; 2, *Arenicolites*; 3, *Rusophycus*; 4, *Cruziana*; 5, *Diplocraterion*; 6, *Neonereites*; 7, *Scolicia*; 8, *Daedalus*; 9, *Diplichnites*; 10, *Merostomichnites*; 11, *Arthrophycus*; 12, *Phycodes*.

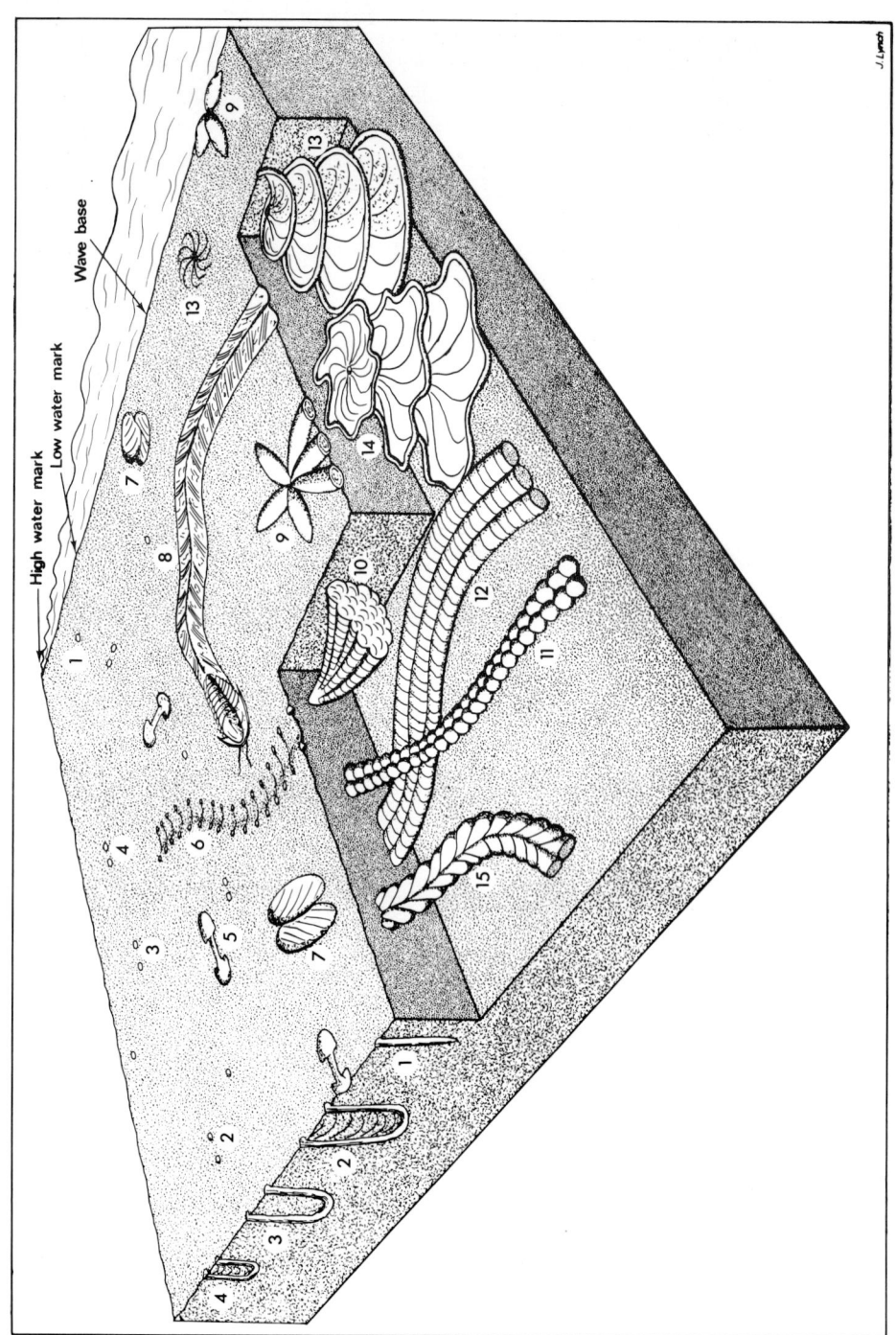

FIG. 3. Schematic representation of trace fossil distribution on the sea floor in North Africa during the Devonian (included for completeness and to show comparison with Lower Palaezoic sandstones). 1, *Skolithos*; 2, *Diplocraterion*; 3, *Arenicolites*; 4, *Corophioides*; 5, *Bifungites* (? section of U-tube); 6, *Myrianites*; 7, *Rusophycus*; 8, *Cruziana*; 9, *Asterosoma* 10, *Arthrophycus*; 11, *Neonereites*; 12, *Scolicia*; 13, *Spirophyton*; 14, *Zoophycos*; 15, *Nereites*.

fossils, in particular those produced by trilobites, represent the activities of marine animals. This has already been used to subdivide the thick Lower Palaeozoic sandstones in several parts of North Africa into fresh water and marine, and papers by DESIO (1940), PICARD (1942), BENDER (1963), KLITZSCH (1963), and BEUF and others (1971) can be cited. The most detailed analysis is that of SELLEY (1970), who investigated the relationship between trace fossils and the depositional environment within the Ordovician sediments of part of the Southern Desert of Jordan. He suggested that the lowest 700 metres of the investigated succession could be identified as of fluvial origin, both by the contained sedimentary structures and the absence of trilobite traces. Marine influence was first detected at the top of this succession by the occurrence of *Cruziana* in silt-filled, abandoned channels, probably produced under fresh water conditions. The overlying succession consists of 250 metres of well sorted sandstones with thin silt sheets. These sediments contain abundant *Arthrophycus*, *Cruziana*, and *Skolithos*, indicating shallow marine conditions. The following sequence consists of shales with turbidites, suggesting a deeper water environment, but so far no trace fossils have been found. The highest beds revert to channel sands with *Skolithos*, *Rusophycus*, *Rouaultia*, *Diplichnites*, and *Merostomichnites* and were apparently deposited in a deltaic environment.

Much work remains to be done, but even now the available data can be used to erect an environmental model to represent trace fossil distributions on the sea floor in North Africa for at least two systems (Figs. 2 and 3). Comparisons of trace fossil assemblages with these models will normally allow a realistic appraisal of the depositional environment, but the most satisfactory interpretation can only follow consideration of all available evidence, including lithology and inorganic sedimentary structures.

7. Conclusions

The main conclusions to be drawn from this review are:

(a) Trace fossils are abundant at many levels within the Lower Palaeozoic rocks of North Africa and also occur at these horizons in South Africa.

(b) Forms with restricted time ranges, such as certain species of *Cruziana* and *Rusophycus*, have great potential for correlating otherwise unfossiliferous strata, particularly in North Africa.

(c) Trace fossils are valuable environmental indicators and can, for example, be particularly useful in distinguishing between freshwater and marine sandstone.

References

BENDER, F. (1963). Stratigraphie der 'Nubischen Sandsteine' in Süd-Jordanien. *Geol. Jb.*, **81**, 237.

BEUF, S., BIJU-DUVAL, B., DE CHARPAL, O., ROGNON, P., GARIEL, O., and BENNACEF, A. (1971). *Les grès du Paléozoïque inférieur au Sahara*. Publ. Inst. Fr. Pétrol., No. 18, Editions Technip, Paris.

BEUF, S., BIJU-DUVAL, B., MAUVIER, A., and LEGRAND, P. (1968). Nouvelles observations sur le 'Cambro–Ordovicien' du Bled el Mass (Sahara central). *Publ. Serv. Carte Géol. Algérie*, **38**, 39.

COCKS, L. R. M., BRUNTON, C. H. C., ROWELL, A. J., and RUST, I. C. (1970). The first Lower Palaeozoic fauna proved from South Africa, *Q.J. geol. Soc. Lond.*, **125**, 583.

CRIMES, T. P. (1968). *Cruziana*: A stratigraphically useful trace fossil. *Geol. Mag.*, **105**, 360.

CRIMES, T. P. (1969). Trace fossils from the Cambro-Ordovician rocks of North Wales and their stratigraphic significance. *Geol. J.*, **6**, 333.

CRIMES, T. P. (1970a). Trilobite tracks and other trace fossils from the Upper Cambrian of North Wales. *Geol. J.*, **7**, 47.

CRIMES, T. P. (1970b). The significance of trace fossils in sedimentology, stratigraphy, and palaeoecology with examples from Lower Palaeozoic strata, *in* CRIMES, T. P. and HARPER, J. C. (Eds.), *Trace fossils*. *Geol. J. Spec. Issue*, No. **3**, 101, Seel House Press, Liverpool.

CRIMES, T. P. and CROSSLEY, J. D. (1968). The stratigraphy, sedimentology, ichnology and structure of the Lower Palaeozoic rocks of part of northeastern County Wexford. *Proc. R. Ir. Acad.*, **67B**, 185.

DESIO, A. (1940). Vestigia problematiche paleozoiche della Libya. *Ann. Mus. Libico. Stor. Nat.*, **2**, 47.

DUBOIS, B. and LESSERTISSEUR, J. (1964). Note sur *Bifungites*, trace problématique du Dévonien du Sahara. *Bull. Soc. Géol. Fr.*, **6**, 626.

FREY, R. W. and HOWARD, J. D. (1970). Comparison of Upper Cretaceous ichnofaunas from siliceous sandstones and chalk, Western Interior Region, U.S.A., *in* CRIMES, T. P. and HARPER, J. C. (Eds.), *Trace fossils. Geol. J. Spec. Issue*, No. 3, 141, Seel House Press, Liverpool.

FRITEL, P. H. (1925). Végétaux paléozoïques et organismes problématique de L'Ouadoü. *Bull. Soc. Géol. Fr.*, **25**, 33.

FURON, R. (1957). *Le Sahara*. Payot, Paris.

FURON, R. (1963). *Geology of Africa*. Oliver and Boyd, Edinburgh.

GERMS, G. J. B. (1972). Trace fossils from the Nama Group, southwest Africa. *J. Paleontol.*, **6**, 864.

HELAL, A. H. (1965). Stratigraphy of outcropping Paleozoic rocks around the northern edge of the Arabian shield (within Saudi Arabia). *Z. Deut. Geol. Ges.*, **117**, 506.

JACQUÉ, M. (1963). Reconnaissance géologique du Fezzan oriental. *Notes et Memoires, Compagnie Française des Pétroles*, No. 5, Paris.

KINAHAN, G. H. (1878). *Manual of the Geology of Ireland*. Kegan Paul, London.

LESSERTISSEUR, J. (1971). L'énigme du *Daedalus* (*Daedalus* Rouault, 1850). Ichnofossilia. *Bull. Mus. Nat. Hist. Nat.*, **20**, 37–72.

OSGOOD, R. G. (1970). Trace fossils in the Cincinnati area. *Palaeontol. Am.*, **6(41)**, 281.

PICARD, L. (1942). New Cambrian fossils and Palaeozoic problematica from the Dead Sea and Arabia. *Bull. Geol. Dep. Hebrew Univ.*, **4**, (1), 1–18.

SEILACHER, A. (1960). Lebenspuren als Leitofossilien. *Geol. Rdsch.*, **49**, 41.

SEILACHER, A. (1964). Biogenic sedimentary structures, *in* IMBRIE, J. and NEWELL, N. D. (Eds.), *Approaches to Paleoecology*. Wiley, New York, 296.

SEILACHER, A. (1967). Bathymetry of trace fossils. *Mar. Geol.*, **5**, 413.

SEILACHER, A. (1969). Sedimentary rhythms and trace fossils in Paleozoic sandstones of Libya, *in* KANES, W. H. (Ed.), *Geology, Archaeology and Prehistory of the southwestern Fezzan, Libya*. Petroleum Exploration Society of Libya, Tripoli, 117.

SEILACHER, A. (1970). *Cruziana* stratigraphy of 'non-fossiliferous' Palaeozoic sandstones, *in* CRIMES, T. P. and HARPER, J. C. (Eds.), *Trace fossils. Geol. J. Spec. Issue*, No. 3, 447. Seel House Press, Liverpool.

SELLEY, R. C. (1970). Ichnology of Palaeozoic sandstones in the Southern Desert of Jordan: a study of trace fossils in their sedimentologic context, *in* CRIMES, T. P. and HARPER, J. C. (Eds.), *Trace fossils. Geol. J. Spec. Issue*, No. 3, 477. Seel House Press, Liverpool.

SIMPSON, S. (1970). Notes on *Zoophycos* and *Spirophyton*, *in* CRIMES, T. P. and HARPER, J. C. (Eds.), *Trace fossils. Geol. J. Spec. Issue*, No. 3, 505. Seel House Press, Liverpool.

SOUGY, J. (1956). La stratigraphie du 'Cambro–Ordovicien de Semmour' (Sahara occidental). *Bull. Soc. Géol. Fr.*, **6**, 99.

Lower Palaeozoic of the Middle East, Eastern and Southern Africa, and Antarctica
Edited by C. H. Holland
© 1981 John Wiley & Sons Ltd.

LOWER PALAEOZOIC PALAEOCLIMATOLOGY

Nils Spjeldnaes

Palaeoecology Department, Aarhus University, Denmark

Contents

1. Introduction

1.1. Previous work, scope, and definitions

The classical work on Early Palaeozoic palaeoclimatology is by SCHWARTZBACH (1950), and numerous papers have dealt either with the climate of part of the period (BAIN, 1960, 1965; BORISOV, 1965; SPJELDNAES, 1961), or with special problems and events.

Although many data and many interesting hypotheses have been published, the great advance in the study of Early Palaeozic palaeoclimatology has come with the breakthrough of the theory of plate tectonics. Since the configuration and relative position of the continents are of major importance for palaeoclimatic reconstruction, there is a fruitful feedback between palaeogeographical reconstruction given by plate tectonics, and palaeoclimatological studies. In many cases climatic studies may supplement the tectonic evidence, and their main purpose in this connection is to provide an independent check on the mainly geophysical evidence upon which the plate tectonic reconstruction rests.

At the present stage it is therefore impossible to divorce palaeoclimatology from palaeogeography, both physical geography and biogeography. In this review, some emphasis is put on these subjects, which are necessary in order to study palaeoclimatology.

Recent progress in palaeoecology (e.g. ZIEGLER, 1965), and in theories of evolution—partly related to plate tectonics and biogeography (DURAZZI and STEHLI, 1972; RAUP, 1972; VALENTINE and MOORES, 1970; VALENTINE, 1973; and VINE, 1973), has widened the base for palaeoclimatic studies, but unfortunately most of the observational data available do not have the precision required for being utilized with the modern theories. This also applies to much of the geophysical evidence and, in order to arrive at a realistic and moderately precise interpretation, it is necessary to re-evaluate many of the primary data, be they biological, sedimentological, or geophysical. This will be long and time-consuming work and is outside the scope of this paper. The picture given here will be one obtained from the material available at present, and it must be considered that it may change considerably both in local detail and in more fundamental aspects, when more of the observational data are revised to the standards required for modern theory.

Because of the considerable plate movements which have taken place since the Early Palaeozoic, it is of course unrealistic, and not very enlightening, to plot observational data on maps giving the recent position of the plates. There have been several attempts to make reconstructed geographies for part of whole of the Early Palaeozoic (SMITH, BRIDEN, and DREWRY, 1973; WHITTINGTON and HUGHES, 1972, 1973; BRIDEN, DREWRY, and SMITH, 1974). Such maps, based mainly on palaeomagnetic evidence, suffer from considerable uncertainties in the areas where the coverage of reliable determinations are scarce, and much work is required before satisfactory maps can be reconstructed. For plotting the data used in this paper, a modification of WHITTINGTON and HUGHE's (1972) map is used in Figs. 2, 3, and 4. The main change from the original is that more of the southern hemisphere is included in order to show relevant data from Africa and South America.

The outlines of the continents have purposely been made crude, but the general shape has been retained in order to make identification easy. Because so much continental crust has been lost, the real outlines of the Early Palaeozoic continents were probably rather different from those in Figs. 2, 3, and 4.

A parsimonious solution has been selected for the amount of plate movement. Only the minimum necessary to accommodate the palaeoclimatic observations has been used. This, and the special map projection used, make the maps rather different from those of SMITH, BRIDEN, and DREWRY (1973), and BRIDEN, DREWRY, and SMITH (1974), and maps derived from these. Other maps by ZIEGLER and others (1977), ZIEGLER and others (1979), and SCROTESE and others (1979) appear more realistic, and are also closer to the reconstructions used here. Since this manuscript was effectively completed early in 1976, the information from these newer maps has not been incorporated.

1.2. Materials and methods

In order to achieve palaeoclimatological reconstructions it is necessary to integrate several types of materials and methods. Geophysical (mostly palaeomagnetic), sedimentological, and biological material must be used. These types of material, and the methods involved in their use, not only supplement one another but also provide independent checks, which is valuable considering the many uncertainties involved.

It must also be noted that geophysical methods give latitudes, whereas sedimentological and biological methods give temperature, or other direct indications of climate. Since the isotherms were probably never parallel to the latitudes this discrepancy must be considered. In Fig. 1 by comparison with the 'ideal' earth circulation pattern—without obstructing continents—the

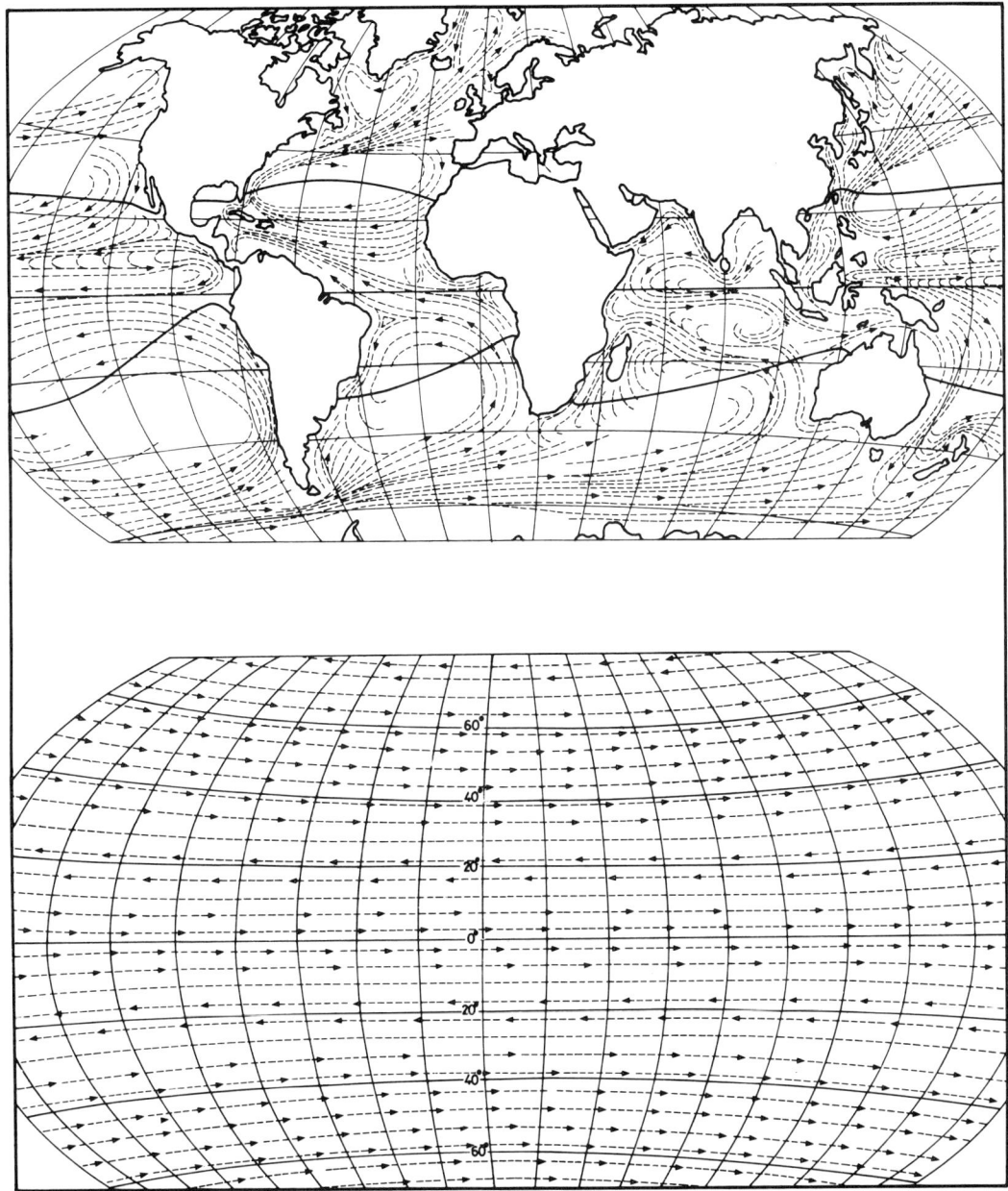

FIG. 1. The oceanic circulation pattern. *Above.* The present circulation pattern, showing deviations due to the shape and latitudinal position of the continents. The heavy line is the 25° February isotherm, which also marks the approximate boundary of coral-algal bioherms today. *Below.* A hypothetical circulation pattern if there were no continental blocks.

effect of continental plates on the circulation pattern is obvious. When plate reconstructions are made, it will often be extremely difficult to interpret the effects of the plate movement on the earth's atmospheric and oceanic circulation and the resulting isotherm pattern. It will depend both on the size and shape of the continental parts of the plates and their latitude. The displacement of the isotherms will be greatest by movements across the latitudes, and at low latitudes.

The effect will also depend on the intensity of the circulation, which will depend largely on the presence or absence of polar ice-caps. Therefore, this effect will make detailed reconstructions difficult, because small changes in parameters which are difficult to estimate will cause large effects. On the other hand, this situation may, under favourable circumstances, be used to improve models of plate and climatic reconstructions.

An example is shown in analysis of plate movements in the North Atlantic during the Ordovician. The North American plate had a uniform climatic pattern, and did therefore not move, at least not across the latitudes, whereas the European plate showed increasingly warmer climates (= lower latitudes) from the low middle Ordovician (Llandeilo), accompanied by an increase of American (= warm water) elements in the fauna, and a general increase in exchange of faunal elements with time, until the fusion was complete in the low Silurian.

Because of this climatic development, it may be concluded that the European plate moved across the latitudes towards the comparatively stable, equatorial American plate. In this way, both the timing and direction (strictly the north–south component of it) of plate movements can be identified. In this particular case, some of the complicated circulation patterns and their biogeographical effects have been analysed by WILLIAMS (1969) and ROSS (1975), who demonstrated the use of climatic and circulatory patterns in palaeobiogeographical analysis and reconstruction of ancient plate movements.

One of the great difficulties in palaeoclimatic interpretations in the Palaeozoic is the absence of modern equivalents and the fact that some processes—in seeming defiance of the principle of uniformitarianism—occurred in a different manner in the Palaeozoic to the present time.

Most of these apparent deviations from the principle of uniformitarianism are connected with vegetation. The absence of land vegetation in the pre-Devonian must have caused profound differences in weathering and sediment transport, to a degree which makes environmental reconstructions based on clay-mineral composition, degree of weathering, and the like unreliable.

CAWLEY, BURRUSS, and HOLLAND (1969) suggested that chemical weathering would be not less than one fifth of the normal rate in an unvegetated area, based on studies from Central Iceland. Their suggestion, that chemical weathering was not very different with or without plant cover, is based on the assumption that all bicarbonate in their samples was due to chemical weathering, and that the partial pressure of carbon dioxide in the pre-Devonian was five times higher than today. The latter would have led to abnormalities in the carbonate balance in the oceans and possible climatic effects.

The impact on run-off will also alter the type of structures and sediments of fluvial deposits profoundly.

The general tendency is to make terrestrial sediments (including fluvial, lacustrine, and estuarine) look as if they came from a regime with a colder climate or greater relief. Certain sedimentary environments characteristic of special climatic conditions (such as the laterite environment) can hardly be conceived without a vegetation cover, and their climatic equivalents in the Early Palaeozoic are unknown.

In the marine realm, vegetation is also highly important, both in creating new ecological niches, and thereby a higher species diversity, and by influencing sedimentation processes. This is well known from studies of the recent *Zostera* and *Laminaria* communities, but the extent to which analogues existed in the Palaeozoic, and how they may have differed from recent ones, are not known.

The non-calcareous marine algae are not important for climatic zoning, but presence or absence of an algal vegetation will profoundly influence species diversity, both in the whole fauna and in specific animal groups. Since specific diversity has been much used in palaeo-climatological analysis (cf. STEHLI, 1973, and others), the bias introduced by algal vegetations must be taken into consideration. The absence of algal vegetation would lead to lower specific diversity, simulating colder climates.

The interface between marine and terrestrial realms is also often marked with special vegetation types, such as mangrove swamps and sediment-binding grasses. They have a marked climatic differentiation, and play an important role in restricting sediment transport and in creating ecological niches. It is unlikely that such vegetation types existed in the Lower Palaeozoic, and these climatically characteristic biotopes have therefore not been found. This absence of biotopes with high diversity may lead to an interpretation in the direction of a colder climate than the real one.

Another type of deviation from the principle of uniformitarianism comes from quantitative changes in the physical parameters of the environment with time. Some of these changes are also related to the presence or absence of vegetation, such as increased aridity and aeolian transport due to absence of vegetation. On a non-vegetated continent, the absence of plant cover would lead to increased aeolian transport, both directly and by increased wind velocity close to the ground. Because of rapid sediment movement, due to faster surface run-off of water, the weathered rocks would be transported faster to their final place of sedimentation. The result would be faster erosion, and mechanical weathering would dominate over chemical weathering. It would also include faster peneplanation, and the character of the resulting sediments would appear more immature than if a plant cover were present.

Land plants were present, and perhaps even locally abundant, at least in the Silurian (GRAY, LAUFELD, and BOUCOT, 1974). It is assumed that these plants lived in or close to water and that they did not form a vegetation cover affecting the climate by increasing the chemical weathering, retaining precipitation, and decreasing the albedo.

In the Early Palaeozoic, there are few well identified terrestrial sediments, and they are not climatically significant. The effects of the absent plant cover are also felt in the marine sediments, which again will have a tendency to be less mature than those from younger periods, when land vegetation was well developed.

This effect of the absent plant cover has been known for a long time and has been much discussed. Considering the theoretical magnitude of its impact, it is remarkable how similar Lower Palaeozoic sediments are to recent ones. This is partly due to frequent recycling of sediments, and other maturation processes in the marine regime. Still one might speculate over the possibility that there was some sediment-binding factor operating in the absence of a normal vegetation cover. Without any factual evidence it is impossible to define this; it may have been chemical, diagenetic, such as caliches, or a specialized algal mat in or at the sediment surface.

Changes in the tides, due to the smaller distance between the moon and the earth, would have profound ecological effects. There is now both astronomical and palaeontological evidence indicating that the distance between the earth and the moon was shorter in the Palaeozoic than it is today (cf. CLOUD, 1968; RUNCORN, 1971; GOLDREICH, 1972). Modern studies on this problem (e.g. LAMBECK, 1978; PIPER, 1978; and especially SCRUTTON, 1978) concentrate on whether the change in distance was a gradual or episodic one. They all indicate that the earth–moon distance during the Early Palaeozoic was considerably smaller than the present one. This would result in more days in the year, and therefore more frequent and stronger tides.

The shorter days may have pronounced biological effects, as indicated by LOVENBURG and others (1972). Even a minor change in the distance to the moon will have had a marked physical effect, and ALFVÉN and ARRHENIUS (1969) have vividly described the extreme geological effects of a close proximity of the moon to the earth. As they aptly remark, such effects would not have escaped the notice of geologists, even if they occurred in the Pre-Cambrian, and it may safely be concluded that there is no geological evidence to indicate that the moon was even close to the Roche limit, where it would have disintegrated owing to the gravitational forces of the earth.

Increased tidal forces would also have caused a heat effect due to stronger friction, but this effect cannot be traced in the geological record and does not appear to have had climatological consequences, at least not in the Early Palaeozoic.

KLEIN (1971, 1972, 1972a) has tried to determine the tidal range in the geological past, and concluded that there was no appreciable change even from the Lower Cambrian. Other workers, such as MERIFIELD and LAMAR (1970) and OLSON (1970), found indications of considerably higher tidal amplitudes in the late Pre-Cambrian and Cambrian, up to 300 metres. The absence of sedimentary structures formed under upper flow regime seems to indicate that the tidal forces in the Early Palaeozoic were not (at least not by a magnitude or more) stronger than at present. The problem of palaeotidal ranges is best studied in sandy sediments with little or no palaeoclimatic significance, but the question is of considerable importance, because an increased tidal range would also have profound effect on littoral marine life, because it would widen and deepen the littoral zone and increase its energy level.

Some of the very widespread, but somewhat atypical, tidal-flat sandstones from the Lower Cambrian and Ordovician may, as mentioned by RUTTEN (1952), have been formed under non-actualistic tidal conditions. The palaeoclimatological effect of this, as well as the other deviations from the principle of uniformitarianism, is that of a 'noise,' which blurs the data (signals) used in interpreting the climate, and tends to make the sediments, and the faunas, less indicative than they would have been under the present conditions.

1.2.1. Sedimentological methods

Sedimentological methods have been much used and discussed in palaeoclimatology. Clastic, terrigenous sediments do normally not give much climatological data. The maturity of such sediments expressed in feldspar content, and the presence of unstable heavy minerals, cannot be used with confidence in the Early Palaeozoic because of the absence of terrestrial vegetation, which will make such sediments appear more immature (= 'colder') than their recent counterparts.

Present day oceanic clay sediments do show distinct latitudinal distribution patterns (LISITZIN, 1972), but the use of clay mineral distributions for palaeoclimatological analysis in the Early Palaeozoic is restricted, partly by the absence of vegetation, which would have limited certain types of weathering considerably, and also by the fact that the regular distribution pattern shown by LISITZIN (1972) is mostly restricted to oceanic sediments.

Pre-Cretaceous oceanic sediments are rare and often strongly deformed, and the shelf sediments, from which most of our evidence comes, are subject to great variation because of local factors. BJØRLYKKE's (1974) studies indicate that the clay mineral content of Ordovician shelf sediments may be more influenced by regional geology than by general climatic patterns. Structures in terrigenous clastics may be used for interpreting climatic parameters other than temperature, such as aridity. Again, this is of restricted value in the Early Palaeozoic, since the absence of terrestrial vegetation would cause desert-like conditions also in areas with moderate or even high seasonal precipitation. It would also have a profound effect on the discharge, sediment load, and run-off of rivers, and the interpretation of pre-Devonian fluviative sediments will therefore be rather uncertain.

Biogenic structures are mostly referred to in Section 1.2.2, but some cases, where the biological interpretation of the structure is doubtful, may be mentioned here. Vertical, tube-like trace fossils of *Skolithos* or similar types are very common in the Lower Cambrian, and middle Ordovician. They are often so crowded that they dominate the rock, and are found over wide areas. They give the impression of representing dense, but almost monospecific, infaunal populations. This can be interpreted as indicating high organic productivity, but very low diversity. This may reflect low temperature, but may also indicate other restricting ecological factors, such as high turbulence and rapid sediment movement, such as the recent *Arenicola* associations on tidal flats. Similar, shallow-water marine environments which by direct evidence can be classified as warm do in most cases have a much more diverse trace-fossil assemblage. LINDSTRÖM (1972) has observed indications of the presence of ice in a sedimentary series which includes *Skolithos* sandstones, and this, together with the widespread occurrence of such rocks in the cold-water Ordovician of North Africa (BEUF and others, 1971), must be taken as an indication that this type of trace fossil may be used as an indicator of cold water.

Evaporites show, as already indicated by LOTZE (1938) and further elaborated by DREWRY, RAMSAY, and SMITH (1974), a latitudinal zonation, with a statistical maximum in the trade-wind belts. Exceptions are known and special types of evaporitic sediments do also occur outside these zones. This issue is also complicated by the discovery of oceanic salt deposits, both in the Mediterranean and in other oceans. Their climatic implications are at present unknown. The well known Early Palaeozoic evaporites (in the midwestern United States, Arctic Canada, and the Siberian Platform) are all consistent with the model proposed by LOTZE, in the palaeogeographical reconstruction given here.

Glacial sediments are important and much used for palaeoclimatological reconstructions. Since direct evidence of mountain glaciations have only very remote chances of preservation, especially from the Early Palaeozoic, the available material will be from low-land glaciations, and mostly from marine glacial beds (cf. SPJELDNAES, 1973). The identification of glacial sediments is often difficult, as demonstrated by CROWELL (1964), SCHERMERHORN (1974, 1975), and HARLAND, HEROD, and KRINSLEY (1966). It may be especially difficult to identify terrestrial glacial deposits (moraines) with certainty, because of the resemblance to mud-flows, solifluction soils, and other unsorted sediments. Since most of the terrestrial tillites have been altered, by sliding, tectonization by later ice movements, or by out-wash, interpretations of one individual rock may often be doubtful. A study of the whole sedimentological and palaeogeographical context may help to solve the problem. A good example is the Ordovician glacial beds in Central Sahara, described by BEUF and others (1971). Here there are few and atypical tillites, but the association with widespread ice-striated pavements, glacially eroded valleys, typical sandur-plain glacio-fluvial drainage systems, and laminated silts with ice-dropped exotic boulders, many of which are ice-scratched, give entirely convincing evidence for glacial conditions.

The use of the whole regional, sedimentological, and stratigraphical context is therefore the most effective way of proving the glacial origin of a sequence. If only a single locality, with a restricted assemblage of rocks and structures, is available, the origin will often be in doubt. Some of the details regarding the distribution and identification of glacial sediments and structures are discussed under the description of the basal Cambrian and the upper Ordovician glaciations (Sections 2.1 and 3.2).

Because carbonates are more soluble in cold than in warm seawater, the presence of carbonates has been much used in palaeoclimatic reconstructions. As mentioned by CLOUD

(1961), the presence and volume of carbonate sediments cannot itself be used as a direct indicator of palaeoclimate. It is necessary to study the type of carbonate in order to identify the typical warm-water associations (micrite–oolith–bioherms), and to eliminate the exceptions. TEICHERT (1958) described several of these, but it is easy to identify his cold-water reefs and avoid confusion with warm-water bioherms. The *Lophohelia–Amphihelia* coral thickets and their fossil equivalents have characteristic fossil assemblages and matrices which make their identification as deep- and cold-water structures quite evident.

Our knowledge of cold to temperate water carbonate have also increased considerably (cf., e.g. LEES and BULLER, 1972; LEES, 1975) and, even if the main importance of the carbonates in palaeoclimatology still is as indicators of tropical conditions, it may be possible to use them as indicators also of intermediate climates. This has been demonstrated by JAANUSSON (1972) for the lower Ordovician carbonates of the Balto-Scandian region.

The typical warm-water carbonates are probably the best climatic indicators known. They are based both on sedimentological (chemical) and biological evidence which has been repeatedly tested, and they are widespread, both in space and time, being among the most common sediments on the shelf in the whole tropical region. They are also known almost without interruption from the Pre-Cambrian to the present. In the Early Palaeozoic, typical warm-water carbonates have been identified repeatedly and this constitutes, together with the concurrent biological evidence, one of the best methods of defining the palaeoequator in this time interval.

An interesting observation, which is not yet sufficiently supported by quantitative evidence to be regarded as fully established, is that the tropical zone, as defined by the biological–sedimentological criteria described here, does not appear to be wider in periods with generally warmer climates, or when the climatic zoning was less sharp, due to absence of polar ice-caps. This is also shown in the well known equatorial distribution of the rudist reefs in the Cretaceous, which cover a narrower zone than the tropical hermatypic corals today. During periods with warmer climates, the important change seems therefore to have been a polewards expansion of the subtropical and boral regions, rather than a widening of the tropics. This (and the scarcity of precise data) is illustrated in Fig. 4, where it is shown that the Silurian reefs, in spite of the generally warmer climate, did not occur at latitudes higher than those of the recent hermatypic corals.

1.2.2. Biological methods

In recent geological periods (mainly the Pleistocene), the use of biological data in palaeo-climatology is based on a knowledge of the ecological and climatological requirements of recent species. This method does not work in the Early Palaeozoic and the biological methods used in this time interval will therefore have to be based on more indirect evidence. The use of taxonomic units higher than species is doubtful, as the precision decreases in most cases, and even the general life habit and requirements of whole groups may have changed with time, as demonstrated with the crinoids. The biological data used for climatological analysis in the Early Palaeozoic must therefore partly be based on assumptions of constant requirements over long time intervals. These assumptions can, in favourable cases, be checked by assuming that the general physiological conditions have been unchanged, and partly against chemical–physical evidence (sedimentological and geophysical) from the same environment.

As mentioned in Section 1.2.1, some important restrictions in the applications of the principle of uniformitarianism are linked with the development of the biosphere, and these must be taken into consideration using biological evidence for climatic analysis in this time interval.

In the last 10 years or so (VALENTINE and MOORES, 1970; DURAZZI and STEHLI, 1972; RAUP, 1972; HALLAM, 1967, 1973; VALENTINE, 1973; and VINE, 1973) there has been an interesting development in theoretical studies on patterns of evolution and diversity related to plate tectonics, which are highly relevant to palaeoclimatic studies, but they are difficult to utilize at present since the field tests of most of these theories have not yet been adequately made. As indicated by STEHLI (1973), quantitative data are absolutely necessary in order to obtain generally valid results and to test these theoretical models.

An analysis of some of the possibilities of using ordinary qualitative biogeographical material for palaeoclimatological reconstructions was made by SPJELDNAES (1961, 1967). Most of the individual features used in such methods are open to exceptions and only when a considerable number of features all pointing in the same direction are found, and checked against the chemical–physical parameters, can they be used as a basis for reconstruction. Much 'noise' is found in the biological as well as in the sedimentological material. Tectonism and plate movements will change the biogeographical pattern and create complications which may simulate climatic changes, even if the climate is stable. All of these difficulties make biologically based palaeoclimatic reconstructions crude and imprecise in details.

When using methods from recent zoogeography, it should be noted that the distribution of sediments in detail is normally not climatically directed (with the notable exceptions of glacial and some carbonate environments). THORSON (1957) has indicated that most of his animal communities are dependent more on bottom type and bathymetry than temperature. This is particularly the case with the soft-bottom communities, which may be found in all climates with similar association of species. ZIEGLER's (1965) marine communities, which are conceptually different from THORSON's, are also mainly related to bathymetry and sediments, and only remotely and indirectly to climate. In some extreme cases the systematic absence of communities may be used, such as the absence or feeble development of the littoral communities in polar regions, where the shores are covered with ice for large parts of the year.

Faunal diversity has been used to determine palaeolatitudes, based on the observation that diversities normally increase with lower latitudes, e.g. by FISHER (1960). Ideally, the diversity index is the number of species per 1000 individuals. Owing mainly to lack of adequate studies and the difficulty of obtaining precise datings, this method can normally not be used on Palaeozoic material. In many cases the simpler method of listing the number of coexisting species within a group in a biotope is used. When working with a taphocoenosis, there are several sources of error as the available material may be mixtures not only of infauna and epifauna but may also involve planktonic and far-drifted nektonic forms. The presence of algal vegetation increases the number of microhabitats, and therefore also the diversity, especially at the species level. In order to obtain reliable diversity figures from fossil material, it is important to be able to sort out the various elements, not only the autochtonous from the allochtonous, but also the infaunal elements from the various epifaunas.

The fact that the fossil material does not give a full representation of the biota, because many forms lacked hard parts or because they had an aragonitic skeleton, will not in itself invalidate diversity studies, provided that equivalent materials are compared. When faunas with differing proportions of infauna and epifauna, and in different habitats, are compared this lack of selective preservation may introduce considerable bias. For fossil material the number of genera in a certain area in a certain time interval is normally used. This is obviously a crude method, because of the sources of error involved in the taxonomic problems and the incompleteness of our knowledge of the fauna in most of the Palaeozoic. In spite of this, the method may give acceptable results when applied to tolerably well described groups, such as the brachiopods and trilobites of the Ordovician and Silurian.

A critical review of the data in the best surveys with this method (cf. STEHLI and WELLS, 1971; BOUCOT, 1975; WHITTINGTON and HUGHES, 1972, 1973; WILLIAMS, 1969, 1973) indicates that the method works well at moderate diversities ($n \geqslant 10$). At low diversities ($n < 10$) the statistical uncertainties become too large and the 'noise' from faunas exposed to ecological stresses other than cold climate may mask the latitudinal effect. At very high diversities, biological and longitudinal factors may distort the results (cf. STEHLI and WELLS, 1971). With the limitations mentioned, faunal diversity may therefore be a valuable indicator of palaeo-climate, especially if used together with other methods which give some mutual control.

THORSON (1957) has postulated that the latitudinal change in diversity is due mainly to changes in the diversity of the epifauna, whereas the diversity of the infauna does not change much with latitude or temperature. Modern studies (cf. THORSON, 1966) have shown that infaunas also change more with latitude than was originally supposed by THORSON (1957), but he was right as to the uniformity of the dominating species in the infaunal communities. By combination with information from studies on the structure of the communities, climatically significant observations may be obtained. By using comparable and regionally covering data (cf. STEHLI, 1973), it may be possible to obtain interesting results in individual cases even by semi-quantitative or qualitative methods.

An example of this is the Llanvirn–Llandeilo of south-western Europe, where faunas are dominated by infaunal elements, mainly molluscs, and where the epifauna, although present, has a remarkably low diversity. Primitive taxodont bivalves make up a considerable part of the fauna (BABIN, 1966). In recent environments the same type of bivalves are also dominant in soft bottom communities in cold, arctic water, and in the equally cold abyssal environments. Even if there are complications in the diversity pattern of recent polar bivalve faunas (NICOL, 1970), it therefore appears reasonable to assume that these Ordovician faunas lived in a cold climate. This assumption can be checked sedimentologically, and the absence of carbonate and other features support the biological interpretations. So also does the position in the general pattern of climatic zoning observed for Europe in this time interval (DEAN, 1966; SPJELDNAES, 1961, 1967).

Below, a short review is given of the usefulness of the various animal and plant groups for palaeoclimatological studies in the Lower Palaeozoic.

Calcareous algae are mostly regarded as indicators of warm water. Generally this is true, but there are striking exceptions. *Lithothamnion* crusts and rhodolites are sediment-forming today in Spitsbergen, and are therefore rather deceptive (cf. TEICHERT, 1958). If such sediments are preserved, they will easily be identified as being 'non-warm'. The Dasycladaceans are also normally regarded as indicators of warm water, even if a restriction to 15° off the equator may be an exaggeration. The deviations mentioned by NITECKI (1972) are partly due to a single reference to a receptacultitid from North Africa. The original reference (TERMIER and TERMIER, 1950) refers to a *Receptaculites* sp. of unknown stratigraphical provenance, from the plain of Rehamna in Morocco. Later references are apparently secondary, applying to the same fossil. The receptaculitid in question may easily be from the upper Silurian or Devonian, and cannot be used without reservation as evidence for the distribution of the dasycladaceans in the Ordovician. Generally, the occurrences of diversified calcareous algal floras in the Early Palaeozoic are accompanied by sediments (bahamites, oolites, and bioherms) indicating climatic conditions similar to those in the present tropical zone. Some special occurrences, such as the massive, monospecific algal floras known from certain beds in the middle Ordovician of Scandinavia (JOHNSON, 1954) and Great Britain (SPJELDNAES, 1955), may indicate habitats resembling those of the *Lithothamnion* floras in cold waters of today. They were probably not really arctic, but occurred—judging both from other biological, and geophysical evidence—at least 40° off the equator.

Corals are useful, mostly by their absence in presumed cold-water areas. Hermatypic corals were, as in recent times, restricted to a narrow, equatorial belt, and the other corals, mostly rugose ones, are found in rapidly decreasing number and diversity from higher palaeolatitudes.

Bivalves are not too well known in the Lower Palaeozoic, and it is difficult to assess their usefulness as much of the information may be masked by the lack of adequate description from large regions. The available evidence seems to indicate that a diversity gradient is found from the supposed cold-water faunas to the warmer ones. There are, on the qualitative level, several parallels in taxonomic position, sculpture, and shell thickness with the climatically dependent variables in recent faunas (cf. BABIN, 1966).

The Lower Palaeozoic *gastropods* are not too well known, and they are at present of little use as climatic indicators. Modern gastropod faunas are often climatically significant, but the old ones are difficult to use because of their aragonitic skeleton, which makes preservation incomplete. A few exceptions exist, such as the *Maclurites–Ceratopea* association which appears to be restricted to warm-water sediments.

Cephalopod faunas are distinctly more diversified in warm water than in cold waters in the Lower Palaeozoic. The data should be used with great care in detail because the nautiloid cephalopods, which are the only group under consideration in this interval, are likely to have drifted over long distances after death, and may have ended up in a different environment from that in which the animal lived. This is well known from the living *Nautilus*, and the same mechanism was certainly operating in the Early Palaeozoic also.

The *brachiopods* are well differentiated into faunal provinces which are climatically dependent. Due to the numerous modern monographs the Early Palaeozoic brachiopods (WILLIAMS, 1962, 1963, 1974; HAVLIČEK, 1958, 1967, 1971; BASSET and COCKS, 1974; BOUCOT, 1963, 1968, 1970; and others) give some of the best material, both for diversity data and for taxonomically equivalent regional coverage, in this time interval. In some cases this can be used even at the species level, and in at least some cases for the Ordovician and Silurian the best material on which many of the reconstructions are based comes from brachiopod evidence.

The *bryozoan* faunas are not well known, but they appear to be well differentiated into climatic zones, and show a smaller change in diversity (= number of genera or species present) going from the pole towards the equator than do many other invertebrate groups. This is also found in recent bryozoans.

Trilobites are among the best known Early Palaeozoic animals and their distribution is well known. The data from Ordovician trilobites have been extensively used by WHITTINGTON (1969) and WHITTINGTON and HUGHES (1972, 1973) for reconstruction both of faunal provinces and of their latitudinal position. SHAW and FORTEY (1977) and FORTEY and BARNES (1977) have recently refined the trilobite faunal provinces, based on modern biogeographical principles. Silurian trilobites are fewer and not too well differentiated. Cambrian trilobites are differentiated into faunal provinces, but there is some doubt as to whether the Cambrian provinces are related to climate only, or to bottom condition and bathymetry (PALMER, 1972; LOCHMAN–BALK, 1970, 1971).

Ostracodes are reasonably well known from the Ordovician and Silurian, even though data from supposed cold-water regions are meagre. Because of this, little is known about their potential as palaeoclimatic indicators, even if the Ordovician ostracodes seems to be differentiated into faunal provinces which compare well with those of better known organisms.

The *conodonts* are also rather cosmopolitan, and the newly suggested faunal provinces do not give much information on the climatic factors in their distribution. It must be mentioned that in several regions, especially where the climate presumably was extreme, conodonts are few, or not well described. This may explain why conodonts are not at present very useful for climatic considerations. In some publications, such as in LINDSTRÖM and PELHATE (1971) and LINDSTRÖM *in* JULIVERT and TRYUOLS (1973), conodont datings of carbonate beds from the mediterranean Ordovician are at variance with the conventional ones. More studies both on the conodonts themselves and on other faunal elements are necessary to solve these discrepancies.

Recent studies (e.g. BERGSTRÖM, 1977) have shown that conodont faunal provinces with a distribution pattern similar to those of graptolites and trilobites could be discerned in the Ordovician.

Chitinozoans and *acritarchs* are not yet sufficiently known regionally, but some results by CRAMER (1971, 1973) indicate that they show a latitudinal distribution in the Silurian.

The *echinoderms* appear to have a high specific diversity in several climatic zones, but quantitatively dominance is found mainly in warm-water and intermediate sediments. It is remarkable that some of the temporary incursions of warm-water faunas into areas of colder climate are dominated by echinoderms, often in low or variable specific diversity. Examples of this are the upper Caradoc–Ashgill cystoid fauns and the *Scyphocrinites* bloom close to the Silurian–Devonian boundary in Southern Europe, North Africa, and Eastern North America. The presence of many of the echinoderms, especially the rarer types, as mass occurrences in infrequent beds ('starfish beds') may be a collection and preservation artefact, and without climatic significance. The widespread occurrence of unidentified fragments of echinoderm skeletons also in localities where there are no formal records of echinoderm faunas may indicate that the echinoderms were much more diverse and widespread than can be gathered from the palaeontological literature only. The high endemism, which partly may be a monographic artefact, also impairs the use of echinoderms as climatic indicators in the Lower Palaeozoic. In some cases, such as in the Ordovician of Europe, it is possible to discern faunal provinces, which may be explained as climatically induced.

The *graptolites* are mainly planktonic, and therefore less effective as indicators of climate than the benthonic faunal elements. Even if the data are few and fragmentary, there may be a climatic dependence in the distribution of their faunal provinces as suggested by SKEVINGTON (1973, 1974). In the presumably 'cold' regions in Southern Europe and North Africa, there is a remarkably low diversity in the post-Arenig Ordovician, which can be interpreted as due to the cold climate. The scarcity of biserial graptolites in the cold-water low middle Ordovician is a good example of this. In the Silurian most of the graptolite assemblages are cosmopolitan, and there is little in the distribution of the comparatively few endemic forms which can be climatically interpreted (Section 4.1).

The *vertebrates* are rather few, especially in the Ordovician, and their use as palaeoclimatic indicators is impaired by the rapid expansion of the group in the upper Silurian, which results in a sharply increasing diversity, which probably is environmental rather than induced by climatic factors. Both the few Ordovician vertebrates and the richest Silurian faunas are from warm-water environments, and there are no records of vertebrates from cold-water environments in the Lower Palaeozoic.

1.2.3. Geophysical methods

The most important geophysical method for palaeoclimatological work is palaeomagnetism. A determination of the ancient pole position will place a locality in relation to longitude and latitude, and since the climatic zones—with some deviations (cf. Fig. 1)—are roughly parallel to the latitudes, this gives a hint of the climate. Palaeomagnetic results are especially important in palaeoclimatology as they represent an independent check on the biological and sedimentological evidence. Being a purely physical parameter, the direction of the magnetic field is not likely to be influenced either by surface temperature or by other climatic parameters, and no interference should occur from biological effects. If, therefore, the biological–sedimentological and palaeomagnetic evidence point to the same latitude, this is a strong argument in favour of the correctness of the interpretation, and a wide discrepancy must lead to a reconsideration of the latitudinal determination.

The palaeomagnetic latitude determination of climatically typical faunas and sediments is also important in determining the general global climate. If the climate was generally warmer, warm water sediments and faunas and floras would occur at higher latitudes than today, and similarly the occurrence of indicators of cold climates at palaeomagnetically low latitudes will indicate a generally colder climate. It is difficult to express these changes in general climate quantitatively by using only biological and sedimentological evidence, as this will involve estimates of area, size, and shape of crustal plates which have changed considerably both in size, shape, and position during geological time. Palaeomagnetic pole determinations are subject to a number of sources of error, which must be considered when evaluating the results for palaeoclimatological purposes. Firstly, a large number of samples are necessary in order to eliminate statistically the secular variation of the dipole field.

Secondly, it is often extremely difficult to identify the original and relevant remanent magnetism. Especially in old rocks it is buried under considerable 'noise' created by different types of re-magnetizations. Some of these are due to chemical changes in the rock, and the character of the rock is therefore of great importance. Basic igneous rocks have a complicated mineralogy of their magnetic minerals, which are likely to change by reheating and weathering in a way which changes the magnetic properties fundamentally, and may erase the original remanent magnetism completely.

The methods for finding the deeply buried remains of the original remanent magnetism varies, and, especially in Palaeozoic rocks, different laboratories may get discrepant results. The most reliable results seem to come from sediments containing haematite, formed shortly after the consolidations of the rock (Løvlie and Kvingedal, 1975). Few such determinations have been made on Lower Palaeozoic rocks.

Thirdly, the deviations of the palaeopoles from the present ones are the result of an interaction between two different types of movements. The polar shift is the result of movements of the crust as a whole in relation to the earth's dipole field, and superimposed on this are the relative movements of the crustal plates. The polar shift is about 90° in about 450 m.y., giving an average velocity of approximately 22 millimetres per year. This figure is not much smaller than that suggested for plate movements, and it may be difficult to discriminate between the two types of movements.

In the Mesozoic and Cainozoic this poses only small problems, as the polar shift is well known, and seems to follow a smooth curve at uniform velocity. In the Lower Palaeozoic, there is no agreement as to whether the movement was uniform and smooth, or not. Storetvedt (1967, 1970) suggested that there was no appreciable polar shift between the base of the Cambrian and the Lower Carboniferous. In this case all of the changes in pole positions in the Lower Palaeozoic would be due to plate movements. Others, such as McElhinny and Briden (1971), advocate large and fast changes in the polar positions during the Lower Palaeozoic. If this were true, it would complicate palaeomagnetic reconstructions considerably. The movements are not on a smooth curve, and the suggested velocities are considerably higher than those suggested for plate movements.

A further, practical complication is that most palaeomagnetic measurements are made on rocks which are not too well dated, at least with a precision lower than the suggested climatic changes, and often from rocks only indirectly correlated with the beds from which the sedimentological and biological evidence has been obtained. A critical examination of the available material of Lower Palaeozoic palaeomagnetic pole determinations reveals that they are very few in number and that they are not only inequally distributed, but also that many of them are suspect in several ways. The data for the North European plate (including the Russian Platform) are the most numerous (cf. CREER, 1970; STORETVEDT, 1968, 1970; KHRAMOV, 1973) and indicate the position of a pole about 15° S, and 20° W. Also, the North Siberian Plate seems to be well placed (KHRAMOV, 1973) but the available data from North America, South America, and Africa are few, and subject to the above-mentioned reservations.

The results of McELHINNY and others (1968, 1971) are based on four sets of samples. One is from an unpublished study of basaltic lavas in south Morocco. These lavas, of Lower Cambrian to basal Ordovician age are deeply weathered and can be suspected of having had a complicated magnetic history. The pole position obtained is in discord with the faunal and sedimentological evidence from contemporaneous beds in the same region, especially the archaeocyathid reefs. The other two sets are from deep-seated alkaline igneous bodies on the African Shield, without direct contact with the surface stratigraphy. The estimate of their age is based on isotopic dating only. Since the cooling histories of these bodies are not adequately known, it is not immediately evident that the isotope ages are well correlated in time with the magnetic history, and none are closely connected with the history of surface geography. The fourth set, from the Ordovician of Table Mountain, South Africa, deviates both from the other African ones, and from the current reconstructions of Early Palaeozoic plate positions. This illustrates the difficulty of getting magnetically reliable, and precisely dated, samples from the Lower Palaeozoic of Africa.

It should be noted that the current interpretation of the wide gap between the European and the African plates in the Lower Palaeozoic (cf. BRIDEN, DREWRY, and SMITH, 1974) is based on these four sets of measurements.

Results from South America have recently been summarized by THOMPSON (1973). His data which are different from those previously published (cf. CREER, 1970), have been obtained from rocks which, according to their lithology, should give reliable results, and which are tolerably well dated. The Cambrian data (THOMPSON, 1973, Table 3, Fig. 7) mostly fall on the Russian Platform (No. 4 is widely off the others, being placed in Alaska). Climatologically this is not altogether impossible as there was a glaciation in the area in the basal Cambrian (cf. Section 2.1), and indicators of cold climate have been reported from the Baltic Lower Cambrian (LINDSTRÖM, 1972). The results are, however, at variance with the local palaeomagnetic results. The Cambro–Ordovician results fall consistently to the east of the comparable African ones (Nos. 6 and 7) from McELHINNY and others (1968, 1971), and No. 8 is placed very close to the shallow water Ordovician carbonates of the Baltic. No. 9 is between the Ordovician pole for Early Palaeozoic Europe, and that suggested by McELHINNY and BRIDEN (1971) for Africa. It is to the west of the pole suggested by SPJELDNAES (1961) on climatological evidence. The resulting picture is rather incoherent and the data are too few for more definite conclusions but, supposing that there were no or only small polar shifts in the Early Palaeozoic, the data may be interpreted to indicate that South America moved a considerable distance eastwards to be fused to Africa about Silurian times. This would parallel the closing of the North Atlantic about the same time.

The present problem of interpreting the available palaeomagnetic data for the Early Palaeozoic is therefore not that inherent in the method—it is certainly a potentially highly useful tool for latitude determination and general palaeogeographical work, even in the Early Palaeozoic—but, owing to the scarcity of reliable, well cleaned and dated material from beds of that age, until many more such observations, preferably from continuous sequences of well dated, tectonically undisturbed rocks have been made available, the potential of this method will not be fully exploited in palaeoclimatology.

When using palaeomagnetic results for palaeogeographical and palaeoclimatic reconstructions, it should be borne in mind that the numerous sources of error and methodological pitfalls make the basic assumption, that the directly observed remanent magnetism faithfully records the magnetic field at the moment when the rock was formed, appear to be a somewhat naive first approximation. Only data from samples, the magnetization history of which has been carefully studied, and where the results have been checked by different methods and preferably different laboratories, should be used. The author's rule-of-thumb is that if the palaeomagnetic results are supported by coinciding results from biological (ecological) and sedimentological studies, they are regarded as good ones. The chances that two so different methods would give the same erroneous result is small and, within the limits of error, such results may be regarded as usable. If there is a marked discrepancy between the palaeomagnetic results and those obtained by the other methods, both sets of data should be critically re-examined.

Other geophysical methods can be arranged under the theory of plate tectonics. Although there are some unexplained features both in the magnetics and rheology involved in the mechanisms required by this model, it has been most useful in reconstructing ancient geography, and thereby also the movements of crustal plate segments. There is a tendency to explain all deviations from a smooth and uniform distribution of sediments and fossils in terms of plate movements, and this over-extension of the theory may lead to an undeserved distrust in the plate tectonic model itself.

Geophysical studies of ancient plate movements sometimes give models which are difficult to explain in the light of biological and sedimentological evidence. The present opening of the South Atlantic may be a deceivingly simple case, both geologically and because it is parallel to the latitudes in its climatic effects. Early Palaeozoic plate movements may have been geologically more complicated, and have had a more profound climatological effect, especially in diverting a latitudinal pattern of climatic zoning by plates moving more or less obliquely in and out of the equatorial region.

At present there are not too many data for a stringent and detailed analysis of the climatic effect of Early Palaeozoic plate movements, but theoretically, and potentially, the climatological data may be used as independent controls. In combination with palaeomagnetic and other geophysical studies, they will—in the future, when more and more precise data are available— allow more refined reconstructions of ancient geography.

2. Cambrian climates

2.1. The basal Cambrian glaciation

The basal Cambrian glaciation was originally discovered by REUSCH in North Norway (for references to the history, *see* BJØRLYKKE, 1967, and SPJELDNAES, 1964) and, later, glacial sediments from the Cambrian–Pre-Cambrian boundary interval have been reported from all continents except Antarctica. Recent summaries of the literature and distribution of these sediments are given e.g. by HARLAND and RUDWICK (1964), SCHERMEHORN (1974, 1975), ČUMAKOV (1964), ČUMAKOV and CAILLEUX (1971), TROMPETTE (1972), and BIJU-DUVAL and GARIEL (1969). The distribution is outlined on Fig. 2. The basal Cambrian glaciation has been suggested as an appropriate lower boundary for the Cambrian and the whole Phanerozoic (HOLTEDAHL, 1961; DUNN, THOMSON, and RANKAMA, 1971), and good arguments may be advanced for the use of the onset of such a sharp event for a stratigraphical boundary. Regardless of whether the relevant international bodies will come to use this boundary, it makes

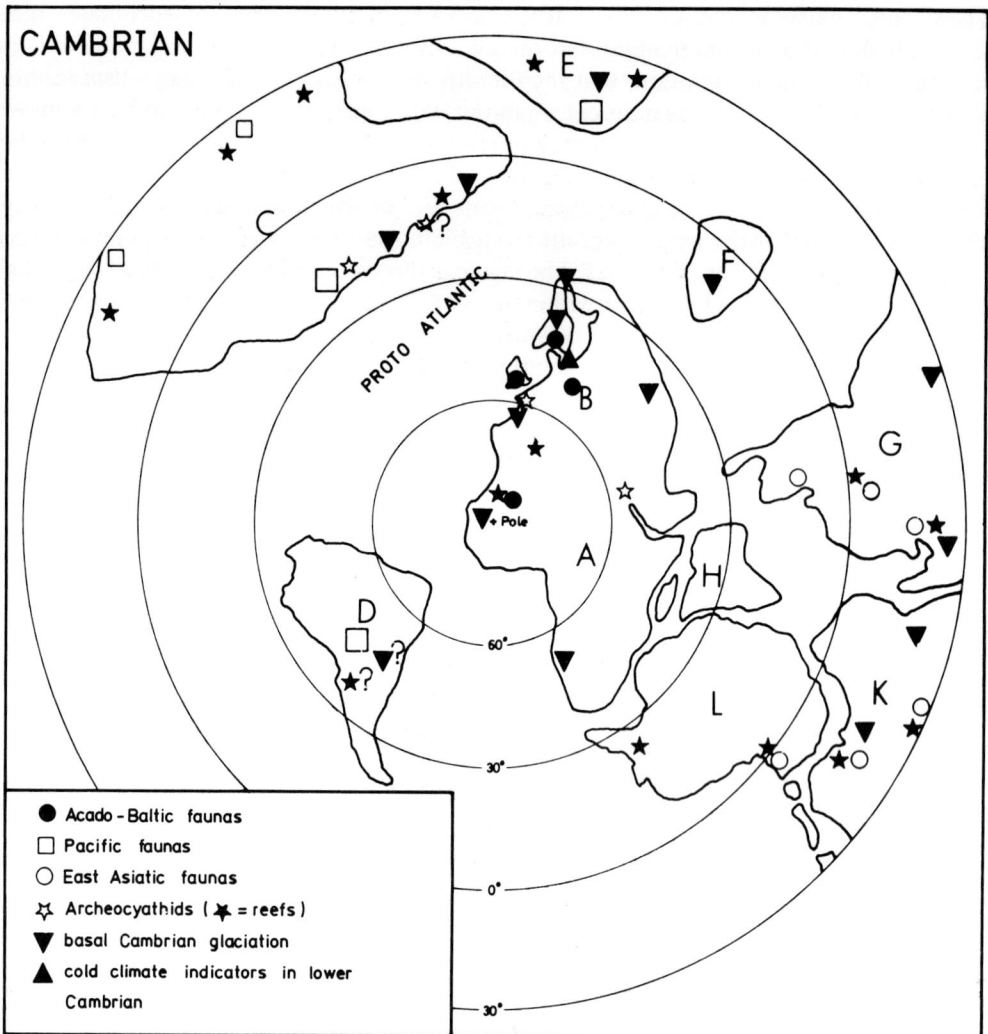

FIG. 2. A reconstruction of the palaeogeography and climatic zoning of the Cambrian. As mentioned in the text, the area in Europe (parts of Ireland, Scotland, and Norway, and Spitsbergen) which belonged to the American side of the Proto-Atlantic, and subsurface Florida and the other regions in America which belonged to the European–African plate have been placed on the correct side, joined to the continents where they belonged. The shape of the continents, especially the equatorial ones, are purposely figured rather crudely as the real shape of these continental blocks in the Early Palaeozoic is not known in detail. This figure (and Figs. 3 and 4) is modified from SPJELDNAES (1961) and WHITTINGTON and HUGHES (1972). Continental plates: A: Africa, B: Europe, C: North America, D: South America, E: North-east Asia, F; Kazakhstan, G: South-east Asia, H: India, K: Australia, L: Antarctica.

an excellent starting point for discussion of Lower Palaeozoic climates, and it is therefore included here.

The exact dating of the beds in many cases poses problems as indicative fossils are scarce, and the glacial beds are often limited by distinct discontinuities in the sedimentary record. It may be that the various glacial beds are not absolutely contemporaneous, even if the difference may be smaller than the time resolution normally obtained in beds of this age. In most areas, the evidence is largely of unsorted sediments, with large, angular clasts. In some cases the lower, and even the upper, boundary is erosional and there are often two or more discrete tillite horizons, with differing lithologies and boulder content. Striated boulders are not common but have been reported (e.g. BJØRLYKKE, 1974). The tillites are often associated with, and may grade into, varved sediments with drop-stones. Even if the tillites have a normal, gradual transition to the surrounding sediments, this does not disprove a glacial origin, as it is to be expected (cf. SPJELDNAES, 1973) that the majority of preserved fossil tills are of marine (or at least aquatic) origin, being dropped from floating ice. This also explains the wide and uniform distribution of many of the tillites. Marginal and bottom moraines formed by ice movement are susceptible to erosion and normally change rapidly in both thickness and character laterally.

Alternative explanations have been suggested for the genesis of the tillites (e.g. slide conglomerates, solifluction soils) by CROWELL (1963), SCHERMERHORN (1974, 1975), and others, but, even if it may be difficult to discriminate between such rocks and glacially formed tillites, the likelihood of preservation of rocks formed by the alternative processes is much smaller than for marine tillites.

There are also, among the beds described as glacial, a number where the sedimentological evidence for a glacial origin is not absolutely convincing, and there are numerous doubtful cases, especially among metamorphic rocks. The criteria for identification of glacial beds, and their classification, has been much debated (HARLAND, HEROD, and KRINGSLEY, 1966) and much of the material on fossil glaciations comes from studies on the basal Cambrian glaciation.

Even if most of the beds described as glacial from this interval are accepted as such, there have been critical voices, such as those of CROWELL (1963), CROWELL and FRAKES (1970), and SCHERMERHORN (1974, 1975), who prefer alternative explanations and refer to the supposed glacial beds as diamictites or mixtites.

The basal Cambrian glacial beds deviate from recent analogues both in depositional environment and presumed latitude. The comparatively few palaeomagnetic latitude determinations of these beds (BIDGOOD and HARLAND, 1961; TARLING, 1974) give consistently very low (0–15°) latitudes. A plot of their occurrences (Fig. 2) on a reconstructed world map shows most of them in tropical latitudes. The only exception is the occurrence in the Eglab massif in Sahara, described by BIJU-DUVAL and GARIEL (1969), one of the best documented cases, which, regardless of which pole position for the Cambrian is used (cf Section 1.2.3), must have been formed at high latitudes.

The idea that the glaciation was total and both hemispheres were covered with ice, leaving only a narrow equatorial zone with drift-ice-covered oceans, is also untenable. In some of the non-glaciated areas (e.g. the Russian Platform) the unbroken sequence does not show breaks which could be referred to a large erosive glaciation and the sediments, some of which are contemporaneous with the glaciated beds, do not indicate a particularly cold climate.

The glacial beds are in most cases found resting on, and in some instances even interbedded with, massive carbonates. These sediments sometimes contain oolites, oncholites, and stromatolites, indicating a warm-water origin. Such carbonates often form a conspicuous part of the boulder content of some of the tillites.

This problem is extremely difficult to solve in terms of modern climatology, and there is at present no solution which can explain the situation adequately. G. E. WILLIAMS (1972, 1975) has suggested that the abnormal situation was caused by a much higher obliquity of the earth. His strongest argument is that the varves and other features associated with the glacial beds are indicative of seasonality of the climate, a feature which is much less pronounced at the equator under the present, low obliquity. There may be a gradual change from the clearly abnormal basal

Cambrian glaciation, through the Ordovician one which is much more 'normal', but still shows some of the same deviations from the actual patterns (*see* Section 3.2), to the Permo-Carboniferous and Pleistocene glaciations, which appear to be perfectly normal in this respect. At present there is not enough information and model calculations to justify the acceptance of WILLIAMS' hypothesis and it must be considered as an interesting idea, which must be taken into account when the anomalies of the basal Cambrian glaciation are explained. Anyhow, there can be no doubt that glacial conditions were present, closely intertwined with equally good evidence for warm-water condition in the oceans, at the beginning of the Phanerozoic.

2.2. Archaeocyathid reefs

One of the conspicuous features of the Lower Cambrian is the widespread occurrence of archaeocyathid reefs (Fig. 2). Since the group is totally extinct, no direct biological evidence bearing on palaeoclimatic conclusions can be drawn from their presence, but studies of their occurrence and similarity to hermatypic corals, their association with algae, and above all the nature of the associated carbonates, strongly indicate a warm-water origin for the archaeo-cyathid reefs. This is also supported by the general occurrence of micritic carbonates, and in some cases evaporites, associated with the archaeocyathids.

Studies made by ZHURAVLEVA (1966) (see *also* HILL, 1972, pp. E27–29) clearly indicate that features normally associated with warm-water carbonates are found in the archaeocyathid reefs, such as micrites, oolites, and oncholites, as well as algal structures similar to those found today in shallow, warm water. The detailed analysis of depth distribution given by ZHURAVLEVA (1966) may indicate that the archaeocyathid reefs grew deeper than modern coral bioherms, and that they were intermediate between these and carbonate mounds in origin and structure. This is supported by the author's studies of some archaeocyatid reefs, which show some of the features characteristic of some early Palaeozoic carbonate mounds (Plate 1).

The present distribution of archaeocyathid reefs (Fig. 2) is almost world-wide and shows no relation to the present climatic zones. The best known occurrences are in Siberia, Australia, Morocco, and North America, where they are met with both in the eastern, and western sides of the continent (COWIE, 1971; NORTH, 1971; PALMER, 1971). The distribution of single archaeocyathids is wider than that of reefs, both in area and in presumed latitudinal distribution. Also in this respect the archaeocyathids resemble the recent scleractinian corals in their zoogeographical distribution. This is supporting evidence for interpreting the archaeocyathids climatologically, in spite of their unknown biology.

In addition to the distribution reported by HILL (1972) and ZHURAVLEVA (1968), there are some new records from marginal areas. OMARA (1972) reported archaeocyathids from a limestone horizon at the base of the Nubian sandstone in Sinai. This occurrence is especially interesting in that the associated algae seems to be different from those normally found in association with archaeocyathids. HOLLAND and STURT (1970) reported colonial archaeo-cyathids from northernmost Norway. Their material is metamorphic (it comes from carbonate xenolites in a gabbro) and is difficult to identify. Besides its interest in extending the geographical distribution of the archaeocyathids considerably, this find is also important as it may indicate that considerable parts of the metamorphic carbonate sequences in Northern Norway are in Pacific facies, and that this part of Europe therefore belonged to the other side of the Proto-Atlantic during the Cambrian.

An interesting correlation is found between the occurrence of archaeocyathid reefs and the occurrence of the basal Cambrian glaciation (Fig. 2). In the case of the glacial beds, it is perplexing that they occur regularly in very low latitudes, but this becomes natural for the archaeocyathid reefs. Regardless of which continental reconstruction is used, the archaeo-cyathid reefs do have a wider latitudinal distribution than the Ordovician and Silurian reefs, and the climatic zoning in the Lower Cambrian appear to have been rather indistinct. It should be noted that this deviation is mainly due to the high latitudes of the Moroccan and other Mediterranean occurrences, which may be related to an atypical distribution of climatic zones.

2.3. Black shales

Black shale is one of the characteristic sediments of the Cambrian. In many cases the fauna includes forms which normally are regarded as benthonic, particularly trilobites and brachio-pods, but the ecology of these shales may be complex. Some certainly include a truly benthonic element, in addition to a nektonic one, but in other cases there is evidence pointing to a purely nektonic and epiplanktonic mode of life of all the fossils found. Shales with high contents of organic carbon, sulphides, and heavy metals indicate that the bottom was stagnant, anoxic, and therefore unsuited for benthonic life. Supporting evidence for this is found in the slight fragmentation, indicating the absence of bottom scavengers, and laminated sediments, indicat-ing the absence of infauna. Such shales are very common in the Cambrian and, although some of the most spectacular examples of unfragmented faunas without traces of infaunas, such as the Burgess Shale fauna, are due to special circumstances, the accumulated evidence indicates that large areas of the ocean bottom including the shelves were anoxic during the Cambrian (STRØM, 1962).

In some well studied cases, such as North America, the anoxic facies is found particularly at the margins of the continent, whereas ventilated facies are more common in the central part of the craton. This is contrary to what would be expected if the anoxic conditions were due to local eutrophication, which would more likely occur in the shallow water on the shelf, where the restricted water volume, high organic production, and topographic features may induce localized areas with anoxic conditions. The fact that the craton is regularly surrounded by black shale facies does indicate that the oceans, or at least large parts of them, were anoxic.

In Europe and North Africa the conditions are more complex and the black shale facies seems to have penetrated deeply into the craton also. The Upper Cambrian Alum Shale in Scandinavia is a typical example. Its high content of organic carbon and hydrocarbons, as well as the content of sulphur and heavy metals, strongly suggests that the shale was deposited under anoxic conditions. This is also supported by the fine laminations and almost total absence of bioturba-tion. In spite of this, there are several indications that these shales (and the Middle Cambrian ones in the same area) were deposited at moderate depth. Considering the wide extension of the black shale facies, this indicates that the regime of oceanic and atmospheric circulation was different in the Cambrian to the present.

The fauna of the black shales is dominated by certain groups of trilobites: olenids and agnostids. The agnostids are especially common in the Middle Cambrian and the olenids, which dominate in the Upper Cambrian, are known (through the admittedly Ordovician *Triarthrus*) to have large gills, similar to arthropods living in the oxygen-deprived waters just above the oxygen–hydrogen sulphide interface in recent waters with anoxic bottom conditions. This may indicate that the olenids were specialized to live in oxygen-deprived waters, and their biogeo-graphical distribution also indicates that anoxic bottom conditions were more widespread in the Cambrian than they are today. The olenids therefore must be expected to have lived nektoni-cally with no, or at least very rare, contact with the poisonous bottom. This is supported by the very rare observations of tracks of olenids in bottom sediments (HENNINGSMOEN, 1957, Fig. 9).

Here the sediment is unusual (silt) and it can be shown that the trilobites just skirted the bottom surface, and took off.

By analogy, it may be supposed that the fauna which accompany the olenids also lived above the anoxic bottom water, as nekton or epiplankton. This is easily understandable in the case of the numerous inarticulate, phosphate-shelled brachiopods, which are known also from other black shales. It probably also was the case with the articulate brachiopods (*Orusia*) which are met with as mass occurrences in the olenid shales. HAVLIČEK (1967, pp. 21–22) has demonstrated that similar thin-shelled brachiopods were originally attached to floating sea-weeds. The presence of moderately sized epiplankton indicates that primary production in the Upper Cambrian oceans was not restricted to microscopic algae, but included floating plants large enough to serve as support for the epiplankton.

No remains of predators are found in the environment, as the trilobites were probably scavengers, feeding on dead organic material sifting down into the oxygen-deficient water where they lived. Their unspecialized mouth appendages prevented them from being active predators. The gills of the agnostids are unknown, and we have no definite clues as to their feeding and respiration mechanism and possible adaptations to low oxygen levels. The brachiopods were certainly filter-feeders, indicating the presence of microscopic plants and/or animals besides the larger algae to which they were attached.

The facts that the olenids were specialized for their rather stressed environment—which had the important advantage of giving good protection against predators, and that non-olenid (and non-agnostid) trilobites are conspicuously uncommon, do indicate that there must have been predators, or at least effective competitors, present. They have left no fossils and the fact that the brachiopods show no signs of fragmentation which can be ascribed to predation indicates that the predators present did not have hard parts for crushing. The most likely suspect for the role as predator or competitor in this biotope may be among the Scyphozoa.

The anoxic environment has, in itself, no great climatic significance. It may be found in any climate. At high temperature the solubility of oxygen is lower and the rate of oxidation is higher, but in modern marine environments anoxic conditions are found also in rather cold water, where the bathymetric and hydrographic conditions are appropriate.

The wide distribution of Cambrian black shales and the indicators of a shallow depth both for the bottom, and for the oxygen–hydrogen sulphide interface, indicate that conditions were different from those at present. A storm of only trivial size according to the modern circulation pattern would have oxygenated large parts of the 'Olenid Sea'. The distribution and depth conditions of the Upper Cambrian may therefore be interpreted to indicate a considerably lower rate of atmospheric and oceanic circulation. This may indicate the absence of polar ice-caps and a warmer and more uniformly distributed climate, with less pronounced zoning.

Similar conditions (generally warm climate, indistinct climatic zoning) prevailed also in the Silurian and the Jurassic–Cretaceous. Most of the sediments preserved from these periods are from the continental shelves and the craton, and they do not give direct information of the conditions in the oceanic basins. The few real deep-water sediments found on land have originated in trenches or other tectonically active environments which may not be typical of the oceanic environment in general. Deep sea drillings have revealed that the Cretaceous oceanic sediments to large degree were anoxic. This supports the hypothesis advanced here, that the periods in which polar ice-caps were absent and the climate was warm with diffuse zoning, also had stagnant oceanic basins with anoxic sediments encroaching upon the shelves and submerged parts of the craton.

The amount of black shales increases markedly through the Cambrian. In the Lower Cambrian these are rare and the typical sediments are well aerated, more or less bioturbated. In the Middle Cambrian black and dark grey shales are increasingly common, but specialized faunas are still few, except for the agnostids. In the Upper Cambrian and Tremadoc, olenid facies is found over a considerable area, from northernmost Norway to northern and west-central Argentina. In many cases this black shale environment passes up into the Tremadoc, which climatologically resembles the Upper Cambrian (cf. Section 3.3).

2.4. Cambrian faunal provinces

Cambrian faunal provinces have been the subject of much discussion, and have been summarized by COWIE (1960, 1974), ÖPIK (1957), KOBAYASHI (1972), PALMER (1972, 1973), and SDZUY (1972). JELL (1974) has attempted to study the provinciality by cluster, principal component, and discriminative analysis of the available faunal data for the Middle Cambrian.

Most authors divide the Cambrian faunal provinces into three, with various subdivisions. PALMER (1972) uses four cratonic provinces and two oceanic ones. The Acado-Baltic province is centered about North-West Europe (Britain and Scandinavia) and the north-eastern part of the United States and adjoining parts of Canada, including Western Newfoundland. The East-Asiatic province (also called the Redlichia, or Tethys, province) includes most of China, the Far East, Australia, and Antarctica. The Pacific province covers most of North and South America. The Siberian shield is referred partly to the East-Asiatic and partly to the Pacific province, and sometimes (e.g. by PALMER, 1972) regarded as a special province. The basis for the provincial subdivision is almost exclusively trilobite faunas. Archaeocyathids are used to a minor extent in the Lower Cambrian, and some few articulate brachiopods and early molluscs are sometimes included.

The areas covered by the various provinces varied with time. During the Middle Cambrian, both North Africa and the north-western part of Siberia were undoubtedly included in the Acado-Baltic province, whereas the whole Mediterranean region belonged to the East-Asiatic Province in the Upper Cambrian. North Africa (and adjoining parts of Southern Europe) was somewhat diffuse during the Lower Cambrian. Its fauna was dominated by Redlichids and there were spectacular archaeocyathid reefs, which normally characterizes the East-Asiatic and Pacific provinces. On the other hand, there were, especially in the upper part, a number of forms (Protolenids) regarded as indicative of the Acado-Baltic province, and it is therefore regarded (e.g. by SDZUY, 1972) as belonging to this province. The limitation of the provinces is also somewhat loose, and many important genera, such as *Fallotaspis*, are found in several provinces and must be regarded as cosmopolitan.

The gradual transitions between the provinces are important in supplying evidence for continuity of plates in the Cambrian. In Europe and North Africa there are gradual transitions from the East-Asiatic influence in Morocco to the pure Acado-Baltic in Southern Scandinavia. The boundary fluctuated and the accumulated evidence indicates that there was a close communication through the area in the Lower and Middle Cambrian. This contradicts the split of Europe seen in many reconstructions (e.g. SMITH, BRIDEN, and BREWRY, 1973; BRIDEN, DREWRY, and SMITH, 1974) based on palaeomagnetic data. On the other hand, the sharp boundaries between the areas with Acado-Baltic and Pacific fauna found partly in eastern North America and North-West Europe indicates that these areas were separated by a Proto-Atlantic ocean during the Cambrian. In this case the biogeographical information is in accordance with the palaeomagnetic and other geophysical evidence.

The gradual transition from the Siberian province (or sub-province) to the Pacific one through Alaska can also be used as an indication of proximity of the plates. Similarly, the close relationship between the faunas of China, Australia, and Antarctica indicates that these areas were close enough to permit free faunal migrations all through the Cambrian. The climatic implications of the Cambrian faunal provinces are not immediately striking. It is generally assumed that the Acado-Baltic province had a colder climate than the others, based mainly on lithology, absence of Archaeocyathids, and scarcity of endemic forms. This seems to be true at least for the Lower Cambrian, but for the Middle and Upper Cambrian the wide distribution of the black shale facies in the Acado-Baltic province introduces an ecological bias which may have obscured some underlying climatic effects. The agnostids, which are the dominant trilobites in the Acado-Baltic province, are a cosmopolitan group. They were probably pelagic forms, which are best preserved in black shales. The reports of 'Acado-Baltic elements' in the other provinces refer in most cases to this cosmopolitan group, which is really not significant for any province. Since many of the agnostids were first described from localities in the Acado-Baltic province, they have been linked with this province without much biogeographical evidence.

In the Upper Cambrian, the olenids are even more dominating in the Acado-Baltic province than the agnostids in the Middle Cambrian. The olenids are not equally cosmopolitan in the Upper Cambrian, even if they are met with especially in the East-Asiatic province, but they become much more so in the Tremadoc, which climatically and biogeographically resembles the Cambrian more than the Ordovician.

If these two, basically cosmopolitan, ecologically specialized groups are subtracted, the Upper Cambrian Acado-Baltic faunas are extremely meagre, and less so but still of much lower diversity than the other provinces in the Middle Cambrian. This low diversity is certainly to great extent caused more by the special ecological conditions than by a marked difference in climate. There are no good indications of cold climates in the Acado-Baltic Middle and Upper Cambrian; the available evidence rather points to a warm climate, though perhaps less warm than in other provinces.

In the Lower Cambrian, there is more evidence for climatic zoning, based on lithology, presence of bioherms and biostromes, and on general faunal diversity. This indicates a colder climate in the Acado-Baltic province than in the others. The East-Asiatic province may have had a warmer climate than the Pacific one, though widespread thick carbonates also in the latter indicate that the difference was not great, and for both provinces the climate would correspond to that found in low latitudes today.

2.5. Summary of the Cambrian climate

The widespread indications of stagnant oceans and the few, and diffusely limited, faunal provinces suggest a uniformly warm climate throughout the Middle and Upper Cambrian. The differences in distribution of the faunal provinces, such as the occurrence of the Acado-Baltic *Paradoxides* fauna in Northern Siberia and North Africa, and the East-Asiatic fauna in the latter region in the Upper Cambrian, may indicate that the climate was warmer in the Upper than in the Middle Cambrian, but the evidence which is based on the assumption of a comparatively cold climate in the Acado-Baltic province is not strong enough for definite conclusions to be drawn.

Conditions in the Lower Cambrian are considerably more complex and controversial. The Basal Cambrian glaciation is found in regions which later in the Lower Cambrian had faunas belonging to all three faunal provinces. It may be an indication of an undifferentiated cosmopolitan low Lower Cambrian fauna that *Fallotaspis tazemourtensis* is found not only in Morocco, but also in Siberia and California. The presence of archaeocyathid reefs in both the Pacific and East-Asiatic provinces (and in North Africa) indicates a mild climate, but this does not make sense when compared with reconstructions of palaeolatitudes (cf. Fig. 2). Even if a nonactualistic explanation (like that of S. E. WILLIAMS, 1974, 1975) is resorted to, it must be admitted that at present we do not have a good explanation for the distribution of climatic zones in the Lower Cambrian. In the Acado-Baltic province there are indications of cold climates (LINDSTRÖM, 1972). The observations reported do not indicate that the area (Scania) was glaciated, but only that sea ice was present. Such sedimentary structures are formed every hard winter in the Baltic and southern part of the North Sea, and do not necessarily indicate an arctic climate. Compared with the currently accepted palaeolatitudes for the Cambrian, it is striking that the cold-water indicators from Sweden are at lower latitudes than the presumably warm-water carbonates of the Mediterranean area.

3. Ordovician climates

3.1. Ordovician faunal provinces and plate movements

Ordovician faunal provinces have been identified and much discussed during the last 100 years. Recent summaries which cover the fauna in general have been given by SPJELDNAES (1961), BURRETT (1973), and ROSS (1975), and detailed studies covering brachiopods have been given by WILLIAMS (1969, 1973) and trilobites by WHITTINGTON and HUGHES (1972, 1973). Even if the Ordovician faunal provinces are easier to identify and more distinct than those of the Cambrian and Silurian, the uneven quality of the data makes a quantitative analysis which would relate to recent faunal provinces and their climatological and ecological background rather difficult. Even major groups are not well known, and the level of taxonomic sophistication in descriptions from different areas is too different for raw data, such as the genera described in the *Treatise on Invertebrate Paleontology*, to be used directly. Even such simple discrimination as between infauna and epifauna—highly necessary for biogeographical analysis—is not obtainable for large elements of the Lower Palaeozoic invertebrates. Although there are interesting ecological and faunal analyses by some authors (cf. BRETSKY, 1970; ZIEGLER, 1965), such modern studies cover only a small fraction of the areas and faunas in question.

An analysis of the Ordovician (and other Lower Palaeozoic) faunal provinces must therefore be based partly on quantitative studies of selected animal groups, or parts of them, which have been studied in a modern way on a world-wide basis. This will have to be supplemented and extrapolated with semi-quantitative methods (biological and sedimentological) of the type used by SPJELDNAES (1961) until more satisfactory data are available. The picture presented here is therefore provisional, and much detail will certainly be added to it in further palaeontological and palaeoecological studies.

In determining the direction of faunal migration, the central problem is to find the origin of characteristic genera and families. There has been a tendency to regard a taxonomic unit as having originated in the area where it was first described—a fact which may easily lead to serious misinterpretations. There is also a biologically plausible hypothesis (NEUMAN, 1972) which suggests that many of the new genera and families of Ordovician organisms originated around tectonically unstable volcanic islands in the Proto-Atlantic. These 'centres of origin' were small areas to begin with and may easily have disappeared completely. Their remains are normally small, isolated, and often found in metamorphic terrain, where precise correlations will be difficult.

In some cases this could be helped by tracing the evolutionary and distributional history of critical genera or families. This method works well in some cases (cf. SPJELDNAES, 1967), but new data from previously unknown areas may cause considerable changes. In the examples provided by SPJELDNAES (1967), where the brachiopods *Christiana* and *Porambonites* were used, the *Christiania* history still holds well, but more recent studies have shown that the Porambonitids did perhaps not originate in the Baltic, but rather on one of the volcanic islands in the Proto-Atlantic (NEUMAN, 1972).

3.1.1. The North Atlantic

The best known, and best defined, Ordovician faunal provinces are the ones on both sides of the North Atlantic. The Anglo-Baltic one may be divided into a Baltic subprovince, based mainly on the shallow-water carbonates of the platform, and the Anglo-Scandic with a more 'geosynclinal' environment with more soft-bottom faunas and higher rates of sedimentation. In North America, there is a similar division between the Appalachian faunas, where soft-bottom forms are important, and the cratonic faunas from the continental interior. In both continents an internal gradation due to slight differences in latitudes and climate can be seen.

Therefore, it may be possible to identify and estimate the rate of exchange of faunas between provinces and plates fairly exactly, but it is normally very difficult to define, and especially to quantify, the direction of movements. It is now well documented (WILSON, 1966; HARLAND, 1967; HARLAND and SAYER, 1972) that there are parts of the original plates on the 'wrong' side of the present Atlantic, because the recent, post-Jurassic split did not follow the trace of the old Proto-Atlantic. In North America, western Newfoundland and other parts of the Maritime Provinces and parts of north-eastern New England belong to the Anglo-Baltic province (with certain admixtures from what are supposed to be isolated volcanic islands acting as centres of dispersal). Florida, parts of Newfoundland (DEAN and MARTIN, 1978), and parts of the metamorphic Piedmont in south-eastern North America belong to the Mediterranean province. The amount of 'European' plate material in this region is not well known because of the metamorphism.

In North-West Europe, it has long been known that most of Scotland, and parts of the Trondheim region, North Norway, and Spitsbergen, belonged to the American plate, due to their Pacific-American faunas in the Cambrian and Ordovician.

Even if the displaced areas are well documented, their precise delineation and size is still very much in doubt, and the exact position of the old suture between the plates is difficult to find. This may partly be due to extensive metamorphism, but also to the possibility that the rather unmetamorphic, and hence dateable, sediments 'spilled over' from one plate to another by large-scale thrusting or by décollement folding.

The closing of the Proto-Atlantic is nicely demonstrated by the rate of faunal exchange between the plates. As mentioned above (Section 2.4) the provinces were well isolated during the Cambrian and the same is the case in the lower Ordovician. Not only are the faunal provinces rather distinct, but—as in the Cambrian—the North American plate shows also sedimentary evidence indicating an equatorial or at least low latitude climate, whereas the North European plate is estimated to have had a boreal climate. As indicated by JAANUSSON (1972), even the widespread Baltic lower Ordovician carbonates do indicate a moderate climate, and not a warm one. LINDSTRÖM (1972a) has reported ice-marked sand grains from the lower Ordovician of Scandinavia, but there are no other indications of a really cold climate in the region. The separation in faunal composition is found not only in the sessile benthos (brachiopods, bryozoans) but also in the vagile benthos and nektonic–planktonic forms (trilobites, cephalopods, and graptolites).

The isolation holds into the low middle Ordovician (Llanvirn–Llandeilo), and the first period of extensive exchange is found in *gracilis–peltifer* time. The cephalopods start earlier (SWEET, 1958)—as could be expected for organisms which were mobile, not only living, but also in terms of post-mortem dispersal. As far as we know, the exchange was mainly one-sided, with

American forms appearing as small numbers of individuals in Europe, but very few European forms coming into North America. The cephalopods continue to show this distribution pattern with only minor changes in rate through the rest of the Ordovician. The maximal exchanges of benthonic faunas is found later and the first arrivals are, as usual, difficult to date precisely. The characteristic faunas in Europe are those of the Derfel Limestone in Wales, the lower Chasmops Shale (4bα) in the Oslo Region, and the Kukruse in Estonia. The trilobites are almost completely exchanged, even if some endemism is still found. The two plates support only one trilobite faunal province for the rest of the middle Ordovician (the Remopleurid province of WHITTINGTON, 1973). The exchange of trilobites seems to be a two-way one, but with dominance of American forms migrating to Europe.

The brachiopod faunas show the same general picture, and, even if most of the exchange is an influx of American forms into Europe, there are also some genera which from their biogeographical history may be judged to be European genera migrating to North America. The best documented examples are *Christiania* and *Kullervo*.

In the middle Cardoc there was a short period of increasing endemism and, in *clingani* time, a new wave of exchange is found. In contrast to the first one, which may be termed the Kukruse wave, this later episode is definitely marked, both in number of genera and even more in the number of individuals, by an invasion of American forms into Europe. The reason for this is that this wave (the Vasalemma wave) not only comprises an exchange of mainly soft-bottom and vagile forms, as did the Krukuse wave, but it also marks the introduction of the carbonate facies with a much greater variety of epifaunas. This wave is therefore marked not only by an increased number of 'American' trilobites and brachiopods, but even more by bryozoans, and the first real occurence of compound corals and stromatoporoids. The calcareous algae, which until this time were rather restricted both taxonomically and in diversity in Europe, also expanded greatly, but the origin of these floras is not known in detail. They, or their ancestors, have not been described from the low or low middle Ordovician of North America.

In the low upper Ordovician, there appears to be another lapse into increased endemism, but less marked than before. The Porkuni wave, the third wave in Europe, at the very end of the Ordovician is remarkable in its increased diversity, high number of 'American' forms, and numerous bioherms and carbonate mounds with high diversities, often of endemics, and also in that the American faunas in the same time interval do not show any marked increase in diversity, and that the same period was marked by cold climate in the Mediterranean province, in spite of the indicators of warm climate in North Europe.

In the Ashgill, a large-scale faunal exchange between the Mediterranean and the North European provinces took place. It is best documented by the trilobite faunas (KIELAN, 1958) but can be identified also in other animal groups. This 'cosmopolitanization' of the faunas, which is completed in the low Silurian, may be related to the late Ordovician glaciation and is discussed in Section 4.1.

In the basal Silurian, there are virtually no differences between the faunas on both sides of the Proto-Atlantic, even at the specific level. This does not indicate that the continental plates were in actual contact, but from this time the Proto-Atlantic ceased to be a barrier for the distribution of shallow-water marine faunas. The rate and degree of exchange between the two plates can best be explained by supposing that the North American plate was permanently in a postion under the equator, whereas the European plate moved towards it from higher latitudes.

The fact that the exchange seems to have taken place in three waves, rather than continuously, was certainly not due to pulsations in the size of the Proto-Atlantic. A more reasonable explanation is that the waves are due to a complex interplay between climatic changes and alternations in the pattern of oceanic circulation, due to the different position of the continents (mainly on the European plate) in relation to the general circulation pattern. The details of this are difficult to make out at present (although a bold attempt was made by Ross, 1975), as our knowledge of the precise directions of plate movements is incomplete and the sizes and shapes of the continental parts of the plates are not known with certainty.

3.1.2. Europe and North Africa

The relations between North Europe and the Mediterranean province, including Africa, have been differently reconstructed. Most of the attempts have relied heavily on palaeomagnetic data. As discussed in Section 1.2.3, these data are rather uncertain and may be grossly in error. In most of these attempts (BRIDEN, DREWRY, and SMITH, 1974; SMITH, BRIDEN, and DREWRY, 1972; WHITTINGTON and HUGHES, 1973), North Europe is separated from the Mediterranean region by an ocean (Proto-Tethys) and, in some reconstructions, an ocean is also introduced between the Mediterranean region, and the African continental block.

Throughout most of the Ordovician there was little faunal exchange between North Europe and the Mediterranean region, and this may have contributed to the suggested separation. A more detailed study of the boundary between the two faunal provinces and their development in time (DEAN, 1966; SPJELDNAES, 1967) reveals that the boundary does not indicate that the provinces were geographically separated, but that the boundary between them was a climatic one. Especially in the Llanvirn and Llandeilo, there are many Mediterranean forms in Wales and adjoining parts of England, with decreasing number in younger beds and northwards. In a monograph on the brachiopods of the Shelve inlier (WILLIAMS, 1974), eight genera of Mediterranean affinities are recorded.

In Belgium, the Llanvirn and Llandeilo are typically Mediterranean, with the *Neseuretus tristani* fauna and a similarity with Bohemia down to specific level in the trilobites. In the Caradoc and Ashgill the fauna is entirely of Anglo-Baltic type, as evidenced by the brachiopods, bryozans, and trilobites (SHEEHAN, 1975; and the author's unpublished studies). It is highly unlikely that a Belgian 'microcontinent' moved from the Mediterranean region to North Europe between the Llandeilo and Caradoc, and the obvious conclusion seems to be that the two regions were part of the same plate or association of plates. As indicated by SPJELDNAES (1967), the movement in the Tethyan area does not seem to resemble that of large-scale plate movements, but does fit much better into the model suggested by ILLIES (1974), with small-scale movements (including rotation) of crustal units, one or two orders of size smaller than the ordinary plates.

The conditions in the Carnic alps are complicated by tectonics, but the Ordovician is found in different facies, some of which appear to be dominantly Bohemian and others with relationships to the Mediterranean (especially Sardinia, South France, and parts of Spain), and also a number of Anglo-Baltic elements which are otherwise rare in the Mediterranean province, such as *Anisopleurella* and *Christiania*. The peculiarities of the Carnic Ordovician can only be solved by a detailed analysis of the numerous faunas which have been recovered during the intensive work in the region. Three models can be suggested. One is that the various facies areas have been tectonically telescoped, but this does not solve the problem of the appearance of the Anglo-Baltic forms. It may be that there was an interfingering of beds of slightly differing age, similar to the situation in Belgium, or, thirdly, a Gulf-Stream situation, where a current of warm water brought in the Anglo-Baltic faunal elements and partly separated the Bohemian region from the rest of the Mediterranean one.

Graptolite faunas of definite Scandinavian stamp have been found in Rügen (JAEGER, 1967) and typical shelly Baltic faunas are met with in Poland (MODLINSKI, 1973) and adjoining parts of Russia. In Podolia (NIKIFOROVA and PREDTECHENSKIJ, 1968; SOKOLOV, 1972; SHULGA, 1972) a very reduced Ordovician consisting mainly of calcareous sandstone is met with, but the

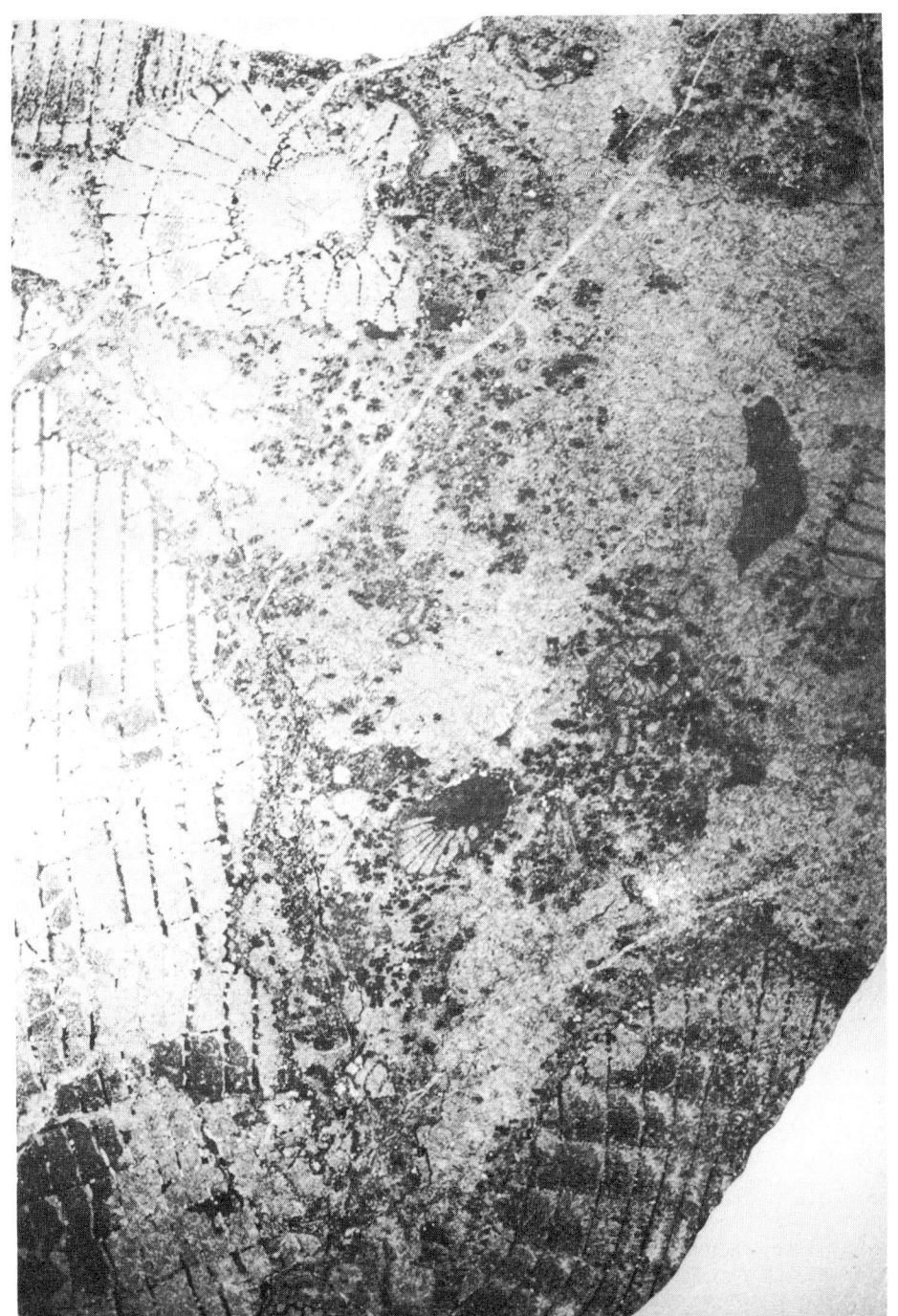

Plate 1. Carbonate structure of an Archaeocyathid reef. The thin section (magnification is 2.1 ×) shows typical carbonate mound structures, indicating deposition in warm water. From a small bioherm at the village of Amagour, Southern Morocco. The locality has been described by CHOUBERT and DEBRENNE (1964), and a section close by is given in CHOUBERT (1952).

Plate 2. Glacial striation from the upper Ordovician of southern Algeria. The striation is part of a very large surface, with striae from south-east to north-west (from bottom to top in picture). The locality is in the western part of Tassili de Tafassasset, the east-north-eastern side of the Hoggar Massif, and has been described by IAP (1970, p. 94) as no. J 3.

fauna ia definitely Baltic. It is interesting that this development is found on the southern side of the Ukrainian shield, and, according to PATRULIUS and IORDAN (1974), it continues into Roumania. In Bulgaria, the Ordovician is developed in typical Mediterranean facies (SPASOV, 1958; TENCOV, SPASOV, and JANEV, 1963) and there are no intermediate faunas known in this part of Europe. In Turkey, (DEAN, 1970–73, 1975) there was an influx of Baltic forms during the Arenig, but otherwise this area, including the island of Kos (DESIO, 1930), has only Mediterranean faunas in the Ordovician.

An analysis of the boundary relation between the Mediterranean and the North European provinces indicates that the relationship was very different from that between the two plates in the North Atlantic (Proto-Atlantic). The obvious explanation of the differences seems to be that Africa and Europe were part of the same plate, or complex of plates, and that there was no oceanic barrier between the regions throughout the Early Palaeozoic. The sharp boundary between the two provinces in Eastern Europe, which contrasts with the interfingering found in Western Europe and Turkey, may be due to some plate movements (crustal shortening or lateral movements) in connection with the Hercynian or Alpine movements in Europe.

It should be noted that Bohemia, which is the classical area (cf. HAVLIČEK and MAREK, 1973) for both fauna and stratigraphy in the Mediterranean province, really is a marginal area with a comparatively low faunal diversity. The apparently high number of species is due, as is the well known and detailed stratigraphy, to diligent work by palaeontologists and stratigraphers in the area, rather than to an original richness of the faunas. The lack of epifaunas is especially striking and is particularly well seen in the upper Caradoc, where numerous carbonates are found in much of the Mediterranean province accompanied by rich epifaunas (bryozoans, brachiopods, and echinoderms). The corresponding beds in Bohemia (Bohdalec beds) also contain bryozoans, but most of them are specialized forms, adapted to life on soft bottoms and with low diversity. When, and if, the other regions in the Mediterranean province are studied in the same detail, and with the same precision, as the Bohemian one, the number of species will probably be much higher.

In the central part of the Mediterranean province, new studies by DESTOMBES (1963, 1966, 1967, 1971) on stratigraphy and trilobites and by HAVLIČEK (1971) on brachiopods from Morocco have made this region the type area for the Mediterranean Ordovician. In Spain and Portugal, numerous workers (HAMMANN, 1971, 1974, on trilobites; HARTEVELDT, 1970, from the Pyrenees; PUSCHMANN, 1968, from Catalonia; SACHER, 1966, and CARLS, 1974, from Aragonia; TAMAIN, 1971, 1973, TAMAIN and others, 1970, CHAUVEL and others, 1969, CARRÉ and others, 1970, and SAUPÈ, 1971, from southern Spain; GIL CID, 1972 from Montes de Toledo; RADIG, 1962, JULIVERT and others, 1968, and JULIVERT, MARCOS and TRUYOLS, 1973 from Asturias; and ROMARIZ and DINIZ, 1962 from Portugal), have established the well known pattern of an Ordovician sequence, basically starting with a supposedly Arenig sandstone series (the Armorican Quartzite), followed by a thick, and sedimentologically varied shale–sandstone series with the *Neseuretus tristani* fauna, mainly of Llanvirn–Llandelio age but possibly reaching up into the Caradoc. On top of this a shale-carbonate sequence is found, locally with a rich fauna of bryozoans, brachiopods, and echinoderms, indicating an upper Caradoc and Ashgill age (HAFENRICHTER, 1979). A peculiarity of this series in the Iberian Peninsula is that it is accompanied by mineralization in several areas (antimony in Galicia, GUILLOU, 1969; mercury—in the beds immediately above—in Almaden, Andalucia). Some of the mineralizations are probably later, due to the filter effect of the carbonate sediments, but some of them may be contemporaneous (SAUPÉ, 1967; MAUCHER and SAUPÉ, 1967). Some indication of the upper Ordovician glaciation is also found at or near the Ordovician–Silurian boundary (cf. Section 3.2). The Silurian follows rather abruptly, often with a distinct hiatus, on the Ordovician and consists mostly of dark shales with normal Llandovery graptolite faunas.

In recent years this picture has been supplemented on three points. Tremadoc fossils have been identified from Sierra de la Demanda by COLCHEN (1968) and by COLCHEN and HAVLIČEK (1968), and from the Iberian Mountains by JOSOPAIT (1972) and HAVLIČEK and JOSOPAIT (1972). The brachiopods reported show strong resemblance to those of Bohemia and Morocco, though the trilobites are less typical, some, such as *Ceratopyge*, being known also from the Baltic region. From Portugal MITCHELL (1974) reported that the lower part of the volcano-detritic and carbonate sequence—which previously was held to be equivalent to the upper Caradoc and Ashgill of the rest of the Mediterranean province—contains fossils of lower Caradoc age. This is of considerable climatological interest as it would indicate that the warm period, indicated by rich and diverse shelly faunas, started earlier and lasted longer, at least in Portugal. It would also indicate that this, comparatively, warm-water facies spread slowly from the north and north-west as there are no signs of such a warming in the well known Moroccan Caradoc. This will necessiate a re-study of the lower part of the shale–carbonate sequence all over the Mediterranean region, in order to see if these faunas have a wider time span than supposed.

Finally, JULIVERT and TRUYOLS (1972) reported Silurian conodonts (determined by M. LINDSTRÖM) from the upper part of the carbonate series at Cabo Peñas in Asturia. Since the lower part of the carbonate sequence here is undoubtedly Ordovician, this may indicate that carbonate deposition continued uninterruptedly during the upper Ordovician glaciation in the surrounding areas. The limestone resembles the Calcaire de Rosan from Brittany, both in lithology and stratigraphical setting, and contains a rich and diversified fauna including rugose and tabulate corals, especially in its lower (Ordovician) part. The fauna is undoubtedly the 'warmest' known from the Mediterranean upper Ordovician and has a considerable climatological interest.

In Brittany and Normandy, the Armorican Quartzite is followed by a series of dark shales and sandstones, with a great variety of sedimentary structures. The fossils occur mostly in thin bands and lenses, and are often much influenced by specialized environments. This makes detailed correlation difficult, but work by BABIN (1966), BABIN and others (1968), BABIN and MELOU (1972), DORÉ (1972), HENRY (1966, 1969), LINDSTRÖM, RACHEBEUF, and HENRY (1974), LINDSTRÖM and PELHATE (1971), MELOU (1971, 1973, 1975), PHILLIPOT (1963), PILLET and ROBARDET (1969), ROBARDET (1973), and others has cleared up this difficult stratigraphy in an admirable way, as well as that of the younger Ordovician beds and the transition to the Silurian. Above the dark shale–sandstone complex there is a series with shale, sands, and some carbonates, with an upper Caradoc to Ashgill fauna. In Normandy there are glacial beds in this sequence (cf. Section 3.2) and in Brittany there are basaltic volcanic beds, partly as extrusives and partly as very shallow intrusives (BARROIS, 1890; LUCAS, 1940). The tuffs of this volcanic series are partly highly fossiliferous and on top of the series the Calcaire de Rosan is found. This is a massive limestone, with a fauna including bryozoans, brachiopods (including *Nicolella*), corals, and conodonts (LINDSTRÖM and PELHATE, 1971). The age is regarded as upper Ordovician, and it is overlain by Silurian graptolite shales which belong to the upper Llandovery.

The type of fauna found in the shale–sandstone sequence indicates comparatively cold water (cf. SPJELDNAES, 1961) but the Calcaire de Rosan indicates warmer water. In spite of its lens-like shape the limestone does not consist of bioherms and there are no reports of definite indicators of warm-water carbonate (micrite or oolites). The limestone is rather recrystallized, but both the massive limestone and the carbonate cemented parts of the tuffs seem to be composed of biogenic carbonate sands, which are rather insignificant climatologically, even if they do not appear to occur under arctic temperatures (cf. LEES, 1975; LEES and BULLER, 1972).

The peculiar concurrence of comparatively 'warm' carbonates with glacial beds (ice-rafted material) in the same general area is typical of the upper Ordovician glaciation (cf. Section 3.2), and so is the frequent occurrence of carbonate boulders in the glacial beds. Fossiliferous carbonate boulders have been reported from the glacial beds in Normandy by DORÉ and LE GALL (1972), but, since the fauna has not been described in detail, it is not known whether the carbonates are related to the Calcaire de Rosan or other known local carbonates, or if they came from faunal assemblages unknown in the region, such as that described by BLUMENSTENGEL (1965) from the German Lederschiefer.

3.1.3. Other regions

The Middle East is an important transitional region. A summary of the older literature is given by FLÜGEL (1964) and a good, recent review is that of DEAN (1975). As mentioned in Section 3.1.2, Turkey has a Mediterranean fauna, except for an incursion of typical Baltic forms in the Arenig. The route of immigration of these forms is unknown, and in the area with Baltic fauna closest to Turkey—Podolia and Roumania—the Arenig is missing.

In Syria, Mediterranean shelly faunas are also found, but to the south mainly unfossiliferous sandstones, with trace fossils and very sparse faunas are met with, cf. BENDER (1963) and SELLEY (1972) from Jordan, HELAL (1964, 1965) and POWERS and others (1966) from Saudi Arabia, WEISSBROD (1969) from Israel and Sinai, and TSCHOPP (1967) from Oman. The sediments are close to those found in the African shield, and the area must be regarded as an extension of the southern margin of the Mediterranean province.

It is interesting that the sedimentary structures in these sandstone sequences are differently interpreted. SELLEY (1972), in the typical British tradition, interpreted the sedimentary environment as mainly a fluviatile one. Both the trace fossil assemblages and the sedimentary structures are very close to those reported from Sahara, where BEUF and others (1971) have interpreted the same structures to indicate a shallow marine (partly tidal) environment, and use of the trace fossils including the trilobite tracks (*Cruiziana*) as supporting evidence for this. Similar discrepancies between the schools of sedimentology are reflected in the interpretation of the Old Red Sandstone facies, where most British sedimentologists such as ALLEN, DINELEY, and FRIEND (1967), ALLEN (1974), and FRIEND (1965, 1966, 1969) regard most of the beds as fluviatile–estuarine, whereas BURROLET, BYRAMJEE, and COUPPEY (1969) interpret the Old Red Sandstone of north Scotland as deposited in a large marine or brackish water body.

These discrepant interpretations may in part be due to the difficulties caused by the non-actualistic conditions in the Early Palaeozoic (lack of terrestrial vegetation, and possible deviation in the circulation pattern and obliquity), which make it difficult to reconstruct the environments because of the absence of precise recent analogues.

Farther east in Iran (cf. VATAN and YASSINI, 1969; DEAN, 1975) the faunas are mixed, and also not very well known. The Cambrian shows relations to the Mediterranean region and this is also true for the Tremadoc (especially as demonstrated from Afghanistan by WOLFART, 1970, 1970a), where there are also some elements from the East Asiatic province. Baltic elements are found not only in northern Iran but also in the east of the country and in Karakorum (GORTANI, 1934). The fragmentary information given by HUCKRIEDE and others (1962) from the Kerman region (southern Iran) does also suggest Baltic affinities in the Ordovician, and there is definitely no Mediterranean aspect of the faunas.

Generally, there seems to be a gradual transition from the pure Mediterranean fauna in the west, through those with Baltic elements, into the East Asiatic province. In the area described by RUTTNER and others (1968), the trilobites are a peculiar mixture of Baltic, Mediterranean, and Chinese forms. The brachiopods and bryozoans, which have been studied by the author, also indicate a mixed fauna. The brachiopods consist partly of endemics with Baltic affinities, cosmopolitans (*Nicolella*), and Mediterranean forms. The bryozoans are typically Baltic–American, belonging to the 'Vasalemma' wave. It is interesting that the bryozoans are entirely different from the Mediterranean ones, but this may be due to ecological differences, and perhaps also to slight differences in age (the maximal development of the Mediterranean bryozoans is found in the Ashgill, rather than in the Caradoc).

Because of the few, and only partly described, faunas involved, it is premature to express the climatological and biogeographical information into arguments for plate movements in this region. It should be noted that some of the 'Baltic' elements may also have come by the way of Kazakhstan.

In Kazakhstan (KELLER, 1954–61; NIKITIN, 1972–73; APPOLONOV, 1968, 1975; APPOLONOV and others, 1968; GNILOVSKAJA, 1972), the Ordovician shows relations both to the Baltic region and even more to North America. Because of the varied facies and the detailed studies in this region the periods of faunal exchange may be outlined, but the routes of migration are still unknown. One of the crucial questions is whether the Baltic, and partly Appalachian, elements found in the Middle East, Himalaya, Karakorum, and Altai-Tien-Shan (SOKOLOV and YOLKIN, 1978) have arrived via Kazakhstan or along an entirely different path.

The Siberian Platform (SOKOLOV and TESAKOV, 1975) and the Far East (of the Soviet Union) (CHUGAEVA, 1964, 1973; ROZMAN and others, 1970) show especially relations to the equatorial part of North America (ASTROVA, 1965; ROZMAN, 1968), and seem to be linked to it through Alaska and Arctic Canada. There are few modern surveys of China, although SOKOLOV (1962, 1963) has summarized the older literature and BURRETT (1974) has attempted an analysis of the faunas and their relationships. MU and others (1973) and other more recent Chinese literature add to the endemism of the area, and bring some interesting information on the boundaries of the faunal provinces, especially in the Himalayan region.

Australia apparently had warm climates all through the Ordovician and the faunas are closely related to those of China, but also to North America, and, to lesser degree, to Kazakhstan (WEBBY, 1971). Most of the 'Baltic' elements reported seems to be cosmopolitan ones, labeled Baltic because they were first described in North-West Europe. In all these regions (Siberia, Kazakhstan, Australia, and China and other parts of the Far East) a warm climate seems to have prevailed during the Ordovician, and they may appropriately be termed the equatorial continents, in contrast to Europe, Africa, and South America. It is just the uniformity of climate which makes it difficult to identify migration routes, and to indicate the position of these continents in relation to one another and to the other continental masses in the Ordovician.

In conventional reconstructions (e.g. SPJELDNAES, 1961; WHITTINGTON and HUGHES, 1972) almost all continents are placed in one hemisphere, with the other covered with oceans. The possible presence of an antiboreal fauna in the Klamath Mountains of California (see below) may complicate this issue considerably, and the longitudes assigned to the equatorial continents in Figs. 2, 3, and 4 are little better than arbitrary, and may be considerably modified when more detailed information from these continents can be assembled and analysed.

From the Klamath Mountains in California, ROHR, BOUCOT, and POTTER (1974) have reported an Ordovician fauna of great importance. The age is given as Ashgill, although an upper Caradoc age is not excluded for part of the assemblage. The fauna (mainly brachiopods and trilobites) is typical Anglo-Baltic. Some of the genera do occur in the Appalachian middle or upper Ordovician, but are unknown in central North America. The presence of such a fauna can be explained in two ways. Either the rocks (at present only found as boulders in a conglomerate) are draped around North America from the present eastern side by a complex series of plate movements, or they indicate the existence of an Ordovician 'antiboreal' fauna, on the opposite side of the equator from the typical Anglo-Baltic faunas.

The latter hypothesis is strengthened by the occurrence of some of the same brachiopods (e.g. *Christiania*) in Alaska (ROSS and DUTRO, 1966; BRABB, 1967; PATTON and DUTRO, 1969), but until the fauna has been studied in greater detail it is impossible to reach a more precise conclusion. This problem is of great importance for Ordovician biogeography, as an 'antiboreal' fauna of this type would lead to more possibilities both for explaining plate movements in the other hemisphere and as an alternative source for Anglo-Baltic faunal elements on other plates, such as the Australian, South Asiatic, and Kazakhstanian ones. The presence of antiboreal faunas would indicate that an equatorial convergence was in existence during the Ordovician. This requires that the deep waters of the oceans were cold, and therefore that polar ice-caps were present.

The immediate aspect of the Klamath Mountain fauna, and those reported by ROSS and DUTRO (1966), BRABB (1967), and PATTON and DUTRO (1969) from Alaska, indicates that the difference between it, and the 'boreal' Ordovician was less—in terms of generic and specific differences—than that between the recent boreal and antiboreal faunas, but this may be an artefact, due to the lack of detailed studies of the Klamath fauna. At present the lack of data from this fauna is the greatest single source of uncertainty as to the position and inter-relationship of the equatorial and 'antipodal' continents of the Ordovician. A detailed analysis of the material is greatly needed.

In west-central Argentina, the Cambrian belongs to the Pacific province (cf. Section 2.4 and BORELLO, 1971) and in the lower Ordovician the faunas have relationships to both the North European and Australian provinces. This is distinctly shown in the trilobites, where WHITTINGTON and HUGHES (1972) do not report any good representatives of the *Selenopeltis* province from this continent and SERPAGLI (1974) reports conodonts of Baltic affinities from Argentina. SERPAGLI (1973) also reports micritic carbonates, which are interpreted as indicators of a warm climate. His rocks are certainly very different from the few carbonates met with in the Mediterranean lower and low middle Ordovician. The receptaculid reported by NITECKI and FORNEY (1978) also supports a warm-water interpretation of the lower Ordovician sediments from parts of South America.

Especially the faunas described from Venezuela (SHELL and CREOLE, 1964) indicate a fairly warm climate and affinities with the Acado-Baltic Province, and even more with the Appalachian one. The faunas described range into the Caradoc, or younger, and are spectacularly different from anything else in South America.

This distribution of the faunas does not correspond well with the current reconstructions of Gondwanaland. In most such reconstructions, the pole will fall somewhere in the northern part of the Brazilian shield, and would therefore put all the Lower Palaeozoic of South America in high latitudes.

From the uppermost Ordovician (or Silurian) the Zapla tillite is described as a glacial sediment. The only described Ashgill fauna from Argentina, by LEVY and NULLO (1974), is also a cold-water type, with a strong resemblance to the Mediterranean Ashgill faunas, even if an endemism is reported, at both the species and generic level. The contrast between this fauna and the roughly contemporaneous warm to boreal fauna from Venezuela (SHELL and CREOLE, 1964) is striking, and may indicate large-scale discordances in palaeogeographical position.

If the admittedly somewhat meagre faunal evidence from South America should be interpreted in terms of plate tectonics, it would indicate that South America was distinctly separated
from Africa in the Cambrian and lower Ordovician, and that the distance decreased during the
Ordovician, until it was at fairly high latitudes in the upper Ordovician (BIGARELLA, 1973;
BERRY and BOUCOT, 1972). DEVINGE's (1971, 1972) studies support such movements, even if
they are only marginally based on evidence from the Ordovician. The model outlined here
resembles that of the North Atlantic, but the evidence for timing is much more diffuse due to the
lack of Early Palaeozoic sediments along most of the adjoining parts of the two continental
blocks (South America and Africa). The considerable structural similarity even in the Pre-
Cambrian, and the late consolidation of great parts of the African shield, may indicate that the
original continents were rather different in size and shape from the present ones. Due to the lack
of observations it is, at least at present, extremely difficult to obtain a more precise picture of
these movements.

The fact that upper Ordovician glacial beds are found in South America also does not
necessarily indicate that the two continental blocks were in contact. It is difficult to envisage how
a glaciation could be established climatologically in the interior of a continent of the size of
Gondwanaland, and it may very well be that the southern Proto-Atlantic acted as an activator
and source of moisture for the upper Ordovician glaciation, in a similar way to the function of the
North Atlantic and Arctic Ocean during the Pleistocene glaciations in the Northern Hemisphere. This would correspond to the presence of synchronous, but separated, ice sheets in
North Europe and North America during the Pleistocene. The distribution of the marine faunas
of the Malvinocaffric faunal province in the Silurian and Devonian (cf. BOUCOT, 1975) also
indicates the presence of marine seaways across the postulated Gondwana continent, even if
there is no good evidence to indicate whether these seaways were oceanic or intercratonal.
TEICHERT (1974) also argues that 'Gondwanaland' was not a continuous continent even if it was
a single plate.

The suggested movements of South America (Figs. 2, 3, and 4) are not, in contrast to the more
recent ones, in unison with those between North America and Europe. In the Ordovician,
Europe moved from higher to lower latitudes, towards North America, which was comparatively stationary at equatorial latitudes, whereas South America may have moved from low to
high latitudes, towards Africa, which was at polar latitudes. The geometrics of this clearly
indicate that the movements had a considerable component of rotation.

The movements suggested here are not in contradiction to the few palaeomagnetic results
from South America. The best ones, by THOMPSON (1973), suggest that, on the basis of a
conventional reconstruction of Gondwanaland, the poles were far to the east of those suggested
for Africa. Thus a polar curve would indicate a polar movement from the north-east towards the
'normal' position of the pole at the end of the Lower Palaeozoic. Even if both the palaeomagnetic and biogeographical data are very imprecise in this area and time interval, the available
information does not contradict a movement of South America of 40–50° of latitude towards the
pole from the Lower Cambrian to the Silurian.

3.2. The upper Ordovician glaciation

Early work on climatic zoning (SPJELDNAES, 1961) and palaeomagnetic work (RUNCORN,
1959) gave clear indications that a pole existed in the central part of Gondwanaland in the
Ordovician, and that polar ice-caps were to be expected because of the pattern of climatic zoning
(SPJELDNAES, 1961). Shortly after (SOUGY and LÉCORCHÉ, 1963; DEBYSER and others, 1965;
BEUF and others, 1966) glacial deposits were identified in various parts of Sahara. The

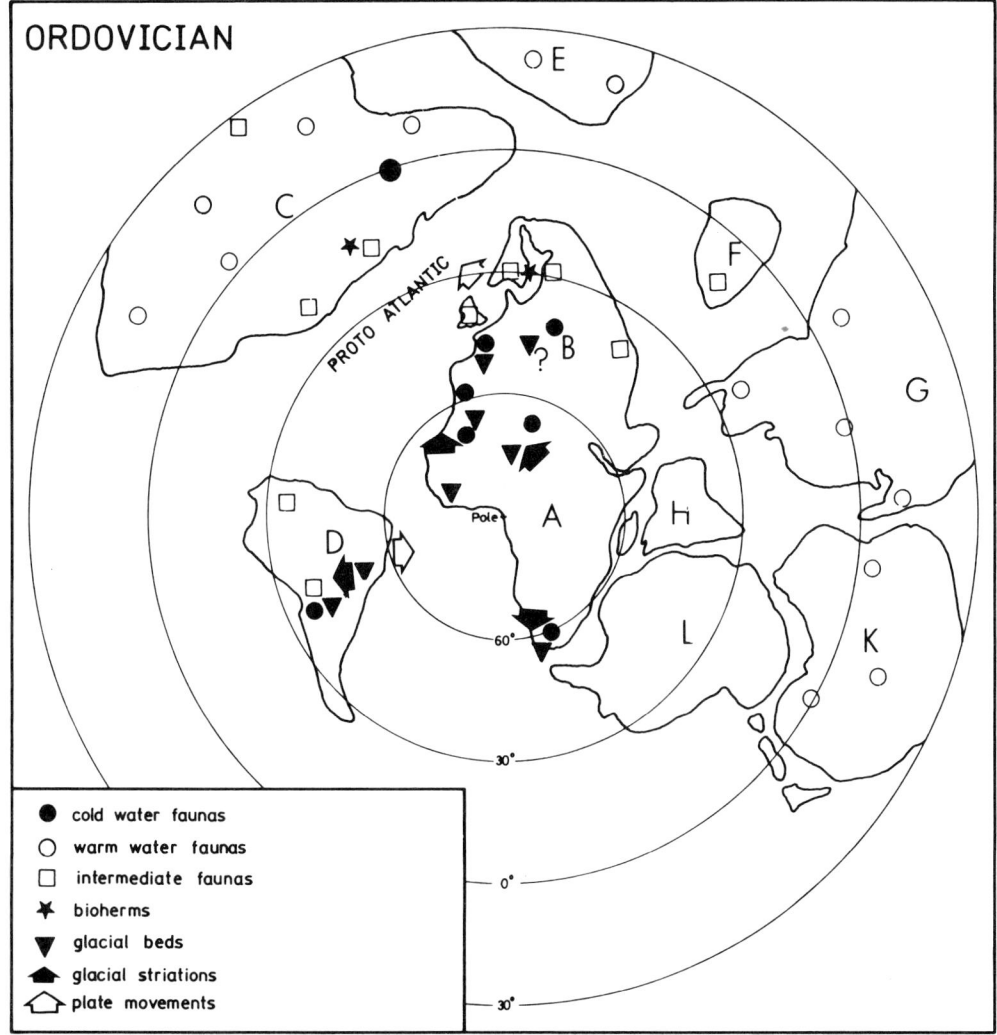

FIG. 3. A reconstruction of the palaeogeography and climatic zoning in the Ordovician. A double
signature indicates change in climate, not mixed faunas. For further explanation, see Fig. 2.

monograph of BEUF and others (1971) gives a rather complete picture of this glaciation in
central Sahara. Some details have been added by BIJU-DUVAL and others (1974) and a recent
summary has been given by ALLEN (1975). The evidence for glacial conditions is impressive,
with a wide array of criteria from glacial striations (frontispiece), through a variety of glacial
sediments, to large-scale glacial erosion and ice tectonics. Typical moraines are missing, due
to the uniform character of the underlying Ordovician sandstones and the very smooth
topography. Erratic boulders are mainly found in the ice-drop tillites in the glacial valleys (silte
vert). It is the marvellous exposures, and the excellent studies by BEUF and others (1971), which
make central Sahara the type area for this glaciation. The Ordovician glaciation has also been
identified in the Eglab Massif by ROGNON, BIJU-DUVAL, and DE CHARPAL (1972); in
Mauritania by GEVIN and others (1968), DEYNOUX and others (1972), and TROMPETTE (1972);
and in the Ougarta Mountains by ARBEY (1968). The studies by DESTOMBES (1968) in Morocco

are especially interesting as a precise dating (upper Ashgill) is given for the glacial beds. Glacial beds have also been reported by TUCKER and REID (1973) from Sierra Leone, where they record ice-drop tillites with carbonate boulders. Good evidence is given for the glacial nature of the sediments, but the dating is not too good. Because of the lithological similarity, and correlation with beds in Guinea which are overlain by graptolite shales (Llandovery to Wenlock), they are referred to the upper Ordovician. The glacial beds reported by DOW and others (1971) and BEYTH (1972) from Ethiopia are of uncertain age and might as well be Permo-Carboniferous.

In South Africa, the tillite from the Pakhuis Formation in the Table Mountain Group has been well documented by RUST (1969), and has been discussed later by e.g. BIGARELLA (1973), LOCK (1973), and VISSER (1974). Striated pavements and striated boulders are met with, but the most spectacular feature of the glaciation is the ice-tectonics indicating a movement from north to south. The age of the tillites is given by the fauna found in the overlying Cedarberg Formation, which is also interpreted as glacial (COCKS and others, 1970). The precise age is somewhat in doubt, either uppermost Ordovician or basal Silurian. The main reason for the uncertainty of the age determination is that the fauna is an endemic one, which in some respects foreshadows the Malvinocaffric fauna found in the southern continents in the Silurian and Devonian (cf. BERRY and BOUCOT, 1973, pp. 59–60).

In South America glacial beds are known from Northern Argentina, South Peru, and Bolivia (for detailed references, see BERRY and BOUCOT, 1972, and BIGARELLA, 1973). The evidence for a glacial origin of the Zapla tillite in southern Bolivia seems to be good, even if detailed and modern descriptions are lacking. Similar beds, also suggested to be glacial, partly by correlation with the Zapla tillite and partly by the presence of striated boulders, striated pavements, and general lithology, have been reported from northern Argentina, Paraguay, and Brazil. A summary of these sediments and a discussion of their origin was given by BIGARELLA (1973). It is not known whether the ice sheets reported from South Africa and South America were continuous with the one found in Sahara. The transport directions reported by RUST (1969) and BIGARELLA (1973) do not contradict this, but there is also no direct evidence for a continuous ice sheet and it may well be that there were several isolated ice masses. This would be easier to explain climatologically since the continuous ice sheet would be considerably larger than the present Antarctic one. It is difficult to envisage a meteorological model which could supply sufficient precipitation to the central part of a glaciated continent of this size. There may therefore have been isolated ice sheets in Africa and South America, separated by an ocean or an intercratonal sea from which the necessary precipitation was generated.

Glaciomarine deposits probably referable to this glaciation have been met with in Europe. The Orea Shale, which is common in parts of Spain was described by GREILING (1967) as a non-glacial sediment due to mudflows or similar processes. Others, such as PFEIFFER (1972), have classified it as glacial, probably more on stratigraphical position and petrology than by detailed sedimentological analysis. In hand specimens, the rock has many of the features of an ice-drop tillite, such as lack of bedding, dispersed clasts, and bad sorting. It also resembles many well documented Ordovician tillites in having clasts of partly fossiliferous dolomite from the underlying Ashgill carbonates. As pointed out by GREILING (1967), it lacks one important criterion for ice-drop tillites as it is not particularly stratigraphically continuous. The clast content varies considerably over short distance and it grades laterally into shales without clasts. The contact with the underlying carbonates is normally erosive but no signs of glacial erosion are found. The details of the contact do suggest a karstic erosion of the carbonate before the deposition of the Orea Shale (HAFENRICHTER, 1979). Glacial striation has not been found on the clasts, but most of them are also too small to develop striae. It cannot be excluded that the

Orea Shale is of glacial origin, and both its stratigraphical level—at the Ordovician–Silurian boundary—and its texture does indicate this. Definite proof for the glacial origin is lacking, however, and for the time being it must be listed as one of the doubtful cases.

The glacial striations reported from Sierra Morena by ARBEY and TAMAIN (1971) and TAMAIN (1972) are also somewhat doubtful. Striations alone are reported, with no supporting evidence. SCHERMERHORN (personal communication), who has seen the locality, is of the opinion that the striations are of tectonic origin. It is often extremely difficult to interpret striations of this kind, and if they are not completely unweathered and typically glacial, such as those in Plate 2, or supported by other evidence (tillites, ice erosion, ice tectonics) they must be looked upon with reservation.

The glacial beds from Normandy are typical ice-drop tillites, at least partly with gradual lower and upper contacts. The detailed analysis of the whole sedimentary setting by DANGEARD and DORÉ (1971) and ROBARDET (1973), and the discovery of glacial striations on clasts in the tillite by DORÉ and LE GALL (1972), provides convincing documentation for the glacial origin of the beds. Also in this region, fossiliferous carbonates are a conspicuous part of the boulder content of the tillite and it is not known whether the carbonates come from limestones in the region (correlatives of the Calcaire de Rosan), or if they belong to otherwise unpreserved formations.

The Lederschiefer in the German Democratic Republic has a texture and distribution similar to that of the 'Tillite de Feuguerolles' in Normandy, but the lack of good exposures and a slight metamorphism make it difficult to interpret the formation and its contacts. GREILING (1967) regarded it as non-glacial, but KATZUNG (1961) and PFEIFFER (1972) supplied evidence for a glacial origin. Both the stratigraphical position (just below the Ordovician–Silurian boundary) and the lithological similarities to well documented tillites tend to support a glacial origin, but it is not beyond doubt. The Lederschiefer also carries carbonate clasts. BLUMENSTENGEL (1965) has described part of the fauna, and it is interesting to note that the ostracodes show definite relations to Scandinavia and very little to the Mediterranean region, including Bohemia, which is rather close. This may have some palaeobiogeographical significance, but may also be due to our lack of knowledge of the Mediterranean upper Ordovician ostracodes. The only well known fauna, from the upper Caradoc Bohdalec Beds in Czechoslovakia, is somewhat older than the Lederschiefer.

In the Maritime Provinces of Canada there have been reports of glacial sediments, in Nova Scotia by SCHRENK (1972) and Newfoundland by MCCANN and KENNEDY (1974). Because of the complicated tectonics and some metamorphism, neither the age nor the glacial nature of these beds can be regarded as well established. In reconstructions of the Proto-Atlantic, the regions involved would be placed opposite Ireland or northern France and, although they may have been at lower latitude, the distance from Normandy with the well documented Tillite de Feuguerolles (DORÉ and LE GALL, 1972) is not too large. Mediterranean type faunas have also been reported from Newfoundland, e.g. by DEAN and MARTIN (1978). There is no reason to suggest special, large-scale plate movements and involve North Africa in these reconstructions. The areas where one might expect to find glacial beds, because of their proximity to North Africa and the Iberian Peninsula during the Ordovician, are in the subsurface of Florida and in the continuation of this region into the metamorphic terrain to the north.

No signs of a glaciation are reported from Australia (SHERWIN, 1975) or from the other 'equatorial' regions. A warm climate seems to have existed into the Silurian but there are often breaks in the stratigraphical sequence at or near the boundary.

The Ordovician glaciation was probably of antarctic type, based on a large, peneplaned continent under the pole (SPJELDNAES, 1973). In the central part of the glaciated area there are few or no sediments because the glaciation was mainly erosive and the low topographic relief resulted in a low sediment production.

Most of the glaciated continent was covered with mature Early Palaeozoic sediments, and most of the glacial sediments were formed by reworking of these. It is therefore in many cases difficult to identify them as glacial sediments, because they have an inherited maturity which is rather unusual in sediments of this origin.

One special feature of the Ordovician glaciation is the occurrence of carbonate rocks, sometimes of decidely warm-water type, at the same time and partly also in the same regions as the glaciated ones. Typically this is found in northern Europe, where biostromal limestones with warm-water carbonates are common in the uppermost Ordovician (the Porkuni beds in Estonia, bioherms in stage 5b in the Oslo Region, and the Keisley and Kildare Limestones in the British Isles). In southern Europe and North Africa the only horizon where carbonates are common is the uppermost Ordovician. In some of the glaciomarine beds carbonate boulders (DORÉ and LE GALL, 1972 from Normandy; BLUMENSTENGEL, 1965, from the German Democratic Republic) are a conspicuous part. The most striking example is in south-eastern Morocco (the Erfoud district) and parts of the subsurface in Algeria and Libya. Here biostromal, quartzose limestones with a very rich fauna, mainly of bryozoans, are found as 'islands' or 'driftless' areas in the otherwise glaciated region. This recalls the explanation given by G. E. WILLIAMS (1974) for the climatic peculiarities of the basal Cambrian glaciation, even if the discrepancies are less pronounced in the upper Ordovician. Other possibilities may be a rather long duration of the Ordovician glaciation. The sections with Ordovician glaciations may have considerable gaps and there may have been several episodes which will be difficult to discriminate.

The dating of the Ordovician glaciation is still somewhat doubtful. As shown by DESTOMBES (1968) from Morocco, the age in the peripheral part of the glaciated area is Ashgill, perhaps even upper Ashgill. The ice-drop tillites in Normandy and the German Democratic Republic are also young Ordovician, even if their age cannot be determined with the same precision as the Moroccan ones. The glacial beds in South Africa have been referred both to the Ashgill and to the lower Llandovery, and the South American ones have been listed under the lower Silurian by BERRY and BOUCOT (1973). The available evidence seems to indicate that all these glacial beds are roughly contemporaneous. The time resolution in the Ordovician is under normal circumstances not better than ±one million years and within this margin there may have been more than one glacial episode. The possibility that there were several glacial episodes therefore cannot be excluded and the glacial beds preserved in the various districts may also differ slightly in age.

In the central part of the glaciated region, in the area around the Hoggar massif (cf. BOEUF and others, 1971), the glaciation was mainly erosive and comparatively few sediments are preserved. Fossils are found in nodules in the green silts in the tunnel valleys. These fossils are trilobites, which, according to Dr J. DESTOMBES (personal communication), are indicative of a Caradoc age, and large phosphatic brachiopods. Such brachiopods are characteristic elements of the faunas both of the Caradoc and Ashgill of North Africa, as shown by HAVLIČEK and MASSA (1973). The rest of the fauna (bivalves, gastropods, cephalopods, and a few articulate brachiopods) do not give a more precise age. Because of the well known trilobite stratigraphy in Morocco (DESTOMBES, 1971) it can be stated that the dominating elements in the fauna are Caradoc, but the possibility of Ashgill admixtures cannot be excluded.

The problem remains as to whether the fossils are autochtonous or derived. In the same beds there are numerous ice-dropped boulders. Many of them are quartzites from the underlying Ordovician beds and some of them are fossiliferous, with brachiopods and trilobites indicating a Llandeilo and lower Caradoc age. The younger fossils mentioned above are found in small ironstone nodules, which may also have been ice-dropped. The fact that the same fauna has been found in three separate localities is also not significant as that would also have been expected if the material was long-distance ice-rafted. The matrix of the nodules contains iron ooliths, which are not found in the surrounding sediment, and the shape and size of the sand grains are also different in the nodules and the green silt. This indicates that the fossiliferous nodules are ice-rafted and the age of the glacial beds must therefore be younger than the age of the fossils found: Caradoc (probably upper Caradoc).

The fauna found in the nodules corresponds to that reported from the Caradoc Melez–Chograne Formation of Libya by HAVLIČEK and MASSA (1973). None of the elements of the younger fauna, reported from the Ashgill Memouniat Formation of Libya, has been found in the nodules, but, since all the fossils have been ice-rafted, the age of the glacial beds may well be Ashgill.

In the Central Sahara (both in Algeria and Libya) there are also limestones preserved as small erosional remains, sometimes only as pebbles in conglomerates. These limestones contain a fauna mainly of bryozoans which strongly recalls that from the Ashgill of the Erfoud area in Morocco.

Although the complete history of the Ordovician glaciation is far from known, it may tentatively be concluded that the available evidence suggest a model where a polar ice-cap was present over the suggested pole from the Arenig onwards, and that this ice-cap expanded drastically in the uppermost Ordovician into one or more glacial episodes reaching rather low latitudes (Fig. 3). The glaciation was accompanied by climatic changes involving the deposition of biostromal carbonate at high latitudes by a mechanism which is not yet fully understood.

3.3. Summary of the Ordovician climate

The outstanding feature of the Ordovician climate is the presence of polar ice-caps in Africa and South America, culminating in a glaciation reaching, in its marginal effects, to fairly low latitudes. This resulted in a strong atmospheric and oceanic circulation, which is reflected in easily identifiable, and often sharply limited, faunal provinces and climatic belts, also marked by sediment distribution.

A development can be traced, from the Tremadoc, which climatically is very similar to the Upper Cambrian, through the Arenig, Llanvirn, and Llandeilo, where faunal provinces are sharply separated and endemism therefore high. In the Caradoc an exchange of faunas took place in a series of pulses where plate movements can be suspected, and as undulations of boundaries between provinces where stable biogeographical conditions may have prevailed. The second type of interchange may be related to pulsation in the extent of the polar ice-caps.

There are generally no difficulties in identifying faunas and lithologies as being either cold, intermediate (boreal), or warm. One of the crucial questions, which is still unsolved, is whether the boreal-looking assemblages found along the western side of North America reflect anti-boreal faunas or more complicated plate movements. If antiboreal faunas existed, they would indicate the presence of an equatorial submergence, with cold and oxygenated deep ocean waters, and this would also open possibilities for reconstructing the position of the equatorial continents.

In the Ashgill, the neat pattern of sharply limited faunal provinces breaks down and a 'cosmopolitanization' is found, by which faunas from the previously cold areas (predominantly trilobites) migrate into the warmer regions and *vice versa*. This occurs in connection with a sharpening of the climatic contrasts, whereby the cold areas show signs of glaciation and warm-water carbonates and bioherms became more widespread in the previously warm and boreal regions. The combination of isostatic sea-level changes and rapid changes in climate may have provided a 'pumping mechanism' of increased selection pressure, especially on the shallow-water benthos. This may have increased the evolution rate and lowered local diversity, but also aided in migration compared with the more stable conditions in the earlier Ordovician.

The increased circulation, which is regarded as the main cause of the sharp climatic zoning both physically and biogeographically, also increases the difficulties of a more detailed palaeo-geographical reconstruction. A stronger circulation would also increase the deviation from a smooth pattern due to the position and shape of the land masses (cf. Fig. 1) just as it does today. Under such circumstances, it may be very difficult to identify the causes of rapid changes in climate. If, for example, the change in littoral fauna at Cape Cod on the east coast of North America (cf. FISCHER, 1960) was observed in a palaeontological context, with very fragmentary preservation both of the biological material and its large-scale regional setting, it would be extremely difficult to interpret. It would not be easy, from the local data, to envisage that the faunal contrasts was due to the Isthmus of Panama acting as a barrier, which, together with the Carribean geography, deflected the warm water (the Gulf Stream) northwards. This would be aggravated by the presence of the eccentric ice-cap on Greenland, which is found at latitudes down to 60°N, far south of North Norway where boreal faunas are found at the same time even up to 71°N. Similar situations were certainly present in the Ordovician, and they can only be properly identified by a large-scale areal analysis of global scale.

Attempts to reconstruct the Ordovician circulation patterns and their deviation can be made with some confidence for small, well known areas as shown by WILLIAMS (1969), but large-scale ones will have to be less rigid. The problem can be approached in two ways, either by measuring the distance between areas in numbers of common forms, as done by BURRETT (1973), or more intuitively by assuming shape and position of the continental landmasses, as demonstrated by ROSS (1975). With increasingly reliable (taxonomically and stratigraphically) data, and with computer simulation of circulation patterns with different models for continental positions and shapes, progress will certainly be made. The fact that many of the fossiliferous sediments have disappeared, either by plate consumption or by metamorphism, especially in the critical regions of plate contact, will put constraints on the possibilities to make precise and detailed reconstructions of Ordovician climate and geography. The main outlines seem to be well documented, and it may be concluded that the Ordovician climate was more similar to the present one than to those of the Cambrian and Silurian.

4. Silurian climates

4.1. Silurian faunal provinces

The problem of the existence of faunal provinces in the Silurian has been discussed by HOLLAND (1971) and an enormous amount of factual and theoretical material has been assembled by BOUCOT (1975, and previously). The general concensus is that the Silurian faunas are considerably more cosmopolitan than the Ordovician ones (as indicated by BOUCOT and JOHNSON, 1973), and that the boundaries between biogeographical regions of distribution are more likely to be geographical and environmental than climatic.

Three main environmental regimes are found from the Silurian. The shallow-water, mainly mud-carbonate environment generally has a highly diverse fauna; the evaporitic facies has rather restricted faunas; and the basinal, graptolite shale facies has planktonic–nektonic faunas. Most of the discussion about faunal assemblages has been concentrated on the first of these regimes, which in some atypical regions and tectonically unstable areas (like Wales) also may include deeper water.

The faunal assemblages described by ZIEGLER (1965), and elaborated by many others such as BOUCOT (1975), are mainly indicators of bathymetry, just as are the Ordovician ones described by FORTEY (1975). In order to use this information for climatological interpretation it is necessary to delimit the faunas, not only according to bathymetry and bottom sediment, but also by sorting out infaunas and epifaunas. Even if THORSON's (1957) blunt statement that the latitudinal change in diversity is due solely to the epifauna must be modified (cf. THORSON, 1966), information is needed about which forms were epifaunal in order to interpret the ecology in climatic terms. The presence or absence of algal vegetation is also of great importance because it will partly increase diversity by creating niches for epiphytic animals and partly reduce animal biomass by competing for space. Very few studies of faunal assemblages from the Silurian do separate the faunas in a way which is helpful for climatological analysis, and although much information is available it does not solve the problems of climatic zoning. As a result of this, the Silurian faunal provinces are vague, but can, according to BOUCOT (1975), be outlined as a North European–North American region, which, in spite of minor differences, seems to form a unit, which is well separated from a Malvinocaffric region, found in South America and South Africa. In the Devonian, this region also covered Australia, but in the Silurian it was more restricted.

The origin of the Malvinocaffric fauna is in doubt and so is the age of its first representatives. The oldest faunas follow above the glacial beds (the Pakhuis tillite in South Africa, Zapla tillite in South America). BOUCOT (1975) and BERRY and BOUCOT (1972, 1973) regard these faunas as probably lower Llandovery in age, whereas COCKS and others (1970) regard the South African (and by extrapolation also the South American ones) as Ashgill.

This problem, which is important for the stratigraphy of the Ordovician–Silurian boundary, is of minor importance climatologically as all the faunas involved are judged to be cold-water ones, both the basal one and the later Silurian and Devonian faunas from the same areas. If BOUCOT's (1975) dating is correct, this may indicate that the glaciations were not exactly synchronous in Sahara and in the southern continents, but the difference will be small and the time resolution is in any case not very good in this time interval. A similar problem concerns the *Hirnantia* fauna, which is found close to the Ordivician–Silurian boundary. The stratigraphical position has been much debated by LESPÉRANCE (1974), COCKS and PRICE (1975), and JAEGER, HAVLIČEK, and SCHÖNLAUB (1975) and seems to span the boundary and not to be restricted to the Ashgill as previously supposed. Climatologically, the fauna is a cold-water one in its type region (Morocco and Bohemia) and also occurs in the Malvinocaffric region and China. In Great Britain the sediments do not contradict a cold-water connection, but in Scandinavia elements of the fauna do occur in sediments which interfinger with typical warm-water carbonates and bioherms. This may be explained in terms of increasingly stressed environments due to the glaciation, which reduced the diversity but also caused a selection pressure in the direction of eurythermic forms, which could expand their area of distribution.

In accordance with this, the lower Llandovery show a rather low diversity and most of the forms found are left-overs from the Ashgill. Only well after this glacially induced crisis (cf. AMSDEN, 1971; SHEEHAN, 1972, 1975) is a rapid increase in diversity found, especially in the shallow-water benthonic faunas. To begin with, the diversity in each region is mostly due to the presence of several, environmentally separated assemblages, each of which has a low diversity. Later, with increased maturity and stability, the diversity of the individual assemblages increases, especially in the uppermost Llandovery and Wenlock. It is not yet clear to what extent this increase is due to establishment of new niches among the existing faunas, or to what extent it is due to a change in the quantitative relation between infauna (which appear to dominate in the early assemblages) and epifauna. The increase in both biomass and diversity of bryozoans, corals, and stromatoporoids from the lower Llandovery into the Wenlock may indicate increase in epifauna, which suggests a warming of the general climate.

In addition to the well studied regions, there are minor differences in those where the Silurian shelly faunas are not known in detail. In most cases the material is insufficient for a complete analysis, and in any case the fragmentary data do not supplement the climatic information. These provinces have been discussed by BOUCOT (1975) and by COCKS and McKERROW (1973).

The evaporitic facies can be subdivided into the deep basins with saline deposition and the more widespread and often thin dolomite sequences, with a characteristic fauna dominated by low diversity, absence of stenohaline forms, and presence of many eurypterids and other forms which are suspected of having been euryhaline. All these faunas indicate a warm and dry climate but are rather undifferentiated. One of the interesting points mentioned by BOUCOT (1975) is the salinity problem on the central parts of the very wide cratons. Under a regime of low circulation and high temperature one would expect the parts of the craton which were farthest from the open ocean to become evaporitic. This seems to hold in the upper Silurian, but in the older Silurian such beds are remarkably scarce. In the lowest Silurian this could be explained by low evaporation due to a cold climate, but the condition prevails far into the definitely warm part of the Silurian. Since there were few land areas from which fresh-water could be supplied, this phenomenon may.be a clue to the Silurian atmospheric circulation pattern.

The graptolite shale facies is widespread in the Silurian, as reflected in the standard stratigraphical divisions being based on graptolite zones. The common occurrence of black shales is in itself a sign of a low rate of atmospheric and oceanic circulation, and therefore probably the absence of polar ice-caps and the presence of a warm climate. The scarcity of real deep-water sediments precludes direct evidence for stagnant oceans in the Silurian, but the fact that almost all non-evaporitic basins and most cratonic margins seem to have had black sediments points in this direction. The graptolite shales grade from condensed, laminated very black shales, with high content of organic material and sulphides (alum shales), to thick, basinal sequences, where alternations between black and grey muds are common and gradations into normal bioturbated shallow-water sediments are found. One of the best examples of this is found in the Baltic, where the graptolite sequence is over 1000 metres thick and shows signs of shallow water (mud-cracks) as well as interfingering with shelly faunas.

Most of the faunas are cosmopolitan (a fact which is utilized in Silurian stratigraphy) but some deviations are found. In the basal Silurian the faunas are comparatively few as a hiatus is often found between the Ordovician and Silurian. The first graptolite faunas have often a low diversity, which may be due either to cold climate or to other environmental conditions, such as high influx of clastic material. They also show more endemism than the later ones, and this does cause difficulties in correlation and tracing the early Silurian transgressions as shown by e.g. BEUF and others (1966) and LEGRAND (1971).

The first two graptolite zones of the Silurian are dominated by these rather sparse faunas of diplograptids, climacograptids, and dimorphograptids. The monograptids appear only in the *atavus* Zone, and after the almost explosive development of this group the faunas became much more diverse and cosmopolitan, as demonstrated by HUTT and RICKARDS (1970).

The endemism of the later Silurian graptolite fauna is normally very low. One of the exceptions is the Mediterranean subfauna, with gigantic forms described originally by GORTANI (1922) from Sardinia and discussed later by GUIERARD and others (1970) and ROMARIZ and others (1971). These gigantic forms, which appear in the upper Llandovery and Wenlock, are difficult to interpret climatologically. The large size does not have to be related to a warm climate or rich supply of nutrients. It may be an adaption to individual (colonial) longevity in an area of temporarily restricted nutrient supply. If the size is due to rich food supply (presumably plankton), this is found most frequently in the zones of high organic productivity, which at present are located at high latitudes and in areas of local upwelling. We know very little about where the zones of highest planktonic productivity were under the conditions suggested for the Silurian, and no climatological conclusions can be drawn from the Mediterranean graptolite subfauna.

Together with the graptolites, this facies also contains a characteristic fauna of molluscs (cf. BERRY and BOUCOT, 1967a). Nautiloid cephalopods are often dominant, and diverse, but their potential for correlation and ecological studies has not been much utilized (cf. HOLLAND, 1971). The nautiloids are nektonic, but some of the other representatives in this fauna pose ecological problems, such as the comparatively thick-shelled bivalves like *Cardiola*. They might have been attached to sea-weeds, or had a special adaptation for floating, such as had the crinoid *Scyphocrinites*.

The records of plate movements in the Silurian are rather meagre. According to McKERROW and ZIEGLER (1972) the final closing of the Proto-Atlantic took place in the early Silurian. As mentioned above, the Proto-Atlantic ceased to be a barrier for benthonic organisms already at the Ordovician–Silurian boundary, or before, and from biogeographical evidence it is therefore difficult to date the precise time of physical closing of the ocean. The development of the Old Red Sandstone facies (cf. Section 4.3) indicates that the continents were in contact, and that tectonic processes were active in the region of suture at least from the middle Silurian to late into the Devonian.

The diffuse boundaries between climatic zones and faunal provinces in the Silurian, and the apparent scarcity of plate movement, have led BOUCOT (1975), BERRY and BOUCOT (1967), and BERRY, BOUCOT, and JOHNSON (1968) to adopt a somewhat critical and reserved attitude towards plate tectonics. The last paper contains a good review of the discussion on this subject, and also an excellent summary of faunal distributions in the Silurian.

Near the end of the Silurian there are some changes in the faunal pattern. In the Arctic, and parts of North America, the Urghan-Nevandan province with a distinctive brachiopod fauna is found. The graptolite faunas have very low diversity, foreshadowing the extinction of the group rather than ecology, and they also occur in a greater variety of sediments than before. As indicated by BERRY and BOUCOT (1972a), the depth range of the Silurian graptolites decreases with time, and the last, uppermost Silurian and Devonian representatives—although they are cosmopolitan—are restricted to shallow water. In addition, the changes in environment connected with the Old Red Sandstone facies (Section 4.3) have profound influence on the faunal distribution in some of the best known areas. Where normal marine conditions prevail, there is a smooth gradation into the Lower Devonian, with no indication of sharp changes either in faunal distribution or in climate.

4.2. Silurian reefs

The first reefs (bioherms) reported are—besides some stromatolite build-ups—formed by archaeocyathids in the Lower Cambrian (cf. Section 2.2). Both in carbonate structure and especially in animal and plant life these bioherms deviated strongly from the present coral–algal bioherms, even if their climatic significance may be similar. Bioherms of a more modern type occur first in the Ordovician. The oldest ones are the Chazy, in the north-eastern United States, of low middle Ordovician age (KAPP, 1975). They are mostly patch reefs, with an initially simple structure and stromatoporoids as frame builders. In the middle Ordovician (upper Caradoc), reefs are known not only in North America but also in Estonia (Vasalemma) and the Oslo

Region in Norway. The framework is made of stromatoporoids in all the real bioherms, but coral banks are also found on a small scale. Many of the so-called Ordovician reefs are really carbonate mud banks, such as the Kullsberg and Boda Limestones in Sweden, the Keisley Limestone in England, the Kildare Limestone in Ireland, and numerous examples in North America.

In the uppermost Ordovician bioherms are found frequently in the supposed equatorial zone (cf. Fig. 3), but very few and small ones are met with in the basal Silurian (lower Llandovery). In the upper Llandovery reef formation increases and it seems to reach a peak in the Wenlock and low Ludlow. The decline in the later Ludlow may be apparent, due to the facies changes found to saline or clastic environments in the best studied areas.

One of the type areas for Silurian reefs is the central part of North America, where they are well known through the classical papers of LOWENSTAM (1948, 1950, 1957), and later information has been added by CROWLEY (1973).

The other important regions is the island of Gotland. Here MANTEN (1971) has reviewed the older literature and given an exhaustive description of the reef topography. For the structure of the reefs, HADDING (1941) is still the most important source, even if much later information has been supplied by MORI (1969, 1970). The reefs range in size from single-colony patch-reefs to very extensive barrier reefs and the framework builders are largely stromatoporoids. Detailed studies of the framework and the stromatoporids involved have been given by MORI (1968, 1970) and ST. JEAN (1971).

In some cases, such as the bioherms in stage 9c at Holmestrand, Norway (upper Wenlock age), tabulate corals of the genus *Thecia* are frame builders, and small scale build-ups with tabulate corals, colonial rugose corals, and algae are known. Coral thickets, formed by colonial rugose corals and bryozoans, are also widespread, but mostly on a small scale. They are considered not to be of climatic significance.

The reefs and their surroundings support rich faunas of specially adapted forms, especially brachiopods, cephalopods, and some molluscs and echinoderms. These faunas contribute much to the high species diversity found in the Silurian. Crinoids were often common around reefs, which often grade into crinoidal debris limestones, especially on the exposed side of the reefs. Algae played a more restricted role on the Silurian reefs than on the recent ones, but they are locally common. Most of them are encrusting and oncolite-forming simple forms, but *Halimeda*-like types have been reported already from the middle Ordovician by HØEG (1927), and they occur also in the Silurian. As indicated by LOWENSTAM (1957) the crinoids substituted for the jointed *Halimeda*-like algae in the Early Palaeozoic. Also in Gotland this seems to be the case as the exposed side of the reefs normally grades into crinoidal limestones, whereas the sheltered, lagoonal side often contains algal floras. The general position of the reefs seems to be as fringes, either (North America) between the stable shallow-water craton and an evaporitic basin, or (the Baltic), between the shore-line and a basin filled with thick graptolite shales.

Silurian reefs are also found in Kazakhstan (BANDALETOV, 1969) and in Australia, but there are few detailed descriptions. The reefs appear to be rather similar to those described from the classical regions. In some 'warm' regions reefs are conspicuously absent. In some cases, like Podolia (NIKIFOROVA and PREDTECHENSKIJ, 1968) there is no obvious reason for this, as the fauna otherwise is very similar to that of the Baltic. In other cases, such as north-eastern North America and the Siberian Platform, the reefs seem to be replaced by evaporitic shallow-water dolomites and other sediments indicating a hot climate.

TEICHERT and STAUFFER (1965) reported Silurian reefs from Pakistan. They may be related to the Siluro-Devonian carbonates described from Afghanistan by DÜRKOOP (1970) and indicate that this part of Pakistan did not belong to the Indian part of Gondwanaland. Except for

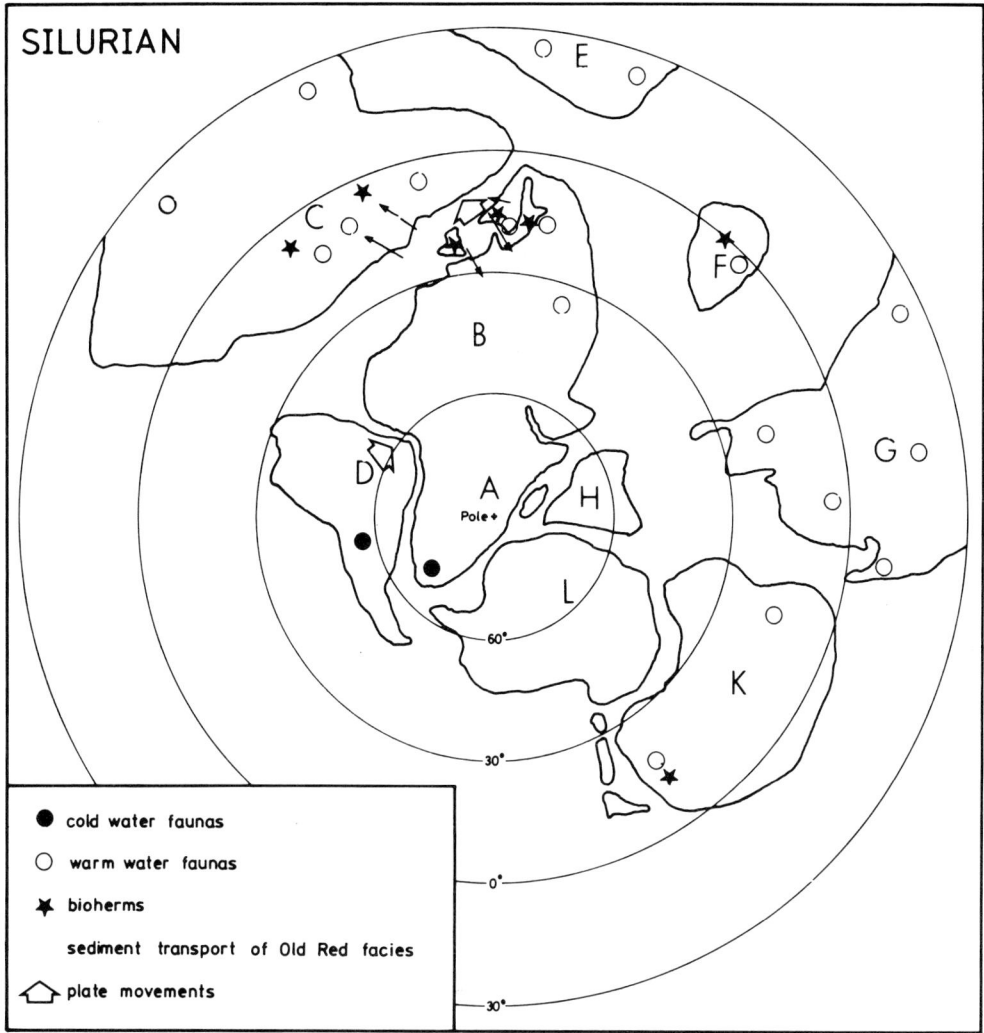

FIG. 4. A reconstruction of the palaeogeography and climatic zoning in the Silurian. For further explanations, see Fig. 2.

the Pakistanian reefs, all others conform without difficulty to the suggested equatorial zone for the Silurian (cf. Fig. 4).

4.3. The Old Red Sandstone facies

Traditionally the Old Red Sandstone facies has been regarded as Devonian, although it has long been known that this facies straddled the Siluro–Devonian boundary in some regions. Since the base of the *Monograptus uniformis* Zone has been accepted as the boundary, even more rocks in the Old Red Sandstone facies have become Silurian. This facies is known especially from north-eastern Europe and the north and eastern parts of North America. Similar faunas, floras, and lithologies are found also in other continents, but these occurrences are small, not very well known, and most of them appear to be of Devonian age. The present comments are therefore concentrated on the classical regions.

The rocks of the Old Red Sandstone facies are mostly clastics: shale, sandstone, and conglomerate. Caliche-type carbonates are also often met with and an impressive suite of sedimentary structures has been described. The red colour—which as shown by FRIEND (1966) and TURNER (1974) is due to haematite—is not always present; many of the sandstones are grey, green, or dark, and in some cases the red colour is evidently secondary (such as in the top of the Ringerike Sandstone at Kolsås in South Norway, where it is due to Permian weathering), and in some cases the original red colour has been changed, due to high content of organic material or the influence of volcanic gases.

The sediments differ from recent ones in some instances. The high content of unaltered mica gives the sandstone an immature aspect, which suggests high topography and/or arid conditions. As mentioned above (Section 1.2) this may to a certain extent be due to the absence of terrestrial plant cover in the Early Palaeozoic. The late Silurian and early Devonian floras seem to have been concentrated in water or in moist, bog-like places, which were low in the landscape and did not influence weathering as does the recent terrestrial plant cover. The frequent occurrence of caliche- or sebka-like carbonates, but the scarcity of real aeolian sediments, also indicates that the climate was one of ample but seasonal rainfall, and rapid run-off due to lack of vegetation. Under such circumstances it is difficult to reconstruct the topography and detailed sedimentary environments on the Old Red Sandstone continent. Topographically the Old Red Sandstone basins can be divided into internal and external types. The former, which mostly are tectonically bounded inter-montane basins, indicate a strong relief and a 'live' tectonics (cf. BRYHNI, 1964). These basins are mostly, if not wholly, found in the Devonian and are therefore not considered further here. The external basins all seem to have graded into marine environments and therefore to have been formed at about sea level and close to the coast line (not necessarily on the ocean coast; in many cases the sea was a shallow, intercratonal one). The best example is probably the development in the Ludlow district in Britain, where typically marine faunas grade into brackish water ones, with lower diversity, as the basin was filled by an increased influx of clastics. The faunal changes, from echinodern–brachiopod faunas, through mollusc-dominated ones, to vertebrate–arthropod faunas are what one would expect under these ecological conditions, but it is very difficult to pin-point the precise transition from marine to fresh water. The general reconstruction of the basins and the sedimentological environments have been studied in detail by ALLEN (1974), ALLEN and TARLO (1963), FRIEND (1965, 1966, 1969), and others. They all base their reconstruction on the typical fluviatile model of the British School of sedimentology. An alternative explanation had been given by BURROLET and others (1969), who favour deposition in, or close to, large bodies of water (salt or brackish to fresh). This discrepancy not only illustrates the different ways of thinking in the two schools of sedimentology, but also reflects the inborn difficulties in environmental reconstruction caused by the somewhat non-actualistic conditions on the Old Red Sandstone continent, especially the lack of dense terrestrial plant cover.

It is difficult to tell if, and how, the external basins were related to sea level, and to what degree they communicated with the sea. In some cases, such as the type Ludlow in Britain, there was a prolonged time with gradual decreasing salinity and shallowing water. In other areas, like the Oslo Region, the condition suggests rapidly prograding deltas as described by RAMBERG and SPJELDNAES (1978). In most cases the proximity to the sea is easily seen in the marginal parts of the 'Old Red Sandstone Continent'; but there are some exceptions, such as Hidra in West Norway. Here the beds are dated as Downtonian by some rather incompletely preserved fossils. The dating is based on a comparison with the Stonehaven beds in Scotland, and it may be severely doubted where they should be placed in relation to the Siluro–Devonian boundary. ALLEN, DINELEY, and FRIEND (1967) discuss the age of the Hidra beds and suggest Ludlow. This may partly be based on the scanty fossil evidence, but even more on the 'marine look' of the beds. They differ considerably in the presence of dark siltstones and other sediments from the other Devonian areas in western Norway and they do also show a variety of sedimentary structures, which may be interpreted as belonging in a fluvial–estuarine environment as suggested by SIEDLECKA and SIEDLECKI (1972), but they may also be interpreted as indicating an estuarine, deltaic environment with some tidal influence.

In the Middle or Lower Devonian Grey Hoek series in Spitsbergen, the environment was probably brackish marine according to the descriptions by FRIEND (1961) and WORSLEY (1972). The vertebrate faunas have a typical Old Red Sandstone aspect, but the invertebrate faunas described by QUENSTEDT (1926, bivalves) and SOLLE (1935, ostracodes) are definitely marine, euryhaline forms. The leperditids, which are the dominant ostracode group, are common in Silurian sediments which are regarded as hypersaline and there is nothing to indicate that this fauna was a fresh-water one. This may lead to reconsideration of the environment of some of the Old Red Sandstone vertebrate faunas. It is remarkable that the pattern of dispersal and the diversity and endemism in the vertebrate faunas do not change markedly during the transition from the marine carbonate-shale facies to the Old Red Sandstone facies. This is well illustrated in the analysis given by HALSTEAD and TURNER (1973). If most of the Old Red Sandstone vertebrate faunas were fresh-water ones, one would have expected a much higher degree of local endemism when new pioneer groups invaded a number of virgin, but isolated habitats.

Zoogeographically, the external basins of the Old Red Sandstone behaved as a marine area and not as an arid continental mass. This may be explained partly by the selective preservation of low marginal areas in a period of rising of the continental mass. The more centrally and higher placed areas would leave sediments only in localities with strong tectonic subsidence as in the model described by BRYHNI (1966, 1974) and BRYHNI and SKJERLIE (1976). In this respect, the Hidra area is interesting because it is centrally located in the Old Red Sandstone continent and still shows the characteristic features of an external basin, possibly in contact with the sea. As it is also the oldest one in the region, it may indicate that large parts of the Old Red Sandstone province were really shallow inland seas in the Silurian, which only gradually—in the Devonian—became more terrestrial. The onset of the Old Red Sandstone facies seems to have progressed from the central parts towards the periphery of the distribution area. In southern Britain the uppermost Silurian (Downtonian) is in Old Red Sandstone facies and in the Ardennes an interfingering of typical marine sandstones and Old Red Sandstone lithologies is found in the Lower Devonian. In Podolia, ZYCH (1927) suggested that an extensive interfingering existed between the Old Red Sandstone, and the marine carbonate facies, but according to more modern studies by NIKIFOROVA and PREDTECHENSKIJ (1968) and SOKOLOV (1972), the transition seems to be restricted to a short time all over the whole area, even if some small-scale 'climbing' may occur, and the change was well above the Siluro–Devonian boundary. In Estonia the Old Red Sandstone facies is also all Devonian.

In Scotland, and other more centrally placed areas, the influx of clastics started much earlier and the first sediments and faunas which may be classified as of Old Red Sandstone facies occur already in the Wenlock and lower Ludlow. ALLEN, DINELEY, and FRIEND (1967) do not include these oldest beds in the Old Red Sandstone sequence. This is understandable as there are considerable local differences, including a disconformity. The evidence from other areas indicates that the definition of the Old Red Sandstone facies should be extended to include these beds because of their vertebrate and merostome faunas and their lithology, indicating an influx of clastic material of the Old Red Sandstone type.

Although the fossil evidence is very poor, both the Stonehaven and Hidra beds may be dated as upper Silurian (Ludlow or Downtonian). In the Olso Region, the Marienberg Formation of low Wenlock age provides the oldest red sandstones and the Bruflat Sandstone, which is also low

Wenlock, is not red but has trace fossils indicating a rich merostome fauna of the same type as that in the typical Old Red Sandstone facies. Later, in the upper Wenlock and lower Ludlow, the Ringerike Sandstone occurs in part of the southern Oslo Region. It has well known vertebrate and merostome faunas in its lower part and there are indications of a gradual expansion towards the south and south-east by deltaic beds (SPJELDNAES, 1966; RAMBERG and SPJELDNAES, 1978). The youngest beds in the southernmost localities indicate that the top of the Ringerike Formation reaches into the lowest vertebrate zones of the typical Downtonian.

In Sweden the unfossiliferous Orsa Sandstone may be an atypical representative of the Old Red Sandstone facies and the Burgsvik Sandstone on Gotland has all the characteristics of a marginal, fully marine tongue of the Old Red Sandstone facies. In spite of a very mature granulometry it has a high content of mica. It has merostome trace fossils and spores of land-plants (GRAY, LAUFELD, and BOUCOT, 1974) in addition to a typical marine, shallow water fauna.

In Scania, the Øved–Ramsåsa beds show an interfingering between red beds and marine carbonates, with an increase of red beds towards the top. The age of the onset of red bed sedimentation is later than in the Oslo Region, and at about the same time as the sedimentation of the Burgsvik Sandstone on Gotland. In Scania there is no reversal to typical marine conditions such as on Gotland.

In the subsurface of Denmark (North Jutland Basin) CHRISTENSEN (1973) has reported red beds, intermediate in lithological character between the Ringerike Formation and the Øved–Ramsåsa beds. The change from fully marine graptolite shales to red beds with fossiliferous carbonates is above the lower Ludlow (*M. colonus* beds).

STØRMER (1967) has shown that in Norway the Old Red Sandstone facies rests on metamorphic and folded older beds (lower Wenlock or older) in the central parts of the Caledonides. In the peripheral areas, the Old Red Sandstone beds rest conformably on the marine Silurian carbonates. This indicates that normal marine sedimentation ceased in the central Caledonides just after the Llandovery–Wenlock boundary, due to uplift and orogenic movements probably in connection with the final phases of the fusion of the North Atlantic continents. At about the same time, and in connection with the same processes, an influx of fresh clastic material in the marginal areas initiated the Old Red Sandstone facies sedimention as defined here. During the rest of the Silurian this facies spread towards the south and east and reached its maximal distribution in the Lower Devonian.

Judging from the interfingering and bordering marine sediments, and the admittedly scanty direct evidence, the climate of the Old Red Sandstone continent was warm. The area was probably a rising one, which may have had considerable topography, but most of the rocks preserved are from marginal areas, deposited at or below sea level. The sediments formed well above sea level are, as usual, removed by later erosion, except where they are preserved in tectonically formed traps, mostly as steep-sided intermontane basins. There is no good evidence for arid conditions. The whole sedimentary and biological picture seems to indicate ample, but perhaps seasonal, rainfall and rapid run-off, with immature sediments due to the absence of a terrestrial plant cover.

4.4. Summary of the Silurian climate

The Silurian climate was generally warm. The evidence for this interpretation is fairly extensive and includes not only the sedimentological data given by the black shales together with the widespread carbonates and evaporites, but also the distribution of high diversity faunas. A trend can be discerned from the rather cold basal Silurian, which resembles the Ordovician rather than the rest of the Silurian, to the stable warm climate, which appears to have been established during or before the upper Llandovery.

Only very few local changes are known from the Silurian. One of them is the comparatively warm-water Llandovery reported by BOUCOT (1975) from northernmost and westernmost South America. Both the upper Ordovician and later Silurian faunas are of cold-water type. Owing to a lack of precise information, it is difficult to interpret this observation in terms of large scale climatic or palaeogeographical developments. The Malvinocaffric Province which has been described especially by BOUCOT (1975) is definitely a cold-water one, at least in comparison with the other Silurian faunas. As indicated by BOUCOT (1975), almost all of the Malvinocaffric faunas belong in shallow-water assemblages. Judging from both the lithologies and the brachiopod assemblages, many of them were littoral or high sublittoral. In spite of this, there are no signs of glacial activity or presence of ice, and the fact that shallow-water faunas on sandy bottoms are highly developed indicates that the climate was not really arctic. Modern arctic faunas are mostly found in soft clay sediments, and have a high proportion of infaunal elements. The few extensive shallow sandy bottoms carry a very restricted fauna as these environments are covered by ice during a large part of the year. The ice not only freezes the sediments and disturbs them (as demonstrated by LINDSTRÖM, 1972), but effectively restricts circulation with the atmosphere and damps the wave-generated turbulence which is necessary to maintain a clean sand environment. The Silurian Malvinocaffric faunas therefore must be assumed to have lived in a cold climate, but not cold enough to have winter ice as a regular feature. This is also in accordance with a climatic model of the Silurian, without polar ice-caps.

References

ALFVÉN, H. and ARRHENIUS, G. (1969). Two alternatives for the history of the moon. *Science*, **165**, 11.

ALLEN, J. R. L. (1974). Sedimentology of the Old Red Sandstone (Siluro-Devonian) in the Clee Hills area, Shropshire, England, *Sed. Geol.*, **12**, 73.

ALLEN, J. R. L., DINELEY, D. L., and FRIEND, P. F. (1967). Old Red Sandstone basins of North America and Northwest Europe, *in* OSWALD, D. H. (Ed.), *International Symposium on the Devonian System, Calgary*, **1**, 69.

ALLEN, J. R. L. and TARLO, L. B. (1963). The Downtonian and Dittonian Facies of the Welsh Borderland. *Geol. Mag.*, **100**, 129.

ALLEN, P. (1975). Ordovician Glacials of the central Sahara, *in* WRIGHT, A. E. and MOSELEY, F. (Eds.), *Ice Ages: Ancient and Modern. Geol. J.*, Spec. Issue, **6**, 275.

AMSDEN, T. W. (1971). Late Ordovician–Early Silurian Brachiopods from the Central United States. *Mem. BRGM*, **73**, 19.

APPOLONOV, M. K. (1968). Zonalnaja shkala srende- i verchneordovikskich otlozhenii Kazachstana. *Mezh. Geol. Kongr. XXIII, Probl.*, **9**, 78.

APPOLONOV, M. K. (1975). Ordovician trilobite assemblages in Kazakhstan. *Fossils Strata*, **4**, 375.

APPOLONOV, M. K., BANDALETOV, S. M., NIKITIN, N. F., and TSAI, D. T. (1968). Ordovikskie i silurijskie otlozhenija kazachstana i ich korreljatsija s evropeiskimi razrezami. *Mezh. Geol. Kongr. XXIII, Probl.*, **9**, 111.

ARBEY, F. (1968). Structures et dépôts glaciaires dans l'Ordovicien terminal des chaînes d'Ougarta (Sahara algérien). *C.R. Acad. Sci.*, **266**, 76.

ARBEY, F. and TAMAIN, G. (1971). Existence d'une glaciation siluro-ordovicienne en Sierra Morena (Espagne). *C.R. Acad. Sci.*, **272**, 1721.

ASTROVA, G. G. (1965). Morfologija, istorija pazvitija i sistema ordovikskich i siluriiskich mshanok. *Tr. Paleont. Inst. Akad. Nauk SSSR*, **56**, 432 pp.

BABIN, C. (1966). Les mollusques bivalves en tant qu'indicateurs écologiques. Application au paléozique armoricain. *91. Congr. Soc. Savantes, Rennes*, **2**, 327.

BABIN, C. (1966a). *Mollusques bivalves et céphalopodes du paléozoïque armoricain. Impr. Comm. Adm., Brest*, 472 pp.

BABIN, C., CHAUVEL, J., CHAUVEL, J.-J., HENRY, J.-L., LE CORRE, C., MORZADEC, P., NION, J., PHILIPPOT, A., PLUSQUELLEC, Y., and RENAUD, A. (1968). le Palćozoique antécarbonifere de Bretagne (France). Résultat recent et problèmes actuels. *Casopis Min. Geol.*, **13**, 261.

BABIN, C. and MELOU, M. (1972). Mollusques bivalves et brachiopodes des "schistes de Raguenez" (Ordovician supérieur du Finistère): conséquences stratigraphiques et paléobiogeographiques. *Ann. Soc. Geol. Nord.*, **92**, 79.

BAIN, G. W. (1960). Climatic zones of the Paleozoic era. *Rep. 21st Int. Geol. Congr.*, Sect. XII, 84.

BAIN, G. W. (1965). Climatic zones throughout the ages, *in Polar Wandering and Continental Drift.* SEPM, 100.

BARROIS, CH. (1890). Memoire sur les eruptions diabasiques siluriennes du Memez-Hom (Finistère). *Bull. Serv. Carte Geol. Fr.*, **7**.

BASSETT, M. G. and COCKS, L. R. M. (1974). A review af Silurian brachiopods from Gotland. *Fossils Strata*, **3**, 1–56.

BENDER, F. (1963). Stratigraphie der 'Nubischen Sandsteine' in Süd-Jordanien. *Geol. Jb.*, **81**, 237.

BERGSTRÖM, S. M. (1977). Early Paleozoic conodont biostratigraphy in the Atlantic Borderlands, *in* SWAIN, F. M. (Ed.), *Stratigraphic Micropaleontology of Atlantic Basins and Borderlands.* Elsevier, Amsterdam, 85–110.

BERRY, W. B. N. (1972). Early Ordovician bathyuid province lithofacies, biofacies, and correlations—their relationship to a proto-Atlantic Ocean. *Lethaia*, **5**, 69.

BERRY, W. B. N. and BOUCOT, A. J. (1967). Pelecypod–Graptolite Association in the Old World Silurian. *Bull. Geol. Soc. Am.* **78**, 1515.

BERRY, W. B. N. and BOUCOT, A. J. (1967). Continental stability—A Silurian point of view. *J. Geophys. Res.*, **72**, 2254.

BERRY, W. B. N. and BOUCOT, A. J. (1972). Silurian graptolite depth zonation. *Proc. 24th Int. Geol. Congr. Sect.* **7**, 59.

BERRY, W. B. N. and BOUCOT, A. J. (Eds.) (1972). Correlation of the South American Silurian Rocks. *Spec. Pap. Geol. Soc. Am.*, No. 133.

BERRY, W. B. N. and BOUCOT, A. J. (1973). Correlation of the African Silurian Rocks. *Spec. Pap. Geol. Soc. Am.*, No. 147.

BEUF, S., BIJU-DUVAL, B., STEVAUX, J., and KULBICKI, G. (1966). Ampleur des glaciations 'siluriennes' du Sahara: leurs influences et leurs conséquences sur la sédimentation. *Rev. Inst. Pet. Fr.*, **21**, 363.

BEUF, S., BIJU-DEVAL, B., CHAPARAL, O. DE, ROGNON, R., GARIEL, O., and BENNACEF, A. (1971). *Les Grès du Paléozoïque inférieur au Sahara.* Technip, Paris.

BEYTH, M. (1972). Paleozoic–Mesozoic Sedimentary of Mekele Outlier, Northern Ethopia. *Bull. Am. Ass. Petrol. Geol.*, **56**, 2426.

BIDGOOD, D. E. T. and HARLAND, W. B. (1961). Palaeomagnetism in some East Greenland sedimentary rocks. *Nature*, **189**, 633.

BIGARELLA, J. J. (1973). Paleocurrents and the problem of continental drift. *Geol. Rundsch.*, **62**, 447.

BIJU-DUVAL, B., DEYNOUX, M., and ROGNON, P. (1974). Essai d'interprétation des 'Fractures en gradins' observées dans les formations Glaciaires précambriennes et ordviciennes du Sahara. *Rev. Géogr. Phys. Géol. Dyn.*, **16**, 503.

BIJU-DUVAL, B., and GARIEL, O. (1969). Nouvelles observations sur les phénomènes glacieres 'Éocambriens' de la bordure Nord de la synéclise de Taoudeni, entre El Hank et le Tanezrouft, Sahara Occidental. *Palaeogeogr. Palaeoclimatol. Palaeoecol.*, **6**, 283.

BJØRLYKKE, K. (1967). The Eocambrian 'Reusch Moraine' at Bigganjargga and the geology around Varangerfjord, Northern Norway. *Norges Geol. Unders.*, **251**, 18.

BJØRLYKKE, K. (1974). Geochemical and mineralogical influence of Ordovician Island Arcs on epicontinental clastic sedimentation. A study of Lower Palaeozoic sedimentation in the Oslo Region, Norway. *Sedimentology*, **21**, 251.

BJØRLYKKE, K. O. (1974). Glacial striations on clasts from the Moelv Tillite of the Late Precambrian of Southern Norway. *Am. J. Sci.*, **274**, 443.

BLUMENSTENGEL, H. (1965). Zur Ostracodenfauna eines Kalkgerölls aus dem Thüringer Lederschiefer (Ordovizium). *Freiberger Forschungsh.* **C182**, 63.

BONDAREV, V. I., BURSKI, A. Z., and NEKHOROSHEVA, L. V. (1968). Skhema stratigrafii ordovika arktiskeskich raionov Uralo-Novozemeljskogo skladchtoi oblasti i ee sopostavljenie so skhemami ordovika Severnoi Evropi. *Mezh. Geol. Kongr. XXIII, Probl.*, **9**, 86.

BORELLO, A. V. (1971). The Cambrian of South America, *in* HOLLAND, C. H. (Ed.), *Lower Palaezoic Rocks of the World.* Vol. 1. Wiley, New York, 385.

BORISOV, A. A. (1965). *Paleoklimati Territorii SSSR.* LOLGY, Leningrad.

BOUCOT, A. J. (1963). The Eospiriferidae. *Palaeontology,* **5**, 682.

BOUCOT, A. J. (1975). *Evolution and extinction rate controls.* Elsevier, Amsterdam.

BOUCOT, A. J., BERRY, W. B. N., and JOHNSON, J. G. (1968). The crust of the earth from a Lower Paleozoic point of view, *in* PHINNEY, R. A. (Ed.), *The History of the Earth's Crust.* Princeton University Press, 208.

BOUCOT, A. J., GAURI, K. L. and SOUTHARD, J. (1970). Silurian and Lower Devonian Brachiopods, structure and stratigraphy of the Green Pond outlier in Southeastern New York. *Palaeontographica, A,* **135**, 1.

BOUCOT, A. J. and JOHNSON, J. G. (1973). Silurian Brachiopods, *in* HALLAM, A. (Ed.), *Atlas of Palaeobiogeography.* Elsevier, Amsterdam, 59.

BRABB, E. E. (1967). Stratigraphy of the Cambrian and Ordovician Rocks of East-Central Alaska. *Prof. Pap. U.S. Geol. Surv. No.* 559-A, 30 pp.

BRETSKY, P. W. (1970). Upper Ordovician Ecology of the central Appalachians. *Peabody Mus. Bull.,* **34**.

BRIDEN, J. C., DREWRY, G. E., and GILBERT SMITH, A. (1974). Phanerozoic equal-area world maps. *J. Geol.,* **82**, 555.

BRYHNI, I. (1964). Migrating basins in the Old Red Continent. *Nature,* **202**, 384.

BRYHNI, I. (1974). Old Red Sandstone of Hustadvika and an occurrence of dolomite at Flatsjer, Nordmøre. *Norges Geol. Unders.,* **311**, 49.

BRYHNI, I. and SKJERLIE, F. S. (1976). Syndepositional tectonism in the Kvamshesten district (Old Red Sandstone), Western Norway. *Geol. Mag.,* **112**, 583.

BURRETT, C. (1973). Ordovician biogeography and contenental drift. *Palaeogeogr. Palaeoclimatol. Palaeoecol.,* **13**, 161.

BURRETT, C. F. (1974). Plate tectonics and the fusion of Asia. *Earth Planet. Sci. Lett.,* **21**, 181.

BURROLET, P. F., BYRAMJEE, R., and COUPPEY, C. (1969). Contribution a l'étude sédimentologique des terrains Dévoniens du Nord-Est de l'Écosse. *Not. Mem. CFP,* **9**, 83 pp.

CARLS, P. (1975). The Ordovician of the Eastern Iberian Chain near Fombuena and Luesma (Prov. Zaragoza, Spain). *Abh. Neues Jb. Geol. Paläont.,* **150**, 127.

CARRÉ, D., HENRY, J.-L., POUPON, G., and TAMAIN, G. (1970). Les Quartzites Botella et leur faune trilobitique. Le problème de la limite Llandelien–Caradocien en Sierra Morena. *Bull. Soc. Geol. Fr.,* (7), **12**, 774.

CAWLEY, J. L., BURRUSS, R. C., and HOLLAND, H. D. (1969). Chemical weathering in Central Iceland: an analog of Pre-Silurian weathering. *Science,* **165**, 391.

CHAUVEL, J., DROT, J., PILLET, J., and TAMAIN, G. (1969). Précisisons sur l'Ordovicien moyen et supérieur de la 'série-type' du Centenillo (Sierra Morena orientale, Espagne). *Bull. Soc. Geol. Fr.,* (7), **11**, 613.

CHRISTENSEN, O. B. (1973). The Rønde and Nøvling Formations (Silurian) in Nøvling No. 1, *in* RASMUSSEN, L. B. (Ed.), The deep test well Nøvling No. 1 in Central Jutland, Denmark. *Danmarks Geol. Unders.,* **40**, 150 [in Danish, with summaries in English].

CHOUBERT, G. (1952). Livret guide de l'excursion A 36. Anti-Atlas Occidental. *Int. Congr. Geol. XIX, Ser. Maroc,* **10**.

CHOUBERT, G. and DEBRENNE, F. (1964). Sur la Paléogéographie des calcaires a archéocyathes dans l'Anti Atlas oriental. *C.R. Acad. Sci.,* **258**, 2122.

CHUGAEVA, M. N. and ORADOVSKAYA, M. M. (1973). Biostratigrafija nizhnei chasti ordovika severo-vostoka SSSR i biogeografija kontsa rannego ordovika. *Tr. Geol. Inst. Akad. Nauk SSSR.,* No. 213.

CHUGAEVA, M. N., ROZMAN, CH, S., and IVANOVA, V. A. (1964). Sravniteljnaja biostratigrafija ordovikskich otlozhenii Severo-Vostoka SSSR. *Tr. Geol. Inst. Akad. Nauk SSSR.,* No. 106, 262pp.

CLOUD, P. E. (1961). Paleobiogeography of the Marine Realm, in *Oceanography.* AAAS, **151**.

CLOUD, P. E., JR. (1968). Atmospheric and hydrospheric evolution on the primitive earth. *Science,* **160**, 729. (See also discussion of this paper in *Science,* **161**, 1364).

COCKS, L. R. M., BRUTON, C. H. C., ROWELL, A. J., and RUST, I. C. (1970). The first Lower Palaeozoic fauna proved from South Africa. *Q. J. geol. Soc.,* **125**, 583.

COCKS, L. R. M., and McKERROW, W. S. (1973). Brachiopod distribution and faunal provinces in the Silurian and Lower Devonian, *Spec. Pap. Palaeont.,* **12**, 291.

COCKS, L. R. M. and PRICE, D. (1975). The biostratigraphy of the upper Ordovician and lower Silurian of south-west Dyfed, with comments on the *Hirnantia* fauna. *Palaeontology,* **18**, 703.

COOPER, G. A. (1956). Chazyan and related Brachiopods. *Smithsonian Misc. Coll.*, **127**.

COLCHEN, M. (1968). Le Cambrien et ses limités dans la Sierra de la Demanda (Burgos–Logrono, Espagne). *C.R. Som. Soc. Géol. Fr.* 180.

COLCHEN, M. and HAVLIČEK, V. (1968). Le niveau à *Billingsella* cf. *lingulaeformis* Nikitin du Cambrien de la Sierra de la Demanda (Logrono, Espagne). *Bull. Soc. Géol. Fr.*, (7), **10**, 133.

COWIE, J. W. (1960). Notes on the Lower Cambrian stratigraphy of the boreal regions. *Rep. XXI Int. Geol. Congr.*, **8**, 57.

COWIE, J. W. (1971). The Cambrian of the North American Arctic Regions, *in* HOLLAND, C. H. (Ed.), *Lower Palaeozoic Rocks of the World*, Vol. 1. Wiley New York, 325.

COWIE, J. W. (1974). The Cambrian of Spitsbergen and Scotland, *in* HOLLAND, C. H. (Ed.), *Lower Palaeozoic Rocks of the World*, Vol. 2. Wiley New York, 123.

CRAMER, F. H. (1971). A Palynostratigraphic model for Atlantic Pangea during Silurian time. *Mem. BRGM*, **73**, 229.

CRAMER, F. H. (1973). Middle and Upper Silurian chitinozoan succession in Florida subsurface. *J. Paleontol.*, **47**, 279.

CREER, K. M. (1970). Review and interpretation of Palaeomagnetic data from the Gondwanic continents. *2nd Gondwana Symp. CSIR*, 55.

CROWELL, J. C. (1964). Climatic significance of sedimentary deposits containing dispersed megaclasts, *in* NAIRN, A. E. M. (Ed.), *Problems in Palaeoclimatology*. Interscience, New York, 86.

CROWELL, J. C. and FRAKES, L. A. (1970). Phanerozoic glaciation and the causes of Ice Ages. *Am. J. Sci.*, **268**, 193.

CROWLEY, D. J. (1973). Middle Silurian patch reefs in Gasport Member (Lockport Formation), New York. *Bull. Am. Ass. Petrol. Geol.*, **57**, 283.

ČUMAKOV, N. M. (1964). Präkambrische tillitähnliche gesteine der Sowietunion. *Geol. Rundsch.*, **54**, 83.

ČUMAKOV, N. and CAILEUX, A. (1971). Glaciation et eolisation dans l'Est et lé Nord de l'Europe a l'Éocambrien. *Rev. Géomorphol. Dyn.*, **20**, 1.

DANGEARD, L. and DORÉ, F. (1971). Faciès Glaciaires de l'Ordovicien Supérieur en Normandie. *Mem. BRGM*, **73**, 119.

DEAN, W. T. (1966). the Lower Ordovician stratigraphy and Trilobites of the Landeyran Valley and the neighbouring district of the Montaigne Noire, South-Western France. *Bull. Brit. Mus. (Nat. Hist.) Geol.*, **12**, 247.

DEAN, W. T. (1970–73). The Lower Palaeozoic stratigraphy and faunas of the Taurus Mountains near Beysehir, Turkey. I–III. *Bull. Brit. Mus. (Nat. Hist.) Geol.*, **19**, No. 8; **20**, No. 1; **24**, No. 5.

DEAN, W. T. (1975). Cambrian and Ordovician correlation and trilobite distribution in Turkey. *Fossils Strata* **4**, 353.

DEAN, W. T. and MARTIN, F. (1978). Lower Ordovician acritarchs and trilobites from Bell Island, Eastern Newfoundland. *Bull. Geol. Serv. Can.*, **284**, 19 pp.

DEBYSER, J., DE CARPAL. O., and MERABET, O. (1965). Sur le caractere glaciale de la sédimentation de l'Unite IV au Sahara Central. *C.R. Acad. Sci.*, **261**, 5575.

DESIO, A. (1930). Sulla presenza del siluriano fossilifero nell'isola di Coo (Egeo). *Rendic. R. Acad. Naz. Lincei. Cl. Fis. Mat. Natural, XI Ser.*, **6**, 1020.

DESTOMBES, J. (1963). Quelques nouveaux Phacopina (trilobites) de l'Ordovicien supérieur de l'Anti-Atlas (Maroc). *Notes Serv. Géol. Maroc.*, **23**, 47.

DESTOMBES, J. (1966). Quelques Calymenina (Trilobitae) de l'Ordovicien moyen et supérieur de l'Anti-Atlas (Maroc). *Notes Serv. Géol. Maroc.*, **26**, 38.

DESTOMBES, J. (1967). Quelques trilobites rares (*Lichas, Amphytrion, Dionide*) de l'Ashgill (Ordovicien supérieur) de l'Anti-Atlas, Maroc. *Ann. Soc. Géol. Nord.*, **87**, 123.

DESTOMBES, J. (1968). Sur la nature glaciare des sédiments du groupe du 2.e. Bani, Ashgill Supérieur de l'Anti-Atlas (Maroc). *C.R. Acad. Sci.*, **267**, 565.

DESTOMBES, J. (1971). L'Ordovicien au Maroc. Essai de synthèse stratigraphique. *Mem. BRGM*, **73**, 237.

DÉVIGNE, J.-P. (1971). L'Afrique a-t-elle dérivé vers l'Amérique du sud au Précambrien supérieur et au Paléozoique inférieur? 6. *Colloque Geol Africaine, Leicester, 14–17 Avril 1971*, 4 pp.

DÉVIGNE, J.-P. (1972). Age protérozoïque de la disjonction des paléocontinents africain et sud-américain. Les données paléoclimatiques. *C.R. Acad. Sci., Ser. D.*, **275**, 1589–1592.

DEYNOUX, M., DIA, O., SOUGY, J., and TROMPETTE, R. (1972). La glaciation 'fini-Ordovicienne' en Afrique de l'Ouest. *Bull. Soc. Geol. Min. Bretagne*, **C4**, 9.

DORÉ, (1972). La trangression majeure du paléozoique inferieur dans le nord-est du Massif Armoricain. *Bull. Soc. Géol. Fr.*, (7), **14**, 79.

DORÉ, F. and LE GALL, J. (1972). Sédimentologie de la 'Tillite de Feuguerolles' (Ordovicien supérieur de Normandie). *Bull. Soc. Géol. Fr.*, (7), **14**, 199.

DOW, D. B., BEYTH, M., and HAILU, T. (1971). Palaeozoic glacial rocks recently discovered in Northern Ethiopia. *Geol. Mag.*, **108**, 53.

DREWY, G. E., RAMSAY, A. T. S., and GILBERT SMITH, A. (1974). Climatically controlled sediments, the geomagnetic field and trade wind belts on Phanerozoic time. *J. Geol.*, **82**, 531.

DUNN, P. R., THOMSON, B. P., and RANKAMA, K. (1971). Late Pre-Cambrian glaciation in Australia as a stratigraphic boundary. *Nature*, **231**, 498.

DURAZZI, J. T. and STEHLI, F. G. (1972). Average generic age, the plantary temperature gradient, and pole location. *System. Zool.*, **21**, 384.

DÜRKOOP, A. (1970). Brachiopoden aus dem Silur, Devon und Karbon in Afghanistan. *Palaeontographica*, A, **134**, 153.

FISCHER, A. S. (1960). Latitudinal variation in organic diversity. *Evolution*, **14**, 64.

FLÜGEL, H. (1964). Die entwicklung des vorderasiatischen Paläozoicums, *in* STILLE, H. and LOTZE, F. (Eds.), *Geotektonische Forschungen*, 18.

FORTEY, R. A. (1975). Early Ordovician trilobite communities. *Fossils Strata*, **4**, 331.

FORTEY, R. A. and BARNES, C. R. (1977). Early Ordovician conodont and trilobite communities of Spitsbergen. Influence on biogeography. *Alcheringia*, **1**, 297–309.

FRIEND, P. F. (1961). The Devonian Stratigraphy of North and Central Vestspitsbergen. *Proc. Yorks. geol. Soc.*, **33**, 77.

FRIEND, P. F. (1965). Fluviatile sedimentary structures in the Wood Bay Series (Devonian) of Spitsbergen. *Sedimentology*, **5**, 39.

FRIEND, P. F. (1966). Clay fractions and colours of some Devonian red beds in the Catskill Mountains, U.S.A. *Q. J. geol. Soc.*, **122**, 273.

FRIEND, P. F. (1969). Tectonic features of Old Red sedimentation in North Atlantic borders. *Mem. Am. Ass. Petrol. Geol.*, **12**, 703.

GEVIN, P. and MONGEREAU, N. (1968). La tillite de la Gara Assaba (bordure sédimentaire sud-orientale de l'Eglab, Sahara occidental). *Bull. Soc. Géol. Fr.*, (7), **10**, 89.

GIL CID, M. D. (1972). Nota sobre la fauna de Trilobites del Ordovicico de los Montes de Toledo (España). *Bol. R. Soc. Española Hist. Nat. (Geol.)*, **70**, 55.

GNILOVSKAJA, M. B. (1972). *Izvestkovie Vodoroslii Srednego i Pozdnego Ordovika Vostoshnogo Kazachstana*. Nauka, Leningrad, 196 pp.

GORTANI, M. (1922). Faune paleozoiche della Sardegna. *Paleont. Ital.*, **28**, 41 and 85.

GORTANI, M. (1934). Fossili Ordviciani del Caracorum. *Spedizione Ital. de Fillippi nel'Himalaia, Caracorum e Turchestan Cinese* (1913–1914), (2), **5**, 1.

GRAY, J., LAUFELD, S., and BOUCOT, A. J. (1974). Silurian trilete spores and spore tetrads from Gotland: their implications for land plant evolution. *Science*, **185**, 260.

GREILING, L. (1967). Der Thüringische Lederschifer. *Geol. Palaeontol.*, **1**, 3.

GUEIRARD, S., WATERLOT, G., GHERZI, A., and SAMAT, M. (1970). Sur l'âge Llandovérien supérieur a Tarranonien inférieur des schistes à Graptolites du Fenouillet, massif des Maures (Var.). *Bull. Soc. Géol. Fr.*, (7), **12**, 195.

GUILLOU, J.-J. (1969). Contribution a l'étude des minéralisations ordoviciennes en antimon de la Sierra de Caurel (provinces de Lugo et d'Orense). *Sci. Terre*, **14**, 5.

HADDING, A. (1941). The pre-Quarternary sedimentary rocks of Sweden. 6. Reef limestones. *Lunds Univ. Årskr.*, (2), **37**, 1.

HAFENRICHTER, M. (1979). Paläontologisch–ökologische und Lithofazielle Untersuchungen des 'Ashgill-Kalkes'' (Jungordovizium) in Spanien. *Arb. Paläont. Inst. Würzburg*, **3**, 139 pp.

HALLAM, A. (1967). The bearing of certain Paleozoogeographic data on Continental Drift. *Palaeogeogr. Palaeoclimatol. Palaeoecol.*, **3**, 210.

HALLAM, A. (1972). Continental drift and the fossil record. *Sci. Am.*, **227**, 56.

HALLAM, A. (1973). Distributional patterns in contemporary terrestrial and marine animals. *Spec. Pap. Palaeont.*, **12**, 93.

HALSTEAD, L. B. and TURNER, S. (1973). Silurian and Devonian ostracoderms, *in* HALLAM, A. (Ed.), *Atlas of Palaeobiogeography*. Elsevier, Amsterdam, 67.

HAMMANN, W. (1971). Stratigraphische einteilung des spanischen ordoviziums nach dalmanitacea und cheirurina (trilobita). *Mem. BRGM*, **76**, 265.

HAMMANN, W. (1974). Phacopina und Cheirurina (Trilobita) aus dem Ordovizium von Spanien. Senckenbergiana. *Lethaia*, **55**, 1.

HARLAND, W. B. (1967). Early history of the North Atlantic Ocean and its margins. *Nature*, **216**, 464.

HARLAND, W. B. and GAYER, R. A. (1972). The Arctic Caledonides and earlier oceans. *Geol. Mag.*, **109**, 289.

HARLAND, W. B. and RUDWICK, M. J. S. (1964). The great Infra-Cambrian ice age. *Sci. Am.*, **211**, 28.

HARLAND, W. B., HEROD, K. N., and KRINSLEY, D. H. (1966). The definition and identification of tills and tillites. *Earth Sci. Rev.*, **2**, 225.

HARTEVELDT, J. J. A. (1970). Geology of the upper Segre and Valira valleys, Central Pyrenees, Andorra, Spain. *Leidse Geol. Meded.*, **45**, 167.

HAVLIČEK, V. (1958). Ramenonžci českého ordoviku. *Rozpr. U.U. Geol.*, **13**.

HAVLIČEK, V. (1967). Brachiopods of the Suborder Strophomenidina in Czechoslovakia. *Rozpr. U.U. Geol.*, 33.

HAVLIČEK, V. (1971). Brachiopodes de l'Ordovicien du Maroc. *Not. Mem. Serv. Geol.*, 230.

HAVLIČEK, V. and JOSOPAIT, V. (1972). Articulate brachiopods from the Iberian Chains, Northeast Spain (Middle Cambrian–Upper Tremadoc). *Neues Jb. Geol. Paläont. Abh.*, **140**, 328.

HAVLIČEK, V. and MAREK, L. (1973). Bohemian Ordovician and its international correlation. *Casopis Min. Geol.*, **18**, 225.

HAVLIČEK, V. and MASSA, D. (1973). Brachiopodes de l'Ordovicien supérieur de Libye occidentale. Implications stratigraphiques régionales. *Geobios.*, **6**, 267.

HELAL, A. H. (1964). On the occurrence of lower Paleozoic rocks in Tabuk area, Saudi Arabia. *Monatsh. Neues Jb. Geol. Paläont.*, 391.

HELAL, A. H. (1965). General geology and litho-stratigraphic subdivision of the Devonian rocks of the Jauf area, Saudi Arabia. *Monatsh. Neues Jb. Geol. Paläont.*, 527.

HENNINGSMOEN, G. (1957). The Trilobite Family Olenidae. *Skr. Vid. Akad. Oslo, Mat. Nat. Kl.*, No. 1.

HENRY, J.-L. (1966). Sur un nouveau Phacopina (Trilobite) de l'Ordovicien de Bretagne. *Bull. Soc. Géol. Fr.* (7), **7**, 558.

HENRY, J.-L. (1969). Données stratigraphiques sur l'Ordovicien de Bretagne et de Normandie. *Bull. Soc. Géol. Min. Bretagne*, **1**, 11.

HILL, D. (1972). Archeocyatha, *in* TEICHERT, C. (Ed.), *Treatise on Invertebrate Paleontology, Part E*, Geological Society of America and University of Kansas, Boulder, Colarado, and Lawrence, Kansas.

HØEG, O. A. (1927). *Dimorphosiphon rectangularis*. Preliminary note on a new *Codiacea* from the Ordovician of Norway. *Avh. Vid. Akad. Oslo, Mat.-Nat. Kl.*, No. 4.

HOLLAND, C. H. (1971). Silurian faunal provinces, *in* MIDDLEMISS, F. A., RAWSON, P. F., and NEWALL, G. (Eds.), *Faunal Provinces in Space and Time*, Seel House Press, Liverpool, 61.

HOLLAND, C. H. and STURT, B. A. (1970). On the occurrence of archaeocyathids in the Caledonian metamorphic rocks of Sørøy and their stratigraphic significance. *Norsk Geol. Tidsskr.*, **50**, 341.

HOLTEDAHL, O. (1961). The 'Sparagmite Formation' (Kjerulf) and 'Eocambrian' (Brøgger) of the Scandinavian Peninsula. Symp. El Sistema Cambrico su paleogeografia y el problema de seu base. *XX Congr. Geol.*, **3**, 9 (for date of publication, *see* SPJELDNAES, 1964, 43).

HUCKRIEDE, R., KÜRSTEN, M., and VENZLAFF, H. (1962). Zur Geologie des Gebietes zwischen Kerman und Sagand (Iran). *Beih. Geol. Jb.*, 51.

HUTT, J. and RICKARDS, R. B. (1970). The evolution of the earliest Llandovery monograptids. *Geol. Mag.*, **107**, 67.

INSTITUT ALGERIAN DU PÉTROLE. (1970). *Paleozoique Inférieur du Sahara (Voyage d'Étude Sédimentologique)*. IAB Algiers.

ILLIES, H. (1975). Intraplate tectonics in stable Europe as related to plate tectonics in the Alpine system. *Geol. Rundschau*, **64**, 677.

JAANUSSON, V. (1972). Aspects of carbonate sedimentation in the Ordovician of Baltoscandia. *Lethaia*, **6**, 11.

JAEGER, H. (1970). Ordoviz auf Rügen. *Ber. Dt. Ges. Geol. Wiss. A, Geol. Paläont.*, **12**, 165.

JAEGER, H., HAVLIČEK, V., and SCHÖNLAUB, H. P. (1975). Biostratigraphie der Ordovicium/Silur-Grenze in den Südalpen—Ein Beitrag um die Hirnatia-Fauna. *Verh. Geol. Bundesanst. Wien*, 271.

JELL, P. A. (1974). Faunal provinces and possible planetary reconstructions of the Middle Cambrian. *J. Geol.*, **82**, 319.

JOHNSON, J. H. (1954). An introduction to the study of rock building algae and algal limestones. *Q. J. Colorado School Mines*, **49**, no. 2.

JOSAPAIT, V. (1972). Das Kambrium und das Tremadoc von Ateca (Westliche Iberische Ketten, NE-Spanien). *Münster Forsch. Geol. Paläont.*, 23.

JULIVERT, M. and TRUYOLS, J. (1972). La coupe du Cabo Peñas, une coupe de référence pour l'Ordovicien du Nord-Ouest de l'Espagne. *C.R. Somm. Soc. Géol. Fr.*, 241.

JULIVERT, M., MARCOS, A., and TRUYOLS, J. (1973). L'evolution paleogeographique du nord-ouest de l'Espagne pendant l'Ordovicien-Silurien. *Bull. Soc. Géol. Min. Bretagne, Ser. C.*, 4, 1.

JULIVERT, M., MARCOS, A., PHILIPPOT, A., and HENRY, J. L. (1968). Nota sobre la extension de las pizarras ordovicicas al E. de la cuenca carbonifera central de Asturias. *Brev. Geol. Asturica*, 12, 1.

KAPP, U. S. (1975). Paleoecology of Middle Ordovician stromatoporoid mounds in Vermont. *Lethaia*, 8, 195.

KATZUNG, G. (1961). Die Geröllführung lederschiefers (Ordovizium) an der SE-Flanke des Schwartzburger Sattels (Thüringen). *Geologie*, 10, 778.

KELLER, B. M. and others (1954–61). Ordovik Kazachstana. *Tr. Inst. Geol. Nauk Akad. SSSR.*, 1, 9, 18, 200 + 232 + 152 pp.

KHRAMOV, A. N. (1971). Palaeomagnetic directions and palaeomagnetic poles, *in* McELHINNY, M. W. and BROWN, D. A. (Eds.), (1973). *A Compilation of Palaeomagnetic Results from the USSR*. Research School of Earth Sciences, Canberra, Publ. 990.

KIELAN, Z. (1958). Upper Ordovician trilobites from Poland and some related forms from Bohemia and Scandinavia. *Paleont. Polonica*, 11.

KLEIN, G. DEV. (1970). Tidal origin of a Precambrian quartzite—The lower fine-grained quartzite (Middle Dalradian) of Islay, Scotland. *J. Sed. Petrol.*, 40, 973.

KLEIN, G. DEV. (1971). A sedimentary model of determining paleotidal range. *Bull. Geol. Soc. Am.*, 82, 2585.

KLEIN, G. DEV. (1972). Sedimentary model for determining paleotidal range: reply. *Bull. Geol. Soc. Am.*, 83, 539.

KLEIN, G. DEV. (1972). Determination of paleotidal range in clastic sedimentary rocks. *Proc. 24th Int. Geol. Congr., Sect.* 6, 97.

KOBAYASHI, T. (1972). Three faunal provinces in the Early Cambrian period. *Proc. Jap. Acad.*, 48, 242.

LAMBECK, K. (1978). The earth's palaeorotation, *in* BROSCHE, P. and SÜNDERMANN, J. (Eds.), *Tidal Friction and the Earth's Rotation*. Springer, Berlin, Heidelberg, 145.

LEES, A. (1975). Possible influence of salinity and temperature on modern shelf carbonate sedimentation. *Mar. Geol.*, 19, 159.

LEES, A. and BULLER, A. T. (1972). Modern temperate-water and warm-water shelf carbonate sediments contrasted. *Mar. Geol.*, 13, M67.

LEGRAND, PH. (1971). Les couches a Diplograptus du Tassili de Tarit (Ahnet, Sahara algerien). *Bull. Soc. Hist. Nat. Afr. Nord*, 16, 3.

LESPÉRANCE, P. J. (1974). The Hirnantian fauna of the Perce area (Quebec) and the Ordovician–Silurian boundary. *Am. J. Sci.*, 274, 10.

LEVY, R. and NULLO, F. (1974). La fauna del Ordovicico (Ashgilliano) de Villicun, San Juan, Argentina ('Brachiopoda'). *Ameghinia*, 11, 173.

LINDSTRÖM, M. (1972). Cold Age sediments in Lower Cambrian of South Sweden. *Geol. Palaeont.*, 6, 9.

LINDSTRÖM, M. (1972a). Ice-marked sand grains in the Lower Ordovician of Sweden. *Geol. Palaeont.*, 6, 25.

LINDSTRÖM, M. and PELHATE, A. (1971). Présence de Conodontes dans les calcaires de Rosan (Ordovicien moyen à supérieur, Massif Armoricain). *Mem. BRGM*, 73, 89.

LINDSTRÖM, M., RACHEBOEUF, P. R., and HENRY, J.-L. (1974). Ordovician condonts from the Postolonnec Formation (Crozon peninsula, Massif Armoricain) and their stratigraphic significance. *Geol. Palaeont.*, 8, 15.

LISITZIN, A. L. (1972). Sedimentation in the world oceans. *Spec. Pub. Soc. Econ. Paleont. Min.*, 17.

LOCHMAN-BALK, C. (1970). Upper Cambrian faunal patterns on the Craton. *Bull. Geol. Soc. Am.*, 81, 3197.

LOCHMAN-BALK, C. (1971). The Cambrian of the Craton of the United States *in* HOLLAND, C. H. (Ed.), *Lower Palaeozoic Rocks of the World*, Vol. 1 Wiley, New York, 79.

LOCK, B. E. (1973). The Ordovician Ice Age in South Africa. *Geol. Mag.*, 110, 372.

LOTZE, F. (1938). Steinsalz und Kalisalze, Geologie, *in* STUTZER, O. (Ed.), *Die Wichtigsten Lagerstätten der 'Nicht-Erze'*, 3, Teil 1. Geb. Borntraeger, Berlin.

LØVLIE, R. and KVINGEDAL, M. (1975). A palaeomagnetic discordance between a lava sequence and an associated interbasaltic horizon from the Faeroe Islands. *Geophys. J. R. Astron. Soc.*, 40, 45.

LOWENSTAM, H. A. (1948). Biostratigraphic studies of the Niagaran inter-reef formations of Northeastern Illinois. *Sci. Pap. Ill. State Mus.*, **4**.

LOWENSTAM, H. A. (1950). Niagaran Reefs in the Great Lakes Area. *J. Geol.*, **58**, 430.

LOWENSTAM, H. A. (1957). Niagaran Reefs in the Great Lakes Area, *in* HEDGPETH, J. W. (Ed.), *Treatise on Marine Ecology and Paleoecology. Mem. Geol. Soc. Am.*, **67-II**, 215.

LUCAS, G. (1940). Contribution a l'étude de Silurien de la presque ila de Crozon. *Bull. Soc. Min. Bretagne, n.s.*, **95**.

MANTEN, A. A. (1971). *Silurian Reefs of Gotland.* Elsevier, Amsterdam.

MAUCHER, A. and SAUPÉ, F. (1967). Sedimentärer Pyrit aus der Zinnober-Lagerstätte Almadén, (Provinz Ciudad Real, Spanien). *Minerali. Depos.*, **2**, 312.

MCCANN, A. M. and KENNEDY, M. J. (1974). A probable glacio-marine deposit of Late Ordovician–Early Silurian age from the north central Newfoundland Appalachian Belt. *Geol. Mag.*, **111**, 549.

MCELHINNY, M. W., BRIDEN, J. C., JONES, D. L., and BROCK, A. (1968). Geological and geophysical implications of paleomagnetic results from Africa. *Rev. Geophys.*, **6**, 201.

MCELHINNEY, M. W. and BRIDEN, J. C. (1971). Continental drift during the Palaeozoic. *Earth Planet. Sci. Lett.*, **10**, 407.

MCKERROW, W. S. and ZIEGLER, A. M. (1972). Silurian Paleogeographic Development of the Proto-Atlantic Ocean. *Proc. 24th Int. Geol. Congr.*, Sect., **6**, 4.

MELOU, M. (1971). Nouvelle espece de Leptestiina dans l'Ordovicien supérieur de l'Aulne (Finistère). *Mem. BRGM*, **73**, 93.

MELOU, M. (1973). Le genre *Aegiromena* (Brachiopode, Strophomenida) dans l'Ordovician de Massif armoricain (France). *Ann. Soc. Géol. Nord*, **93**, 253.

MELOU, M. (1975). Le genre Heterorthina (Brachiopoda, Orthida) dans la Formation des schistes de Postolonnec (Ordovician), Finistère, France. *Geobios*, **8**, 191.

MERIFIELD, P. M. and LAMAR, D. L. (1970). Paleotides and the geologic record, *in* RUNCORN, S. K. (Ed.), *Palaeogeophysics.* Academic Press, New York, 31.

MITCHELL, W. I. (1974). An outline of the stratigraphy and palaeontology of the Ordovician rocks of Central Portugal. *Geol. Mag.*, **111**, 385.

MODLINSKI, Z. (1973). Stratigraphy and development of the Ordovician in North-Eastern Poland [in Polish, with summary in English]. *Prace Inst. Geol.*, **72**.

MORI, K. (1968–70). Stomatoporoids from the Silurian of Gotland. I–II. *Stockholm Contrib. Geol.*, **19**, 1 and **22**, 1.

MU, A.-T., WEN, S. H., WANG, Y.-K., and CHANG, P. K. (1973). Stratigraphy of the Mount Jolmo Lungma region in Southern Tibet, China. *Sci. Sin.*, **16**, 96.

NEUMAN, R. B. (1972). Brachiopods of Early Ordovician Volcanic Islands. *Proc. 24th Int. Geol. Congr.*, Sect. **7**, 297.

NICOL, D. (1970). Antarctic Pelecypod faunal peculiarities. *Science*, **168**, 1248.

NIKIFOROVA, O. I. and PREDTECHENSKIJ, N. N. (1968). *A Guide to the Geological Excursion on Silurian and Lower Devonian Deposits of Podolia.* VSEGEI, Leningrad.

NIKITIN, I. F. (1972–73). *Ordovik Kazakchstana* 1–2. Akad. Nauk KazSSR Alma-Ata. 242 and 99 pp.

NITECKI, M. H. (1972). The Paleogeographic significance of the Receptaculitids. *Proc. 24th Int. Geol. Congr.*, Sect., 7, 303.

NITECKI, M. H. and FORNEY, G. G. (1978). Ordovician *Receptaculites camacho* n. sp. from Argentina. *Feildiana, Geol.*, **37**, 93–110.

OLSON, W. S. (1970). Tidal amplitudes in geologic history. *Trans. N.Y. Acad. Sci.*, (2), **32**, 220.

OMARA, S. (1972). An Early Cambrian outcrop in Southwestern Sinai, Egypt. *Monatsh. Neues Jb. Geol. Paläont.*, 306.

ÖPIK, A. A. (1957). Cambrian Palaeogeography of Australia. *Bull. Bur. Min. Res.*, **49**, 239.

PALMER, A. R. (1971). The Cambrian of the Great Basin and adjacent areas, Western United States, *in* HOLLAND, C. H. (Ed.), *Lower Palaeozoic Rocks of the World*, Vol. 1. Wiley, New York, 1.

PALMER, A. R. (1971a). The Cambrian of the Appalachian and Eastern New England Regions, Eastern United States, *in* HOLLAND, C. H. (Ed.), *Lower Palaeozoic Rocks of the World*, Vol. 1. Wiley, New York, 169.

PALMER, A. R. (1972). Problems of Cambrian biogeography. *Proc. 24th Int. Geol. Congr.*, Sect. 7, 310.

PALMER, A. R. (1973). Cambrian trilobites, *in* HALLAM, A. (Ed.), *Atlas of Palaeobiogeography*, Elsevier, Amsterdam, 3.

PATRULIUS, D. and IORDAN, M. (1974). Asupra prezentei pogonoforuli Sabadellites cambriensis Ian. şi a 'algei' Vendotaenia antiqua Gnil. in depozitele detritice presiluriene din podisұl Moldovenesc. *Dări de Seamă Inst. Geol. 4 (Stratigrafie)*, **60**, 1.

PATTON, W. W., JR., and DUTRO, J. T., JR. (1969). Preliminary report on the Paleozoic and Mesozoic sedimentary sequence on St. Lawrence Island, Alaska. *Prof. Pap. U.S. Geol. Surv.* **650-D**, D138.

PFEIFFER, H. (1972). Zur Bildungsgeschichte von Hauptquarzit und Lederśchiefer (Ordovizium, Saxo-thuringikum). *Geologie*, **21**, 763.

PHILIPPOT, A. (1963). Remarques sur la sédimentation de l'Ordovicien supérieur et de l'Ordovicien moyen dans la presqu'île de Crozon (Finistère). *Bull. Soc. Géol. Min. Bretagne*, **1–2**, 133.

PILLET, J. and ROBARDET, M. (1969). Les 'Schistes à Trinucleus' de la Sangsuriere (Manche). *Bull. Soc. Linn. Normandie*, (10), **9**, 66.

PIPER, J. D. A. (1978). Geological and geophysical evidence relating to continental growth and dynamics and the hydrosphere in Precambrian times: a review and analysis, *in* BROSCHE, P. and SÜNDERMANN, J. (Eds.), *Tidal Friction and the Earth's Rotation*. Springer, Berlin, Heidelberg, 197.

POWERS, R. W., RAMIREZ, L. F., REDMOND, C. D., and ELBERG, E. L. (1966). Geology of the Arabian Peninsula: sedimentary geology of Saudi Arabia. *Prof. Pap. U.S. Geol. Surv.* **560-D**.

PUSCHMANN, H. (1968). Stratigraphische Untersuchungen im Paläozoicum des Montseny (Katalonien/Spanien). *Geol. Rundschau*, **57**, 1066.

QUENSTEDT, W. (1926). Mollusken aus den Redbay- und Greyhookschichten Spitzbergen. *Skr. Svalbard Ishavet*, **11**.

RADIG, F. (1962). Ordovizium/Silurium und die Frage prävariszischer Faltungen in Nordspanien. *Geol. Rundschau*, **52**, 346.

RAMBERG, I. B. and SPJELDNAES, N. (1978). The tectonic history of the Oslo region, *in* RAMBERG, I. B. and NEUMANN, E. R. (Eds.), *Tectonics and Geophysics of Continental Rifts*. Reidel, Dordrecht, 167.

RAUP, D. M. (1972). Taxonomic diversity during the Phanerozoic. *Science*, **177**, 1065.

ROBARDET, M. (1973). *Évolution géodynamique du nord-est du Massif Armoricain au Paléozoique*. Thèse a l'Université de Paris, Reg. No. CNRS A0 8533.

ROGNON, P., BIJU-DUVAL, B., and DE CHARPAL, O. (1972). Modelés glaciaires dans l'Ordovicien Supérieur saharien: phases d'érosion et glaciotectonique sur la bordure Nord des Eglab. *Rev. Géogr. Phys. Géol. Dynam.*, **14**, 507.

ROHR, D. M., BOUCOT, A. J., and POTTER, A. W. (1974). Age corrections for some Silurian localities in northern California. *J. Paleont.*, **48**, 413.

ROSS, R. J., JR. (1975). Early Paleozoic trilobites, sedimentary facies, lithosphere plates and ocean currents. *Fossils Strata*, **4**, 307.

ROSS, R. J. and DUTRO, J. T. (1966). Silicified Ordovician Brachiopods from East-Central Alaska. *Smithsonian Misc. Coll.*, **149**.

ROMARIZ, C. and DINIZ, F. (1962). Alguns aspectos petrograficos dos calcarios ordovicico-siluricos portugueses. *Rev. Fac. Cienc. Lisboa.*, *2. Ser.*, **10**, 55.

ROMARIZ, C., ARCHE, A., BARBA, A., GUTIERREZ ELORZA, M., and VEGAS, R. (1971). The mediterranean graptolitic fauna of the wenclokian in the Iberian Peninsula. *Bol. Soc. Geol. Portugal*, **18**, 57.

ROZMAN, CH. S. (1968). Jarusnoe rarasclenenje verchnego ordovika i biogeograficheskie osobennosti razvitija pozdneordovikskoi fauni. *Mezh. Geol. Kongr. XXIII, Probl.*, **9**, 95.

ROZMAN, CH. S., IVANOVA, V. A., KRASILOVA, I. N., and MODZALEVSKAJA, E. A. (1970). Biostratigrafija verchnego ordovika severo-vostoka SSSR. *Tr. Geol. Inst. Akad. Nauk SSSR.*, **205**, 318 pp.

RUNCORN, S. K. (1959). Rock magnetism. *Science*, **129**, 1002.

RUNCORN, S. K. (1971). Marine life and the rotation rates of the earth and moon, *in* MAXWELL, A. E. (Ed.), *The Sea*, Vol. 4. Wiley, New York, 759.

RUST, I. C. (1969). The Western Cape some 450 million years ago. *S. Afr. J. Geogr.*, **3**, 351.

RUTTEN, M. G. (1953). Shallow shelf sea sedimentation during non-glacial and a-tectonic times in geologic history. *Proc. 19th Int. Geol. Congr.*, Sect. 4, 119.

RUTTNER, A., NABAVI, M., and HADJIAN, J. (1968). Geology of the Shirgesht area, Tabas area, East Iran. *Rep. Geol. Surv. Iran*, **4**.

SACHER, L. (1966). Stratigraphie und Tektonik der nordwestlichen Hesperidischen Ketten bei Molina de Aragon/Spanien. *Neues Jb. Geol. Paläont. Abh.*, **124**, 151.

SAUPÉ, F. (1967). Note preliminaire concernant la genese du gisement de mercure d'Almadén, Province de Ciudad Real, Espagne. *Mineral. Depos.*, **2**, 26.

SAUPÉ, F. (1971). La série ordovicienne et silurienne d'Almadén (Province de Ciudad Real, Espagne), points des connaissances actuelles. *Mem. BRGM*, **73**, 355.

SCHERMERHORN, L. J. G. (1974). Late Precambrian mixtites: glacial or and/or non-glacial? *Am. J. Sci.*, **274**, 673.

SCHERMERHORN, L. J. G. (1975). Tectonic framework of Late Precambrian supposed glacials, *in* WRIGHT, A. E. and MOSELEY, F. (Eds.), *Ice Ages: Ancient and Modern. Geol. J. Spec. Issue*, **6**, 241.

SCHRENK, P. E. (1972). Possible Late Ordovician glaciation in Nova Scotia. *Can. J. Earth Sci.*, **9**, 95.

SCHWARZBACH, M. (1950). *Das Klima der Vorzeit*, (3. Auft. 1974), F. Enke Verlag, Berlin.

SCROTESE, C. R., BAMBACH, R. K., BARTON, C., VAN DER VOO, R., and ZIEGLER, A. M., (1979). Paleozoic base maps. *J. Geol.*, **87**, 217.

SCRUTTON, C. T. (1978). Periodic growth features in fossil organisms and the length of the day and month, *in* BROSCHE, P. and SÜNDERMANN, J. (Eds.), *Tidal Friction and the Earth's Rotation*. Springer, Berlin, Heidelberg, 154.

SDZUY, K. (1972). Das Kambrium der Acadobaltischen Faunenprovinz. *Zbl. Geol. Paläont.*, **II**, 1.

SELLEY, R. C. (1972). Diagnosis of marine and non-marine environments from the Cambro-Ordovician sandstones of Jordan. *J. geol. Soc.*, **128**, 135.

SERPAGLI, E. (1973). Carbonati de tipo bahamitico nell ordoviciano inferiore delle precordillera argentina e relative osservationi paleoclimato logiche. *Atti Soc. Nat. Mat. Modena*, **104**, 239.

SERPAGLI, E. (1974). Un momento della storia geologica dell' Ordoviciano atlantico: Conodonti baltici nella Precordillera argentina. *Acc. Naz. Lincei, Rendic. Sci., Fis., Mat. Nat., Ser. VIII*, **55**, No. 5.

SHAW, F. C. and FORTEY, R. A. (1977). Middle Ordovician facies and trilobite faunas. *Geol. Mag.*, **114**, 409–443.

SHEEHAN, P. M. (1972). The relation of Late Ordovician glaciation to the Ordovician–Silurian change-over in North American brachiopod faunas. *Lethaia*, **6**, 147.

SHEEHAN, P. M. (1975). Brachiopod synecology in a time of crisis (Late Ordovician–Early Silurian). *Paleobiology*, **1**, 205.

SHEEHAN, P. M. (1975a). Late Ordovician Brachiopods from Belgium. *Abstr. Progr. Geol. Soc. Am.*, 1267.

SHELL and CREOLE (1964). Paleozoic rocks of Merida Andes, Venezuela. *Bull. Am. Ass. Petrol. Geol.*, **48**, 70.

SHERWIN, L. (1975). Glaciation and the Benambran Orogeny. *Q. Notes Geol. Surv. N.S. Wales*, **18**, 13.

SHULGA, P. L. (Ed.) (1972). *Stratigrafija URSR, III, Kembrii, Ordovik*. Nauknova Dumka, Kiev.

SIEDLECKA, A. and SIEDLECKI, S. (1972). A Contribution to the Geology of the Downtonian Sedimentary Rocks of Hitra. *Norges Geol. Unders.*, **275**, No. 3.

SKEVINGTON, D. (1973). Ordovician graptolites, *in* HALLAM, A. (Ed.), *Atlas of palaeobiogeography*. Elsevier, Amsterdam, 27.

SKEVINGTON, D. (1974). Controls influencing the composition and distribution of Ordovician graptolite faunal provinces, *in* RICKARDS, R. B., JACKSON, D. E., and HUGHES, C. P. (Eds.), *Graptolite Studies in Honour of O.M.B. Bulman. Spec. Pap. Palaeont.*, **13**, 59.

SMITH, A. G., BRIDEN, J. C., and DREWRY, G. E. (1973). Phaenerozoic world maps. *Spec. Pap. Palaeont.*, **12**, 1.

SOKOLOV, B. S. (1962). *Paleogeograficheskii Atlas Kitaya. Izdateljstvo inostrannoi Literaturi*. Moscow, 119 pp.

SOKOLOV, B. S. (1963). *Regionalnaja stratigrafija Kitaya. Izdateljstvo Inostrannoi Literaturi*. Moscow, 274 pp.

SOKOLOV, B. S. (Ed.) (1972). Opornji razrez silura i nizhnego devona podolii. *Tr. Mezh. Strat. Kom. SSSR*, 5.

SOKOLOV, B. S. and TESAKOV, YU. I. (Eds.) (1975). Stratigrafija ordovika sibirskoi platformi. *Tr. Inst. Geol. Geofiz. Akad. Nauk SSSR, Sib. Otdel.*, **200**.

SOKOLOV, B. S. and YOLKIN, E. A. (Eds.) (1978). Pogranitsie sloj ordovika i silura Altae-Sajanskogj oblasti i Tian Shan. *Tr. Inst. Geol. Geofiz. Sibir. Otdel. Akad Nauk SSSR*, **397**.

SOLLE, G. (1935). Die Devonischen Ostracoden Spitzbergens. *Skr. Svalbard Ishavet.*, 64.

SOUGY, J. and LÉCORCHÉ, J.-P. (1963). sur la nature glaciare de la base de la serie de Garat el Hamoueïd (Zemmour, Mauritanie serpentrionale). *C.R. Acad. Sci.*, **256**, 4471.

SPASOV, H. R. (1958). *Fosilite na Bulgarija. I. Paleozoi*. Bulg. Akad. na Naukite, Sofia.

SPJELDNAES, N. (1955). *Coelosphaeridium* (Chlorophyta Dasycladacea) from the Caradocian beds of N. Wales. *Norsk Geol. Tidsskr.*, **35**, 151.

SPJELDNAES, N. (1961). Ordovician Climatic Zones. *Norsk Geol. Tidsskr.*, **41**, 45.

SPJELDNAES, N. (1964). The Eocambrian glaciation in Norway. *Geol. Rundschau*, **54**, 24.

SPJELDNAES, N. (1966). Silurian tidal sediments from the base of the Ringerike Formation, Oslo Region, Norway. *Norsk Geol. Tidsskr.*, **46**, 497.

SPJELDNAES, N. (1967). The Palaeogeography of the Tethyan Region during the Ordovician. *Publ. Syst. Ass.*, **7**, 45.

SPJELDNAES, N. (1973). Moraine stratigraphy, with examples from the basal Cambrian ('Eocambrian') and Ordovician glaciations. *Bull. Geol. Inst. Uppsala*, **5**, 165.

ST. JEAN., J., JR. (1971). Paleobiologic Considerations of Reef Stromatoporoids. *Proc. North. Am. Paleont. Conv.*, *J*, 1389.

STEHLI, F. G. (1973). Review of Paleoclimate and Continental Drift. *Earth Sci. Rev.*, **9**, 1.

STEHLI, F. G. and WELLS, J. W. (1971). Diversity and age patterns in hermatypic corals. *System. Zool.*, **20**, 115.

STORETVEDT, K. (1967). A synthesis of the Palaeozoic palaeomagnetic data for Europe. *Earth Planet. Sci. Lett.*, **3**, 444.

STORETVEDT, K. (1970). The Devonian palaeomagnetic field for Europe, *in* RUNCORN, S. K. (Ed.), *Palaeogophisics*. Pergamon Press, Oxford, 247.

STØRMER, L. (1967). Some aspects of the Caledonian geosyncline and foreland west of the Baltic Shield. *Q. J. geol. Soc.*, **123**, 183.

STRØM, K. (1962). Meere der Vorzeit als Bildungsräume des Schwarzschiefer. *Natur Museum*, **92**, 415.

SWEET, W. C. (1958). The Middle Ordovician of the Oslo Region, Norway. 10. Nautiloid Cephalopods. *Norsk Geol. Tidsskr.*, **38**, 1.

SYLVESTER-BRADLEY, P. C. (1971). Dynamic factors in animal palaeogeography, *in* MIDDLEMISS, F. A., RAWSON, P. F., and NEWALL, G. (Eds.), *Faunal Provinces in Space and Time. Geol. J. Spec. Issue*, **4**, 1.

TAMAIN, G. (1971). L'Ordovicien est-marianque (Espagne), sa place dans la province mediterraneenne. *Mem. BRGM*, **73**, 403.

TAMAIN, G. (1972). *Recherches géologiques et Minières en Sierra Morena Orientale (Espagne)*. Thèse a l'Université de Paris/Orsay, Reg. No. 990.

TAMAIN, G., OVTRACHT, A., CARRÉ, J., HELOIR, J.-P., PERAN, M., and POUPON, G. (1970). L'Ordovicien de la Sierra Morena Orientale (Espagne). *C.R. 94. Congr. Nat. Soc. Savantes., Sect. Sci.*, **2**, 275.

TARLING, D. H. (1974). A palaeomagnetic study of Eocambrian tillites in Scotland. *J. geol. Soc.*, **130**, 163.

TEICHERT, C. (1958). Cold- and Deep-water coral banks. *Bull. Am. Ass. Petrol. Geol.*, **42**, 1064.

TEICHERT, C. (1974). Marine sedimentary environments and their faunas in Gondwana area. *Mem. Am. Ass. Petrol. Geol.*, **23**, 361.

TEICHERT, C. and STAUFFER, K. W. (1965). Paleozoic reefs in Pakistan. *Science*, **150**, 1287.

TENCOV, J., SPASOV, HR., and JANEV, SL. (1963). Sur la stratigraphie du Paléozoïque en Bulgarie Occidentale. *5. Congr. Ass. Geol. Carpato-Balkanique*, **3/2**, 245.

TERMIER, G. and TERMIER, H. (1950). *Paléontologie Marocaine, II, Invertébrés de l'Ere Primaire*. Fasc. 1–4. Hermann et Cie, Paris.

THOMPSON, R. (1973). South American Palaeozoic palaeomagnetic results and the welding of Pangaea. *Earth Planet. Sci. Lett.*, **18**, 226.

THORSON, G. (1957). Bottom communities, *in* HEDGPETH, J. W. (Ed.), *Treatise on Marine Ecology and Paleoecology. Mem. Geol. Soc. Am.*, **67-1**, 461.

THORSON, G. (1966). Some factors influencing the recruitment and establishment of marine benthic communities. *Neth. J. Sea Res.*, **3**, 267.

TROMPETTE, R. (1972). Le Précambrian supérieur et le Paléozoïque inférieur de l'Adrar de Mauritanie. *Trav. Lab. Sci. Terre. Marseilles*, No. 6.

TSCHOPP, R. H. (1967). The general geology of Oman. *Proc. 7th Petrol. Congr. Mexico*, **16**, 231.

TUCKER, M. E. and REID, P. C. (1973). The sedimentology and context of Late Ordovician glacial marine sediments from Sierra Leone, West Africa. *Palaeogeogr., Palaeoclimatol., Palaeoecol.*, **13**, 289.

TURNER, P. (1974). Origin of red beds in the Ringerike Group (Silurian) of Norway. *Sedimentary Geol.*, **12**, 215.

VALENTINE, J. W. (1973). Plates and provinciality, a theoretical history of environmental discontinuities. *Spec. Pap. Palaeont.*, **12**, 79.

VALENTINE, J. W. and MOORES, E. M. (1970). Plate tectonic regulation of faunal diversity and sea level: a model. *Nature*, **228**, 657.

VATAN, A. and YASSINI, Y. (1969). Les grandes lignes de la géologie de l'Elbourz Central dans la region de Teheran et la Plaine de la Caspienne. *Rev. IFP, Ann. Combust. Liq.*, **24**, 841.

VINE, F. J. (1973). Organic diversity, palaeomagnetism, and Permian palaeogeography. *Spec. Pap. Palaeont.*, **12**, 61.

VISSER, J. N. J. (1974). The Table Mountain Group: a study in the deposition of quartz arenites on a stable shelf. *Trans. Geol. Soc. S. Afr.*, **77**, 229.

WEBBY, B. D. (1971). The trilobite *Pliomerina* Chugaeva from the Ordovician of New South Wales. *Palaeontology*, **14**, 612.

WEISBROD, T. (1969). The Paleozoic outcrops of southwestern Sinai and their correlation with those of southern Israel. *Bull. Geol. Surv. Isr.*, 48.

WHITTINGTON, H. B. (1966). Phylogeny and Distribution of Ordovician Trilobites. *J. Paleont.*, **40**, 696.

WHITTINGTON, H. B. (1973). Ordovician Trilobites, *in* HALLAM, A. (Ed.), *Atlas of Palaeobiogeography*. Elsevier, Amsterdam, 13.

WHITTINGTON, H. B. and HUGHES, C. P. (1972). Ordovician geography and faunal provinces deduced from trilobite distribution. *Phil. Trans. Roy. Soc.*, **263**, 235.

WHITTINGTON, H. B. and HUGHES, C. P. (1973). Ordovician trilobite distribution and geography. *Spec. Pap. Paleont.*, **12**, 235.

WILLIAMS, A. (1962). The Barr and Lower Ardmillan Series (Caradoc) of the Girvan District, South-West Ayrshire, with descriptions of the Brachiopoda. *Mem. Geol. Soc. Lond.*, 3.

WILLIAMS, A. (1963). The Caradocian Brachiopods faunas of the Bala district, Merionethshire. *Bull. Brit. Mus. (Nat. Hist.) Geol.*, **8**, 327.

WILLIAMS, A. (1969). Ordovician faunal provinces with reference to Brachiopod distribution, *in* WOOD, A. (Ed.), *The Pre-Cambrian and Lower Palaeozoic Rocks of Wales*. University of Wales Press, Cardiff, 117.

WILLIAMS, A. (1973). Distribution of Brachiopod assemblages in relation to Ordovician palaeogeography. *Spec. Pap. Palaeont.*, **12**, 241.

WILLIAMS, A. (1974). Ordovician Brachiopoda from the Shelve District, Shropshire. *Bull. Brit. Mus. (Nat. Hist.) Geol.*, Suppl. **11**, 163 pp.

WILLIAMS, G. E. (1972). Geological evidence relating to the origin and secular rotation of the solar system. *Mod. Geol.*, **3**, 165.

WILLIAMS, G. E. (1975). Late Precambrian glacial climate and the Earth's obliquity. *Geol. Mag.*, **112**, 441.

WILSON, J. T. (1966). Did the Atlantic close, and then re-open? *Nature*, **211**, 676.

WOLFART, R. (1970). Fauna, Stratigraphie und Paläogeografie des Ordoviziums in Afghanistan. *Beiheft. Geol. Jb.*, **89**.

WOLFART, R. (1970a). The age of the Early Tremadocian and of the Saukia Zone and the boundary between Cambrian and Ordovician. *Newsl. Stratgr.*, **1**, 10.

WORSLEY, D. (1972). Sedimentological observations on the Grey Hoek Formation of northern Andree Land, Spitsbergen. *Årbok Norsk Polarinst.*, **1970**, 102.

ZHURAVLEVA, I. T. (1966). Rannekembriyskie organogennye postroyki na territorii Sibirskoy platformy, *in* GEKKER, R. F. (Ed.), *Organizmi i Sreda v Geologicheskom Proshlom*. Akad. Nauk SSSR, Nauka, Moscow, 61.

ZIEGLER, A. M. (1965). Silurian marine communities and their environmental significance. *Nature*, **207**, 270.

ZIEGLER, A. M., HANSEN, K. S., JOHNSON, M. E., KELLY, M. A., SCROTESE, C. R., and VAN DER VOO, R. (1977). Silurian continental distributions, paleogeography, climatology, and biogeography. *Tectonophysics*, **40**, 13–51.

ZIEGLER, A. M., SCROTESE, C. R., MCKERROW, W. S., JOHNSON, M. E., and BAMBACH, R. K. (1979). Paleozoic paleogeography. *Ann. Rev. Earth Planet. Sci.*, **7**, 473–502.

ZIEGLER, A. M. and MCKERROW, W. S. (1975). Silurian marine red beds. *Am. J. Sci.*, **75**, 31.

ZYCH, W. (1927). Old-Red Podolski. *Prace Polsk. Inst. Geol.*, **2**, No. 1.

Lower Palaeozoic of the Middle East, Eastern and Southern Africa, and Antarctica
Edited by C. H. Holland

LOWER PALAEOZOIC ROCKS OF ANTARCTICA

M. G. Laird

New Zealand Geological Survey, Department of Scientific and Industrial Research, Christchurch, New Zealand

Contents

1. Introduction

Including its continental shelf, Antarctica is a roughly circular landmass about 4500 kilometres in diameter, its coastal outline broken by the narrow curving Antarctic Peninsula south of South America, and by two deep embayments, the Ross and Weddell Seas (Fig. 1).

Geographically Antarctica is divided into two distinct regions: West Antarctica, lying south of the Americas, and East Antarctica, bounded by the Atlantic and Indian Oceans and separated

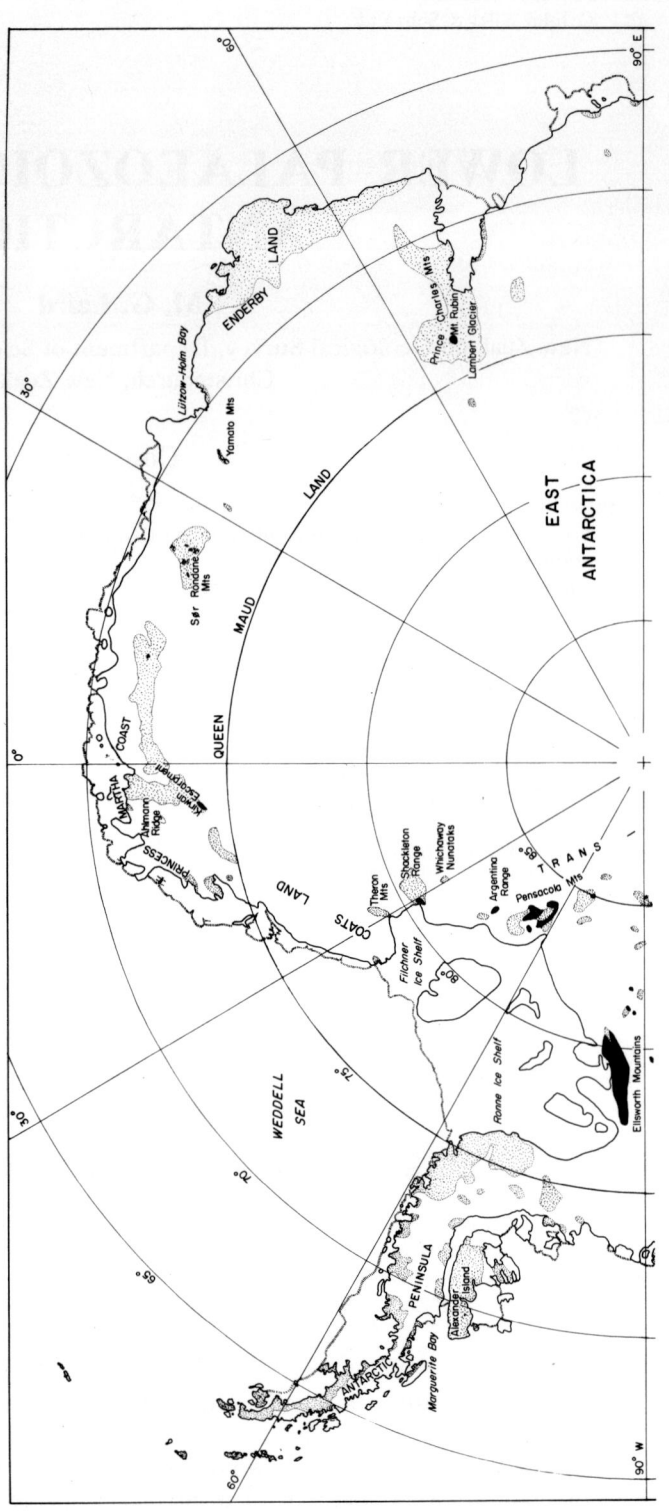

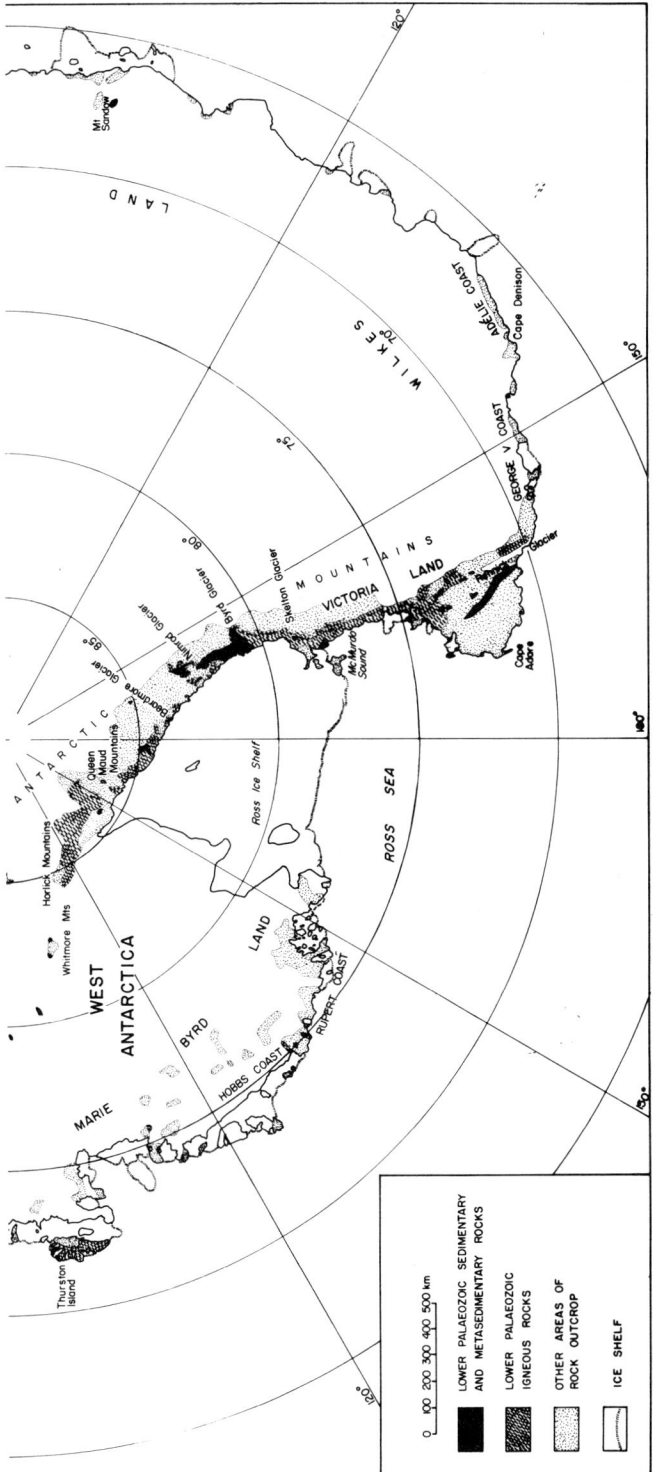

FIG. 1. Map of Antarctica showing distribution of Lower Palaeozoic rocks.

from West Antarctica by a line linking the southern extremities of the Ross and Filchner Ice Shelves. Until comparatively recently this two-fold geographical division was considered to represent also a fundamental geological division, namely, between an ancient stable shield (or craton) and a much younger alpine-type fold-mountain system mainly of late Mesozoic to Cenozoic age. Most geologists now regard this interpretation as an oversimplification.

The Antarctic Ice Sheet covers approximately 98% of the land mass of Antarctica and comprises about 90% of the world's ice. East Antarctica is a high ice plateau rising to 4000 metres in the centre, from which ice drains radially out to the bordering oceans and ice shelves. Rock outcrops occur only on the periphery of the Polar Plateau, and most extensively in the Trans-Antarctic Mountains (Plate 1), an immense mountain range bordering East Antarctica for 3000 kilometres from near Cape Adare in Victoria Land to the Pensacola Mountains flanking the Filchner Ice Shelf. Summit heights reach 4000 metres. Only two other major mountain systems project through the East Antarctic ice cap—that near the Queen Maud Land coast that stretches intermittently in an arc from near the Weddell Sea coast eastwards for 1800 kilometres, and the Prince Charles Mountains flanking the Lambert Glacier. In most other places the ice sheet extends to sea level, unbroken except for nunataks near the coast.

West Antarctica makes up only about one quarter of the total area of Antarctica, and has a generally much lower elevation than East Antarctica. It also is largely ice-covered, with rock outcrops chiefly in ranges near the coast of Marie Byrd Land, scattered along the coast of the Antarctic Peninsula, and in the Ellsworth Mountains bordering the Ronne Ice Shelf.

The harsh climate and the continent's relative inaccessibility have meant that both geographical and geological knowledge of Antarctica remained sketchy until the mid-twentieth century, and vast areas had remained unseen until this time. Great impetus was given to scientific research in Antarctica by the advent of the International Geophysical Year in 1957–58. Permanent bases were set up by many countries and geological and geophysical expeditions sent into the interior. Since the International Geophysical Year, geological programmes have in general been accelerated and few areas now remain unexplored, at least at the reconnaissance level.

1.1. History of Lower Palaeozoic studies in Antarctica

The presence of Lower Palaeozoic rocks in Antarctica was first recorded by GRIFFITH TAYLOR from limestone erratics collected from the Beardmore glacier by F. WILD of Shackleton's 1908–09 expedition (GRIFFITH TAYLOR in DAVID and PRIESTLEY, 1914, pp. 235–240). Some of these samples contained representatives of the key Cambrian Phylum Archaeocyatha. Limestone found *in situ* by Shackleton's party reportedly at the south end of Buckley Nunatak* at the head of the Beardmore Glacier was examined petrographically by SKEATS (1916, p. 196), who noted that 'obscure traces of fossils including Archaeocyathinae can be seen'. Similar samples were re-collected from moraines at the foot of the glacier by the pole party of Scott's last expedition, who dragged them for several hundred kilometres until the party died. Neither Scott's nor Shackleton's party was able to determine the stratigraphical relation between the Cambrian limestone and the Late Palaeozoic Beacon sediments, but it was tentatively suggested that it formed a basal unit under the Beacon Sandstone and above a great erosional unconformity separating the flat-lying Beacon Sandstone from older igneous and folded metamorphic rocks.

The later dredging of similar fossiliferous material from the floor of the Weddell Sea (GORDON, 1920), the discovery of erratics containing archaeocyathid limestone in the

* More likely on Mt. Bowers; *see* p. 274.

Whichaway Nunataks during the British Commonwealth Trans-Antarctic Expedition of 1955–58 (STEPHENSON, 1966, pp. 45–46), and the recording by MAWSON (1940) of numerous erratics of marble, one of which contained probable Archaeocyathinae, at Cape Denison in Adelie Land, showed that the Cambrian limestone was extremely widespread.

The next major discovery was made in the summer of 1960–61 by a New Zealand Geological and Survey party in the Nimrod Glacier area, 200 kilometres north of the Beardmore Glacier. Here, tightly folded archaeocyathid limestone at least 3000 metres thick was observed to overlie greywacke and argillite of probable Late Pre-Cambrian age (Beardmore Group), and to underlie the Beacon rocks unconformably (LAIRD and WATERHOUSE, 1962; LAIRD, 1963). Evidence was thus obtained for the first time that the Cambrian limestone belonged to the folded basement sequence rather than to the flat-lying Beacon Supergroup. Subsequently, fossiliferous Cambrian sediments have been discovered elsewhere in the Trans-Antarctic Mountains and in the Ellsworth Mountains south of the Antarctic Peninsula.

More recent expeditions have shown that, in at least some areas, fossiliferous Cambrian strata rest with marked unconformity on Late Pre-Cambrian greywacke and argillite successions (SCHMIDT and others, 1965; LAIRD, MANSERGH, and CHAPPELL, 1971).

At the time of writing no dated fossiliferous rocks of Lower Palaeozoic age definitely younger than Late Cambrian have been reported, although unfossiliferous formations overlying dated Cambrian sequences in the Ellsworth and Pensacola Mountains are almost certainly of Ordovician–Silurian age, and an Ordovician age is also likely for the Leap Year Group in northern Victoria Land. Radiometric age determinations so far available from igneous and metamorphic rocks show a marked grouping in the 440–540 m.y. range within their total spread from a few million years to 3000 m.y. B.P. Rocks of this age (Late Cambrian and Ordovician) are now known in almost every region for which radiometric age measurements have been recorded, and much of the continent clearly experienced a period of major deformation and plutonic activity, called the Ross Orogeny, during this time.

2. Regional distribution

Sediments of known or probable Lower Palaeozoic age are now known from many parts of Antarctica (Fig. 1). A semi-continuous belt of folded sediments of this age occurs along nearly the whole extent of the Trans-Antarctic Mountains, a distance of 3000 kilometres. Here the Lower Palaeozoic rocks unconformably overlie strongly folded greywacke and argillite sequences of Late Pre-Cambrian age, and are in turn overlain unconformably by the flat-lying or gently dipping Beacon Supergroup of Late Palaeozoic age. At the Weddell Sea end of the Trans-Antarctic Mountains the Pensacola Mountains contain a calcareous sequence with Lower and Middle Cambrian fossils, overlain by volcanic and hypabyssal rocks and by shale and sandstone (SCHMIDT and others, 1965). At the other extremity of the Trans-Antarctic Mountains Late Cambrian fossils occur in a regressive sequence passing upwards from mudstone into fine-grained sandstone, shallow marine red-beds, and then into conglomeratic quartzite (LAIRD and others, 1972). At two localities in between these extremes, in the Nimrod Glacier area (HILL, 1964a) and in the Queen Maud Mountains (PALMER, 1970), fossils of Lower and Middle Cambrian age occur in limestone units. Elsewhere in the range, apparently unfossiliferous rocks of similar lithologies to the fossil-bearing ones are also tentatively correlated, and it is probable that sedimentary rocks of this age occupy discontinuous belts throughout the length of the Trans-Antarctic Mountains. A probable extension of the mountains to the north, which includes the Whichaway Nunataks, Shackleton Range, and Theron Mountains is also suspected to have buried Cambrian limestone in the vicinity (STEPHENSON, 1966), and some of the unfossiliferous sediments of the outcropping rock may also be Lower Palaeozoic in age.

Horizontal or near-horizontal platform deposits of possibly Early Palaeozoic age occur at several localities in the shield area of East Antarctica. At Mt. Sandow in Wilkes Land (Fig. 1) slightly tilted low-grade metasediments occur, containing spores of Late Pre-Cambrian or Early Cambrian age. On the other side of East Antarctica in Queen Maud Land sequences of quartzite, arkose, and conglomerate are of doubtful age, but may be in part Lower Palaeozoic (AUCAMP, WOLMARANS, and NEETHLING, 1972).

In West Antarctica, fossiliferous Lower Palaeozoic sediments are known only from the Ellsworth Mountains, where they occupy part of a folded sequence ranging in age from Late Pre-Cambrian to Permian. Isolated nunataks lying between the Ellsworth and Horlick Mountains contain unfossiliferous sedimentary rocks which may also be Early Palaeozoic in age (CRADDOCK, 1972). Unfossiliferous lightly metamorphosed sandstones and phyllites cropping out in Western Marie Byrd Land may also be of similar age (ADAMS and others, in 1979).

Igneous rocks of Lower Palaeozoic age are extremely widespread throughout East Antarctica and are known in almost every region for which radiometric age measurements have been recorded. Evidently much of the continent experienced a period of major plutonic activity during this time. The Late Cambrian and Ordovician Granite Harbour Intrusive Complex intrudes sediments and metasediments of the Trans-Antarctic Mountains throughout their length. Intrusive rocks of this age also occur in most exposed coastal areas in East Antarctica. In West Antarctica, igneous and high-grade metamorphic rocks of possible Early Palaeozoic age crop out at widely separated localities in the coastal region of Marie Byrd Land, and on Thurston Island (Fig. 1).

3. Regional descriptions of sedimentary rocks

3.1. Lower Palaeozoic sedimentary rocks of West Antarctica

3.1.1. Western Marie Byrd Land

The oldest rock unit of western Marie Byrd Land, the Swanson Group (WADE and LONG, in press), consists of green-schist facies, olive-grey quartzose alternating sandstones and mudstones of minimum thickness 4300 metres. Graded sandstones and siltstones are widespread, but appear to be distributed randomly. Thin fining-upward and coarsening-upward sub-units also occur, and the bulk of the sediments probably accumulated as part of a deep-sea fan complex (ADAMS and others, 1979). The uppermost facies at two localities however appear to have shallow marine and beach characteristics respectively. Palaeocurrent measurements show that there was a bimodal pattern, gravity-induced features suggesting a palaeoslope dipping towards the east, with strong bottom currents flowing both north and south along the slope (ADAMS and other, 1979). Rare volcanics occur interbedded with the clastic sediments: they consist of rhyolitic tuffs and spilitic basalts. Another rock unit in the area, the Fosdick Metamorphic Complex, was considered by WADE and LONG (in press) to be older than the Swanson Group. However, the presence in both units of similar zoned carbonate nodules of unusual appearance has led ADAMS and others (1979) to suggest that the Fosdick Metamorphic Complex is merely a more highly metamorphosed portion of the Swanson group. The group is characterized by close-folds about west-north-west axes and is intruded by granites of earliest Carboniferous age.

The age of the Swanson Group has not been definitely established. ILTCHENKO (1972) recovered an assemblage of acritarchs and other microfossils from the group which indicate a Late Pre-Cambrian–Early Palaeozoic age. Whole-rock K–Ar dating of Swanson Group phyllites resulted in three ages ranging from 445 to 475 m.y. (KRYLOV and others, 1971), which suggests that the main folding event and low-grade metamorphism was early Ordovician. An Early Palaeozoic or perhaps latest Pre-Cambrian age seems likely.

3.1.2. The Ellsworth Mountains

The Ellsworth Mountains (Fig. 1) contain the only fossiliferous Lower Palaeozoic sediments so far described from West Antarctica. Five sedimentary units, totalling approximately 13,000 metres, are exposed (CRADDOCK, 1969) (Fig. 2). The entire stratigraphical sequence has been

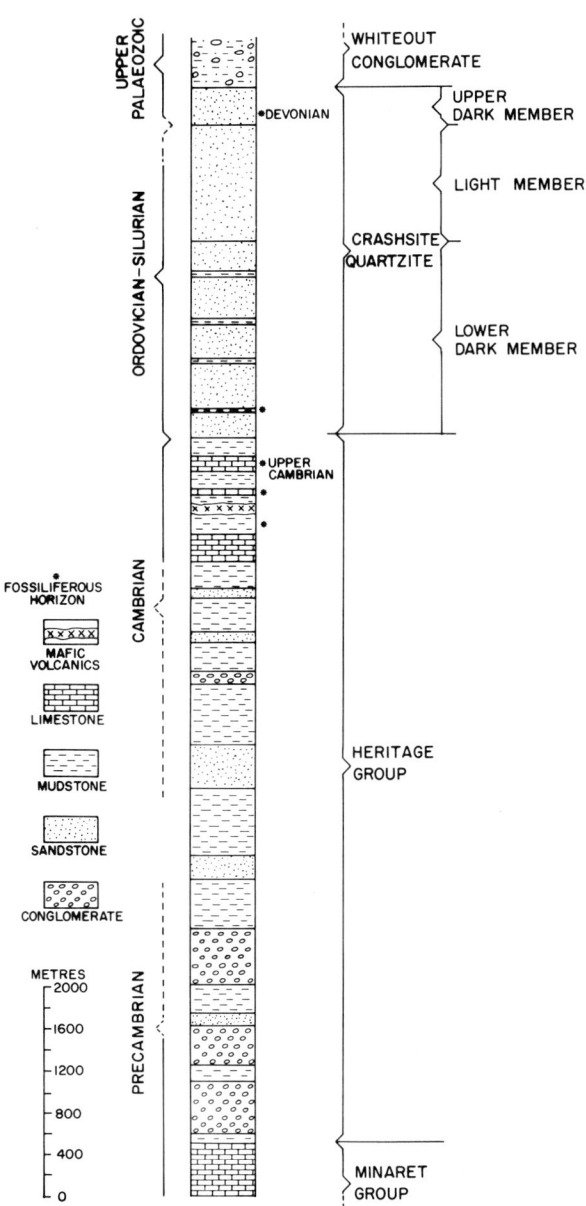

FIG. 2. Columnar section of Lower Palaeozoic rocks of the
Ellsworth Mountains (after CRADDOCK, 1969).

strongly folded about north-west to north-north-west axes, with the majority of dips ranging between 40° and vertical or slightly overturned.

The oldest unit, the Minaret Group, consists of recrystallized carbonate rocks and is probably Late Pre-Cambrian in age. The conformably overlying Heritage Group, which consists dominantly of a variety of clastic rocks with some carbonate units towards the top, contains fossils indicating that it is of Cambrian age. The age of the lower portion of the group is unknown, but it may range down into the Late Pre-Cambrian.

Small-scale structures, apart from bedding, are not preserved in the Heritage Group because of strong folding and shearing. In the lower third of the group massive conglomerate with well rounded pebbles is common. The conglomerate-dominated unit overlies mainly green and red slate and phyllite, and is in its turn overlain by mainly fine-grained clastic rocks, mainly greenish grey phyllite but with some interbedded quartzite and conglomerate. The upper units consist mainly of pelitic rocks interbedded with thick carbonate units.

The red bed sequence at the base of the Heritage Group suggests further shallowing from the probably shallow marine carbonate rocks of the Minaret Group, and may represent tidal flat or marginal marine deposits. Continued regression to a wholly terrestrial environment may be indicated by the dominantly conglomeratic nature of the lowest third of the group. The overlying fine-grained clastics probably represent transgression of the sea, and the highly calcareous nature of the uppermost third, together with the presence of marine fossils, suggests a return to moderately shallow shelf conditions. WEBERS (1972) noted that some of the faunas probably accumulated in a nearshore, possibly sublittoral, high energy environment.

A well preserved Upper Cambrian fauna occurs near the top of the Heritage Group in the north-central Heritage Range, the more southern of the two ranges making up the Ellsworth Mountains. Approximately twelve species of trilobites, eleven species of molluscs, one species of articulate brachiopod, two species of inarticulate brachiopods, one species of archaeocyathid, and probably three species of pelmatozoan echinoderms are present in the fauna (WEBERS, 1972). The fauna is dominated by trilobites, which are mainly Aphelaspids referable to at least three species of *Aphelaspis*. Large pustulose trilobites representing a single species of a new genus close to *Onchopeltis* are abundantly present. Fragmentary pygidia of a large trilobite show affinities to *Eugonocare* and *Olenaspella*, and three species of agnostid trilobites representing *Homagnostus, Pseudagnostus*, and an affaced form probably referable to *Litagnostus* also occur. The homagnostids present are close to *H. tumidosus*, and the pseudagnostids present are close to *P. communis*. Additional trilobite species representing new genera occur but have not yet been described.

In general, the Heritage trilobite fauna exhibits an Asiatic affinity and most clearly resembles Upper Cambrian trilobites described by IVSHIN (1956, 1962) from Kazakhstan. To a lesser degree there are faunal affinities with the Upper Cambrian faunas of Australia.

A well preserved and diverse molluscan fauna is also present, consisting mainly of several species of monoplacophorans and gastropods. The monoplacophorans consist of a high-coned species referable to *Hypseloconus* and others related to *Helcionella, Proplina*, and *Ozarkoconus*. Coiled, asymmetric, hyperstrophic gastropods are abundant in the fauna and are represented by four species referable to the Macluritacea, two of them referable to the Macluritidae. A small species of *Scaevogyra* is abundant in the fauna, and a second species of the Onychochilidae is present as a single specimen possibly referable to *Matharella*. A single species of *Hyolithes* is also present.

One of the most unusual features of the fauna is the presence of specimens of an Irregularian archaeocyathid, a phylum not previously recorded from the Upper Cambrian. The archaeocyathids most closely resemble the genus *Archaeocyathus* and there are also some affinities to *Protopharetra*.

A coarsely ribbed species of *Billingsella* represents the articulate brachiopods in the fauna. Inarticulate brachiopods, referable to *Acrotreta*, also occur. An unidentified lingulid is also present. Scattered pelmatazoan columnals have been recovered.

The precise stratigraphical position of this fauna within the Upper Cambrian is difficult to fix. Trilobites show affinities to described faunas from the Dresbachian and Franconian Stages of Asia and Australia. The molluscs are closest to described faunas from Franconian and younger strata of Minnesota and Wisconsin. The age of the Heritage fauna is probably best considered to be earliest Franconian.

Three other trilobite or brachiopod collections, not yet described, have been made from the upper third of the Heritage group. Some of these fossils are probably as old as Middle Cambrian (CRADDOCK, 1969).

The conformably overlying Crashsite Quartzite (minimum thickness 3200 metres) consists mainly of well sorted medium- to coarse-grained, thick-bedded quartzite. In the Sentinel Range, which forms the northern half of the Ellsworth Mountains, the Crashsite Quartzite has been subdivided into three members. The Lower Dark Member (1830 metres) consists of green and buff quartzite and interbedded argillite; the Light Member (1070 metres) consists of white, grey, and buff quartzite; and the Upper Dark Member (305 metres) comprises brown, micaceous quartzite. A brachiopod fauna collected from the Upper Dark Member indicates a probable Devonian age, but much of the older part of the Crashsite Quartzite may be of Ordovician and Silurian age.

The Crashsite Quartzite consists nearly entirely of quartzite, but minor interbeds of slate, argillite, and quartz, quartzite, or argillite pebble conglomerate are present. The bedding thickness ranges from 3 millimetres to 1.8 metres, and most beds are 0.3 to 0.9 metres thick. The quartzite texture ranges from fine- to coarse-grained with the larger sizes predominating. Scour and fill structures are rare, but cross-bedding is abundant. Current ripples are common, but oscillation ripples occur less frequently.

Thin-section studies of samples from quartzitic horizons (CRADDOCK, ANDERSON, and WEBERS, 1964) show that they are composed predominantly of quartz with minor plagioclase and potash feldspar. Other minerals, usually present only in very small amounts, include garnet, apatite, zircon, clinozoisite, epidote, iron oxides, titanium oxides, and calcite. A few probable volcanic rock fragments also occur in the upper part of the formation.

Sorting is moderate to good; grains are mainly sub-angular, but in some units sub-rounded or rounded. Very finely crystalline quartz is the common cementing agent, but the interstices of most beds contain small flakes of sericite or chlorite.

The good sorting of grains and common occurrence of cross-bedding suggest a moderately active environment of deposition affected by current activity. The presence of trace fossils and, in one locality, of a probable cephalopod, suggests that shallow marine conditions persisted. The occasional occurrence of mud-cracks (WEBERS, 1972) indicates at least local emergence at times.

The only fossils found in the lower (and probably Lower Palaeozoic) part of the Crashsite Quartzite were fragmented inarticulate brachiopods of no age significance.

3.2. Lower Palaeozoic sedimentary rocks of the Trans-Antarctic Mountains

3.2.1. Coats Land

Although not included in the Trans-Antarctic Mountains proper by the majority of writers, the isolated ranges and nunataks extending along the eastern side of the Filchner Ice Shelf into Coats Land form a logical topographical and geological extension of these mountains. Investigations during the British Commonwealth Trans-Antarctic Expedition (1955–58) in this area resulted in the discovery in the Shackleton Range of a sedimentary sequence almost certainly Lower Palaeozoic in age, and in the Whichaway Nunataks, erratics of Cambrian archaeocyathid limestone (STEPHENSON, 1966).

Shackleton Range

The Blaiklock Glacier Group comprises the youngest of three folded sedimentary and metasedimentary sequences exposed in the Shackleton Range (Fig. 3). Originally named the 'Blaiklock Beds' by STEPHENSON (1966), who first described them, the strata were redescribed and redefined by CLARKSON (1972). Cropping out mainly on either side of Blaiklock Glacier in the western part of the Shackleton Range, the Blaiklock Glacier Group comprises a sequence at least 6000 metres thick of sandstone with some conglomerate. It appears to lie on the flanks of an anticline plunging gently southwards, with dips of between 15° and 64° on the limbs. The sediments rest unconformably on the Pre-Cambrian Shackleton Range Metamorphic Complex along the northern margin of the outcrop, but its southern margin (concealed by glaciers) is probably faulted. The group consists of two formations, the Mount Provender and Otter Highlands Formations.

The Mount Provender Formation, the older of the two, is composed almost entirely of feldspathic sandstone with some breccia at the base. The basal member, which unconformably overlies the Shackleton Range Metamorphic Complex, consists of up to 6 metres of breccia or conglomerate containing blocks, derived from the underlying basement rocks, of white marble and dark green gneiss up to 0.8 metres in diameter embedded in a purple–red matrix. This is overlain by red or grey fine-grained, well bedded sandstone containing occasional fine conglomerate layers. Cross-bedding is sometimes present, and fold structures attributable to slumping are numerous at some horizons. Red–grey micaceous shales and slates make up a minor proportion of the formation.

In thin section, the sandstones are well sorted and have a tightly packed texture. The grains are most commonly sub-angular. Quartz is the most common mineral, with feldspar amounting to approximately 10%. Micas, both biotite and muscovite, make up most of the remainder of the grains. Clastic calcite is relatively common in the calcareous sandstone layers which make up a portion of the sequence.

The maximum exposed thickness of the Mount Provender Formation is approximately 760 metres, but, as the top of the formation is not seen, this is a minimum.

The Otter Highlands Formation is exposed mainly on the western edge of the Blaiklock Glacier, where both its upper and lower limits lie beneath glacier ice. The majority of the sediments form a monotonous sequence of hard, grey–green feldspathic grits and sandstones. The sandstones are well bedded and commonly show cross-bedding and intraformational folding. They are generally light coloured, ranging from grey to brown or greenish, but thin dark grey calcareous beds occur in places. In addition to grits and sandstones, the lowest exposed members of the formation comprise compact even bedded conglomerates with pebbles and small boulders of metamorphic rocks (mostly quartzites), thin beds of micaceous shales and sandstones, and, at the bottom of the sequence, thick beds of calcareous grits and sandstones.

The sandstones in thin section are mostly sub-angular in medium-grained lithologies and sub-rounded in others, while all specimens show good sorting. They are mainly medium to coarse-grained feldspathic arenites with less common arkosic arenites. Composition is similar to that of the sandstones of the Mount Provender Formation.

The estimated thickness of the Otter Highlands Formation, discounting any major faulting which might be present, is approximately 5300 metres, which must be a minimum figure as neither the top nor the bottom of the formation is exposed.

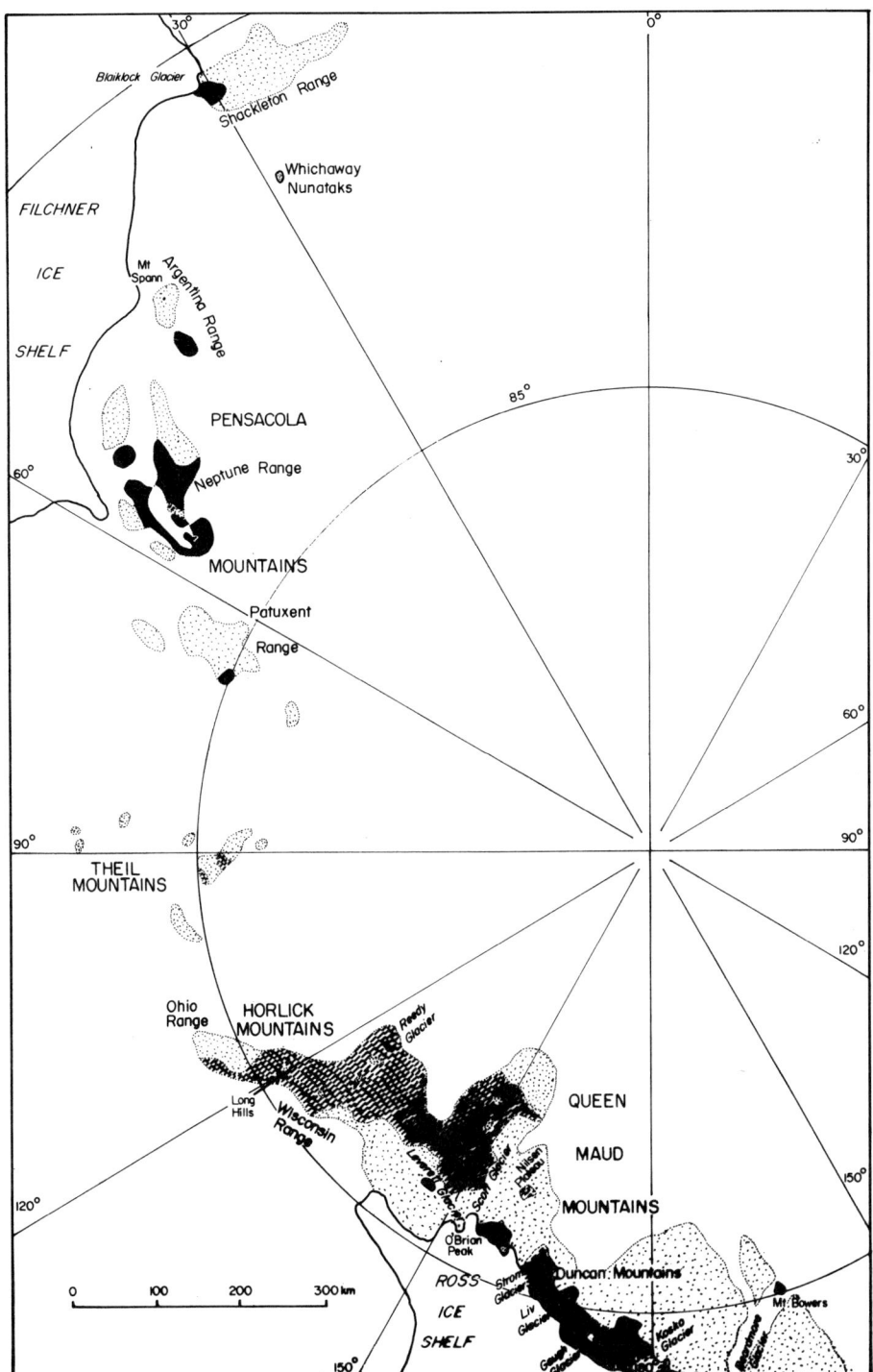

FIG. 3. Geological map of the Trans-Antarctic Mountains, Shackleton Range to Queen
Maud Mountains, showing outcrop areas of Lower Palaeozoic rocks. Legend as for
Fig. 1. (Modified from CRADDOCK, 1972).

Good sorting in the sandstones making up the Blaiklock Glacier Group and their even grain size suggest deposition under shallow water conditions, with considerable reworking. Heavy minerals show high concentrations in some horizons in the Otter Highlands Formation and they may have been 'blacksands' concentrated during reworking by currents. The common red haematitic coloration in the Mount Provender Formation probably indicates accumulation under oxidizing conditions, perhaps in an arid climate.

No direct age data are available for the Blaiklock Glacier Group. Radiometric dating carried out by REX (1972) places the age somewhere between the Late Pre-Cambrian and the Late Carboniferous. The group rests unconformably on the Shackleton Range Metamorphic Complex (which is cut by a dioritic dike dated at 1446 ± 60 m.y.) and is itself cut by a dolerite dike dated at 297 ± 12 m.y. Fossiliferous erratics of silty shale found in moraine flanking the Blaiklock Glacier were assigned by STEPHENSON (1966) to intermediate strata of the Blaiklock Glacier Group assumed to be concealed beneath the glacier. This rock type has not been found *in situ* in the Shackleton Range but its nearest lithological equivalents are found among the sandstones of the Blaiklock Glacier Group (CLARKSON, 1972). The erratics contain inarticulate brachiopods identified as obolids, possibly *Lingulella* (THOMSON, 1972). Obolids range in age from Early Cambrian to late Ordovician, while the genus *Lingulella* ranges no higher than middle Ordovician. Trilobites of Middle Cambrian age were found in similar erratics by the 21st Soviet Antarctic Expedition (SOLOVEV and GRIKUROV, 1978). These consist of *Triplagnostus praecurrens* (Westergand), *Ehmania shackletonia* (Solovev and Grikurov), *Lyriaspis antarctica* (Solovev and Grikurov), and *Elrathina parallela longa* (Solovev and Grikurov). Thus the age of the Blaiklock Glacier Group is likely to be Cambrian if the source of the erratic material has been correctly inferred. The beds are best correlated with the Lower Palaeozoic rocks exposed in the Pensacola Mountains 300 kilometres to the south-east.

Whichaway Nunataks

The Whichaway Nunataks, lying 100 kilometres south of the Shackleton Range, comprise a group of small hills rising to a maximum of 300 metres above the ice surface. They are composed entirely of flat-lying clastic sediments belonging to the Beacon Supergroup and containing plant material of Permian or Carboniferous age (PLUMSTEAD, 1962; STEPHENSON, 1966). However, morainic material which occurs almost up to the summit of the highest nunatak contains numerous fragments of limestone. Some of the limestone boulders contain Archaeocyatha, which have been dated by HILL (1965) as Lower Cambrian. The source of the erratics is not definitely known, as no outcrops of similar limestone occur in the area. No moraines containing similar material occur in the Shackleton Range, and this, combined with the fact that the limestone comprises almost the only exotic material in the moraines at the Whichaway Nunataks, suggests that it is likely to be of local origin. The local ice movement is from the south or south-east, and it is therefore likely that Cambrian limestone lies buried beneath the ice in that direction. One erratic is a limestone breccia of which some fragments contain Archaeocyatha. This breccia is of sedimentary origin with a matrix of poorly sorted lithic sandstone. The finer rock fragments are composed of mosaic quartz and occasional grains of shale, siltstone, granite, and pyritic limestone. Some included quartz grains contain fine acicular inclusions similar to those occurring in the sandstones of the Whichaway Nunataks. From this lithological similarity it could be inferred that the limestone breccia represents the basal member of the Beacon Supergroup in the area. This would indicate that the Beacon Supergroup rests directly on Cambrian limestone in some areas.

3.2.2. Pensacola Mountains

Eleven sedimentary and volcanic rock units (of which seven are probably Lower Palaeozoic) have been mapped in the ranges making up the Pensacola Mountains (SCHMIDT and FORD, 1969), lying south of the Filchner Ice Shelf (Fig. 3). The oldest unit, the Patuxent Formation, consists of isoclinally folded sandstone, slate, and basaltic flows of probable Late Pre-Cambrian age. Overlying the Patuxent Formation with angular unconformity are three conformable formations (Fig. 4): the Nelson Limestone; the Gambacorta Formation, a felsic volcanic unit; and the Wiens Formation, an interbedded siltstone and shale unit (SCHMIDT and others, 1965). These units have been deformed into open, sinuous folds trending northerly or north-north-easterly, with moderately to steeply dipping limbs.

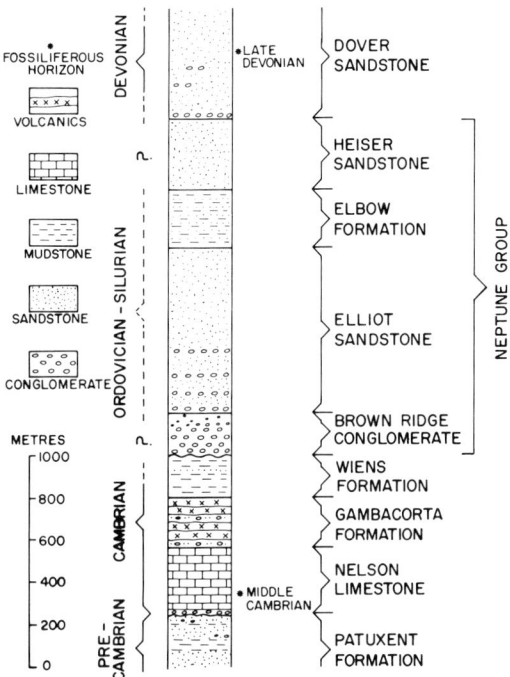

FIG. 4. Columnar section of Lower Palaeozoic rocks of the Pensacola mountains (after SCHMIDT and FORD, 1969).

The Nelson Limestone is exposed in the southern portion of the Patuxent Range and in the Neptune Range. It consists of five members: (1) a basal red quartz–clast conglomerate, 0.3 to 6 metres thick, which is composed largely of erosional detritus of the Patuxent Formation; (2) a local red bed clastic succession which is as much as 18 metres thick; (3) a lower grey limestone, 60 to 90 metres thick, which consists of thin-bedded limestone containing interbedded lamellae of limy shale; (4) a middle grey limestone, 60 to 90 metres thick, which consists of thick-bedded massive limestone; and (5) an upper grey limestone, about 30 to 45 metres thick, which consists of thin-bedded limestone and shaly limestone. The thin-bedded limestones, members 3 and 5, are commonly oolitic, pisolitic, and nodular. Member 4 is commonly bleached white by alteration that probably was associated with intrusion of hypabyssal rhyolite. The Nelson Limestone is 180 to 240 metres thick, both in the type area and elsewhere in the Neptune

Range: in the Patuxent Range it reaches a thickness of at least 550 metres, and is probably considerably thicker (SCHMIDT, DOVER, and BROWN, 1964). The formation evidently represents shallow-water deposits formed during marine transgression.

Two different faunal assemblages of trilobites have been collected from Member 3 of the Nelson Limestone in the Neptune Range. The faunules, which consist of *Amphoton oatesi*, *Chondranomocare australis*, *Trinepea trinodus*, *Kootenia styrax*, *Peronopsis* cf. *P. fallas* (Linnarsson), genus and species undetermined 4, *Nelsonia schesis*, and *Suludella? spinosa*, indicate a medial or late Middle Cambrian age for the oldest limestone beds in the section (PALMER and GATEHOUSE, 1972). In the southern Argentina Range extensive outcrops of unnamed archaeocyathid limestone occur, but the fauna has not been described. At Mt. Spann, about 65 kilometres to the north-east, moraine boulders bear archaeocyathids, mixed archaeo-cyathids and trilobites, and trilobite–mollusc associations which indicate the nearby presence of limestone ranging in age from Early Cambrian to late Middle Cambrian. The trilobite faunules consist of *Australaspis magnus*, *Glabrella? pitans*, *Bathyuriscellus australis*, *Chorbusulina wilkesi*, *Pensacola isolata*, *Redlichia* sp. indet., *Chorbusulina subdita*, *Bathyuriscellus modestus*, *Xystridura glacia*, *X. multilinia*, *Pagetia longispina*, and *Goldfieldia ninguis* (PALMER and GATEHOUSE, 1972). Almost all trilobites in these faunules are close relatives of Australian, Chinese, or Siberian forms. There are no significant relations between any of the faunules and the Cambrian trilobites of South or North America. Algal remains of uncertain age were also collected from the Nelson Limestone in the Patuxent Range (SCHMIDT, DOVER, and BROWN, 1964).

The Gambacorta Formation is named from Gambacorta Peak in the Neptune Range and crops out only in this range. The base and lower part of the formation consists of intensely altered sandstones and conglomerates of volcanic detritus, about 90 metres thick, which conformably overlie member 5 of the Nelson Limestone. The rest of the formation consists of interlayered dark brown, red–brown, and light green rhyolite flows, volcanic breccias, pyroclas-tic deposits, and detrital sandstones and conglomerates. These latter are composed mainly of volcanic clasts. All units are variously and complexly interlayered and thicken towards the south-eastern part of the Neptune Range. Most of the rocks have been intensely altered.

The Gambacorta Formation thins northward and north-westward, from a maximum thick-ness of more than 330 metres in the vicinity of Gambacorta Peak, and completely disappears within 16 to 24 kilometres. Fossils have not been found within the formation, but a radiometric date on interbedded rhyolite gives an age of 563 ± 35 m.y. (EASTIN and others, 1969).

The Wiens Formation is named from Wiens Peak in the Neptune Range, where it conform-ably overlies the Gambacorta Formation. Elsewhere than the type locality, the lower part of the Wiens Formation intertongues with volcanic sedimentary rocks of the upper part of the Gambacorta Formation. It consists of interlayered green and red–brown thin-bedded shale, siltstone, and fine sandstone. Several thin-bedded grey oolitic limestone members occur in the section. The Wiens Formation is less than 330 metres thick in the southern Neptune Range and does not occur in the northern part of the range. Non-diagnostic fucoidal impressions are locally abundant on bedding planes of green shale and are the only evidence in it of former life.

Overlying the Wiens Formation with angular unconformity is the Neptune Group (Fig. 4). It is gently deformed into broad folds trending north or north-north-east with mainly gentle dips. The Neptune Group contains no diagnostic fossils and is therefore of uncertain age. However, the uppermost formation of this group is disconformably overlain by a sandstone (Dover Sandstone) containing plant fossils of probable Late Devonian age (SCHMIDT and FORD, 1969). Much of the Neptune Group is thus likely to be Early Palaeozoic in age, although the upper two formations may be as young as Devonian (NELSON, SCHMIDT, and SCHOPF, 1968). It has been correlated by BARRETT and others (1972) with the Taylor Group of the Beacon Supergroup.

The lowest unit of the Neptune Group, the Brown Ridge Conglomerate, is characteristically red, poorly bedded, and poorly sorted; however, the lower part is green at some localities. Cobbles consist predominantly of sandstone, slate, and vein quartz clasts derived from the Patuxent Formation, but abundant limestone and felsic volcanic clasts are locally present. Larger clasts are commonly 8 centimetres in diameter, but some boulders exceed 1 metre. The formation is distributed discontinuously throughout the Neptune Range. Its thickness is highly variable locally reaching 1000 metres.

The Elliot Sandstone, which overlies the Brown Ridge Conglomerate conformably, consists of a pink to buff coarse-grained cross-bedded sandstone with thick and thin interbedded conglomerate beds and minor thin red shaly beds in the lower half. Volcanic detritus is common in the lower part, and quartz and rock detritus predominate in the upper part. Calcareous cement is characteristic. The formation probably averages about 760 metres in thickness. It is approximately 1500 metres thick near the southern end of the Neptune Range, and thins away to nothing 65 kilometres to the north.

Conformably overlying the Elliot Sandstone is the Elbow Formation, which is a red-bed unit consisting of interbedded red argillaceous siltstone and grey, fine-grained quartzose sandstone in well indurated layers 0.3 to 1.2 metres thick. These layers are made up of laminations 2 to 4 millimetres thick. In overall colour, about half the formation is red and half is light grey. The coarser grained layers are commonly mottled grey or grey with red specks. Cross-bedding is abundant, and ripple-marked bedding planes are common. In some beds, mottled patterns perpendicular to the bedding disrupt laminations and suggest animal burrowings. The facies or character of the red beds changes laterally away from the type section, and the definition of the upper and lower contacts of the Elbow Formation on the southern Neptune Range has not been resolved. It is about 330 metres thick at the type section in the central Neptune Range, and thins towards the south and west; this may correspond in part to a thickening of the Elliot Sandstone in the same direction.

The Heiser Sandstone gradationally overlies the Elbow Formation, and is disconformably overlain by the Dover sandstone of probable Late Devonian age. The Heiser Sandstone consists of a light green to brown quartzose sandstone which is moderately well bedded in beds 0.3 to 1 metre thick. It is characterized by the common occurrence of *Skolithos*-like tubes of lighter coloured quartzite set in a darker, less well indurated, encompassing sandstone. The tubes range from 5 to (rarely) 25 millimetres in diameter, are characteristically perpendicular to the bedding planes, and commonly extend through the thickness of only one bed. At several outcrops, tubes about 25 millimetres in diameter occur, randomly oriented on bedding planes. The *Skolithos*-like tubes are indicative of ancient life. The Heiser Sandstone is about 330 metres thick.

The diversity of lithofacies represented within the Neptune Group reflects varied environments of deposition. The Brown Ridge Conglomerate is considered (WILLIAMS, 1969) to be an orogenic deposit locally derived from the rising fold mountains. The Elliott Sandstone spread as an alluvial apron around the mountains, the relief of which was progressively subdued by erosion. The evenly bedded, well sorted, and rounded sandstones of the Elliott, Elbow, and Heiser Formations indicate a return to more quiescent conditions of sedimentation with some periods of slight uplift in the area of deposition and accompanying local erosion and reworking of the sediments. The Elbow, Heiser, and Dover sandstones were probably deposited at shallow depths in an epineritic environment on a slowly subsiding continental shelf (WILLIAMS, 1969). During the deposition of the Nelson Group, the provenance of the sediments probably changed from a local source, a terrain made up largely of volcanic rocks, to a more distant source, probably the East Antarctic shield, supplying mainly quartz.

3.2.3. Horlick and Queen Maud Mountains

Two areally and lithologically distinct units of metasediments are present in the basement complex of this region (Fig. 3). The older of the two consists mainly of metagreywacke, pelitic hornfels, and schist of the Goldie, Duncan, and La Gorce Formations regarded as of late Pre-Cambrian age (MINSHEW, 1967; McGREGOR, 1965; McGREGOR and WADE, 1969). A metavolcanic unit, the Wyatt Formation (MINSHEW, 1967), which occurs in the Scott Glacier area and appears to interfinger with metasediments of the La Gorce Formation (KATZ and WATERHOUSE, 1970b), has been radiometrically dated at 633 m.y. (FAURE, MURTAUGH, and MONTIGNY, 1968) and thus confirms a late Pre-Cambrian age for both the Wyatt and La Gorce Formations.

The second group of metasediments is less strongly deformed and less metamorphosed than the first group, and is considered to be the younger of the two. The Liv Group (STUMP, in press) comprises a scattering of formations in the Horlick and Queen Maud Mountains, known or thought to be of Cambrian age, in which silicious volcanic rocks are prominent. These include the Taylor, Fairweather, Leverett, and Greenlee Formations. In the Horlick and eastern Queen Maud Mountains these rocks are restricted to a few widely separated outcrops characterized by a high proportion of carbonate rocks. North of the Leverett Glacier a sequence of limestone, shale, sandstone, and rhyolite, making up the Leverett Formation, crops out. It probably exceeds 2000 metres in thickness (MINSHEW, 1967), although neither top nor bottom of the formation is exposed. At the type locality just north of the Leverett Glacier, the Leverett Formation has been divided into seven units (Fig. 5). The oldest unit (A) consists of 300 metres

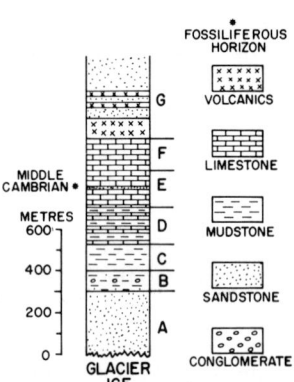

FIG. 5. Columnar section of the Leverett Formation (after MINSHEW, 1967).

of coarse-grained, red, cross-bedded arkosic sandstone, the base of which is not exposed. Rock fragments are abundant, and consist of limestone (about half of the samples), phyllite, siltstone, shale, and volcanics. Unit B, overlying Unit A, is a sequence of poorly sorted, red, massive pebbly mudstone 100 metres thick. It consists of angular to well rounded sand-sized grains of metamorphic and volcanic rock fragments in a red clayey sericitic matrix that makes up about 60% of the samples. Unit C, a thin-bedded shale about 125 metres thick, rests conformably on the pebbly mudstone. It grades upward through a few metres into 150 to 175 metres of interbedded limestone and shale that constitute Unit D. The limestone varies from light to dark grey, weathering to a light pinkish brown. Unit E, lying gradationally above Unit D, is a thin-bedded, dark grey argillaceous limestone at least 175 metres thick. Greyish brown limestone, shale, and volcanic sandstone become abundant in the upper part of the unit. Trilobite fragments are abundant in some of the limestone beds near the middle of the unit. Overlying Unit E is a white massive cliff-forming limestone, Unit F, about 100 to 150 metres thick. Unit G consists of interbedded rhyolite and coarse-grained conglomeratic sandstone, probably a 1000 or more metres thick, gradational with the underlying limestone of Unit F.

The sediments of the Leverett Formation accumulated largely in a shallow marine environment (MINSHEW, 1967). The immaturity of the sandstones suggests that they are first cycle sediments, derived primarily from a granitic and metamorphic terrain but with some intermittent volcanic contributions. The lack of sorting and general coarseness of the sandstones in the lower part of the formation suggests a moderate to high energy environment. The bimodality of the sandy mudstone making up Unit B may indicate emplacement by mass movement. The shale in Unit C probably formed under quiet conditions. A quiet environment is also indicated for much of the time during deposition of the limestones making up Units D, E, and F, although some layers of coarse-grained limestone containing severely broken up trilobite fragments were probably formed under high energy conditions. A high energy environment is indicated for the poorly sorted, coarse-grained conglomeratic sandstones which make up much of Unit G.

The badly deformed and slightly metamorphosed trilobite fragments in the limestone of Unit E at Leverett Glacier are considered to be of Middle Cambrian age (MINSHEW, 1966; PALMER, 1970). Although the trilobites are distorted and broken, three distinct species can be recognized. The commonest has affinities with the genus *Mapania*, a late Middle Cambrian genus found in China and Australia; a second form is close to the genus *Lisamiella*, also of Middle Cambrian age; and the third is a nearly featureless form perhaps representing another Middle Cambrian genus, *Sunaspis*. The trilobite fauna is of general Asiatic aspect.

In the same nunatak, rhyolite stratigraphically above the trilobite-bearing bed gives a radiometric age of 470 ± 11 m.y., and an apparently equivalent rhyolite in the Long Hills, south-eastern Wisconsin Range, gives an age of 498 ± 45 m.y. (FAURE, MURTAUGH, and MONTIGNY, 1968).

Along the strike from the Leverett Glacier succession near O'Brien Peak and in the Duncan Mountains exposures of limestone, marble, and sandstone, although more highly metamorphosed than those north of the Leverett Glacier, may also be correlative (KATZ and WATERHOUSE, 1970a). Similar rocks also occur in an isolated nunatak at the head of the Reedy Glacier in the south-western Horlick Mountains (MIRSKY, 1969).

In the western Queen Maud Mountains the Duncan Formation is overlain by the Fairweather Formation, regarded as of Early Cambrian age (MCGREGOR, 1965), although no fossils are known. The unit is in excess of 3000 metres thick and crops out extensively in the Duncan Mountains. Here the Fairweather Formation is in contact with the Duncan Formation, but both units consist of strongly deformed schists and the stratigraphical relation between the two formations is uncertain (MCGREGOR and WADE, 1969). The Fairweather Formation has thick units of porphyritic silicic volcanic rock throughout. Basaltic rocks are developed towards the bottom of the formation, and chert-like rock is fairly common in the lower and middle portions. The upper portion of the formation contains cross-bedded quartzites, schists, metaconglomerates, and some thin calcareous beds (STUMP, in press). The conformably overlying Henson Marble Member (Plate 2) consists mainly of pure, white, coarsely crystalline marble, with some thin beds of quartzitic and calc-silicate rocks near the base. The full thickness of the member is not exposed, but a minimum thickness is 500 metres.

Separated areally from these formations is a 2500-metre sequence of quartzite, calcareous quartzite, marble, and felsic volcanics (Taylor Formation), which crops out between the Kosco and Gough Glaciers and is considered to be the lateral equivalent of the Fairweather Formation (McGregor and Wade, 1969).

Porphyritic rhyolite with phenocrysts of quartz, plagioclase and sometimes K-feldspar is the most abundant rock type in the formation. Both ash-fall tuffs with welded-shard structure and lavas with flow banding are represented (Stump, 1974, 1976). Spherulites occur in some of these units. On Mt. Greenlee, in what is thought to be the lower portion of the formation, several units of amygdaloidal, basaltic lava crop out. At the Taylor Nunatak locality, the silicic volcanic rocks are associated with carbonate and volcaniclastic rocks. The carbonates are micrites, clastic oosparites, and a breccia from which the Early Cambrian fossil *Cloudinia*? has been found (Yochelson and Stump, 1977). Certain units containing only coarse-grained quartz and plagioclase, some of which is euhedral, seem to be detrital concentrations of phenocrysts. Rocks which appear in the field to be chert, can be seen in thin section to be deposits of fine ash. Microscopic, graded-bedding indicates that these were deposited subaqueously.

The rest of the Taylor Formation to the north on the east side of Shackleton Glacier is distinct from the Taylor Nunatak section. Interbedded with the silicic volcanic rocks in this area are units of argillite and festoon cross-bedded quartzite, probably representing a fluvial or shallow marine environment. Also found toward the northern part of the area are volcaniclastic conglomerates and chert-like deposits which most likely originated as water-laid ash.

The environment during deposition of the Taylor Formation appears to have been one of active volcanic islands, perhaps becoming more emergent through time, as indicated by the cross-bedded quartzites and argillite in the supposed upper portion of the formation.

Lying stratigraphically below the Taylor Formation on Mt. Greenlee and Epidote Peak is a sequence of fine-grained quartzite and phyllite called the Greenlee Formation (Wade, 1974). The sedimentary and volcanic rocks are interbedded in a transition and the Taylor Formation is marked at the lowest thick felsite unit. The thickness of exposed Greenlee Formation is probably not in excess of 1000 metres. Bedding is 5 centimetres to 1 metre thick and even; parallel lamination is the only common sedimentary structure.

No lithological equivalents to the Greenlee Formation are known throughout the Queen Maud Mountains. It is distinct from the volcanic formations of the Liv Group and may precede initiation of this period of volcanism.

3.2.4. Central Trans-Antarctic Mountains—Beardmore Glacier to Byrd Glacier

The area occupied by the central Trans-Antarctic Mountains (Fig. 6) contains the most extensive sequence of fossiliferous Cambrian sediments so far known in Antarctica, and it is also historically important. It was from near the head of the Beardmore Glacier that the first Cambrian fossils to be discovered in Antarctica were found in place by Shackleton's polar party in 1908–09. They occurred in four small limestone outcrops off the extreme south-west of Mt. Bowers (Fig. 3) and separated from it by glacier ice (Young and Ryburn, 1968). The location was previously given as Mt. Buckley, a nunatak lying to the north of Mt. Bowers (David and Priestley, 1914).

The bulk of the Cambrian sediments, however, lie in the region between the Byrd and Nimrod Glaciers (Fig. 6), where a thickness of more than 8000 metres unconformably overlies indurated sandstone, mudstone, quartzite, and rare marble of the Late Pre-Cambrian Goldie Formation. The Cambrian sediments comprise the Byrd Group, which is divided into four formations: the Shackleton Limestone, the Dick Formation, the Douglas Conglomerate, and the Starshot Formation.

The Shackleton Limestone, which comprises the lowermost unit, is also the most widespread and distinctive. In the Holyoake Range, north of the Nimrod Glacier, which contains the type

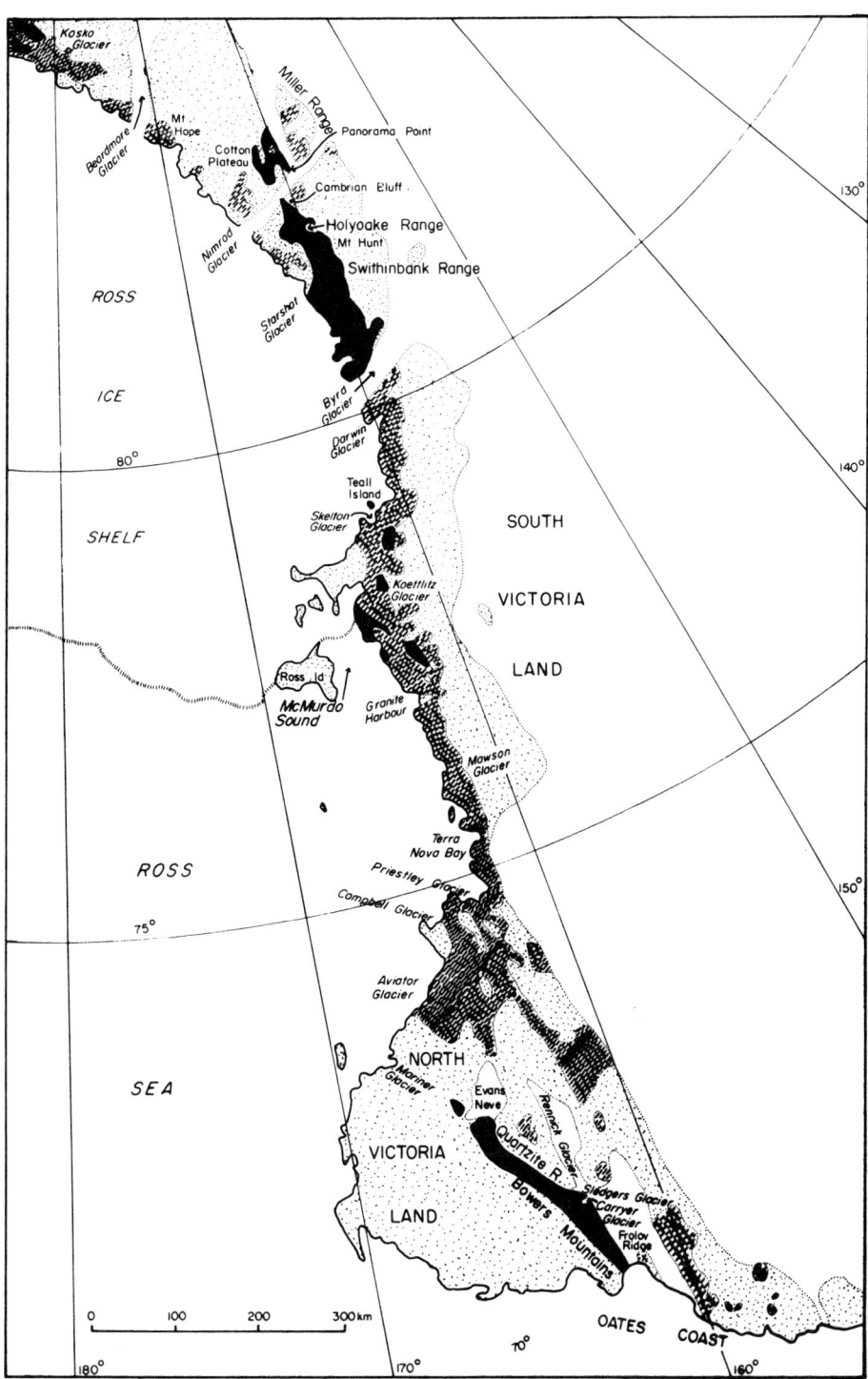

FIG. 6. Geological map of the Trans-Antarctic Mountains from the Beardmore Glacier to
the Oates Coast, showing outcrop areas of Lower Palaeozoic rocks. Legend as for
Fig. 1. (Modified from CRADDOCK, 1972).

section (LAIRD, 1963, 1964), and where almost continuous exposures occur, the formation attains a minimum thickness of 5400 metres (LAIRD, MANSERGH, and CHAPPELL, 1971). The Shackleton Limestone is a highly calcareous unit, consisting dominantly of a pure limestone. The colour is commonly grey, but thick (up to 600 metres) units of cream limestone also occur. At Cambrian Bluff at the southern end of the Holyoake Range, 15 metres of pink limestone crops out, and in the Holyoake and Swithinbank Ranges black limestone, in units up to 300 metres thick, also occurs. The thickest sequence recorded occurs in the Holyoake Range north of Mt. Hunt where there is a well exposed section on the steeply dipping, in part overturned eastern limb of a major north-north-west trending syncline (Fig. 7); the base is obscured by ice

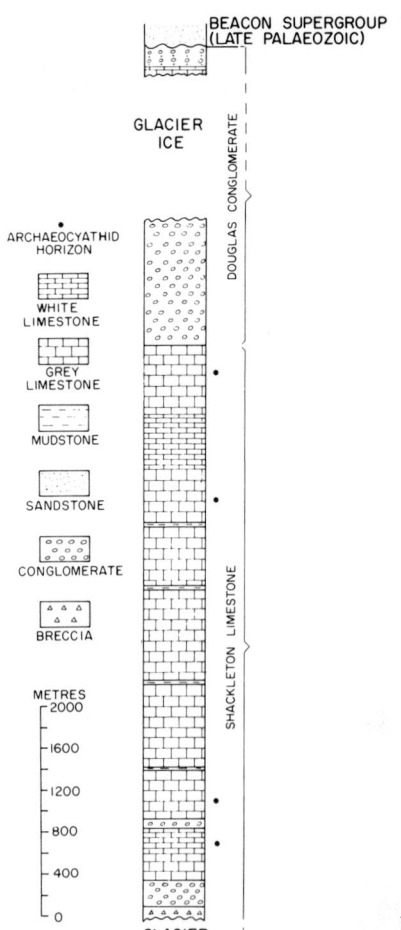

FIG. 7. Columnar section of the Shackleton Limestone, northern Holyoake Range.

and the top of the formation is sharply overlain by the Douglas Conglomerate. The bulk of this sequence consists of limestone sometimes with archaeocyathids, but a smaller proportion of breccia and conglomerate, consisting largely of limestone clasts, also occurs. A 90-metre thick conglomerate, 840 metres above the base of the section, and consisting of grey, rounded limestone clasts averaging 5 centimetres in diameter but reaching 65 centimetres, is unusual in containing large-scale cross-bedding in 65-centimetre to 1-metre sets. A small proportion of pebbles in the conglomerates contain archaeocyathids.

At Cambrian Bluff, approximately 4400 metres of complexly folded Shackleton Limestone rests unconformably on indurated sandstone of the Goldie Formation (Plate 3). The bulk of the sequence consists of grey limestone, commonly oolitic and including a few thin beds of conglomerate and sandstone. Several horizons containing archaeocyathids occur, notably a 15-metre bed of finely bedded pink and grey highly pyritized limestone half-way up the sequence. A cream limestone unit 460 metres thick and whose top is not exposed forms the core of a major syncline (Plate 3) and is correlated with the 600 metres of cream limestone seen in the thickest section at the northern end of the Holyoake Range (Fig. 7).

The basal members of the formation are exposed on the western limb of the syncline on the ridge west of Mt. Hunt where they unconformably overlie marble belonging to the Late Pre-Cambrian Goldie Formation. Here the basal 1200 metres of the Shackleton Limestone is a breccia containing blocks of slightly metamorphosed sandstone up to 65 centimetres in diameter and a small percentage of marble clasts. Shale and sandstone form rare interbeds within the breccia. Breccia and conglomeratic breccia, correlated with the upper part of the basal breccia on Mt. Hunt, are exposed on the western flank of the Holyoake Range at two localities between Mt. Hunt and Cambrian Bluff. These deposits contain limestone clasts and rare pebbles of quartzite, sandstone, and a few conglomeratic boulders derived by local reworking. On the eastern side of the Holyoake Range, at the type locality 15 kilometres south-east of Mt. Hunt, approximately 3500 metres of continuously exposed strata occurs. Although here most of the thickness is made up of limestone, several units of mudstone, sandstone, and conglomerate up to 60 metres thick, consisting of rounded quartz and limestone clasts up to 15 centimetres in diameter, are also present. These conglomerates outcropping on both flanks of the syncline trending along the axis of the Holyoake Range do not appear in the continuous sequence exposed on Cambrian Bluff, and thus are probably lenticular. At Cambrian Bluff, at the type section on Holyoake Range, in the thick sequence exposed north of Mt. Hunt, and in the area between the Starshot and Byrd Glaciers, Archaeocyatha occur sporadically. In the latter area trilobite fragments, green and blue–green algae, brachiopods, and abundant trace fossils also occur (BURGESS and LAMMERINK, 1979). Between the Starshot and Byrd Glaciers the Shackleton Limestone consists mainly of fine-grained, even-textured grey, white, cream, pink and black limestones, although at some localities it is pisolitic, oolitic, algal, or siliceous (SKINNER, 1964, 1965). In the vicinity of Mt. Hamilton over 2000 metres of continuous section of Shackleton Limestone has been recorded (BURGESS and LAMMERINK, 1979). The sequence here includes an anomalous unit of megabreccia, 200 metres in thickness, which consists entirely of Shackleton Limestone fragments most of which were unindurated at the time of erosion. The unit is interpreted as a series of debris-flow deposits resulting from penecontemporaneous uplift and erosion of proximal limestone sequences (BURGESS and LAMMERINK, 1979).

South of the Nimrod Glacier the Shackleton Limestone has been metamorphosed to a coarse-grained marble, containing occasional spectacular crystal roses of wollastonite. No thick sequences have been measured although apparently continuous sections a few thousand metres thick were viewed from a distance (LAIRD, MANSERGH, and CHAPPELL, 1971). Those that were examined indicate a distinct facies change south of the Nimrod Glacier, cream limestone or cream and grey quartzite without conglomerate forming the basal members of the sections. Quartzite is absent on the eastern side of the Cotton Plateau where limestone rests directly on the Goldie Formation, but thickens from an irregularly bedded sequence 5 to 13 metres thick at Panorama Point to a minimum of 500 metres, 16 kilometres to the south. Fossils were found only in one locality and consisted of Archaeocyatha.

The Shackleton Limestone lies unconformably on an unweathered surface cut across beds of the Late Pre-Cambrian Goldie Formation. At Cambrian Bluff, limestone containing sand lenses rests on an undulating surface cut across bedded sandstone and indurated mudstone. At Panorama Point, and on the eastern edge of the Cotton Plateau, metamorphosed Shackleton Limestone rests on an undulating surface with up to 12 metres of relief cut into drag-folded schist. At the former there is an angular discordance of 40°, but at the latter little discordance was noted. Three kilometres south-west of Mt. Hunt, greywacke breccia at the base of the Shackleton Limestone buries a karst-like surface with a relief of up to 9 metres, formed in marble of the Goldie Formation. The angular discordance is about 20°.

The significance of the contacts as part of a regional unconformity between the Lower Cambrian and Upper Pre-Cambrian is emphasised by the abrupt change from probably deep-water, redeposited sediments of the Goldie Formation to shelf limestone, breccia, conglomerate, and quartzite.

At the head of the Beardmore Glacier, Shackleton Limestone appears once again in four small outcrops at the extreme southern point of Mt. Bowers, itself one of a group of three small nunataks (YOUNG and RYBURN, 1968). The limestone here is light or dark grey, fine-grained, moderately pure, and steeply dipping. Scattered Archaeocyatha are present and have been described by GRIFFITH TAYLOR (*in* DAVID and PRIESTLEY, 1914, pp. 235–240).

Although much of the Shackleton Limestone is massive or poorly bedded, some sedimentary structures of assistance in environmental interpretation are present in some localities. Large-scale disharmonic folds, affecting a few hundred metres of limestone but not affecting the beds above or below, and detached balls of limestone, sometimes several metres across, suggest penecontemporaneous slumping. These features occur prominently at Cambrian Bluff and on the eastern flank of the range a few kilometres to the north.

The evidence available suggests that the Shackleton Limestone was laid down in shallow water. Some limestone conglomerate is cross-bedded and shows signs of strong local reworking such as the inclusion of limestone clasts with Archaeocyatha. HILL (1964b) stated that the most favourable depth for their growth was from 20 to 50 metres, and that they are not known in sediments presumed to be deposited below 100 metres. The shelly fauna, algae, and abundant trace fossils recorded south of Byrd Glacier also indicate a shallow shelf environment (BURGESS and LAMMERINK, 1979). A variety of sedimentary features also indicate that the majority of the limestone was deposited in supratidal to very shallow subtidal conditions. South of the Byrd Glacier, desiccated red carbonate mudstones capping lime–mudstone mounds are common features. Birdseye structures (SHINN, 1968) and cargneules (evaporite collapse breccias, WARRAK 1974) are common. Such features are characteristic of exposure in a warm and probably arid environment (BURGESS and LAMMERINK, 1979). In some places oncoliths (LOGAN and others, 1964), formed in a tidal or subtidal environment where relatively strong traction currents operate, are common. Other shallow water subtidal deposits include oolite (relatively common throughout the Nimrod–Byrd Glacier region), skeletal packstones and grainstones, and bioturbated lime mudstones. Scour channels and oscillation ripples in the sand, shale, and quartzite members in the Nimrod Glacier area also imply a shallow water environment.

The facies relationships (Fig. 8) suggest a steep rising coastline not far to the west of Mt. Hunt with less active uplift to the south. The great thickness of shallow-water calcareous sediment on the eastern limb of the syncline implies that deposition more or less kept pace with subsidence of the shelf. The repeated occurrence of reworked materials containing Archaeocyatha indicates

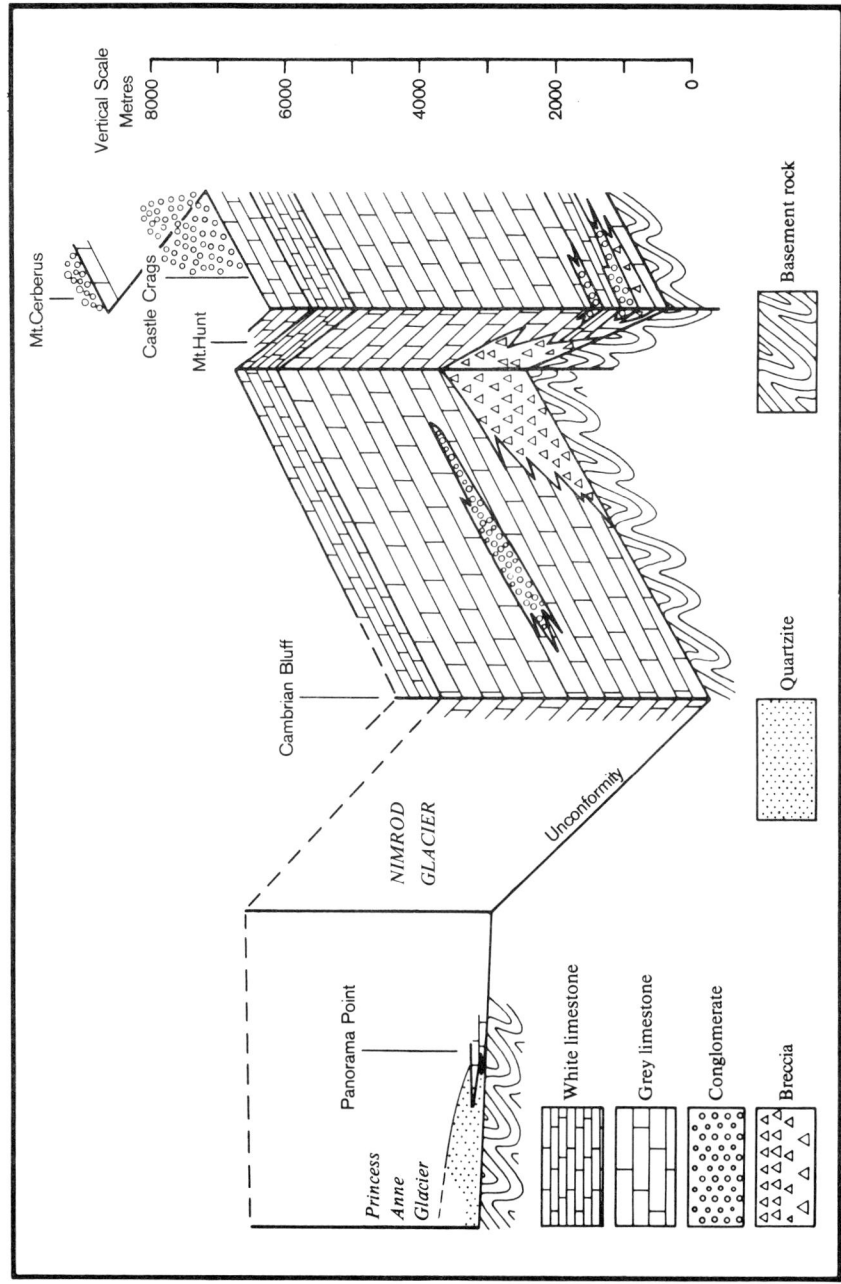

FIG. 8. Reconstruction of the facies relationships in the Shackleton Limestone before folding—Nimrod Glacier area. Horizontal distance between Mt. Hunt and Panorama Point is approximately 80 kilometres (from LAIRD, MANSERGH, and CHAPPELL, 1971, p. 441).

that parts of the inner shelf were continually emerging and being eroded while the basin of deposition was subsiding possibly with outlying reef-building keeping pace with subsidence. The angularity and softness of many of the clasts precludes transport over large distances and they must have been derived from newly emergent land or a nearby growing reef. The picture is thus of an emergent coast rising more rapidly in some places than in others and with an elongate Archaeocyathine bioherm parallel to the coast. On the offshore (eastern) side of the reef downwarp of the basin proceeded, sometimes outpacing sedimentation, accompanied by intermittent slumping and triggering of debris flows.

The fauna of the Shackleton Limestone is dominated by Archaeocyatha, but also includes trilobites, algae, and brachiopods. Archaeocyathine have been recorded in limestone at the head of the Beardmore Glacier, at one locality on the Cotton Plateau, and at numerous localities between the Nimrod and Byrd Glaciers. They occur both as individuals and in colonies most commonly in grey limestone but also in cream limestone. Individuals normally range in size from 1 to 5 centimetres maximum cup diameter, with the stem up to 15 centimetres long. Some particularly large individuals measuring 15 centimetres across at the cup opening and up to 45 centimetres long occur together with possible bryozoans at the head of the Starshot Glacier.

A small collection of Archaeocyatha from beds within 1000 to 1300 metres of the base of the formation near Cambrian Bluff in the Holyoake Range has been identified by HILL (1964a) as belonging to the Lower Lena Stage (late Lower Cambrian). The stratigraphically higher Archaeocyatha were collected from close to the top of the formation, and thus the upper part of the limestone is unlikely to be older than early Middle Cambrian. The fauna comprises *Porocyathus nimrodi*, *Thalamocyathus trachealus* (Taylor), *Cosinocyathus* sp., *Flindersicyathus latiloculatus* Hill, *Protopharetra* sp., and *Bedfordcyathus lairdi* (HILL, 1964a).

Limestone erratics collected by Shackleton's party from the Beardmore Glacier and presumably originating from the outcrops at Mt. Bowers contained an identifiable fauna. An archaeocyathid akin to *Archaeocyathus profundus* occurs, together with a specimen of *Protopharetra* akin to *P. radiata* (Bornemann), *P. rete* (Taylor), and *P. dubiosa*. In addition some algae and fragments of possible trilobites also occur (GRIFFITH TAYLOR, in DAVID and PRIESTLEY, 1914).

Gradationally overlying the Shackleton Limestone between the Starshot and Byrd Glaciers is the Dick Formation (SKINNER, 1964, 1965), a moderate olive–brown siliceous argillite. Thin siltstone beds are often present showing current ripple marks and load casts. Occasional fine conglomerate beds also occur. The finest beds are an irridescent purple colour with a greasy feel and are highly sheared. At one locality a total of 54 metres of impure limestone is interbedded with slightly schistose argillite, and at another cross-bedded sandstone and siltstone are interbedded with thin impure limestones. A flow (9 metres thick) of spilite interbedded with the sequence has also been recorded. The thickness of the Dick Formation varies considerably. Almost 100 metres of argillite overlie the Shackleton Limestone south of Mt. Hamilton (BURGESS and LAMMERINK, 1979), and elsewhere in the region it is at least 150 metres thick (SKINNER, 1964). It is apparently not present in a similar stratigraphical position south of the Starshot Glacier, but similar lithologies from 1 to 20 metres thick occur within the Shackleton Limestone both here and in Byrd Glacier region, and it probably has an interfingering relationship with the limestone.

The Douglas Conglomerate (SKINNER, 1964) overlies the Dick Formation south of the Byrd Glacier with possible slight unconformity (BURGESS and LAMMERINK, 1979). This formation is a polygenetic coarse conglomerate of rounded clasts up to 60 centimetres in diameter with a coarse sand matrix. The most abundant clasts in the Mt. Hamilton area are of Shackleton Limestone, some of which contain Archaeocyatha. Other clasts in the conglomerate are of quartzite, mudstone, argillite, and various igneous and metamorphic rocks surrounded by a coarse sandstone matrix. SKINNER (1964) recorded a thickness of 140 metres of Douglas Conglomerate in his study area, but elsewhere it is much thicker. On Mt. Hamilton the unit is estimated to be at least 1000 metres thick, and, north of Mt. Hunt, a probable correlative of the Douglas Conglomerate consisting of rounded clasts of limestone (some Archaeocyatha-bearing) and argillite resting sharply on Shackleton Limestone is at least 1200 metres, and perhaps up to 2850 metres thick (Fig. 7).

No *in situ* fossils have been recovered from the Dick Formation or the Douglas Conglomerate, although the former, because of its possibly interfingering relationship with the upper part of the Shackleton Limestone, is likely to be of similar age to this unit. The Douglas Conglomerate is certainly younger, and from its stratigraphical position is probably Middle, or perhaps even Late Cambrian in age. It documents renewed and intense tectonic activity with erosion of underlying Pre-Cambrian rocks as well as semi-lithified limestone. The folded Byrd Group sequence is truncated by an erosion surface, on which lie nearly flat-lying sandstones of the Beacon Supergroup.

East of the main outcrop of the Byrd Group are scattered outcrops of clastic sediments referred to the Starshot Formation (LAIRD, 1963, 1964). At the type locality near the mouth of the Starshot Glacier the formation consists of coarse conglomerate, mudstone, and calcareous sandstone, with sandstone as the dominant lithology. The conglomerate occurs in well defined beds up to 3 metres thick, and consists of clasts of well rounded quartz and limestone boulders up to 25 centimetres in diameter in a matrix of sandstone. Load casts occur on the soles of some sandstone beds, and some top surfaces display symmetrical ripple-marks. Several flows of rhyolite, up to 1.5 metres thick, and one of trachyte 6 metres thick, are interbedded with the sedimentary sequence in the type locality. The lateral extent of the Starshot Formation is uncertain. East of Mt. Hunt in the Holyoake Range, beds apparently continuous with the type Starshot Formation, and comprising an alternating sequence of calcareous graded sandstone and mudstone with common coarse sandstone and fine conglomerate layers, were correlated with the Starshot Formation by LAIRD, MANSERGH, and CHAPPELL (1971). Alternatively, they may represent a facies of the Goldie Formation. Because of complex folding no accurate thickness for the Starshot Formation could be determined, but a minimum thickness of 3000 metres was measured at the type locality (Mt. Ubique).

No contact between the Starshot Formation and other units of the Byrd Group has been seen, and a faulted relationship is implied by the field evidence. The petrographic similarity between the sandstones and conglomerates of the Starshot Formation and those of the Shackleton Limestone and the Douglas Conglomerate suggests a common source area and perhaps a similar age. The presence of boulders of limestone, indistinguishable from limestone of the Shackleton Formation, in the Starshot Formation, suggests that the latter must be younger than at least a portion of the Shackleton Limestone. Possibly the Starshot Formation may be a lateral, deeper water equivalent of part of the Shackleton Limestone or the Douglas Conglomerate. The rhyolite and dacite flows present in the formation may be correlatives of the Cambrian acidic volcanics common in the Queen Maud Mountains and Pensacola Mountains.

3.2.5. South Victoria Land

Many of the sediments in this area (lying between the Byrd Glacier and McMurdo Sound) (Fig. 6) are highly metamorphosed, sometimes to amphibolite facies. In the south, where the sediments are only mildly metamorphosed, they are included in the Skelton Group (GUNN and WARREN, 1962), while the northern, more metamorphosed sequences are included in the Koettlitz Group (GRINDLEY and WARREN, 1964; WARREN, 1969). Although the Skelton and

Koettlitz Groups have different metamorphic grades and histories, it is likely that they are in part correlatives.

The mildly metamorphosed sediments of the Skelton Glacier area (Skelton Group) were divided by GUNN and WARREN (1962) into two formations, the Anthill Limestone and the Teall Greywacke. The Teall Greywacke at its type locality (Teall Island) is composed of steeply dipping complexly folded and thermally altered calcareous greywacke and argillite, exposed over a distance of 3 kilometres on the eastern side of the island, at the entrance to the Skelton Glacier. The sequence consists of beds of indurated black mudstone and greenish siltstone varying in thickness up to 1 metre. Groups of graded beds 25 to 50 millimetres thick form ribbon sequences; cross-laminated horizons of the same thickness are also present. Outcrops farther up the glacier at Red Dyke Bluff, which were also included by GUNN and WARREN (1962) in the Teall Greywacke, have subsequently been found to contain pilotaxitic felsic volcanic clasts associated with fine-grained, actinolite-rich tuffaceous beds (SKINNER, in press), and probably belong to a younger unit (see below).

All rocks are traversed by fine, irregular quartz veins and by systems of transverse and horizontal joints which give rise to extensive screes. The strike varies between 074° and 100°, and the beds in the main dip steeply south. Both top and base of the succession at Teall Island are truncated by post-tectonic intrusions, but the thickness of the formation is not less than 1800 metres. Thin-section study showed the rocks to consist dominantly of quartz and feldspar grains, with some clastic grains in a recrystallized matrix.

The Anthill Limestone (GUNN and WARREN, 1962) was defined as those folded rocks of the Skelton Group that consist primarily of white or grey limestone, with minor mudstone, siltstone, and quartzite; they have in the main been altered only by low-grade regional and thermal metamorphism. The type locality is at Ant Hill on the west side of the lower Skelton Glacier, but the limestone beds extend for 8 kilometres along this flank of the glacier and also occur on the east side.

Although the Anthill Limestone is complexly folded, it has been possible to subdivide the formation by use of bottom and top criteria into five lithological units (FLORY and others, 1971). From oldest to youngest these are thin-layered meta-limestone, greywacke–mudstone, quartzite, slate, and meta-limestone. Individual beds in the limestone are usually coarsely crystalline and vary in thickness from a few centimetres to a few metres. The beds are white, grey, green, or yellow–brown, the colour varying with the amount of mud present and the degree of alteration. The limestones, although strongly folded, have been only mildly regionally and thermally metamorphosed, a small amount of fine biotite being found in some rocks. The clastic content of the beds varies considerably in composition, some limestones containing up to 15% of clastic quartz and albite in addition to calcite, while other samples contain high proportions of secondary minerals. The greywacke–mudstone unit is a quartzose greywacke similar to that of Teall Island.

Much of the Anthill Limestone is complexly folded, but the prevailing strike of the bedding is east, this structural trend being disrupted by late north-striking fold systems. The minimum thickness of the formation is 3000 metres.

Recent work (SKINNER, in press) suggests that the Teall Greywacke at its type locality is best considered as a lateral facies of the Anthill Limestone. No fossils have been found in either unit, and their age is in doubt. Both units have been complexly folded during three phases of deformation, a feature common to other basement rocks in the McMurdo Sound region. On degree of metamorphism and general lithology they have some similarities to the Beardmore and Byrd Groups farther south, and are likely to be of Late Pre-Cambrian or Cambrian age.

Plate 1. View up the Nimrod Glacier towards the Polar Plateau. The Ranges form part of the Trans-Antarctic Mountains. Centre right: Holyoake Range, consisting of Cambrian Shackleton Limestone, overlying Pre-Cambrian Goldie formation (near ridge), with Cambrian Bluff extending into the Nimrod Glacier. The ranges in the left foreground and distance consist of folded Pre-Cambrian metasediments overlain by horizontal Late Palaeozoic Beacon Group sediments. (*Photograph by U.S. navy*).

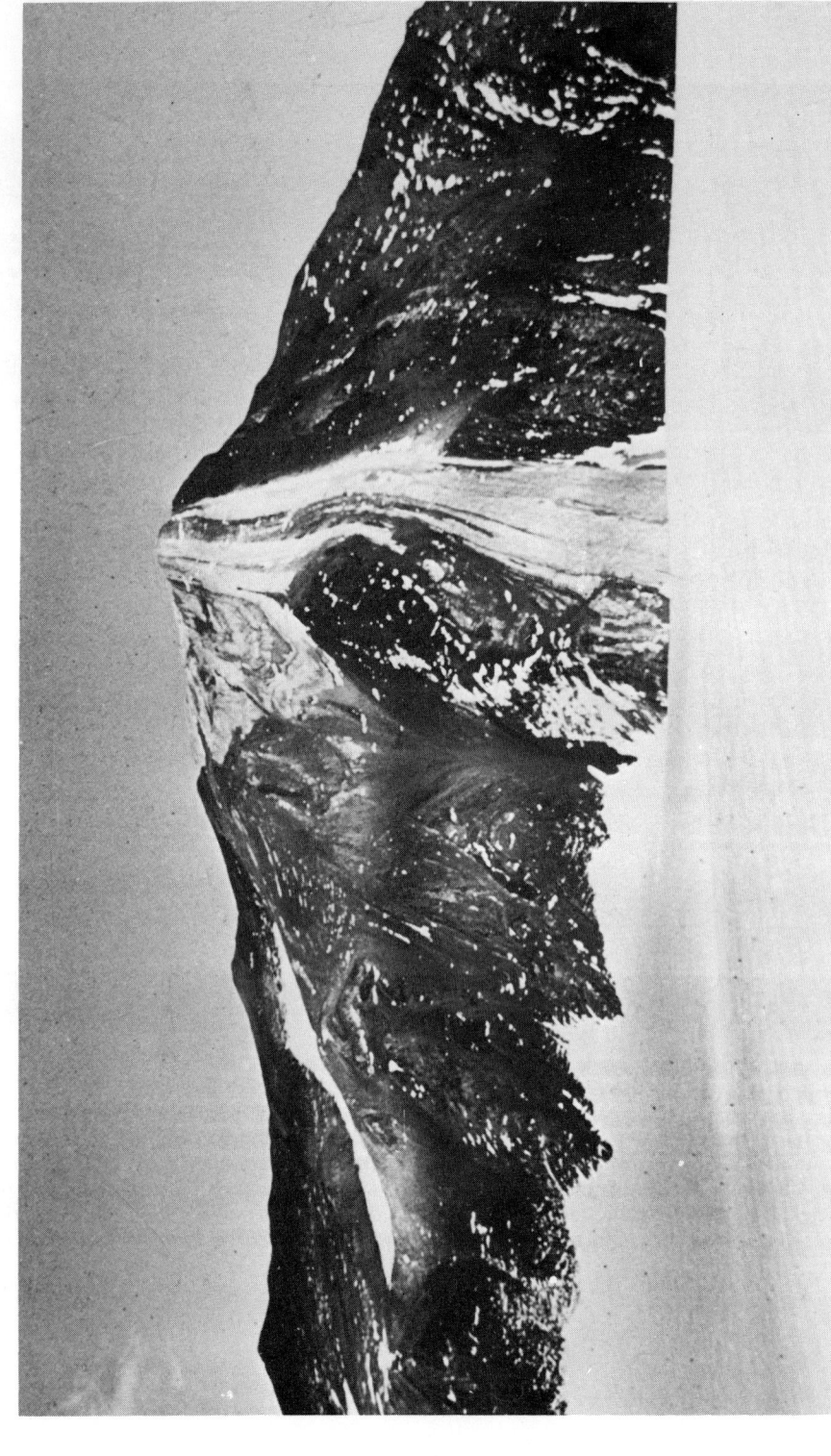

Plate 2. South face of Mt. Henson, type locality of the Henson Marble. Grey and white marble units enclose a small laccolith of actinolite–amphibolite and are faulted against dark hornfelses of the Pre-Cambrian Duncan Formation (right). The face is about 500 metres high. (*Photograph by* BARRETT, P. J.).

Plate 3. South face of Cambrian Bluff (1670 metres) showing complex folding within the Shackleton Limestone. The axis of a major syncline running the length of the Holyoake Range is visible to the left of the photograph. Strata on the eastern limb (right) are slightly overturned.

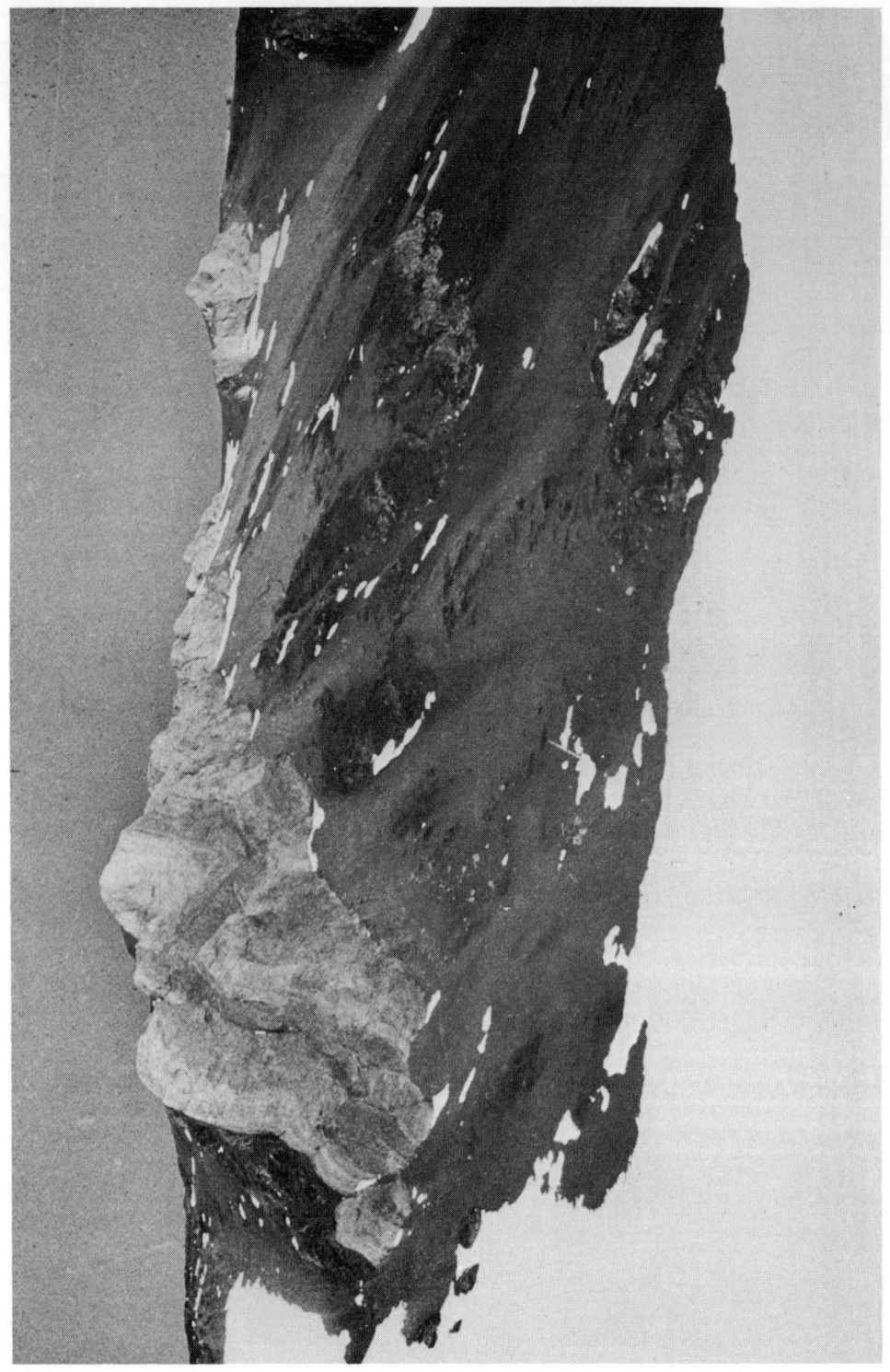

Plate 4. Twenty-metre thick lens of fossiliferous limestone of late Middle Cambrian age in laminated mudstone, Spurs Formation, Reilly Ridge.

Plate 5. Carryer Conglomerate at its type locality, lower Carryer Glacier.

Plate 6. Bluffs formed by Hope Granite, coastal range between the Nimrod and Byrd Glaciers. The low ridges in the middle distance are composed of metasediments of the Pre-Cambrian Goldie Formation.

Overlying the Anthill Limestone with notable angular unconformity in areas to the east of the Skelton Glacier is a unit of metasediments and felsic metavolcanics (Cocks Formation; SKINNER, in press). Locally, channels up to 2 metres deep eroded into complexly folded limestone are infilled by ellipsoidal limestone and metavolcanic clasts up to 35 centimeters in size within a calcareous, tuffaceous, quartz sandstone matrix. Thin lenses of coarse grey limestone are interbedded higher in the local conglomerate and with overlying well bedded volcaniclastic sandstones showing grading and cross-bedding. Penecontemporaneous volcanism is indicated by soft sediment deformation around a 1 metre thick layer of felsic, porphyritic pillow lava along the contact with the limestone at one locality. The thickness of the unit is uncertain, but a minimum is 150 metres. No fossils have been recorded, but the presence of an angular unconformity at its base and the fact that it has been subjected to only two periods of folding indicate that the unit is notably younger than the Teall Greywacke and Anthill Limestone. The felsic volcanics which are a common constituent of the unit invite correlation with Cambrian units elsewhere in the Transantarctic Mountains, (Pensacola Mountains, Queen Maud Mountains) which are also notably volcanogenic.

A fairly regular increase in metamorphic grade can be traced from the comparatively unaltered sediments of the Skelton Glacier area, through the coarse-grained, strongly metamorphosed sillimanite schists and chondrodite marbles of the Koettlitz Glacier and McMurdo Sound areas, to the Granite Harbour region, where only highly migmatized xenoliths are found enclosed in great volumes of granite. In the Koettlitz Glacier–McMurdo Sound region, the Koettlitz Group (GRINDLEY and WARREN, 1964) which consists of schists, marble, gneiss, and rare quartzite, is metamorphosed to the amphibolite facies (BLANK and others, 1963). A central stock of granite divides the metasediments into two belts. The north-eastern belt consists of five formations: the Marshall Formation, 900 metres of schist and gneiss; Miers Marble, 600 metres of coarse-grained marble; Garwood Lake Formation, 760 metres of schist and gneiss; Salmon Marble, 2400 metres; and Hobbs Formation, 1800 metres of schist and conglomerate. The same formations probably occur in the western belt. Cylindrical siliceous objects thought by BLANK and others (1963) to represent Archaeocyatha occur in the Salmon Marble in two localities, but subsequent investigations have shown that they represent inorganic structures (FINDLAY, 1978). No fossils are known from either the Skelton or the Koettlitz Groups, and their age is still in doubt.

Farther north, in the area of the McMurdo Oasis comprising the Dry Valleys, Taylor, Wright, and Victoria, the metasediments of the Koettlitz Group are included in one unit, the Asgard Formation (McKELVEY and WEBB, 1961; McKELVEY and WEBB, 1962; ALLEN and GIBSON, 1962; HASKELL and others, 1965), which consists of schist, marble, and metaquartzite metamorphosed to the amphibolite facies. The Asgard Formation is at least 4500 metres thick, but this thickness may include rocks of Pre-Cambrian as well as Cambrian age.

3.2.6. North Victoria Land

All known Lower Palaeozoic sedimentary rocks of north Victoria Land are included within the Bowers Supergroup (LAIRD and others, in press), which occurs within an infaulted strip 20–25 kilometres wide, extending south-east from the lower Rennick Glacier across the southern Bowers Mountains to at least the upper Mariner Glacier, a distance of 270 kilometres (Figs 6 and 9). It is flanked to the south-west by the Pre-Cambrian Wilson Group, which consists of schists and gneisses of amphibolite metamorphic facies, and to the north-east by the sedimentary Robertson Bay Group of probable Late Pre-Cambrian age (COOPER and others, in press), with which it is in probable fault contact. In its central portion it is overlain by gently dipping rocks of the Beacon Supergroup and Ferrar Group of Late Palaeozoic and Mesozoic

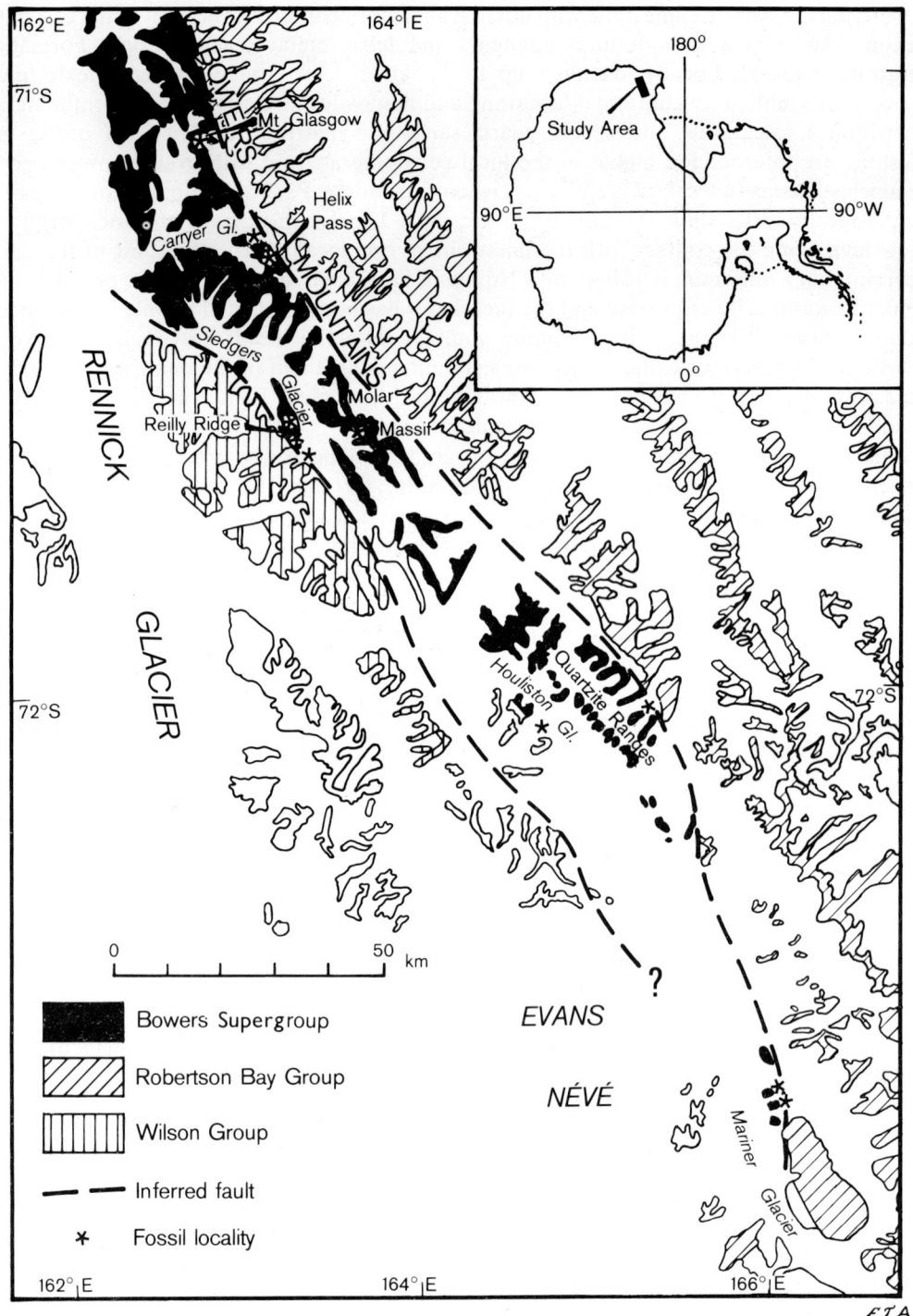

FIG. 9. Map of part of northern Victoria Land showing distribution of Bowers Supergroup and fossil localities (from LAIRD, BRADSHAW, and WODZICKI, 1976, p. 276).

age. The sequence, which dips steeply, is folded about north-west to south-east trending axes into a complex anticline in the west and a major syncline in the east (BRADSHAW and others, in press). The syncline, which extends from near the upper Mariner Glacier north at least as far as Mt. Glasgow, has a nearly vertical axial plane and a gently undulating fold axis. In the west folding is more complex, overturned limbs are common, and fold plunges are steeper. Dips are steep, most ranging between 50° and vertical. Folding of the Bowers Supergroup produced an axial plane cleavage and metamorphism to the prehnite–pumpellyite and pumpellyite–actinolite facies (WODZICKI and others, in press). This folding event occurred during the Borchgrevink Orogeny (CRADDOCK, 1970), K–Ar dated at between 380 and 420 m.y. (Silurian) by ADAMS and others (in press).

Previous descriptions of the succession have included it within the Bowers Group (STURM and CARRYER, 1970), which consisted of three conformable formations (DOW and NEALL, 1972, 1974; and Table 1). However, subsequent field observations and palaeontological study have shown the need for comprehensive stratigraphical revision, including changes in the order of succession (LAIRD and others, 1976; LAIRD and others, in press). The sequence is transected by two important unconformities, and the revised stratigraphy now accommodates four major units, the upper three forming part of a newly defined Bowers Supergroup (Table 1). These units are as follows, from oldest to youngest: the Husky Conglomerate, which is inferred to rest unconformably on the Wilson Group and is of uncertain affiliation; the Sledgers Group of Vendian to Early Cambrian age; the Mariner Group, of late Middle Cambrian to middle or late Late Cambrian age; and the Leap Year Group, of probable late Late Cambrian to Ordovician age. Their stratigraphical relationships are indicated in Fig. 10.

The Husky Conglomerate is a sedimentary succession of dark green amphibolitic conglomerate and minor sandstone and mudstone cropping out on the eastern flanks of the Lanterman Range to the north and south of Husky Pass and west of Reilly Ridge (Fig. 9). At the type locality west of Reilly Ridge, 370 metres of steeply east-dipping dark green conglomerate and breccia, consisting dominantly of amphibolite clasts, with minor sandstone and mudstone, rests on amphibolite of the Wilson Group. Although the basal contact is obscured by snow, an unconformity is inferred. The top of the sequence is separated by a snow gap and probable fault from overturned east-dipping alternating beds of mudstone and volcanogenic sandstone correlated with the Sledgers Group. Conglomerate units, some as thick as 200 metres, make up the bulk of the Husky Conglomerate. The clasts are mainly sub-rounded to sub-angular, reaching 3 metres in diameter. They are often tightly packed in a sparse matrix, but mudstone units up to 95 metres thick with scattered subrounded boulders also occur. Massive breccia units up to 40 metres thick and massive mudstone and sandstone units up to 15 metres form a smaller proportion of the sequence. The clasts at all localities visited consist almost entirely of amphibolite indistinguishable from the Wilson Group amphibolite units on which the Husky Conglomerate rests at localities north of Husky Pass. Rounded pebble-sized clasts of quartz and granite make up less than 1%. Because its upper contact is faulted, and no fossils have been recorded, the stratigraphical relationship of the Husky Conglomerate with other units is uncertain. Its metamorphic rank (green-schist facies; WODZICKI and others, in press) suggests a correlation with either the Robertson Bay Group or the basal part of the Sledgers Group.

Although the Sledgers Group was originally included in the Bowers Group as the Sledgers Formation (CROWDER, 1968), the unit has been given separate group status within the Bowers Supergroup (LAIRD and others, in press). The unit is well exposed in the northern Molar Massif (Fig. 9), where more than 2500 metres of nearly continuous outcrop occurs, and which is taken as the type area. The group crops out throughout the Bowers Mountains at least as far north as Frolov Ridge. It has not been certainly identified in the Evans Neve or farther south, but some of

TABLE 1
Lower Palaeozoic Stratigraphical Units

Dow & Neall, 1972, 1974

Laird, Bradshaw, & Wodzicki (in press)

Group	Formations		Groups	Formations	Thickness (m)	Age
	Camp Ridge Quartzite		Leap Year	Camp Ridge Quartzite / Reilly Conglomerate / Carryer Conglomerate	3000+	Ordovician to ?late Late Cambrian
				HIATUS		
Bowers	Sledgers Formation	BOWERS SUPERGROUP	Mariner	Eureka Formation / Spurs Formation / Edlin Formation	2000+	middle or late Late Cambrian to late Middle Cambrian
				HIATUS		
			Sledgers	Glasgow Volcanics / Molar Formation	3500+	Early Cambrian to Vendian
	Carryer Conglomerate			Husky Conglomerate	370+	?Vendian

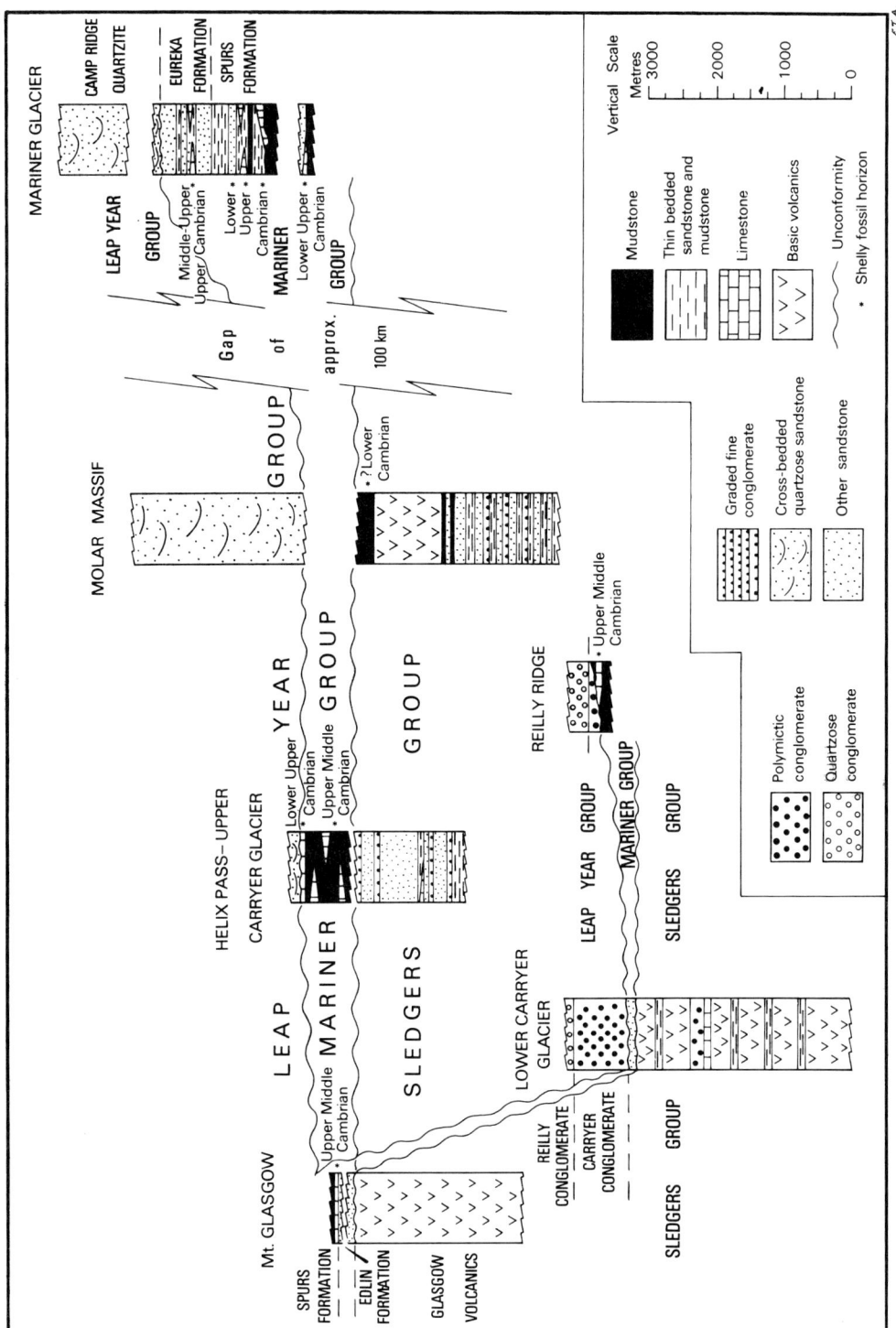

FIG. 10. Correlation of main stratigraphical sections through the Bowers Supergroup (from LAIRD and others, in press).

the rocks lying south and south-east of the Quartzite Ranges and previously included within the Robertson Bay Group may form part of the Sledgers Group. Sediments exposed in isolated nunataks lying west of the Quartzite Ranges in the Houliston Glacier have previously been mapped as Sledgers Formation (now Group) (DOW and NEALL, 1974; COOPER and others, 1976). However, their stratigraphical position is uncertain and the discovery within the sequence by COOPER and others (1976) of Middle Cambrian trilobites suggests possible correlation with the Mariner Group. The group is overlain unconformably by the Mariner Group. Its base is not seen, but the minimum thickness of the Sledgers Group is 3500 metres. It is divided into two formations, the Glasgow Volcanics and the Molar Formation (Table 1).

The Glasgow Volcanics are named for the massive basaltic breccia at its type section in Mt. Glasgow, where the formation is at least 2500 metres thick. The unit occurs as sheets and lenses within the sedimentary phase (Molar Formation) of the Sledgers Group, occupying a high proportion (perhaps more than half) of the succession in the northern part of the area, and apparently wedging out to the south where the formation is unknown south of the Molar Massif. Its northern extent is uncertain, but similar volcanics which are almost certainly part of the same sequence have been reported from Frolov Ridge in the northern Bowers Mountains (STURM and CARRYER, 1970). The base of the succession at Mt. Glasgow was unseen, but the top is marked by a sharp contact with the basal beds of the Mariner Group. The sequence here consists mainly of unsorted angular to subangular basaltic breccia, with minor flows, some containing pillows. In the Molar Massif the formation occurs as one massive sheet-like body up to 1000 metres thick lying between sandstone and mudstone of the Molar Formation near the top of the Sledgers Group. It consists entirely of unsorted angular to subangular basaltic breccia occasionally containing fragments of pillows. In the lower Carryer Glacier area, the formation comprises several thick units composed entirely of volcanic breccia except for 100 metres of reddish weathered lava near the top of the succession. No pillows were observed. The top surface of the uppermost breccia is weathered and overlain abruptly by quartzose sandstone and mudstone of the Mariner Group; the base of the succession has not been seen.

The Molar Formation is named for the sequence of sandstones, mudstones, and minor conglomerates cropping out in the western part of the Molar Massif. The unit is also well exposed in the nunataks lying immediately south of the Molar Massif, and in the mountains between the Sledgers and Carryer Glaciers. The base of the formation is not seen, but at its type locality in the north-western Molar Massif a minimum of 1350 metres of sediments lies below the sharp contact with the Glasgow Volcanics. The sequence here consists dominantly of dark mudstone containing thin beds of well sorted rippled or wavy-bedded sandstone. Intercalated within the mudstones are isolated units of massive or graded-bedded sandstone, lenticular pockets of graded breccia or coarse sandstone, and a 40 metres thick chaotic slide breccia containing angular blocks up to 1 metre in diameter. The sequence is similar throughout the Molar Massif and in the nunataks to the south, although rare thin conglomerate horizons with rounded boulders up to 30 centimetres in size occur in exposures south of the type section. In the Molar Massif and the ridges to the north, up to 350 metres of sediments tentatively included within the Molar Formation occur apparently conformably above the Glasgow Volcanics. They consist of mudstone with thin cross-laminated beds and lenses of calcareous sandstone or limestone occasionally containing burrows. The Molar Formation is generally similar in the area between the Sledgers and Carryer Glaciers, except in the lower Carryer Glacier, where the sediments consist of thin (up to 350 metres) units of conglomerate, sandstone, mudstone, and one bed of limestone 10 metres thick contained within thick sequences belonging to the Glasgow Volcanics (LAIRD and others, in press). The graded coarse sandstone, breccia, and conglomerate beds in the southern part of the succession are poorly sorted; they contain angular to moderately well rounded fragments mainly of volcanic and intrusive rocks (about 70%) and lesser amounts of sandstone, siltstone, and limestone. Basaltic fragments, similar in lithology to the Glasgow Volcanics, comprise about half the igneous clasts; the remainder include andesites, dacites, rhyolites, and intrusive rocks ranging from grandodiorite to granite (LAIRD and others, 1976). Acritarchs extracted from the Molar Formation indicate a Vendian age for the bulk of the sequence (COOPER and others, in press). The following are represented: *Bavlinella faveolata* (Shepeleva) Vidal, *Vendotaenia* sp., *Trachysphaeridium levis* (Lopukhin) Vidal, *Leiosphaeridia* sp., cf. *Stictosphaeridium* sp. (sensu Vidal), *Pterospermopsimorpha* cf. *densicoronata* Vidal, *Chuaria circularis* Walcott, and *Lacunosphaera* cf. *simplex* Pychora. However, at one locality in Molar Massif, above a band of volcanics and representing the highest beds of the Molar Formation, small poorly preserved shelly (chitino-phosphatic) fossils considered to be inarticulate brachiopods are present. Although they occur in association with apparently Vendian acritarchs, they are probably Cambrian, and thus a Vendian to Early Cambrian age for the Molar Formation is likely (COOPER and others, in press).

The Mariner Group (ANDREWS and LAIRD, 1976; LAIRD and others, in press) is well exposed in the upper Mariner Glacier at Eureka Spurs, where 1600 metres occurs, and crops out poorly to the north of this as far as Mt. Glasgow. The base of the group has been seen only at Mt. Glasgow and in the lower Carryer Glacier, where it rests sharply on Glasgow Volcanics. Although the contact is apparently concordant, the weathered nature of the upper surface of the volcanics in the lower Carryer Glacier, the abrupt change in lithology from volcanics and volcanogenic sediments to dominantly quartzose and calcareous sediments, and the lack of fossils older than late Middle Cambrian from the Mariner Group, suggest that a major hiatus representing much of Early and Middle Cambrian time is present between the Mariner and Sledgers Groups. No complete Mariner Group sequence has been seen, and the thickness of the group is unknown but likely to be in excess of 2 kilometres. The group is subdivided into three units: the Edlin, Spurs, and Eureka Formations (LAIRD and others, in press) (Table 1).

The basal unit of the Mariner Group—the Edlin Formation—has been identified only in the lower Carryer Glacier and north of Mt. Glasgow in the Edlin Neve, from which it takes its name. The sediments are characteristically highly quartzose, and represent marginal marine sands transgressive on the underlying Sledgers Group. In the type locality 1 kilometre north of Mt. Glasgow a poorly exposed composite sequence 100 metres thick comprises a basal 8 metres of red–brown muddy sandstone which rests sharply on Glasgow Volcanics, followed by thin-bedded or massive quartzose sandstone. This passes up into limestone and calcareous mudstone of the Spurs Formation. In the lower Carryer Glacier 100 metres of fine-grained, thin-bedded quartzose sandstone sharply but apparently concordantly overlies weathered breccia of the Glasgow Volcanics, and is itself overlain by Carryer Conglomerate. The sandstone, which contains abundant ripple-marks and low-angle cross-bedding, is muddy and notably red–brown in the basal 10 metres, where desiccation cracks and arthropod trails occur. No shelly dossils were seen in the Edlin Formation, but trilobites of late Middle Cambrian age occur in limestone of the Spurs Formation immediately overlying quartzose sandstone of the Edlin Formation north-east of Mt. Glasgow. A Middle Cambrian age is probable.

Named for Eureka Spurs, at the head of the Mariner Glacier, the Spurs Formation is the most widely represented unit of the Mariner Group (see Fig. 10). At the type locality 900 metres of fissile mudstone with thin limestone lenses passes gradationally upwards with increasing sandy layers into wavy-bedded mudstones forming the basal unit of the Eureka Formation. The mudstone contains sparse channels and scattered stratified, parallel-sided massive or parallel-laminated ungraded sandstone beds with occasional flute and groove casts on the soles and symmetrically rippled upper bedding surfaces. Cross-laminated horizons occur rarely. Elsewhere, the Spurs Formation is dominated by mudstone often containing thin lenses and layers of limestone. In the Helix Pass area scattered outcrops of this lithology also include thin beds of quartzose or calcareous sandstone as well as limestone. Thicker limestones (up to 20 metres) occur on Reilly Ridge (Plate 4), where considerable slumping and brecciation are present (BRADSHAW and others, in press), and also at Mt. Glasgow, where a limestone lens 30 metres thick occurs. East of the Mt. Glasgow sequence, at the head of Edlin Neve, several hundred metres of calcareous mudstone containing thick units of poorly sorted quartzo-feldspathic, thinly bedded or massive, medium sandstone exposed in low ridges can probably be correlated with the Spurs Formation. In Houliston Glacier mudstone and conglomerate occurring in isolated nunataks have previously been included within what is now the Sledgers Group (*see* p. 288). The presence of Middle Cambrian fossils, however, shows that the beds are much younger than the youngest known beds (earliest Cambrian) (COOPER and others, in press). Their lithostratigraphical correlation is uncertain, but lithologies similar to those in the Spurs Formation are common (Dr R. A. COOPER, personal communication) and inclusion in this formation is possible. As no continuous section through the Spurs Formation has been recorded, its total thickness is uncertain. The 900 metres of nearly continuous outcrop exposed on Eureka Spurs comprises a minimum, and scattered outcrops of similar lithology lying apparently stratigraphically below this sequence suggest that at least 1500 metres is likely to be present. The Spurs Formation is notably fossiliferous, the main shelly fauna being dominated by trilobites and brachiopods. The oldest fossils occur in an isolated outcrop of uncertain stratigraphical relationship to the ridges forming the type locality, and include *Palaeodotes* cf. *italops* and *Rhodonaspis*, which indicate a Mindyallan age. The stratigraphically lowest fossils collected from the type locality occur approximately 200 metres above the base of the exposed section, and include *Proceratopyge* and *Pseudagnostus*, indicating an early Late Cambrian age. *Pseudagnostus* occurs at several younger horizons along with abundant trilobites, articulate and inarticulate brachiopods, and hyolithids. At least seventeen taxa of trilobites, four of molluscs, and three of brachiopods are present in the upper part of the Spurs Formation, and represent a fauna of Late Cambrian (late Idamean, *Erixianum sentum* Zone; late Dresbachian, late Tuorian) age (SHERGOLD and others, 1976). The trilobites include *Pseudagnostus* sp., *Stigmatoa* sp., *Olentella* cf. *olentensis*, ?*Irvingella* sp., *Pedinocephalus* sp. cf. *P. bublichenkoi*, three unnamed species of the subfamily *Aphelaspidinae*, ?*Talbotinella* sp., an unnamed species of olenid, *Prochungia* sp. aff. *P. granulosa*, *Proceratopyge* sp. cf. *P. lata.*, and three undefined species of Trilobita. The inarticulate brachiopods are represented by *Schizambon reticulata* and *Prototreta* sp., and the articulates by *Billingsella antarctica*. Molluscs present are *Contitheca webersi*, *Hyolithes* sp., *Pelagiella* sp., and *Scaevogyra* sp. The general composition of the fauna resembles that of the Heritage Group of the Ellsworth Mountains (*see* pp. 264–265). It also bears strong affinity with faunas of Australia, China, and Kazakhstan (SHERGOLD and others, 1976), in agreement with affinities indicated by Middle and Early Cambrian trilobite faunas of Antarctica described by PALMER and GATEHOUSE (1972). The limestone at Mt. Glasgow contains a late Middle Cambrian trilobite fauna including *Dorypyge*, a dolichometopid close to *Amphoton*, ?*Centropleura*, and agnostids. At Helix Pass, the lowest fauna, which is underlain by at least 300 metres of unfossiliferous beds, includes *Centropleura* and several agnostids, including *Clavagnostis* and a ptychagnostid indicating a late Middle Cambrian (*Leiopyge laevigata* Zone) age. The youngest fauna, which is 200 metres higher stratigraphically and which lies immediately below the contact with the Camp Ridge Quartzite, includes *Glyptagnostus* ?*stolidotus* and *Aspidagnostus* and is early Late Cambrian in age (probably *Glyptagnostus stolidotus* Zone, Mindyallan Stage). At Reilly Ridge (Plate 4) a rich and well preserved trilobite fauna occurs in mudstone with layers, lenses, and derived blocks of limestone. The fauna includes *Amphoton*, *Dorypyge*, *Corynexochus*, '*Solenopleura*', *Leiopyge*, *Hypagnostus*, and *Ptychagnostus*, and indicates a late Middle Cambrian (*Leiopyge laevigata* Zone) age, thus correlating with the oldest fauna preserved at Helix Pass and perhaps also the Mt. Glasgow fauna. The commonly calcareous nature of the sediments and the relatively prolific shelly fossils strongly suggest that the Spurs Formation was deposited under shallow, open-marine conditions (ANDREWS and LAIRD, 1976).

The Eureka Formation is named from the upper part of the succession at Eureka Spurs. The unit does not occur farther north where it has probably been removed by erosion prior to deposition of the Leap Year Group. At the type locality 700 metres of sediments dominated by quartzose sandstone are well-exposed above the Spurs Formation. The succession has been informally divided into: the basal Wavy-bedded Sandstone Unit (250 metres), which passes gradationally upward from the Spurs Formation; the Limestone Unit (100 metres), which contains lenses of limestone up to 8 metres thick in mudstone; and the Red Bed Unit (350 metres), consisting mainly of brownish grey, very fine or fine-grained quartzose sandstone and minor mudstone with mud-cracks and arthropod tracks (ANDREWS and LAIRD, 1976). The formation appears to be generally shallow marine and regressive upwards. The Eureka Formation is overlain sharply by the basal conglomerate of the Camp Ridge Quartzite (Leap Year Group) which infills small scours eroded into the underlying sediments.

Trilobites were found by COOPER and his associates in a thin limestone layer 200 metres above the base of the formation. Provisional identifications are *Pseudagnostus* (*Pseudagnostus*), *Homagnostus*, *Amorphella* sp., *Kujandaspis* sp., *Proceratopyge*, *Kazelia*? sp., and *Olentalla*? sp. The age is probably early post-Idamean. Trace fossils from the uppermost beds of the Eureka Formation (Red Bed Unit of ANDREWS and LAIRD, 1976) include *Rusophycus*, *Cruziana*, and ?*Diplichnites*.

The Leap Year Group rests on an erosion surface which truncates the Mariner Group. Although locally the two groups appear to be concordant, the Leap Year Group rests on deposits of middle or late Late Cambrian age in the upper Mariner Glacier region, and of early Late Cambrian and late Middle Cambrian age at Helix Pass and Reilly Ridge, suggesting that a considerable thickness of sediments (perhaps in excess of 2 kilometres—*see* Fig. 10) has been eroded from the Mariner Group to the north. The Leap Year Group is exposed in two separate strips of contrasting lithology occupying the eastern and western flanks of a complex anticline with Sledgers Group rocks forming its core. It is divided into three dominantly fluvial formations: the Camp Ridge Quartzite, which is recognized only in the eastern strip of outcrop; and the Carryer and Reilly Conglomerates which crop out in the western strip.

The Camp Ridge Quartzite, which is the most widely represented unit of the Leap Year Group, occupies the core of a major syncline which has been mapped from the upper Mariner Glacier north-westward for 180 kilometres to Helix Pass. At its type locality in the Quartzite Ranges (LE COUTEUR and LEITCH, 1964) it consists of red–brown or yellow quartzose sandstone, quartzose conglomerate, and minor mudstone. Large-scale unimodal cross-bedding is common in the coarser beds, and small channels also occur. Conglomerate pebbles are set in a sandy matrix, are well rounded and up to 8 centimetres in diameter. The lithology is relatively uniform throughout the Camp Ridge Quartzite outcrop, although in exposure north of the Molar Massif conglomerate becomes rare. The basal contact of the formation is known only at Eureka Spurs and 5 kilometres north of Helix Pass. At the former locality it infills small scours eroded in the underlying Eureka Formation (ANDREWS and LAIRD, 1976); at the latter, quartzose sandstone rests sharply and possibly with slight angular discordance on Spurs Formation. The upper limit of the Camp Ridge Quartzite is nowhere seen, and its true thickness is uncertain. However, 2900 metres has been measured at the type locality (LAIRD and others,

1974), and it has been suggested (DOW and NEALL, 1972) that as much as 7000 metres may be exposed 20 kilometres to the north. No shelly fossils have been collected from the Camp Ridge Quartzite, and its age is uncertain. At Helix Pass and in outcrops within and south of the Quartzite Ranges *Skolithos* occurs and at the type locality the lower part of the Camp Ridge Quartzite contains a variety of trace fossils. They include *Daedalus*, *Arthrophycus*, ?*Mono-craterion*, and a U-shaped structure probably related to *Diplocaterion* (LAIRD and others, 1974). A specimen of *Rusophycus* was collected from the West Quartzite Range by DOW and NEALL (1972). None of the trace fossils is of sufficiently restricted time range to be useful in dating. A maximum age is provided by the middle or late Late Cambrian age of the unconformably underlying Mariner Group at Eureka Spurs. A minimum is given by a radiometric age of 421 m.y. (early Silurian) determined from the formation by ADAMS and others (in press). The Camp Ridge Quartzite is thus probably no older than late Late Cambrian and is likely to range up into the Ordovician.

The Carryer Conglomerate (DOW and NEALL, 1972) which is the older of the two divisions of the Leap Year Group occurring in the western strip, is very well exposed in the lower Carryer Glacier (Plate 5) and on Reilly Ridge, and also appears to occupy the south-western flanks of Mt. Soza and much of the massif between the lower Sledgers and Carryer Glaciers. It consists dominantly of polymictic well bedded conglomerate of well rounded clasts, occasionally reaching 3 metres in diameter and consisting mainly of granite, quartz, sandstone, rare volcanics, and locally, limestone. Lenses of cross-bedded pebbly coarse sandstone are common, frequently infilling small channels. The succession at Reilly Ridge, although dominated by polymictic conglomerate, also locally includes units up to 70 metres thick of mudstone and fine-grained sandstone. The basal contact with the Mariner Group is obscured by scree in the lower Carryer Glacier, but appears to be concordant. However, on Reilly Ridge the Carryer Conglomerate appears to infill channels up to 200 metres deep cut into the underlying mudstone and limestone of the Helix Formation (*see* Fig. 1, BRADSHAW and others, in press). At both Reilly Ridge and the lower Carryer Glacier the formation passes upwards with rapid gradation into basal sandstone of the Reilly Conglomerate. It attains its greatest thickness of 800 metres at the type locality. At Reilly Ridge it has a maximum thickness of approximately 300 metres, but thins steadily south-eastwards and is not present in outcrops 6 kilometres south-west of the southern end of Reilly Ridge. No body or trace fossils are known from the Carryer Conglomerate, but a maximum age is provided by the unconformably underlying mudstone and limestone of the Mariner Group, which contains late Middle Cambrian trilobites. A Late Cambrian to Ordovician age is likely.

The Reilly Conglomerate is named for pebbly quartzose conglomerate cropping out on Reilly Ridge and on nunataks to the south. It also crops out in the lower Carryer Glacier where 20 metres of reddish, medium- to coarse-grained quartzose sandstone and quartzose conglomerate rests conformably on Carryer Conglomerate; the unit probably continues southwards as a strip of outcrop to the northern flank of the lower Sledgers Glacier. Isolated outcrops of quartzose pebbly sandstone and conglomerate tentatively correlated with the Reilly Conglomerate also occur 15 kilometres south-east of Husky Pass and in the eastern Evans Neve (LAIRD and others, 1974).

At the type locality at the south end of Reilly Ridge, mudstone and polymictic conglomerate forming the uppermost beds of the Carryer Conglomerate pass gradationally upwards within 2 metres into 10 metres of parallel-laminated, thinly bedded quartzose fine sandstone forming the basal unit of the Reilly Conglomerate. This in turn passes upwards into parallel- and cross-stratified quartzose conglomerate containing well rounded quartz pebbles up to 10 centimetres in diameter. The top of the sequence rests in inferred fault contact with alternating mudstone and volcanogenic sandstone correlated with the Molar Formation. To the south of Reilly Ridge the underlying Carryer Conglomerate thins and finally wedges out, Reilly Conglomerate resting directly on limestone of the Mariner Group 6 kilometres south-east of Reilly Ridge. No upper stratigraphical limit to the Reilly Conglomerate was seen, but a minimum thickness of 300 metres is represented by outcrops of the formation north of the type locality.

No fossils or fossil traces have been recorded from the Reilly Conglomerate, but its age, like that of the conformably underlying Carryer Conglomerate, is probably Late Cambrian or Ordovician. Although coarser grained, the Reilly Conglomerate is otherwise closely similar in lithology to the Camp Ridge Quartzite and occupies a similar stratigraphical position (at least south of Reilly Ridge) overlying rocks of the Mariner Group. The two formations are considered to be stratigraphical equivalents.

The units making up the Bowers Supergroup are lithologically extremely diverse, and reflect a wide variety of depositional environments (LAIRD and others, in press). That of the Husky Conglomerate is difficult to determine because of its limited thickness and the limited number of outcrops mapped. However, its generally coarse nature and particularly the presence of angular, unsorted breccias suggests that the unit was deposited very close to source with little sorting. The composition leaves little doubt that the source lay to the south-west in the Lanterman Range, which probably represented a tectonically active basin margin. The bulk of the sediments of the Sledgers Group are marine and consist of fine-grained deposits emplaced by traction currents moving from north-west to south-east. By contrast, the subordinate slide breccias and conglomerates and lenticular bodies of graded coarse sandstone and fine breccia appear to have been emplaced by downslope mass-transport mechanisms. Sparse directional data suggest that these deposits were derived from the north-eastern flank of the main basin of deposition.

The deposits of the Mariner Group represent a shallow marine environment, with transgressive and regressive phases represented by the relatively coarse-grained deposits of the Edlin and Eureka Formations, respectively. The Spurs Formation and the Wavy-bedded Sandstone Unit of the Eureka Formation are considered to have accumulated under shallow open marine conditions; the Limestone Unit probably represents sedimentation on open marine shoals or on the protected inner shelf; and the Red Bed Unit is inferred to have accumulated intertidally and subtidally in a tidal estuary (ANDREWS and LAIRD, 1976). The generally fine-grained nature of the group indicates tectonic quiescence, although the occurrence of slump horizons and common brecciated limestone at Reilly Ridge suggests the presence of a slope or proximity of an active basin margin. Palaeocurrent data are conflicting, and no provenance direction or basin morphology could be determined. However, the high quartz content and absence of volcanogenic deteritus indicate that the Sledgers Group was not an important source. Derivation from an acid plutonic or high grade metamorphic terrain, such as the Pre-Cambrian Wilson Group now exposed in the Lanterman Range to the south and west, is likely.

All three formations of the Leap Year Sub-Group show characteristics typical of fluvial deposits, although the presence of trace fossils near the base of the Camp Ridge Quartzite indicates the likelihood of some local marine influence. Cross-bedding azimuths from the Camp Ridge Quartzite show a transport direction towards the north-west or north-north-west south of the Molar Massif; north of this there is a swing in direction towards the north-east. The highly quartzose and quartzo-feldspathic nature of the sediments, coupled with the presence of rare pebbles of gneiss and granite, suggest a granitic and gneissic provenance, probably from the Wilson Group or similar rocks. The direction of transport of the Carryer Conglomerate at its type locality is towards the north-east; palaeocurrent directions at Reilly Ridge are sparse and show wide dispersion, but the dominant clasts are compatible with a Wilson Group source. The few current directional readings obtained from the Reilly Conglomerate suggest a transport direction towards the north-west, and the petrography suggests a provenance similar to that of the Camp Ridge Quartzite. The main palaeocurrent trends for the Sledgers and Leap Year Groups are summarized in Fig. 11.

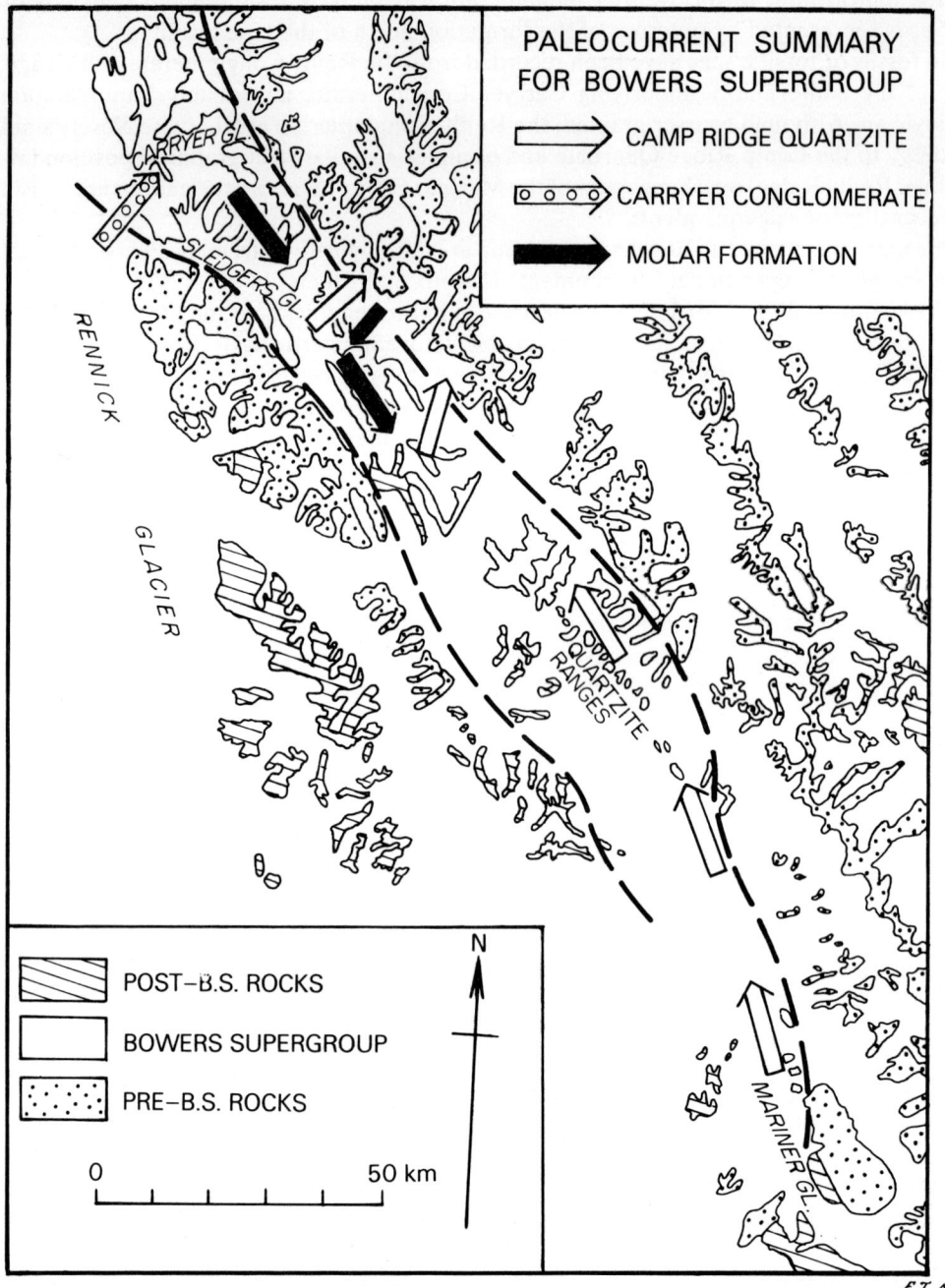

FIG. 11. Sketch map of Bowers Supergroup outcrop showing main palaeocurrent trends of the Molar Formation, Camp Ridge Quartzite, and Carryer Conglomerate (from LAIRD and others, in press).

The sedimentary data suggest that the bulk of the Bowers Supergroup was deposited in a structurally controlled trough-like basin (the Bowers Trough) elongated north-west to south-east lying parallel to and immediately north-east of a Pre-Cambrian crystalline block. Periodic tectonic uplift of the south-western margin of the trough accompanied subsidence and infill of the trough from Vendian to Late Cambrian or Ordovician time. A tectonically active north-eastern margin during Vendian or earliest Cambrian times may be indicated by the presence of north-westerly derived mass flow deposits in the Sledgers Group.

3.3. Possible Lower Palaeozoic sedimentary rocks of the East Antarctic Shield

Several widely separated successions of sub-horizontal or moderately dipping sedimentary rocks overlie dated Pre-Cambrian sedimentary sequences or metamorphic and plutonic complexes of the East Antarctic Shield. They lack fossils or are only sparsely fossiliferous, and correlation with sequences of known Lower Palaeozoic age elsewhere in Antarctica can only be tentative.

3.3.1. Western Queen Maud Land

The region inland from the Princess Martha Coast (Fig. 1) consists almost entirely of metamorphic and igneous rocks of the Pre-Cambrian basement complex. Thick sequences of lightly metamorphosed sedimentary rocks also occur, and, although the age of some is in doubt, a late Pre-Cambrian age has been generally accepted (NEETHLING, 1972).

One exception is the Urfjell Group, which occurs in tectonic contact with Pre-Cambrian gneiss (Sverdrupfjell Group) at the southern end of the Kirwan Escarpment. This sequence, which comprises at least 1680 metres of deformed quartzite and conglomerate, has been divided (AUCAMP, WOLMARANS, and NEETHLING, 1972) into three conformable formations: the Uven Formation (400 metres), the Tunga Formation (1050 metres), and the Urnosa Formation (200 metres). The Urfjell Group is unconformably overlain by at least 100 metres of near horizontal sediments, the Amelang Formation, containing carbonaceous shale with woody tissues dated as Carboniferous or Permian, and correlated with the Beacon Supergroup.

The Uven Formation, the lowest unit of the Urfjell Group, is composed mainly of impure arenite featuring even-bedded, fine- to medium-grained quartzite which is occasionally cross-bedded with rare pebble stringers. A gradation to coarser quartzite is noticeable towards the top.

The predominantly quartzitic rocks of the Uven Formation grade into the conglomerate of the Tunga Formation over a vertical stratigraphical thickness of 80 metres. The lowest exposed 520 metres of the Tunga Formation consists of pale green, massive to cross-bedded, poorly sorted conglomerate and quartzite with occasional interbeds of micaceous quartzite, coarse arkose, and shale. A prominent 3 metres thick, pink quartzite marker occurs approximately 120 metres below the top of this sequence. The upper 450 metres of this formation comprise mainly dark green quartzite and loosely packed lenses of conglomerate with quartzitic pebbles set in a sandy cross-bedded groundmass. The highest exposed part of the Tunga Formation consists of similar but coarser conglomerate.

The overlying Urnosa Formation consists mainly of greenish, medium- to coarse-grained quartzite. Intercalated lenticular beds of darker coloured, fine- to medium-grained quartzite, reddish brown sandy mudstone, and reddish brown micaceous and fissile quartzite are characteristic of this formation. Beds, up to 1.5 metres thick, of very coarse-grained arkose occur. The upper and lower limits of this formation are not exposed in the Kirwan Escarpment and its assignment to the upper part of the Urfjell Group is therefore tentative.

The detrital mineral suite of the Urfjell Group, *viz.* strongly undulose quartz, orthoclase and plagioclase feldspar, relatively abundant garnet, and occasional muscovite with rare zircon, suggests a granitic–metamorphic provenance. The conglomerate pebbles are mainly of fine-grained, brick red to pale green, often fissile quartzite, with subordinate vein quartz and occasional shale fragments. The large-scale cross-bedding and poor sorting of the conglomerate points to a high-energy, near-shore depositional environment. Palaeocurrent measurements indicate a major source area to the south-south-west.

The trend of the near-horizontal fold axes of the Urfjell Group is generally to the south-west or west-south-west. The sediments are mainly gently dipping, but the intensity of deformation increases towards the south, where 800 metres of overfolded sediments occur in fault contact with the Basement Complex.

The age of the Urfjell Group is uncertain. It must be older than the unconformably overlying Carboniferous or Permian Amelang Formation, which is correlated with the Beacon Super-group. It is structurally and lithologically very different from the Proterozoic Ahlmannrygg Group, which also crops out in the vicinity. The most likely correlation is with the Cambro-Ordovician Blaiklock Glacier Group of the relatively nearby Shackleton Range (CLARKSON, 1972), and perhaps also with the folded Lower Palaeozoic formations of the Pensacola Mountains (SCHMIDT and others, 1965).

3.3.2. Prince Charles Mountains

The mountain fringe of the Lambert Glacier (Fig. 1) is composed mainly of gneiss and metasedimentary rocks of Archaean to Middle Proterozoic age overlain by flat-lying beds of arkosic sandstone, mudstone, and coal of Permian age (TRAIL and MCLEOD, 1969b; SOLOVIEV, 1972). Recently, however, problematic organic remains were discovered in limestone boulders collected from a conglomerate horizon within a lightly metamorphosed sedimentary sequence exposed on Mt. Rubin, an isolated nunatak near the head of the Lambert Glacier (G. GRIKUROV, personal communication). Although the organic remains could not be accurately identified, the presence of skeletal material suggests a Phanerozoic age for the host rock. The sequence in which the 30 to 50 metres conglomerate horizon occurs consists of at least 3500 metres of mainly quartzitic sandstones and quartzites with subordinate phyllites, all tightly folded. The degree of deformation and metamorphism suggests an age older than that of the nearby almost flat-lying Permian strata and an Early Palaeozoic, probably Cambrian age is tentatively accepted both for the derived pebbles and the enclosing strata.

3.3.3. Wilkes Land (Mts. Amundsen and Sandow)

These two nunataks (Fig. 1), which rise only 150 metres above the ice surface, occur at the head of the Denman Glacier. The thicker of the two sequences (120 metres), at Mr. Sandow, rests on actinolite–epidote–chlorite schist, which is overlain by a metamorphosed conglomerate containing green-schist pebbles (RAVICH, KLIMOV, and SOLOVIEV, 1968). Higher in the sequence metamorphosed sandstone, siltstone, and claystone are interbedded. The sequence on Mr. Amundsen is thinner, but of similar lithology. The sections on both nunataks appear to belong to the same group, but that on Mt. Amundsen is probably stratigraphically higher (RAVICH, KLIMOV, and SOLOVIEV, 1968). On Mt. Sandow the sediments have a consistent submeridional strike and dip to the east at angles between 25° and 40°. On Mt. Amundsen the dip is to the south-east at an angle of 15° to 20°.

The green-schist forming the basal portion of the sequence at Mt. Sandow appears to be metamorphosed basalt. It is separated by a stratigraphical break from the overlying terriginous rocks, which have also been metamorphosed to the green-schist facies. The sequence overlying the metamorphosed basalt consists mainly of sandstone, often cherry-coloured, with the sand grains showing roundness and good sorting. Argillite is interlayered with the succession of Mt. Sandow, but occurs much less frequently in the Mt. Amundsen section. The metamorphosed conglomerates consist mainly of quartz pebbles, with lesser amounts of quartzite and green-schist. Often the conglomerates occur as lenses.

Few data are available about the environment of deposition, but the nature of the common cross-bedding, the presence of red-coloured sediments, and the presence of lenticular conglomerate bodies, are consistent with aeolian and with fluviatile environments. The sequence is probably continental or coastal-continental.

A radiometric date of 610 m.y. on sericitic schist in the Mt. Sandow section indicates that the sequence was metamorphosed at a time close to the Pre-Cambrian–Cambrian boundary (RAVICH, KLIMOV, and SOLOVIEV, 1968). Determinations on spores found in the sequence give a broad Late Pre-Cambrian–Cambrian age range (KOROTKEVICH and TIMOFEEV, 1959). The relatively gentle dip of these terriginous sediments and their probable Late Pre-Cambrian or Cambrian age suggest that they may represent sediments deposited on the East Antarctic craton at the same time as a marginal geosyncline, developed on the present site of the Trans-Antarctic Mountains, was being filled with thick marine deposits.

4. Regional correlation

Because of the lack of fossils and the almost complete absence of radiometric dates in sequences of possible Lower Palaeozoic age on the East Antarctic craton, no attempt will be made to correlate them with fossiliferous deposits. In the Trans-Antarctic Mountains and the Ellsworth Mountains, however, fossil horizons, although sparse, are sufficient in number to make approximate correlations between rock sequences in the area between the Ross and Weddell Seas.

One of the key units for regional correlation and interpretation of the regional stratigraphy is the Nelson Limestone of the Pensacola Mountains. In the Neptune Range, this is part of a relatively thin Cambrian sequence that has a base no older than medial Middle Cambrian (PALMER and GATEHOUSE, 1972). Farther north, towards the Weddell Sea, the Argentina Range has outcrops of archaeocyathid-bearing limestone that must be older than the base of the Nelson Limestone exposed in the Neptune Range. Closer to the coast, the boulders from Mt. Spann contain trilobite faunules found also in the Nelson Limestone of the Neptune Range, as well as older archaeocyathid faunas. If the Ellsworth Mountains represent a crustal block translocated from its original position as an Atlantic extension of the Trans-Antarctic Moun-tains, as suggested by SCHOPF (1969), the Cambrian section here represents a location still more seaward than Mt. Spann. The thick section of grey pelites with thin limestone interbeds that underlies the limestone unit bearing Late Cambrian trilobites could represent deeper water sediments beyond the main area of carbonate sedimentation. It is important to note, however, that not all writers agree that the Ellsworth Mountains have been translocated, and they may be autochthonous with respect to East Antarctica (FORD, 1972).

In the Ross Sea sector of the Trans-Antarctic Mountains, the section near the Leverett Glacier (Eastern Queen Maud Mountains) is in many respects analogous to the section in the Neptune Range. The trilobite-bearing limestone unit is, at least in part, of Middle Cambrian age and probably correlative with the Nelson Limestone. In both areas, the trilobite-bearing limestones are overlain by a unit of volcanic rocks. Although the terriginous units underlying the limestones in both areas differ in thickness, they appear to be unconformable on older deformed sediments.

In the area between the Beardmore and Byrd Glaciers, the extremely thick carbonate succession (Shackleton Limestone) has archaeocyathids in its lower beds that have a comparable Early Cambrian age to those of the oldest beds dated in the Mt. Spann area. Although unfossiliferous, the Henson Marble to the south can probably be correlated on the grounds of lithology with the Shackleton Limestone. The presence of Early Cambrian fossils in the Taylor Formation of the western Queen Maud Mountains shows that at least part of that unit, also, is a time equivalent of the Shackleton Limestone.

In Northern Victoria Land, the late Middle and Late Cambrian trilobite and brachiopod faunules from the Mariner Group bear a striking similarity to that described from near the top of the Heritage Group in the Ellsworth Mountains, and thus the Mariner Group probably correlates with the upper part of the Heritage Group. It also correlates with the Nelson Limestone, which contains medial or late Middle Cambrian trilobites near its base, and perhaps also with the Gambacorta and Wiens Formations.

Although neither the Crashite Quartzite of the Ellsworth Mountains nor the Leap Year Group of Northern Victoria Land have yielded diagnostic fossils, their stratigraphical position following dated Late Cambrian calcareous units suggests a common post-Cambrian, probably Ordovician age. Both units also consist of terriginous, dominantly quartzitic shallow-water sediments, and they are almost certainly correlatives.

Fig. 12 is an interpretative diagram of the lower Palaeozoic stratigraphical relationships of sedimentary units in the Trans-Antarctic Mountains and the Ellsworth Mountains.

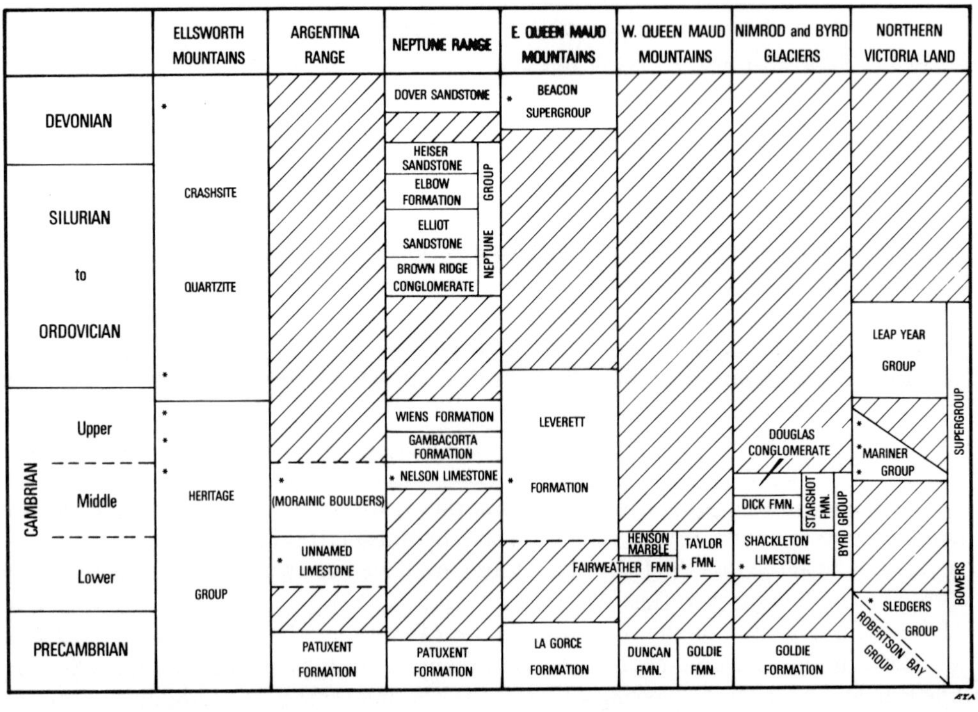

FIG. 12. Correlation of Lower Palaeozoic formations in the Ellsworth Mountains and in the Trans-Antarctic Mountains. The table is diagrammatic and not to scale on a spatial or temporal basis, and the thicknesses of the formations are not represented. Asterisks denote fossil horizons; hatched areas indicate a time gap.

5. Regional descriptions of igneous and high-grade metamorphic rocks

5.1. Lower Palaeozoic igneous and high-grade metamorphic rocks of West Antarctica

Exposures of regionally metamorphosed rocks, widespread in the Antarctic Peninsula, were termed 'Basement Complex' by ADIE (1954) and assigned an (?)Archaean age. However, the only rocks in this whole region which have so far been dated as Pre-Cambrian are those exposed on Clarence Island (ILTCHENKO, 1972), lying north-east of the tip of the Antarctic Peninsula (Fig. 1). The oldest radiometric dates so far recorded from the metamorphic rocks give a late Triassic age (HALPERN, 1972), and they are now considered unlikely to be older than the middle Palaeozoic (SKINNER, 1973). Metamorphosed volcanic rocks which overlie the 'Basement Complex' in Marguerite Bay and adjacent areas of the Antarctic Peninsula (Fig. 1), and granitic rocks which intrude it, were considered by ADIE (1962) to be of Early Palaeozoic age. Radiometric age determinations on granitic rocks from Marguerite Bay give consistent Cretaceous ages (HALPERN, 1972), and there is so far no evidence that rocks older than the middle Palaeozoic crop out anywhere in the Antarctic Peninsula.

Evidently, however, an Early Palaeozoic metamorphic event affected parts of Marie Byrd Land (LOPATIN and ORLENKO, 1972), and Rb–Sr dates of 423 and 473 m.y. on metagabbros and gneisses from the eastern Ruppert Coast indicate an Ordovician or early Silurian metamorphic event (CRADDOCK, 1972). Radiometric Rb/Sr dates of 430 and 502 m.y. have also been recorded from some of the basement rocks of Thurston Island (CRADDOCK, 1972).

5.2. Lower Palaeozoic igneous rocks of the Trans-Antarctic Mountains

Within the Lower Palaeozoic and Pre-Cambrian rocks of the Trans-Antarctic Mountains there are abundant pre-, syn-, and post-tectonic batholiths, stocks, bosses, sheets, and dikes. Although the age range covered by the intrusives is large, radiometric dates show a marked clustering in the range 400 to 540 m.y.

The injection of these Lower Palaeozoic plutonic rocks, now grouped as the Granite Harbour Intrusives (GUNN and WARREN, 1962), was associated with and followed the Late Cambrian and Ordovician Ross Orogeny, whose effect was felt along the length of the present site of the Trans-Antarctic Mountains. The syn-tectonic intrusives are most abundant in the McMurdo Sound region of South Victoria Land from the Darwin Glacier to Terra Nova Bay, but the post-tectonic intrusives are found in all areas from Northern Victoria Land to the Thiel Mountains (Figs. 3 and 6). The Lower Palaeozoic age determinations come from granitic rocks of a wide range of composition and structure—granite in the strict sense, granodiorite, diorite, pegmatite, aplite, and granitic gneiss. Besides these, various schists, paragneisses, and charnockitic granites have also yielded Early Palaeozoic radiometric ages. Most are closely associated with intrusive granitic masses, and the ages they now give are clearly due to metamorphism contemporaneous with, and locally doubtless accentuated by, the intrusion of granitic material.

The rocks of both the Thiel Mountains and Neptune Range of the Pensacola Mountains (Fig. 3), near the Weddell Sea end of the Trans-Antarctic Mountains, were intruded by granitic magma of Early Palaeozoic age. In the Thiel Mountains discordant granitic masses that belong to a single or perhaps several related plutons have intruded a sill-like porphyry body of probable Late Pre-Cambrian age (SCHMIDT and FORD, 1969). Granitic, aplitic, and pegmatitic dikes, apparently related to the larger plutons, also cut the porphyry at several places. The granitic rock is mostly granodioritic or quartz monzonite in composition. Radiometric studies indicate an Early Palaeozoic, probably Late Cambrian to early Ordovician age for the intrusives. In the

Neptune Range, granite magma was intruded as sills and shallow bodies of rhyolite porphyry and as a deep-seated granitic pluton. The radiometric age of this granite (510 m.y.), based on a Rb–Sr whole-rock analysis, indicates that the granite cooled during Late Cambrian time, and was probably related to the intrusive rocks of the Thiel Mountains.

In the Horlick and Queen Maud Mountains region (Fig. 3) metasediments have been invaded by large volumes of calcalkaline intrusive rocks giving radiometric dates of between 400 and 520 m.y., indicating a Late Cambrian to Silurian age (MIRSKY, 1969). Granodiorite and ademellite form the most abundant rock types in the Granite Harbour Intrusives, but there is also some hornblende gabbro, diorite, and granite. The more calcic plutons are mainly older than the less calcic ones. Most of the intrusive rocks make up a large composite batholith, the Queen Maud Batholith, which extends almost continuously across the coastal portion of the Queen Maud Mountains. Small plutons of fine-grained tonalite and ademellite, the latter lithologically similar to the Hope Granite of the area west of the Beardmore Glacier (GUNN and WALCOTT, 1962; GRINDLEY, 1963), cut the metasediments adjacent to the batholith in the eastern Duncan Mountains and elsewhere. Felsic dikes are abundant, and metamorphosed mafic dikes are common in some places. In the Ohio Range (Fig. 3) a grey granodiorite is intruded by a more widespread pink porphyritic quartz monzonite (TREVES, 1965); inclusions of biotite schist, metavolcanics, metadiorite, and metadiabase commonly occur in the granodiorite but are less common in the quartz monzonite. In the Wisconsin Range porphyroblastic quartz monzonite, with biotite and locally abundant xenoliths of probable metasediments, grades into a more equigranular homogeneous granodiorite. Near the head of Reedy Glacier two small masses of hornblendite intrude these rocks. Both the granitic rocks and the mafic masses are intruded by fine- to medium-grained, equigranular granites with muscovite. Along the Scott Glacier the chief intrusive rock is a medium- to coarse-grained, light grey granite and quartz monzonite (DOUMANI and MINSHEW, 1965). Farther west, in the Nilsen Plateau, medium- to coarse-grained, locally porphyritic granodiorite and several quartz monzonite bodies have been identified, mapped, and classified into pre-, syn-, and post-tectonic rocks which form a composite batholith (McLELLAND in MIRSKY, 1969). The youngest intrusive, a quartz monzonite with large pink phenocrysts of potassium feldspar, is similar in appearance and composition to the quartz monzonite in the Ohio Range and to porphyritic granite in the Long Hills and along the north-facing escarpment of the Wisconsin Range. In the Strom Glacier area the intrusive rocks are primarily coarse-grained granodiorite and porphyritic quartz monzonite (McGREGOR, 1965). All granitic rocks are cut by aplites and pegmatites, dated at 477 ± 14 m.y. in the Wisconsin Range (RAY and FAURE, 1974). Black tourmaline is conspicuous in pegmatite dikes in the Wisconsin Range, along the Scott Glacier, and in the Liv Glacier area (MIRSKY, 1969). In the Leverett Glacier area and in the Long Hills, rhyolite occurs (see p. 273) giving radiometric ages of 470 and 498 m.y. (FAURE, MURTAUGH, and MONTIGNY, 1968).

The Hope Granite (Plate 6), an oligoclase–quartz–microcline–biotite–muscovite ademellite or granodiorite, typically coarse-grained, is a common post-tectonic type in the southern part of the central Trans-Antarctic Mountains (GUNN and WALCOTT, 1962; LAIRD, 1964; GRINDLEY and LAIRD, 1969). It obtains its name from Mt. Hope at the mouth of the Beardmore Glacier (Fig. 6), where a large granitic mass is exposed. Other granitic masses occur at several places along the coast between the Beardmore and Starshot Glaciers (Fig. 6), and in the Miller Range, where they have been radiometrically dated at between 450 and 480 m.y. (McDOUGALL and GRINDLEY, 1965). A variety of felsic and mafic dikes, including quartz porphyries, meladiorites, and felsic pegmatites were intruded both before and after the Hope Granite in the Miller Range. Xenolithic tonalite, probably related to the Hope Granite, forms a large intrusion in the

Shackleton Limestone north of the Cotton Plateau (Fig. 6), south of the Nimrod Glacier (LAIRD, MANSERGH, and CHAPPELL, 1971). It contains albite–oligoclase commonly showing oscillatory zoning, biotite, hornblende, and quartz together with minor orthoclase, microcline, and sphene.

The Ida Granite, a name given to intrusions of aplitic quartz monzonite and granite and associated aplite and tourmaline-bearing pegmatite dikes, is apparently intrusive into Hope Granite in the Beardmore Glacier area (GUNNER, 1971). A radiometric age of 463 m.y. on the Ida Granite falls within the range of dates obtained from the Hope Granite, and the former is interpreted to be a late stage differentiate of the magma from which the Hope Granite crystallized.

The granitic intrusive rocks of South Victoria Land (Fig. 6) fall into two groups, both included within the Granite Harbour Intrusives. One comprises intrusions emplaced shortly before or during the main period of deformation of the sedimentary rocks, and the other post-tectonic intrusions. The Larsen Granodiorite includes most of the pre- and syn-tectonic intrusives. This rock is in most places a uniform grey biotite granodiorite and crops out as a major batholith between Terra Nova Bay and the Koettlitz Glacier in the south (WARREN, 1969). It has a distinctly gneissic texture, is intruded by younger granites, and clearly predates at least the later, Ordovician, part of the Ross Orogeny. South of the Mawson Glacier there is uncertainty as to the age of the oldest igneous rocks. Coarse-grained orthogneiss and metadiorite crop out in many areas, in most places in close association with metasedimentary rocks or as inclusions in Larsen Granodiorite. Some are intrusive into the sedimentary rocks, but others, such as the Olympus Granite-Gneiss, which suffered an event dated at 610 m.y. (DEUTSCH and GRÖGLER, 1966), are clearly notably older than the Ross Orogeny. In the Wright Valley, west of McMurdo Sound, granite bodies are cut by lamprophyre and acid porphyry dike swarms, the latter giving Rb–Sr dates of 470 ± 7 m.y. (JONES and FAURE, 1967).

Post-tectonic intrusives range in composition from diorite to alkaline granite. The Skelton Granodiorite (GUNN and WARREN, 1962) is a batholith of hornblende–oligoclase–microcline granodiorite found on both sides of the lower Skelton Glacier (Fig. 6). The Irizar Granite (GUNN and WARREN, 1962) is a name given to all stocks and batholiths of hornblende–biotite–oligoclase–perthite post-tectonic granite found in south Victoria Land. The name is taken from Cape Irizar, just south of Terra Nova Bay. The Irizar Granite (also called Vida Granite; McKELVEY and WEBB, 1962) is the 'pink granite' of the pioneer geologists of the Scott and Shackleton expeditions. It forms steep-sided isolated plutons and bosses characterized by coarse potash feldspars, and hornblende. A number of small intrusive bodies, mainly of undeformed hornblende diorite, have also been mapped.

A granitic batholith complex forms the basement between the Byrd and Skelton glaciers (HASKELL, KENNETT, and PREBBLE, 1965). The Carlyon Granodiorite of the Darwin Glacier area is coarse-grained, foliated biotite–hornblende–andesite granodiorite and may be the dominant intrusive in the batholith. The younger Mt. Rich Granite is a pink, weakly foliated hornblende–biotite–orthoclase granite, resembling the Irizar Granite lying farther to the north.

The Palaeozoic acid igneous rocks of North Victoria Land fall into two distinct groups. The Admiralty Intrusive Group, which lies within the age range 300 to 385 m.y. (Devonian to Carboniferous) is characterized by abundant hornblende and biotite. The Granite Harbour intrusives, lying within the age range of 400 to 530 m.y., in contrast lack hornblende and appear more differentiated than the Admiralty Intrusive Group (NATHAN, 1971a).

The Granite Harbour Intrusives form a belt up to 150 kilometres wide trending north-northwest from Terra Nova Bay and the Aviator Glacier to the Oates Coast (Fig. 6). The granitic rocks form a typical orogenic calc-alkaline suite ranging from biotite–hornblende tonalite through granodiorite and ademellite to leucogranites (GAIR and others, 1969).

Near the Ross Sea coast the intrusive rocks consist of high-level unfoliated post-tectonic plutons of varying size. ADAMSON (1971) recognized three large plutons between the Campbell and Priestley Glaciers, together with minor intrusive complexes. The Dickason Granite is a very coarse-grained porphyritic biotite–orthoclase granite which crops out over a wide area. Contact relationships show it to be the oldest intrusive in the area although it has the relatively young radiometric age of 408 m.y. (NATHAN, 1971b). The Northern Foothills Granite, which intrudes the Dickason Granite, is a pale-coloured even-grained biotite–microcline granite; and the Recoil Granite, which represents the youngest acid intrusion in the area, is a pale-coloured muscovite-bearing biotite granite. Other, smaller intrusives include biotite tonalite and biotite ademellite. Farther north, NATHAN (1971a) distinguished several separate plutons within the Granite Harbour Intrusives. The Aviator Granodiorite, which is coarse-grained white or light grey biotite granite, may be a correlative of the Northern Foothills Granite. The Cosmonaut Granite, which intrudes the Aviator Granodiorite, is a distinctive coarse-grained biotite granite containing abundant megacrysts of potash feldspar. It has been radiometrically dated at 454 m.y. (NATHAN, 1971b).

Granitic, gneissic, pegmatitic, and related rocks occurring near the head of the Campbell and Rennick Glaciers were all included by GAIR (1967) in the Campbell Plutonics, although a radiometric date of 530 m.y. (FAURE and GAIR, 1970), on a post-tectonic granite belonging to the complex, indicates that at least a portion of the plutonics can be referred to the Granite Harbour Intrusives. Intrusions in the area are typically of light-coloured, coarse-grained, muscovite–biotite–hornblende granite, gneissic in places, and frequently intruded by pegmatites.

Farther to the north, west of the lower Rennick Glacier, undeformed granitic and pegmatitic dikes and stocks belonging to the Granite Harbour Intrusives intrude paragneiss and migmatite of the Pre-Cambrian Wilson Group. These intrusives consist primarily of medium-grained or porphyritic biotite granodiorite, but grade to biotite granite with pegmatites close to some of the contacts. Radiometric dates between 420 and 500 m.y. have been obtained from this area (STURM and CARRYER, 1970).

5.3. Lower Palaeozoic igneous and high-grade metamorphic rocks of the Antarctic Shield

Although radiometric dates from igneous and metamorphic rocks of the East Antarctic Shield range up to 3000 m.y., dates falling within the range 440 to 570 m.y. form the most widespread age group (KRYLOV, 1972). These dates do not necessarily indicate a primary age of intrusion of plutonic rocks, but it appears certain that the whole periphery of East Antarctica was affected by an important tectonic or metamorphic event approximately coincident with the Ross Orogeny of the Trans-Antarctic Mountains segment.

5.3.1. Western Queen Maud Land

Most of the crystalline complex constituting this area is likely to be of Pre-Cambrian age, although most radiometric ages fall within the range 400 to 500 m.y. The intrusive rocks of the western portion of the area are all considered to be Pre-Cambrian in age (ROOTS, 1969) although little radiometric dating has been carried out. In the central and eastern portion of the area the majority of over seventy radiometric ages fall within the range of 400 to 500 m.y., the remainder giving late Pre-Cambrian ages. In the central sector, a complex of 'subalkaline granitoids' (RAVICH and others, 1961), predominantly of granosyenite but including syenite, gabbroic rocks, and smaller masses of granite and granodiorite, is intruded into migmatized Pre-Cambrian gneiss and schist. The granitoids are largely undeformed, and the younger

radiometric ages may indicate date of emplacement. Farther east, a large intrusive body of porphyritic hornblende granite (RAVICH, 1959) may be of similar age. Throughout much of the eastern sector, however, no intrusive bodies clearly younger than the main body of ancient gneisses have been described, and the Ordovician and Silurian ages that have been obtained may merely indicate deep-seated metamorphism and metasomatism on a regional scale at this time.

5.3.2. Sør Rondane Mountains

These mountains, located in eastern Queen Maud Land, consist entirely of crystalline rocks which can be divided into two major gneiss assemblages of probable Pre-Cambrian age separated by granitic intrusives (VAN AUTENBOER, 1969). These intrusives, dated at 600 m.y. (VAN AUTENBOER and LOY, 1972), form homogeneous masses of massive, apparently structureless rock. They have sharp contacts with the surrounding gneiss and contain displaced xenoliths, some of which can be identified with one of the gneissic types. The granites are pink to grey, medium-grained rocks consisting of microcline, plagioclase, and biotite, with sphene, allanite, apatite, zircon, fluorite, and opaques as accessory minerals. The structure is often slightly cataclastic. At some localities the granite massif is cut by younger syenites containing xenoliths of both the granite and the gneiss.

A group of intrusive rocks to the north of the main ranges give somewhat younger radiometric ages of 515 to 520 m.y. These Late Cambrian plutons consist of pyroxene–biotite–orthoclase granite, biotite diorite, diopside–biotite–leuconorite, biotite–hypersthene–norite, and amphibole gabbro.

5.3.3. Yamato Mountains (Queen Fabiola Mountains) and Lützow–Holm Bay

The rocks of the region, which can be considered as a single structural unit, consist largely of gneisses, metabasites, and granites referred to the Lützow–Holm Bay System (TATSUMI and KIZAKI, 1969). The majority of the rocks can be correlated on petrographical grounds with the basement complex of East Antarctica, and were almost certainly emplaced during the Pre-Cambrian. However, Rb–Sr and U–Pb age determinations yielding ages of about 500 m.y. suggest that regional metamorphism in this area occurred late in the Cambrian (NICOLAYSON and others, 1961; SAITO, TATSUMI and SATO, 1961; PICCIOTTO and COPPEZ, 1964). In the Yamato Mountains outcrops of migmatitic gneiss, granitic gneiss, and microcline granite form a younger group of rocks resulting from plutonic activity, and an Rb–Sr date of 457 m.y. probably represents the age of emplacement of the bodies in the Late Cambrian or early Ordovician (TATSUMI and KIZAKI, 1969).

5.3.4. Enderby Land to Wilkes Land

The rocks of Enderby Land consist almost entirely of high-grade metamorphic varieties, mainly of the granulite facies, considered to be of Pre-Cambrian age (TRAIL and McLEOD, 1969a; KAMENEV, 1972). The youngest event in the region resulted in the formation of pegmatite dikes which cut all the metamorphic rocks. Radiometric Pb dates of between 465 and 540 m.y. on these pegmatites are considered to be close to their age of emplacement, and probably represent a Middle or Late Cambrian event (KAMENEV, 1972). Ages obtained from metamorphic rocks lying within pegmatite fields are similar to those of the pegmatites and have evidently been affected by their intrusion.

The metamorphic rocks of the Lambert Glacier region are also considered to be Pre-Cambrian in age, although a late metamorphic event occurred at about 480 m.y. (TRAIL, 1964; TRAIL and McLEOD, 1969b).

Rocks outcropping along the 3500 kilometre stretch of coastline between the Lambert Glacier and George V Coast consist almost entirely of charnokites and granulite facies gneisses and schists of Pre-Cambrian age (RAVICH and KRYLOV, 1964). However, a scattering of ages in the range 400 to 630 m.y. at several localities suggests an Early Palaeozoic metamorphic event.

6. Lower Palaeozoic geological history and palaeogeography

A Lower Palaeozoic geological history of Antarctica can describe in reasonable detail only those events affecting the margin of the East Antarctic craton facing the western hemisphere, and the Weddell Sea sector, and sites now occupied by the Trans-Antarctic Mountains and the Ellsworth Mountains respectively. Very little is known of events on the craton itself because of nearly complete snow cover, and an almost complete lack of dates for sedimentary sequences. A similar problem exists in West Antarctica (excluding the Weddel Sea sector).

Cambrian sedimentation along the present trend of the Trans-Antarctic Mountains began on the site of an earlier geosyncline in which a great thickness of sediment, mainly sandstone and mudstone, had accumulated. These sediments, of Late Pre-Cambrian age (Patuxent Formation, Beardmore Group, and their correlatives) were folded and the geosyncline everted during the Late Pre-Cambrian Beardmore Orogeny (GRINDLEY and McDOUGALL, 1969) and subsequently eroded. At the same time continuous geosynclinal deposition was occurring on the site of the Ellsworth Mountains. Here no notable break in sedimentation occurred, although the effects of the Beardmore Orogeny may possibly be represented by regression and deposition of red muds and conglomerates following on the cessation of sedimentation of the calcareous deposits of the ?Pre-Cambrian Minaret Group.

In the new geosyncline Early Cambrian sedimentation in the area of the Ellsworth Mountains was probably represented by deposition of the conglomerate, quartzite, and shale making up much of the middle third of the Heritage Group. Most of the Pensacola Mountains must have been dry land at this time, but marine conditions in the area of the Argentina Range are indicated by the presence of outcrops of archaeocyathid limestone and of morainic boulders of limestone containing Early Cambrian trilobites on Mt. Spann. The sea must have extended to the vicinity of the Whichaway Nunataks where carbonate deposition occurred. From the Horlick Mountains north to the Beardmore Glacier silicic volcanism dominates the Early Cambrian successions, to give way farther north, at least as far as the Byrd Glacier, to limy shallow seas in which archaeocyathid limestone was accumulating. Dating of probable Cambrian sediments is poor in the McMurdo Sound area, but marine conditions also probably prevailed in South Victoria Land with sands, limestones, and volcanogenic sediments being deposited. That the region inundated by the Early Cambrian sea may have extended as far as the coast of eastern Wilkes Land, west of Northern Victoria Land, is hinted at by the discovery of an erratic of marble containing probable archaeocyathids at Cape Denison (MAWSON, 1940). In Northern Victoria Land, the Bowers Trough, a structural depression offset from the main axis of deposition along the Trans-Antarctic Mountains (LAIRD and BRADSHAW, in press) although probably initiated in Vendian times continued to accumulate a mixture of clastics and basic volcanics during at least the early part of the Early Cambrian. Marine incursions may also have occurred in western Marie Byrd Land and in the Prince Charles Mountains. Fig. 13A shows the probable extent and depositional environments of the Early Cambrian sea.

In the late Early Cambrian and Middle Cambrian uplift of Pre-Cambrian basement rocks and previously deposited Cambrian sediments in the central Trans-Antarctic Mountains led to erosion and the formation of conglomerate, breccia, and slump deposits in a subsiding basin

lying to the east of a rising shoreline. Sedimentation of shallow-water carbonate deposits mainly kept pace with subsidence, however, and archaeocyathid bioherms were able to form on the unstable shelf. Sedimentation offshore of this unstable shelf may be represented by the mainly younger but possibly partly contemporaneous Starshot Formation lying to the east of the limestone belt.

By Middle Cambrian times, marine conditions prevailed over almost the entire length of the Trans-Antarctic Mountains and the Prince Charles Mountains, as well as over the site of the Ellsworth Mountains and possibly western Marie Byrd Land (Fig. 13B). A broad carbonate bank, represented by the Nelson Limestone, limestone units of the Leverett Formation, Henson Marble, Shackleton Limestone, Anthill Limestone, and perhaps also marbles of the Koettlitz Group, probably extended from near the site of Whichaway Nunataks in Coats Land to McMurdo Sound. Tectonic activity, uplift, and erosion in the central Trans-Antarctic Mountains, probably occurring during Middle Cambrian times, is indicated by the deposition of thick conglomerates (Douglas Conglomerate) derived from earlier Cambrian and older rocks. Calcareous sedimentation also probably extended to the site of the Ellsworth Mountains during this period, resulting in the deposition of mudstones and limestones containing brachiopods and trilobites of probable Middle Cambrian age. New marine transgression occurred during the later part of the Middle Cambrian in Northern Victoria Land, with deposition first of quartzitic sands and then of shallow marine mudstone and thin limestone units.

Late Cambrian events are less clear because of scant palaeontological control except in the Ellsworth Mountains and Northern Victoria Land. On the site of the Ellsworth Mountains deposition of mudstone and limestone continued from Middle to Late Cambrian times with little variation. A near-shore, possibly sublittoral environment prevailed for most of the time. On the site of the Pensacola Range the extrusion of the rhyolitic lavas and pyroclastic deposits of the Gambacorta Formation indicate contemporary volcanism during late Middle or Late Cambrian times. Acid volcanism, perhaps contemporaneous with this, also occurred in the Late Cambrian or early Ordovician in the Queen Maud Mountains (rhyolites of the Leverett Formation). Rhyolite and trachyte flows of the Starshot Formation in the central Trans-Antarctic Mountains and the felsic volcanics of the Skelton Glacier area may also be correlatives. In northern Victoria Land deposition of mudstone and minor limestone continued into Late Cambrian times, but shallowing of the basin is evident in the middle to late Late Cambrian, with regression causing replacement of mudstone deposition by sedimentation of shallow marine sandstone and then by very shallow water (in part intertidal) red beds. Evidence for this regressive phase elsewhere in the Trans-Antarctic Mountains may be the thick deposits of cross-bedded, coarse-grained conglomeratic sandstone interbedded with and following on extrusion of the rhyolites of the Leverett Formation, Queen Maud Mountains, and a truncated portion may be represented by the shale and fine sandstone with limestone lenses of the Wiens Formation of the Pensacola Mountains. Regression, perhaps coinciding with that of latest Cambrian age in the Trans-Antarctic Mountains, is also evident in the Ellsworth Mountains. It is probably reflected by the transition from the shallow marine shelf environment of the upper part of the Heritage Group, to cross-bedded quartzite (Crashsite Quartzite) of very shallow marine and perhaps in part intertidal environment. The area covered by Late Cambrian seas and lithofacies development is indicated in Fig. 13C.

Whilst largely marine sedimentation was proceeding along the unstable margin of the East Antarctic shield, much of the craton itself was probably above sea level and acting as a source for the geosynclinal deposits flanking it.

Widely scattered remnants of flat-lying or gently dipping sediments of probable Early Palaeozoic age occurring throughout East Antarctica, show, however, that some terrestrial or

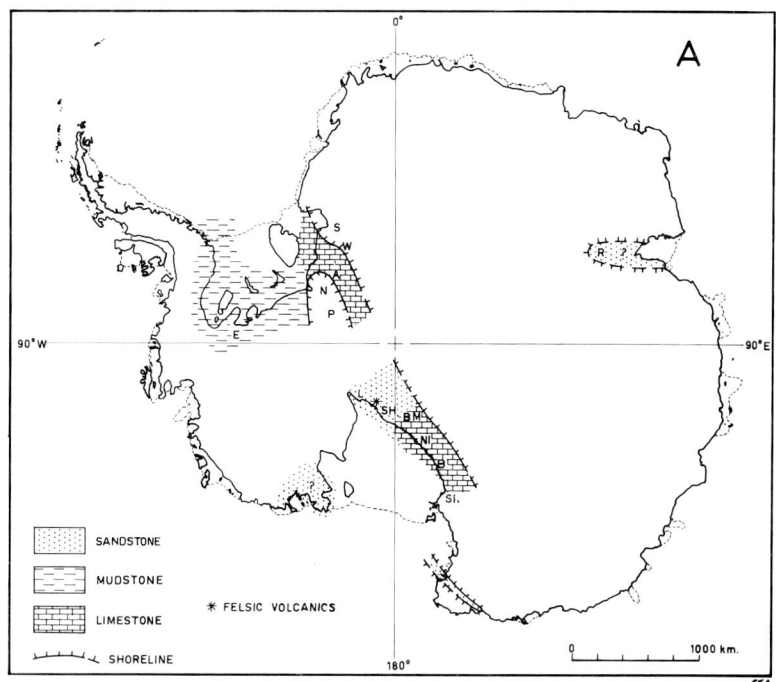

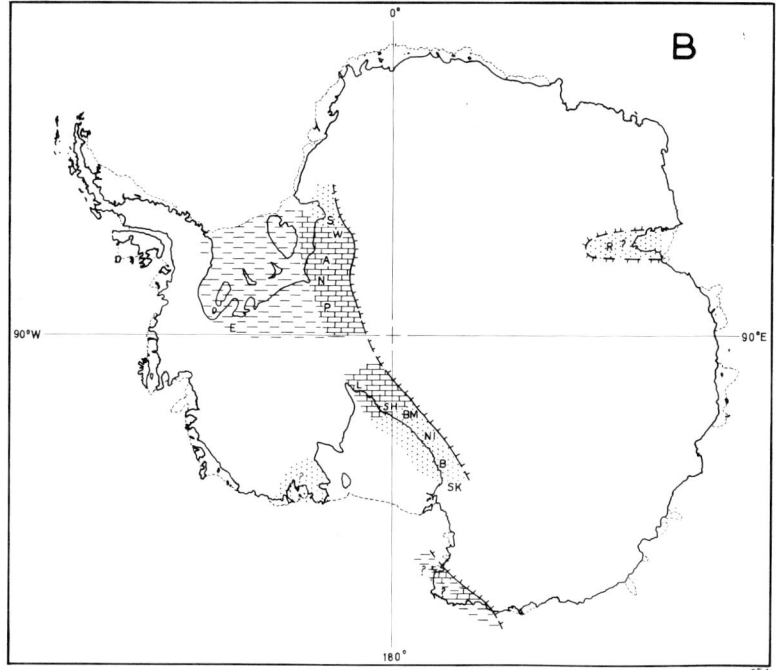

FIG. 13. Cambrian and Ordovician palaeogeography of Antarctica, showing
the probable extent of areas covered by sea and lithofacies distribution in
(A). Early Cambrian, (B) Middle Cambrian, (C) Late Cambrian, and (D)
Ordovician times. Key: S, Shackleton Range; W, Whichaway Nunataks;

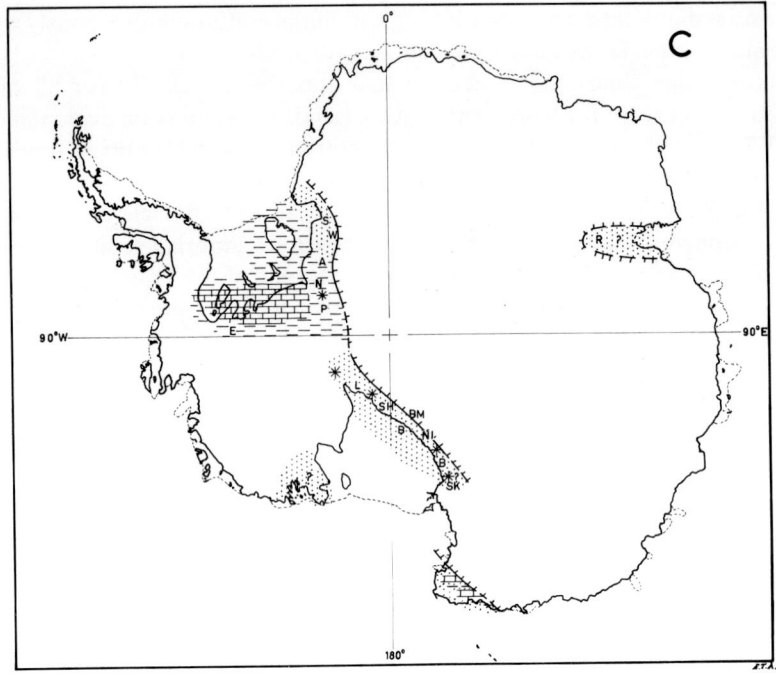

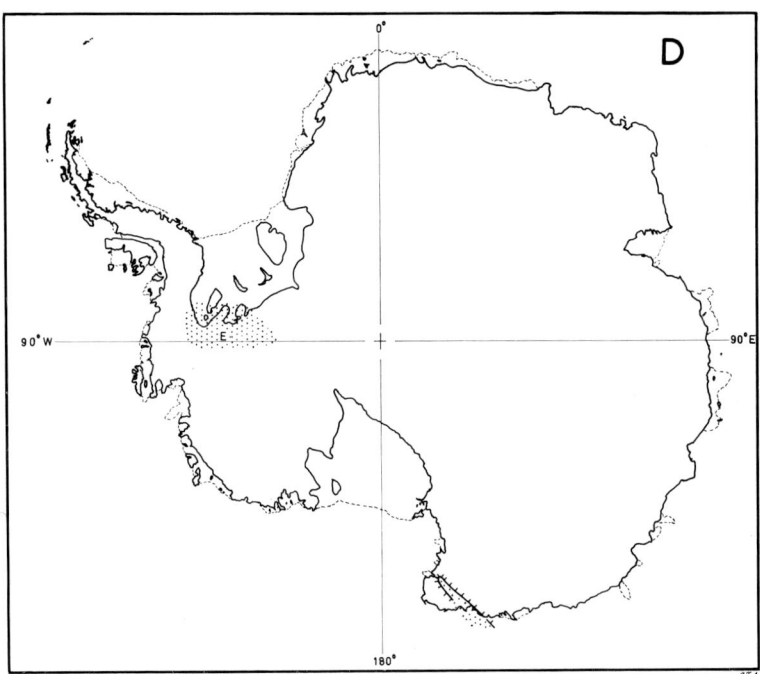

A, Argentina Range; N, Neptune Range; P, Pensacola Mountains; E, Ellsworth Mountains; L, Leverett Clacier; SH, Shackleton Glacier; BM, Beardmore Glacier; NI, Nimrod Glacier; B, Byrd Glacier; SK, Skelton Glacier; R, Mt. Rubin.

marginal marine sedimentation was continuing within the craton simultaneously with sedimentation on the sites of the Trans-Antarctic and Ellsworth Mountains.

By early Ordovician times sedimentation had apparently ceased over all of the Trans-Antarctic Mountains except for Northern Victoria Land. The regression evident along the Ross Sea–Weddell Sea margin of the East Antarctic craton in Late Cambrian times occurred in response to uplift along the site of the Trans-Antarctic Mountains. This uplift heralded the beginning of the Ross Orogeny, a major Late Cambrian to Ordovician tectonic event which resulted in eversion and folding of the Pre-Cambrian and Cambrian sediments of the geosyncline along axes approximately parallel to the present trend of the Trans-Antarctic Mountains.

Uplift and folding was accompanied and succeeded by the intrusion of acid plutons during the Late Cambrian and Ordovician. This magmatic activity was not restricted to what is now the western hemisphere margin of the East Antarctic Shield, and widespread intrusion of acid plutons took place all over East Antarctica. Gneisses cropping out on Thurston Island and in parts of Marie Byrd Land may have formed at this time, and Ordovician radiometric dates on Swanson Group phyllites in western Marie Byrd Land show that metamorphism occurred here at the time of the Ross Orogeny (LOPATIN and ORLENKO, 1972).

While the onset of the Ross Orogeny brought marine sedimentation to a halt in most areas, deposition still occurred on the site of the Ellsworth Mountains and in northern Victoria Land. Apart from the regression noted earlier, the Ross Orogeny appears to have had little effect in the Ellsworth Mountains area and deposition of quartzose sands under shallow marine conditions continued from Late Cambrian until Devonian times. In northern Victoria Land the Ross Orogeny resulted in block faulting and the re-initiation of the Bowers Trough. During probably Ordovician times the trough was infilled with largely non-marine quartzose sands and polymictic gravels, although some local marine influence is evident. The sequence was then folded and eroded in the Silurian to Early Devonian Borchgrevink Orogeny, and finally intruded by Carboniferous to Devonian granites (Admiralty Intrusives). On the site of the Pensacola Mountains, folding and uplift accompanying the Ross Orogeny exposed the Cambrian and Pre-Cambrian formations to erosion. The orogenic Brown Ridge Conglomerate was locally derived from the rising fold mountains, with the Elliott Sandstone spread as an alluvial apron at their foot (WILLIAMS, 1959). Following slow subsidence on an epicontinental shelf, sandstone, probably supplied by the craton lying to the east, was deposited at shallow depths in an epineritic environment until at least Devonian times.

7. Acknowledgements

I wish to record my sincere thanks to Drs G. W. Grindley and P. B. Andrews, New Zealand Geological Survey, for constructive criticism of the manuscript.

A review such as this, which is based on the work of numerous geologists of several different nationalities, would have been extremely difficult to carry out without the willing cooperation of fellow workers. Unpublished material was generously made available by Dr G. Grikurov, Research Institute of the Geology of the Arctic, Leningrad, and by Dr K. B. Spörli, University of Auckland, to both of whom I am very grateful. The manuscript has also benefitted by discussion and correspondence with numerous colleagues, particularly Dr A. B. Ford, Dr C. Craddock, Dr D. S. Soloviev, and Dr O. S. Vialov.

Mrs L. Leonard kindly carried out the final drafting of the majority of the diagrams.

References

ADAMS, C. J., ANDREWS, P. B., BRADDOCK, P., and BRADSHAW, J. D. (1979). Geology of Ford Ranges, western Marie Byrd Land, Antarctica. *N.Z. Antarctic Rec.*, **2**, 30 (Abstract).

ADAMS, C. J., GABITES, J. E., BRADSHAW, J. D., LAIRD, M. G., and WODZICKI, A. (in press). Potassium–argon geochronology of the Precambrian–Cambrian Wilson and Robertson Bay Groups and Bowers Supergroup, North Victoria Land, Antarctica, *in* CRADDOCK, C. and VIERIMA, T. (Eds.), *Antarctic Geoscience* (proceedings of the 1977 Madison Symposium on Antarctic Geology and Geophysics). University of Wisconsin Press.

ADAMSON, R. G. (1971). Granitic rocks of the Campbell–Priestley Divide, Northern Victoria Land, Antarctica. *N.Z. J. Geol. Geophys.*, **14**, 486.

ADIE, R. J. (1954). The Petrology of Graham Land. I. The Basement Complex; early Palaeozoic plutonic and volcanic rocks. *Falkland Islands Dependencies Survey, Sci. Rep.*, No. 11.

ADIE, R. J. (1962). The Geology of Antarctica. *Am. Geophys. Union, Geophys. Monogr.*, No. 7, 26.

ALLEN, A. D. and GIBSON, G. W. (1962). Geological Investigations in Southern Victoria Land, Antarctica. Pt. 6. Outline of the Geology of the Victoria Valley Region. *N.Z. J. Geol. Geophys.*, **5**, 234.

ANDREWS, P. B. and LAIRD, M. G. (1976). Sedimentology of a Late Cambrian regressive sequence (Bowers Group). Northern Victoria Land, Antarctica. *Sediment. Geol.*, **16**, 21.

AUCAMP, A. P. H., WOLMARANS, L. G., and NEETHLING, D. C. (1972). The discovery of a pre-Beacon sedimentary sequence in the South-western portion of the Kirwan Escarpment, Western Queen Maud Land, Antarctica, *in* ADIE, R. J. (Ed.), *Antarctic Geology and Geophysics*. Universitetsforlaget, Oslo, 557.

BLANK, H. R., COOPER, R. A., WHEELER, R. H., and WILLIS, I. A. G. (1963). Geology of the Koettlitz–Blue Glacier Region, Southern Victoria Land, Antarctica.

BRADSHAW, J. D., LAIRD, M. G., and WODZICKI, A. (in press). Structural style and tectonic history in Northern Victoria Land, Antarctica, *in* CRADDOCK, C. and VIERIMA, T. (Eds.), *Antarctic Geoscience* (proceedings of the 1977 Madison Symposium Antarctic Geology and Geophysics). University of Wisconsin Press.

BURGESS, C. J. and LAMMERINK, W. (1979). Geology of the Shackleton Limestone (Cambrian) in the Byrd Glacier area. *N.Z. Antarctic Rec.*, **2**, 12.

CLARKSON, P. D. (1972). Geology of the Shackleton Range: a preliminary report. *Brit. Antarctic Survey Bull.*, No. 31, 1.

COOPER, R. A., JAGO, J. B., MACKINNON, D. I., SHERGOLD, J. H., and VIDAL, G. (in press). Late Precambrian and Cambrian fossils from Northern Victoria Land and their stratigraphic implications, *in* CRADDOCK, C. and VIERIMA, T. (Eds.), *Antarctic Geoscience* (proceedings of the 1977 Madison Symposium on Antarctic Geology and Geophysics). University of Wisconsin Press.

COOPER, R. A., JAGO, J. B., MACKINNON, D. I., SIMES, J. E., and BRADDOCK, P. E. (1976). Cambrian fossils from the Bowers Group, Northern Victoria Land, Antarctica (preliminary note). *N.Z. J. Geol. Geophys.*, **19**, 283.

CRADDOCK, C. (1969). Sheet 4—Geology of the Ellsworth Mountains. Geologic Map of Antarctica 1:1,000,000. *Antarctic Map Folio Series*, Folio 12, Plate IV. (BUSHNELL, V. C., Ed.). American Geographical Society, New York.

CRADDOCK, C. (1972a). *Geologic Map of Antarctica* 1:5,000,000. American Geographical Society, New York.

CRADDOCK, C. (1972b). Antarctic Tectonics, *in* ADIE, R. J. (Ed.), *Antarctic Geology and Geophysics*. Universitetsforlaget, Oslo, 449.

CRADDOCK, C., ANDERSON, J. H., and WEBERS, G. F. (1964). Geologic outline of the Ellsworth Mountains, *in* ADIE, R. J. (Ed.), *Antarctic Geology*. North Holland, Amsterdam, 155.

CRADDOCK, C., RUTFORD, R. H., BASTIEN, T. W., and ANDERSON, J. J. (1965). *Glossopteris* discovered in West Antarctica. *Science*, **148**, 634.

CROWDER, D. F. (1968). Geology of a part of North Victoria Land, Antarctica. *Prof. Pap. U.S. Geol. Surv.*, **600D**, 95.

DAVID, T. W. E. and PRIESTLEY, R. E. (1914). Glaciology, Physiography, and Tectonic Geology of South Victoria Land, with short notes on Palaeontology by G. Griffith Taylor. *Rep. Brit. Antarctic Exped. 1907–09, Geology*, 1.

DEUTSCH, S. and GRÖGLER, N. (1966). Isotopic age of Olympus Granite Gneiss (Victoria Land– Antarctica). *Earth planet Sci. Lett.*, **1**, 82.

DOUMANI, G. A. and MINSHEW, V. H. (1965). General Geology of the Mount Weaver area, Queen Maud Mountains, Antarctica, *in* HADLEY, J. B. (Ed.), *Geology and Paleontology of the Antarctic* (Antarctic Research Series, Vol. 6). American Geophysical Union, Washington, D.C., 127.

DOW, J. A. S. and NEALL, V. E. (1972). A summary of the Geology of the Lower Rennick Glacier, Northern Victoria Land, Antarctica, *in* ADIE, R. J. (Ed.), *Antarctic Geology and Geophysics*. Universitetsforlaget, Oslo, 339.

DOW, J. A. S. and NEALL, V. E. (1974). Geology of the Lower Rennick Glacier, Northern Victoria Land, Antarctica. *N.Z. J. Geol. Geophys.*, **17**, 659.

EASTIN, R., FAURE, G., SHULTZ, C. H., and SCHMIDT, D. L. (1969). Rb–Sr ages of the Littlewood Volcanics and of the acid volcanic rocks of the Neptune Range, Pensacola Mountains, Antarctica. *Geol. Soc. Am., Abstr. 1969*, Pt. 6, 13.

FAURE, G. and GAIR, H. S. (1970). Age determinations from northern Victoria Land. *N.Z. J. Geol. Geophys.*, **10**, 309.

FAURE, G., MURTAUGH, J. G., and MONTIGNY, R. J. E. (1968). The geology and geochronology of the basement complex of the central Transantarctic Mountains. *Can. J. Earth Sci.*, **5**, 555.

FINDLAY, R. H. (1978). Provisional report on the geology of the region between the Renegar and Blue Glaciers, Antarctica. *N.Z. Antarctic Rec.*, **1**, 39.

FLORY, R. L., MURPHY, D. J., SMITHSON, S. B., and HOUSTON, R. S. (1971). Geologic studies of basement rocks in southern Victoria Land. *Antarctic J. U.S.*, **6**, 119.

FORD, A. B. (1972). Fit of Gondwana Continents—drift reconstruction from the Antarctic continental veiwpoint. *Proc. 24th Int. Geol. Congr.*, Sect. 3, 113.

GAIR, H. S. (1967). The geology of the Upper Rennick Glacier to the Coast, Northern Victoria Land, Antarctica. *N.Z. J. Geol. Geophys.*, **10**, 309.

GAIR, H. S., STURM, A., CARRYER, S. J., and GRINDLEY, G. W. (1969). Sheet 13—The Geology of Northern Victoria Land. Geologic Map of Antarctica 1 : 1,000,000. *Antarctic Map Folio Series*, Folio 12, Plate XII. (BUSHNELL, V. C., Ed.). American Geographical Society, New York.

GORDON, W. T. (1970). Cambrian organic remains from dredgings in Weddell Sea (Scottish National Antarctic Expedition 1902–04). *Trans. R. Soc., Edinburgh*, **52**, 681.

GRINDLEY, G. W. (1963). The Geology of the Queen Alexandra Range, Beardmore Glacier, Ross Dependency, Antarctica; with notes on the Correlations of Gondwana Sequences. *N.Z. J. Geol. Geophys.*, **6**, 307.

GRINDLEY, G. W. and LAIRD, M. G. (1969). Sheet 15—Geology of the Shackleton Coast. Geologic map of Antarctica 1 : 1,000,000. *Antarctic Map Folio Series*, Folio 12, Plate XIV. (BUSHNELL, V. C., Ed.). American Geographical Society, New York.

GRINDLEY, G. W. and McDOUGALL, I. (1969). Age and correlation of the Nimrod Group and other Precambrian rock units in the central Trans-Antarctic Mountains, Antarctica. *N.Z. J. Geol. and Geophys.*, **12**, 391.

GRINDLEY, G. W. and WARREN, G. (1964). Stratigraphic nomenclature and correlation in the Western Ross Sea Region, *in* ADIE, R. J. (Ed.), *Antarctic Geology*. North Holland, Amsterdam, 314.

GUNN, B. M. and WALCOTT, R. I. (1962). The Geology of the Mt. Markham Region, Ross Dependency, Antarctica. *N.Z. J. Geol. Geophys.*, **5**, 407.

GUNN, B. M. and WARREN, G. (1962). Geology of Victoria Land between the Mawson and Mulock Glaciers, Antarctica. *Bull. Geol. Surv. N.Z.*, n.s., **71**.

GUNNER, J. (1971). Ida Granite: a new formation of the Granite Harbour Intrusives, Beardmore Glacier region. *Antarctic J. U.S.*, **6**, 194.

HALPERN, M. (1972). Rubidium–strontinum total rock and mineral ages from the Marguerite Bay area, Kohler Range, and Fosdick Mountains of West Antarctica, *in* ADIE, R. J. (Ed.), *Antarctic Geology and Geophysics*. Universitetsforlaget, Oslo, 197.

HASKELL, T. R., KENNETT, J. P., and PREBBLE, W. M. (1965). Geology of the Brown Hills and Darwin Mountains, Southern Victoria Land, Antarctica. *Trans. R. Soc. N.Z.*, **2**, 231.

HASKELL, T. R., KENNETT, J. P., PREBBLE, W. M., SMITH, G., and WILLIS, I. A. G. (1965). The geology of the Middle and Lower Taylor Valley of South Victoria Land, Antarctica. *Trans. R. Soc. N.Z.*, **2**, 169.

HILL, D. (1964a). Archaeocyatha from the Shackleton Limestone of the Ross System, Nimrod Glacier area, Antarctica. *Trans. R. Soc. N.Z.*, (*Geol.*), **2**, 137.

HILL, D. (1964b). The Phylum Archaeocyatha. *Biol. Rev.*, **39**, 232.

HILL, D. (1965). Geology: 3. Archaeocyatha from Antarctica and a review of the phylum. *Trans-Antarctic Exped.*, *1955–58*, *Sci. Rep.*, No. 10.

ILTCHENKO, L. N. (1972). Late Precambrian Acritarchs of Antarctica, *in* ADIE, R. J. (Ed.), *Antarctic Geology and Geophysics*. Universitetsforlaget, Oslo, 599.

IVSHIN, N. K. (1956). Verkhnekembriyskie triloboty Kazakhstana. Chast' I. Kuyandinskii faunisticheskii gorizont mezhdurech'ya Olenty–Shiderty (Upper Cambrian trilobites of Kazakstan. Part I. Kuyandinsk faunal horizon of the river area between Olenta and Shiderta). *Tr. Inst. Geol. Nauk, Alma-Ata*.

IVSHIN, N. K. (1962). Vaerkhenkembriyskie trilobity Kazakhstana. Chast' II. Selentinskii gorizont Kuyandinskogo yarusa tsentral'nogo Kazakhstana (Upper Cambrian trilobites of Kazakstan. Part II. Selentinsk horizon of the Kuyandinsk beds of central Kazakstan). *Tr. Inst. Geol. Nauk, Alma-Ata*.

JONES, L. M. and FAURE, G. (1967). Age of the Vanda Porphyry dikes in Wright Valley, southern Victoria Land, Antarctica. *Earth Plan. Sci. Lett.*, **3**, 321.

KAMENEV, E. N. (1972). Geological Structure of Enderby Land, *in* ADIE, R. J. (Ed.), *Antarctic Geology and Geophysics*. Universitetsforlaget, Oslo, 579.

KATZ, H. R. and WATERHOUSE, B. C. (1970a). Geological reconnaissance of the Scott Glacier area, south eastern Queen Maud Range, Antarctica. *N.Z. J. Geol. Geophys.*, **13**, 1030.

KATZ, H. R. and WATERHOUSE, B. C. (1970b). Geologic situation at O'Brien Peak, Queen Maud Range, Antarctica. *N.Z. J. Geol. Geophys.*, **13**, 1038.

KLIMOV, L. V. (1967). Some results of geological investigations in Marie Byrd Land in 1966–1967. *Sov. Antarctic Exped. Info. Bull.*, No. 66. English translation for Am. Geophys. Union, 555.

KOROTKEVICH, E. S. and TIMOFEEV, B. V. (1959). O vozraste porod Vostochnoi Antarktidy (on the age of East Antarctica). *Byull. Sov. Antarkticheskoi Ekspedi.*, No. 12.

KRYLOV, A. JA. (1972). Antarctic geochronology, *in* ADIE, R. J. (Ed.), *Antarctic Geology and Geophysics*. Universitetsforlaget, Oslo, 419.

LAIRD, M. G. (1963). Geomorphology and Stratigraphy of the Nimrod Glacier–Beaumont Bay region, Southern Victoria Land, Antarctica. *N.Z. J. Geol. Geophys.*, **6**, 465.

LAIRD, M. G. (1964). Petrography of rocks from the Nimrod Glacier–Starshot Glacier region, Ross Dependency, *in* ADIE, R. J. (Ed.), *Antarctic Geology*. North Holland, Amsterdam, 463.

LAIRD, M. G., ANDREWS, P. B., and KYLE, P. R. 1974. Geology of northern Evans Nēvē, Victoria Land, Antarctica. *N.Z. J. Geol. Geophys.*, **17**, 587.

LAIRD, M. G., ANDREWS, P. B., KYLE, P., and JENNINGS, P. (1972). Late Cambrian fossils and the age of the Ross Orogeny, Antarctica. *Nature*, **238**, 34.

LAIRD, M. G. and BRADSHAW, J. D. (in press). Uppermost Proterozoic and Lower Paleozoic Geology of the Transantarctic Mountains, *in* CRADDOCK, C. and VIERIMA, T. (Eds.), *Antarctic Geoscience* (proceedings of the 1977 Madison Symposium on Antarctic Geology and Geophysics). University of Wisconsin Press.

LAIRD, M. G., BRADSHAW, J. D., and WODZICKI, A. (in press). Stratigraphy of the Late Precambrian and Early Paleozoic Bowers Supergroup, Northern Victoria Land, Antarctica, *in* CRADDOCK, C. and VIERIMA, T. (Eds.), *Antarctic Geoscience* (proceedings of the 1977 Madison Symposium on Antarctic Geology and Geophysics). University of Wisconsin Press.

LAIRD, M. G., MANSERGH, G. D., and CHAPPELL, J. M. A. (1971). Geology of the Central Nimrod Glacier area, Antarctica. *N.Z. J. Geol. Geophys.*, **14**, 427.

LAIRD, M. G. and WATERHOUSE, J. B. (1962). Archaeocyathine limestones of Antarctica. *Nature*, **194**, 861.

LE COUTEUR, P. C. and LEITCH, E. C. (1964). Preliminary report on the geology of an area southwest of Upper Tucker Glacier, Northern Victoria Land, *in* ADIE, R. J. (Ed.), *Antarctic Geology*. North Holland, Amsterdam, 229.

LOGAN, B. W., REZAK, R., and GINZBURG, R. N. (1964). Classification and environmental significance of algal stromatolites. *J. Geol.*, **72**, 68.

LOPATIN, B. G. and ORLENKO, E. M. (1972). Outlines of the geology of Marie Byrd Land and Eights Coast, *in* ADIE, R. J. (Ed.), *Antarctic Geology and Geophysics*. Universitetsforlaget, Oslo, 245.

MCDOUGALL, I. and GRINDLEY, G. W. (1965). Potassium–argon dates on micas from the Nimrod–Beardmore–Axel Heiberg Region, Ross Dependency, Antarctica. *N.Z. J. Geol. Geophys.*, **8**, 304.

MCGREGOR, V. R. (1965). Geology of the area between the Axel Heiberg and Shackleton Glaciers, Queen Maud Range, Antarctica. *N.Z. J. Geol. Geophys.*, **8**, 314.

MCGREGOR, V. R. and WADE, F. A. (1969). Sheet 16—Geology of the Western Queen Maud Mountains. Geologic Map of Antarctica 1 : 1,000,000. *Antarctic Map Folio Series*, Folio 12, Plate XV. (BUSHNELL, V. C., Ed.). American Geographical Society, New York.

MCKELVEY, B. C. and WEBB, P. N. (1961). Geological reconnaissance in Victoria Land, Antarctica. *Nature*, **189**, 545.

MCKELVEY, B. C. and WEBB, P. N. (1962). Geological Investigations in Southern Victoria Land, Antarctica. Pt. 3. Geology of Wright Valley. *N.Z. Geol. Geophys.*, **5**, 143.

MAWSON, D. (1940). Part 11. Sedimentary rocks. *Australian Antarctic Expedition 1911–14, Scientific Reports*, Series A, Vol. IV, Geology.

MINSHEW, V. H. (1966). Stratigraphy of the Wisconsin Range, Horlick Mountains, Antarctica. *Science*, **152**, 637.

MINSHEW, V. H. (1967). *Geology of the Scott Glacier and Wisconsin Range Areas, Central Transantarctic Mountains, Antarctica*. Ph.D. Thesis lodged in the Library of the Institute of Polar Studies, Ohio State University.

MIRSKY, A. (1969). Sheet 17—Geology of the Ohio Range, Liv Glacier area. Geologic Map of Antarctica 1 : 1,000,000. *Antarctic Map Folio Series*, Folio 12, Plate XVI. (BUSHNELL, V. C., Ed.). American Geographical Society, New York.

NATHAN, S. (1971a). Geology and petrology of the Campbell–Aviator Divide, Northern Victoria Land, Antarctica. Part 2. Palaeozoic and Precambrian rocks. *N.Z. J. Geol. Geophys.*, **14**, 564.

NATHAN, S. (1971b). Potassium–argon dates from the area between the Priestley and Mariner Glaciers, Northern Victoria Land, Antarctica. *N.Z. J. Geol. Geophys.*, **14**, 504.

NEETHLING, D. C. (1972). Age and correlation of the Ritscher Supergroup and other Precambrian rock units, Dronning Maud Land, *in* ADIE, R. J. (Ed.), *Antarctic Geology and Geophysics*. Universitets-forlaget, Oslo, 547.

NELSON, W. H., SCHMIDT, D. L., and SCHOPF, J. M. (1968). Structure and stratigraphy of the Pensacola Mountains, Antarctica. *Spec. Pap. Geol. Soc. Am.*, No. 115, 344.

NEWELL, N. D., PURDY, E. G., and IMBRIE, J. (1960). Bahamian oolitic sand. *J. Geol.*, **68**, 481.

NICHOLAYSEN, L. O., BURGER, A. J., TATSUMI, T., and AHRENS, L. H. (1961). Age measurements on pegmatites and a basic charnockite lens occurring near Lützow–Holm Bay, Antarctica. *Geochim. Cosmochim. Acta*, **22**, 94.

PALMER, A. R. (1970). Early and Middle Cambrian Trilobites from Antarctica. *Antarctic J. U.S.*, **5**, 162.

PALMER, A. R. and GATEHOUSE, C. G. (1972). Early and Middle Cambrian Trilobites from Antarctica. *Prof. Pap. U.S. Geol. Surv.*, 456-D.

PICCIOTTO, E. and COPPEZ, A. (1964). Bibliographie des mesures d'ages absolus en Antarctique (addendum, aout 1963). *Ann. Soc. Geol. Belg.*, **87**, 115.

PLUMSTEAD, E. P. (1962). Geology: 2. Fossil floras of Antarctica. *Trans-Antarctic Exped. 1955–58, Sci. Rep.*, No. 9.

RAVICH, M. G. (1959). Kratkie Svendeniya o Geologicheskm Stroenii Vostochnoy Chasti Gor na Zemle Korolevy Mod v Vostochnoy Antarktide. *Dokl. Akad. Nauk SSSR*, **128**, 152.

RAVICH, M. G., KLIMOV, L. V., and SOLOVIEV, D. S. (1968). *The Pre-Cambrian of East Antarctica* (English translation). Israel Program for Scientific Translations, Jerusalem.

RAVICH, M. G. and KRYLOV, A. J. (1964). Absolute ages of rocks from East Antarctica, *in* ADIE, R. J. (Ed.), *Antarctic Geology*. North-Holland, Amsterdam, 579.

RAVICH, M. G., SOLOVIEV, D. S., REVNOV, B. I., and SHULYATIN, O. G. (1961). Predvaritelnye Dannye o Geologicheskom Stroenii Tsentralnoy Chasti Gor na Zemle Korolevy Mod (Vostochnaya Antarktida). *Inf. Bull. Inst. Arctic*, **25**, 5.

RAY, F. T. and FAURE, G. (1974). Age of post-tectonic aplite and pegmatite, Wisconsin Range Transantarctic Mountains. *Geol. Soc. Am., Abstra. with Programs (North-Central Section)*, **6**, 539.

REX, D. C. (1972). K–Ar age determinations on volcanic and associated rocks from the Antarctic Peninsula and Dronning Maud Land, *in* ADIE, R. J. (Ed.), *Antarctic Geology and Geophysics*. Universitetsforlaget, Oslo, 133.

RIDDOLLS, B. W. and HANCOX, G. T. (1968). The geology of the Upper Mariner Glacier Region, North Victoria Land, Antarctica. *N.Z. J. Geol. and Geophys.*, **11**, 881.

ROOTS, E. F. (1969). Sheet 6—Geology of Western Queen Maud Land. Geologic Map of Antarctica 1 : 1,000,000. *Antarctic Map Folio Series*, Folio 12, Plate VI. (BUSHNELL, V. C., Ed.). American Geographical Society, New York.

SAITO, N., TATSUMI, T., and SATO, K. (1961). Absolute age of euxinite from Antarctica. *Antarctic Rec.*, No. 12, 1057.

SCHMIDT, D. L., DOVER, J. H., FORD, A. B., and BROWN, R. D. (1964). Geology of the Patuxent Mountains, *in* ADIE, R. J. (Ed.), *Antarctic Geology*. North Holland, Amsterdam, 276.

SCHMIDT, D. L. and FORD, A. B. (1969). Sheet 5—Geology of the Pensacola and Thiel Mountains. Geologic Map of Antarctica 1 : 1,000,000. *Antarctic Map Folio Series*, Folio 12, Plate V. (BUSHNELL, V. C., Ed.). American Geographical Society, New York.

SCHMIDT, D. L., WILLIAMS, P. L., NELSON, W. J., and EGE, J. R. (1965). Upper Precambrian and Paleozoic Stratigraphy and Structure of the Neptune Range of Antarctica. *Prof. Pap. U.S. Geol. Surv.*, 525-D, 112.

SCHOPF, J. M. (1969). Ellsworth Mountains: position in West Antarctica due to sea-floor spreading. *Science*, **164**, 63.

SHERGOLD, J. H., COOPER, R. A., MACKINNON, D. I., and YOCHELSON, E. L. (1976). Late Cambrian brachiopoda, mollusca, and trilobita from northern Victoria Land, Antarctica. *Palaeontology*, London, **19**, 247.

SHINN, E. A. (1968). Practical significance of birdseye fabric structures in carbonate rocks. *J. Sediment. Petrol.*, **38**, 612.

SKEATS, E. W. (1916). Report on the Petrology of some Limestones from the Antarctic. *Rep. Brit. Antarctic Exped. 1907–09, Geology*, **2**, 189.

SKINNER, A. C. (1973). Geology of north-western Palmer Land between Eureka and Meiklejohn Glaciers. *Brit. Antarctic Surv. Bull.*, No. 35, 1.

SKINNER, D. N. B. (1964). A summary of the Geology of the region between Byrd and Starshot Glaciers, South Victoria Land, *in* ADIE, R. J. (Ed.), *Antarctic Geology*. North Holland, Amsterdam, 284.

SKINNER, D. N. B. (1965). Petrographic criteria of the rock units between the Byrd and Starshot Glaciers, South Victoria Land, Antarctica. *N.Z. J. Geol. Geophys.*, **8**, 292.

SKINNER, D. N. B. (in press). Stratigraphy and structure of lower grade metasediments of Skelton Group, McMurdo Sound, *in* CRADDOCK, C. and VIERIMA, T. (Eds.), *Antarctic Geoscience* (proceedings of the 1977 Madison Symposium on Antarctic Geology and Geophysics). University of Wisconsin Press.

SOLOV'EV, I. A. and GRIKUROV, G. E. (1978). First findings of Middle Cambrian trilobites in the Shackleton Range (Antarctica). *Reports–Investigations*, **17**, 187.

SOLOVIEV, D. S. (1972). Geological structure of the mountain fringe of Lambert Glacier and the Amery Ice Shelf, *in* ADIE, R. J. (Ed.), *Antarctic Geology and Geophysics*. Universitetsforlaget, Oslo, 573.

STEPHENSON, P. J. (1966). Geology: 1. Theron Mountains, Shackleton Range, and Whichaway Nunataks. *Trans-Antarctic Exped. 1955–58, Sci. Rep.*, No. 8.

STUMP, E. (1974). Volcanic rocks of the Early Cambrian Taylor Formation, central Transantarctic Mountains. *Antarctic J. U.S.*, **9**, 228.

STUMP, E. (1976). On the Late Precambrian–Early Paleozoic metavolcanic and metasedimentary rocks of the Queen Maud Mountains, Antarctica, and a comparison with rocks of similar age from southern Africa. *Inst. Polar Studies Rep.*, No. 62, 212 pp.

STUMP, E. (in press). The Ross Supergroup in the Queen Maud Mountains, *in* CRADDOCK, C. and VIERIMA, T. (Eds.), *Antarctic Geoscience* (proceedings of the 1977 Madison Symposium on Antarctic Geology and Geophysics). University of Wisconsin Press.

STURM, A. and CARRYER, S. J. (1970). Geology of the region between the Matusevich and Tucker Glaciers, North Victoria Land, Antarctica. *N.Z. J. Geol. Geophys.*, **13**, 408.

TATSUMI, T. and KIZAKI, K. (1969). Sheets 9 and 10—Geology of the Lützow–Holm Bay region and the 'Yamato Mountains' (Queen Fabiola Mountains). Geologic Map of Antarctica 1 : 500,000. *Antarctic Map Folio Series*, Folio 12, Plate IX. (BUSHNELL, V. C., Ed.). American Geographical Society, New York.

THOMSON, M. R. A. (1972). Inarticulate brachiopoda from the Shackleton Range and their stratigraphic significance. *Brit. Antarctic Surv. Bull.*, No. 31, 17.

TRAIL, D. S. (1964). Schist and granite in the southern Prince Charles Mountains, *in* ADIE, R. J. (Ed.), *Antarctic Geology*. North Holland, Amsterdam, 492.

TRAIL, D. S. and McLEOD, I. R. (1969a). Sheet 11—Geology of Enderby Land. Geologic Map of Antarctica 1 : 1,000,000. *Antarctic Map Folio Series*, Folio 12, Plate X. (BUSHNELL, V. C., Ed.). American Geographical Society, New York.

TRAIL, D. S. and McLEOD, I. R. (1969b). Sheet 12—Geology of the Lambert Glacier Region. Geologic Map of Antarctica 1 : 1,000,000. *Antarctic Map Folio Series*, Folio 12, Plate XI. (BUSHNELL, V. C., Ed.). American Geographical Society, New York.

TREVES, S. B. (1965). Igneous and metamorphic rocks of the Ohio Range, Horlick Mountains, Antarctica, *in* HADLEY, J. B. (Ed.), *Geology and Paleontology of the Antarctic* (Antarctic Research Series, Vol. 6). American Geophysical Union, Washington, D.C., 117.

VAN AUTENBOER, T. (1969). Sheet 8—Geology of the Sør Rondane Mountains. Geologic Map of Antarctica 1 : 500,000. *Antarctic Map Folio Series*, Folio 12, Plate VIII. (BUSHNELL, V. C., Ed.). American Geographical Society, New York.

VAN AUTENBOER, T. and LOY, W. (1972), Recent geological investigations in the Sør-Rondane Mountains, Belgicafjella and Sverdrupfjella, Dronning Maud Land, *in* ADIE, R. J. (Ed.), *Antarctic Geology and Geophysics*. Universitetsforlaget, Oslo, 563.

WADE, F. and WILBANKS, J. R. (1972). The Geology of Marie Byrd Land and Ellsworth Lands, Antarctica, *in* ADIE, R. J. (Ed.), *Antarctic Geology and Geophysics*. Universitetsforlaget, Oslo, 207.

WADE, F. A. and LONG, D. R. (in press). The Swanson Formation, Ford Ranges, Marie Byrd Land—Evidence for direct relationship with Robertson Bay Group, Northern Victroia Land, *in* CRADDOCK, C. and VIERIMA, T. (Eds.), *Antarctic Geoscience* (proceedings of the 1977 Madison Symposium on Antarctic Geology and Geophysics). University of Wisconsin Press.

WARRAK, M. (1974). The petrography and origin of dolomitized, veined, and brecciated carbonate rocks, the 'cornielues' in the Frejus region, French Alps. *J. geol. Soc. London*, **130**, 229.

WARREN, G. (1969). Sheet 14—Geology of the Terra Nova Bay–McMurdo Sound Area, Victoria Land. Geologic Map of Antarctica 1 : 1,000,000. *Antarctic Map Folio Series*, Folio 12, Plate XIII. (BUSHNELL, V. C., Ed.). American Geographical Society, New York.

WEBERS, G. F. (1972). Unusual Upper Cambrian fauna from West Antarctica, *in* ADIE, R. J., (Ed.), *Antarctic Geology and Geophysics*. Universitetsforlaget, Oslo, 235.

WILLIAMS, P. L. (1969). Petrology of Upper Precambrian and Palaeozoic sandstones in the Pensacola Mountains, Antarctica. *J. Sediment. Petrol.*, **39**, 1455.

WODZICKI, A., BRADSHAW, J. D., and LAIRD, M. G. (in press). Petrology of the Wilson and Robertson Bay Groups and Bowers Supergroup, Northern Victoria Land, Antarctica, *in* CRADDOCK, C. and VIERIMA, T. (Eds.), *Antarctic Geoscience* (proceedings of the 1977 Madison Symposium on Antarctic Geology and Geophysics). University of Wisconsin Press.

YOCHELSON, E. L. and STUMP, E. (1977). Discovery of Early Cambrian fossils at Taylor Nunatak, Antarctica. *J. Paleontol.*, **81**, 872.

YOUNG, D. J. and RYBURN, R. J. (1968). The geology of Buckley and Darwin Islands, Beardmore Glacier, Ross Dependency, Antarctica. *N.Z. J. Geol. Geophys.*, **11**, 922.

INDEX